DATA ANALYSIS, LEARNING SYMBOLIC AND NUMERIC KNOWLEDGE

INSTITUT DE RECHERCHE
EN INFORMATIQUE ET EN AUTOMATIQUE

DATA ANALYSIS, LEARNING SYMBOLIC AND NUMERIC KNOWLEDGE

(Proceedings of the Conference on Data Analysis, Learning Symbolic and Numeric Knowledge)

Antibes
September 11 - 14, 1989

Edited by E. Diday

INRIA

Nova Science Publishers, Inc.
New York•Budapest

Nova Science Publishers, Inc.
283 Commack Road
Suite 300
Commack, New York 11725

Library of Congress Cataloging-in-Publication Data
available upon request

ISBN 0-941743-64-0

Graphic Design by Elenor Kallberg and Peggy Harvey

Printed in the United States of America

Foreword

The numerical approach in statistics and data analysis like the symbolic approach in Artificial Intelligence and Machine Learning, concerns all the domains of human activity which need to acquire and use knowledge in an automatic process. These Proceedings contain new advances in data analysis: Modeling, multiway tables, textual data, detection of outliers, robustness and quality of solutions etc... and in learning: Abductive recovery and analogy, lattices and formal concept analysis, symbolic approach in time series, automatic construction of a knowledge base etc... The practical examples presented (concerning nuclear plants, document retrieval, biological data etc...) show the wide variety of application domains.

In this symposium, several schools of thought meet: Multidimensional Data Analysis, which has been so successful in processing large data sets with its graphical representations and its possibility of statistical modeling aiming at robustness of the results; Machine Learning in Artificial Intelligence, with the richness of its complex objects, its logic modeling and its symbolic representation, the use of background knowledge and the self-explanatory power of the handled and obtained objects; Neural-Networks approach, with its ability to use architectures on multilevel structures, which are well adapted to the data and problems concerned, owing to their great flexibility.

Both by the concerned problems and the way of solving them these different trends may enrich one another. Such is the wish of the organizers of the symposium.

The Scientific Committee wishes to extend its thanks to all the authors, to the referees who have selected the contributions, to all the scientific specialists for their participation in the symposium and to the Public Relations Department of INRIA which actively contributed to the organizational success of this event.

SPONSORSHIP

ASU	Association pour la Statistique et son Utilisation
EDF	Electricité de France
ISI	International Statistical Institute
IASC	International Association for Statistical Computing
GMFIIS	Groupe des Membres Français de l'Institut International de Statistique
IFCS	International Federation of Classification Societies
RSS	Royal Statistical Society
SFC	Société Francophone de Classification

SCIENTIFIC COMMITTEE

S.	Aivazian	CEMI, Moscow, USSR
H.H.	Bock	Technical University of Aachen, FRG
I.	Bratko	Institute «Jozef Stefan», Ljubljana, Yugoslavia
J.M.	Chambers	Bell Labs, Murray Hill, USA
E.	Diday	Université Paris IX, INRIA-Rocquencourt, France
J.C.	Gower	Rothamsted Experimental Station, UK
W.	Gaul	Karlsruhe University, FRG
J.P.	Haton	INRIA-Lorraine, France
C.	Hayashi	University of the AIR, Chiba, Japan
C.	Lauro	Naples University, Italy
R.S.	Michalski	George Mason University, Fairfax, USA
A.	Rizzi	University of Roma, Italy
M.	Schader	Hamburg University, FRG
R.R.	Sokal	State University of New York, USA
C.	Tricot	Université de Genève, Suisse
N.	Zagoruiko	Inst. Math., Novosibirsk, USSR

PROGRAM COMMITTEES

DATA ANALYSIS

E.	Diday	Université Paris IX, INRIA-Rocquencourt
Y.	Escoufier	Université de Montpellier
M.	Euriat	INSEE, Paris
G.	Govaert	Université de Metz, INRIA-Lorraine
L.	Lebart	ENST, Paris
J.P.	Pages	CEA, Fontenay-aux-Roses
G.	Saporta	CNAM, Paris
Y.	Schektman	Université de Toulouse
R.	Tomassone	INRA, INA-PG, Paris

LEARNING SYMBOLIC AND NUMERIC KNOWLEDGE

E.	Diday	Université Paris IX Dauphine, INRIA-Rocquencourt
Y.	Kodratoff	Université d'Orsay, George Mason University, Fairfax
D.	Sleeman	University of Aberdeen
R.	Wille	Darmstadt University

FORUM COMMITTEE

G.	Govaert	Université se Metz, INRIA-Lorraine
Y.	Lechevallier	INRIA-Rocquencourt
Y.	Schektman	Université de Toulouse

SCIENTIFIC SECRETARIAT

G.	Celeux	INRIA-Rocquencourt
H.	Ralambondrainy	INRIA-Rocquencourt

ORGANISATION

Th.	Bricheteau	INRIA-France
M.	Chazal	INRIA-France
S.	Gosset	INRIA-France
S.	Morel	INRIA-France

TABLE OF CONTENTS

VALIDITY

DATA BASES

DATA ANALYSIS AND MODELS

APPLICATIONS

SMOOTHING DATA

TIME SERIES

GRAPHS AND MODELS

COMPUTER SYSTEMS

PART 2

LEARNING SYMBOLIC AND NUMERIC KNOWLEDGE

RULES GENERATION

SYMBOLIC CLUSTERING

LEARNING SYSTEMS

NEURONAL APPROACH

STRUCTURING AND GENERALIZATION KNOWLEDGE

PART 1

DATA ANALYSIS

THE NEXT FUTURE OF DATA ANALYSIS

Stephen Morgenthaler
Swiss Federal Institute of Technology, Lausanne (Switzerland)
and John W. Tukey*
Princeton University, Princeton, N.J. (USA)

Abstract. Our paper is an attempt to discuss important areas and issues of data analysis as they appear to us today. The philosophical sections of *The Future of Data Analysis* (Tukey, 1962) can continue as they were, and we invite the reader to peruse them. The present paper is divided into three parts. First, we explore basic issues that present themselves in many situations. Then, we go on to offer some comments on a selection of broadly useful tools, and finally, we discuss major kinds of analyses.

1. Introduction

The next future of data analysis begins with a computer – and carries on until it reaches a picture! Neither computer at input nor picture at output was adequately emphasized, almost 30 years ago, when *The Future of Data Analysis* (FDA; Tukey, 1962) was written. Today, neither could possibly be emphasized too strongly.

To an ever increasing degree, the computer is – and will be – the means of placing skillfully tuned computational procedures in the hands of users, whose details the user need not understand, who, however, knows what the overall procedure is intended to do, and a considerable amount of what is known about how well it does what it is supposed to do.

There is no substitute for pictorial output as an effictive mode of communication. We know a variety of styles – scatter diagrams, a curve or a few curves, stem-and-leaf displays, rootograms, etc. We need many more.

Rather less emphasis, than was true for FDA, can now be given to exploration, as contrasted with confirmation. In 1962 being serious about exploration was a novelty. We should, however, expect to need both exploration and confirmation (see Tukey, 1986, Vol. IV, Chapter 22). Indeed, we need to classify confirmatory procedures – depending on their purpose – into detailed categories (see Tukey, 1986, Vol. IV, Chapter 27).

Many methods of data analysis have been developed for smallish data sets. When we apply these methods to large samples, difficulties arise. In a setting where we want to make inferences, we are challenged by the infeasibility of probability models. Our known families of distributions will usually fail to describe reality accurately enough. Testing sharp hypothesis by standard statistical methods becomes impossible. As the sample size grows, variability of estimates becomes relatively unimportant compared to their bias. Yet another example of this phenomenon is the assumption of uncorrelated errors that we almost routinely use. The evidence contained in the data will argue against this assumption with growing sample size. In summary, we can say that finer structures in the data can be seen and become important for large data sets and we will likely want to use new methods developed to describe and detect these details. Nonparametric techniques are the natural candidates.

We also need programs that can analyze automatically gathered data masses, and filter them before more detailed analyses. For this purpose, exploratory techniques and relatively crude descriptions remain very useful.

BEGINNING, COMPLETION AND THREE CHALLENGES

Three issues that deserve special emphasis are *Multiplicity*, *Bias* and *Re-randomization*. Multiplicity has been slowly recognized as a bundle of major issues. We stress focusing attention on fewer than all aspects of what is before us, thus greatly reducing the dilution that would otherwise accompany broader multiplicity. Bias has typically been given inadequate attention. It is harder to attack

* Prepared in connection with research at Princeton University sponsored by the Army Research Office (Durham), DAAL03-86-K-0073

and to measure than variance. Re-randomization offers the highest standard of statistical certainty that we know how to provide.

2. Computers: Flexibility and a Wise Default

Ideas that need to be discussed involve desirable features of statistical packages, both for analysis and graphics, and an approach for keeping statistical software up to date.

A statistician's "expert" system denotes a package that leads the greenhorn by the hand, through a planned sequence of steps, almost always selected by automatic diagnostics from a branching tree. Such systems must be designed to be usable also by a more experienced user who wants to change the path through the guided analysis easily and in a variety of ways. We feel strongly that side calculations made by the system – such as always following the default route – should be reported to the user, whether requested or not.

The graphic capabilities that accompany most statistical packages are not able to produce a substantial fraction of the displays we want. And the very few systems that are flexible enough to do almost anything you want, provide this flexibility by requiring an awful lot of detailed input. Like the "expert" data analysis system, the "friendly" data display system needs to accomodate the user on several levels of experience.

Computer programs come in "releases". What one sees and gets first, is not what one sees and gets later. Consumer datesware (= data analysis techniques embodiable in software) is a way of handling such an approach in statistics. In robustness, for instance, where there are many types of problems, and where we know more about approaching some problems, but less about others, we could put numbers to the types of problems solved, perhaps as follows:

- Robust analogs of classical techniques, involving iterative reweighting (Release 1.x).
- Techniques providing high breakdown points (Release 3.y).
- Techniques allowing specific attention to a much larger number of alternative underlying situations (Release 5.z).

3. Graphics

If we want the results of data analysis to be effectively transmitted we need to present them both in numbers and in pictures. What is most needed is better – and more nearly omnipresent – pictures. The more extensive use of graphical-pictorial display cannot possibly be over-emphasized. Thus our goal should be no less than this: *to parallel every set of numerical results with an appropriate picture.*

Much of our personal experience with plotted points is with points of nearly independent and nearly common variability. We need to take advantage of this experience. As a result, we should tailor our point plots to minimize (and at least localize) dependence and stabilize variability.

More effective comparison with a straight line than with a curve – and more effective comparison with a horizontal line than with a tilted one – seems to be inherent in human psychophysics and perception. Again we should change our point plots to match our capabilities.

If we are showing information about an empirical power spectrum, for instance, we ought at a minimum, plot the *logarithm* of the estimated spectrum. When showing something like a histogram, using rectangles to show counts, we first should plot $\sqrt{\text{counts}}$ rather than counts, thus obtaining a *rootogram*. If we wish to compare with a curve, we should slide the rectangles to have one end on the curve so we can compare the other ends with a horizontal line. (For examples, see Tukey, 1988, Vol. V, chapter 4.)

There is a tendency to regard dynamic graphics as principally a matter of displays of 3-dimensional data shown – in a 2-dimensional view – revolving around some helpful axis. We should not expect this to be the only major dynamic display. Alternagraphic presentation – where two or more images alternate or cycle fast enough for mental superposition but show enough for perceptual separation – deserves careful and extensive investigation.

4. Multiple Comparisons

A man or a woman who sits and deals out a deck of cards repeatedly will eventually get a very unusual set of hands. A report of unusualness would be taken quite differently if we know it was the only deal ever made, or one of a thousand deals, or one of a million deals, etc. Someone who raises 1415 strains of some variety of corn, and measures the yield of each, can make slightly more than a million comparisons of the yield of one strain with another.

We clearly need to think about how to handle such questions of multiplicity. And we cannot expect unique answers. What we think has to depend on how the results are accessed and analyzed.

At one extreme, the 1415 results may be just stored away, available for a small fee. If only an occasional pair of gamblers pay the fees, so they can settle a bet on which of two strains yielded higher, the essential multiplicity will be somewhere between "one" and the number of purchasing pairs. At the other extreme, where an analyst puts the 1415 results in his or her PC, and uses lots of data analysis software, we can be sure that the most unusual of the million comparisons will be found, and, if at all unusual, reported upon. We need to be prepared to deal with either.

One of the easy ways to deal with either is to be careful about the definition of error rate. As a rate, an error rate has a numerator – a number of "errors" – and a denominator – a number of "trials". For one exogenously selected individual comparison, we have "error" equal to getting that particular comparison wrong and "trial" equal to repeating data collection for those two treatments. For exploration of the million comparisons, followed by emphasis on the largest of all, we have "error" equal to getting one or more of the million comparisons wrong and "trial" equal to repetition data collection for all 1415 treatments. For this second case we may even count "error" equal to number of wrong comparisons. If we apply a conventional procedure, with single-application error rate of 5%, we will obtain an error rate of 5% for the individual error rate, $\geq$ 99.9% for the family-wise error rate and 5'000'000% for the Bonferroni error-rate. If either of the latter two are appropriate, we *must* do something.

There are two types of confidence procedures. On one hand we have *confidence directions*, where we are only confident that one yield compared to another is "higher", "lower", or "uncertain". On the other hand, we have *confidence intervals* for the difference in yields. Confidence intervals contain more information than a single test of significance. Information that is needed when we want to put our aquired knowledge to practical use. (It is not enough to know that 500 pounds per acre of a fertilizer will increase the yield. We want to go at least as far as "how much will the extra yield return in monetary terms?".) Confidence intervals also contain an indication of precision, that can be roughly adjusted to other levels of confidence. (If the 95% confidence interval goes from 70 to 130, a 99% confidence interval will be wider, but probably not much outside 60 to 140.)

With what coin do we buy knowledge or belief? At least three different kinds of payment are always important:

1) The care and insight with which the data collection, or the experiments, were planned and performed.
2) The effort involved in collecting enough data.
3) The formal error rate that we are willing to accept for our conclusions.

In the face of variable results we can say little, if anything, without some chance of error. A crucial task is to expend this chance wisely, and to see that how it was spent – as well as how much was spent – is clearly recognizable.

We can thus specify a small number of relatively extreme situations for which we need to be prepared:

1) the data bank, exogenously used, where only a few of the many values will ever be looked at,
2) the clinical trial, focussed, but not exclusively, where a very few prespecified comparisons will be allowed to eat up the available error rate, and the remaining comparisons have the logical status of hints,
3) the full exploration, where we look at all simple comparisons, shrinking our per comparison error rate enough to keep our simultaneous error rate at the desired value.

The most natural way of setting confidence intervals for all comparisons leads to the use of the *Studentized range*. There is, however, a need for several different methods, if we want to be prepared for different stages in the fineness of resolution.

Very early in multiple-comparison history, Henry Scheffé pointed out that using F instead of the Studentized range would give confidence intervals for every contrast (for every $\sum c_i y_i$ with $\sum c_i = 0$) with the property that for any contrast, the simultaneous interval was the same multiple of the individual interval. However,

1) doing this widened the confidence intervals for simple comparisons,
2) it was often hard to find contrasts – other than comparisons – in which there was much interest,
3) most of those found were comparisons of subgroup means, and
4) limits on simple comparisons implied, indirectly, limits on all contrasts.

5. Multiple Covariates and Multiple Responses

We have seen above, that procedures that try to treat everything alike are unsatisfactory. Examples include: all contrasts among treatments, all linear combinations of responses, more coefficients fitted at the same time than the data can support, etc. *It is essential to focus as sharply as we dare.*

Variables, whose direct influences we wish to keep out of the final analysis, are usually correlated, sometimes highly so. It will rarely be most helpful, to discard some of these variables to reduce correlation among the remainder, rather we will do much better to assemble composites – using judgement if necessary – of such variables and then work with these composites. To deal with the desirabilty of making use of many covariates (for avoidance or correction) in the face of limited data we will often have to recognize that, in a particular situation, direct fitting of separate regression coefficients is neither feasible nor desirable. Some form of two-phase fitting is needed, where some kind of a single composite – or a few composites – is formed, based upon one "part" of the data, and a regression on these composites is then calculated from another "part" of the data.

In a clinical trial with "yes or no" response for each patient, the first "part" of the data can consist of all patients analyzed together, while the second "part" consists of the antithesis between treatments. In other problems, it might be reasonable to divide the available data randomly into two parts, one containing about 1/3 of the cases, the other about 2/3 of the cases.

The central point about multiresponse techniques is simple and very pointed: Procedures that are **un**necessarily multirespone are very unsatisfactory. This is a special class of cases within the general scope of a broader principle: equalizing error-rates over all potential results is very unsatisfactory. It is plausible then, that we want either minimum or near-minimum sets of (simple or composite) responses. Ordinarily, the ***minimum*** set of responses, that we are most likely to consider is made up of the individual responses that were originally measured. A near-minimum set might in addition include sums and differences of original responses.

The simplest multiresponse data structure consists of one k-dimensional response vector for each case, no structure for the cases, and k = 2 or 3. The result is an undifferentiated point cloud in 2 or 3 dimensions, which we would like to "simplify" as much as possible, first by modifying the original coordinates and then by mixing them together.

Modification naturally begins by re-expressing each original coordinate to produce approximate symmetry, followed by shift and scaling of each original coordinate to typical location zero and typical spread unity. The mixing of such re-expressed and scale-matched responses naturally proceeds by first looking for principal components (using either Euclidian distance or some robust analog). Using the raw principal components is **un**likely to be desirable, usually because of difficulty of interpretation. In practice, our knowledge of the exact directions of apparent principal components is usually soft enough to allow the selection of very simple sets of coefficients so that the resulting linear combinations remain nearly enough principal.

Just as simple comparisons have come to dominate most contrasts in multiple-comparison situations, so, too, in multiple-response situations, univariate analyses of original coordinates should come to dominate the analyses of most linear combinations of responses. The much greater importance of results that are easier to think about, needs to override any feeling that "treating all on an equal basis" has even token importance. This view says that one *serious problem* with most multivariate analyses is that they are, indeed, multivariate – that they do intend to "treat all linear combinations alike".

Since we are likely to have to pay a Bonferroni factor for the number of original coordinates plus special coordinates, we ought routinely also ask, preferably in advance of seeing the data: "do the original responses come in clusters, within which all correlations are so high as to make replacing the individual responses in the cluster by a single, probably judgement-based composite desirable?" We ought to ask and then *act* on the reply.

Other conventional multivariate techniques like canonical correlations and factor analysis can sometimes be helpful. Clustering, a technique not yet well enough understood will be discussed later. We find it hard to urge many other truly multresponse techniques. Usually a careful selection of univariate responses, followed by appropriate uniresponse analyses will suffice.

6. Bias: Lifting the Rug

Bias comes from many different sources: misperceived methods of measurement, systematic errors of measuring instruments (whether physical devices or questionnaires), naively formulated analytical procedures, to name only three. Analysis itself cannot compensate for misperceived methods, although ideas of statistical design may be effective in ameliorating many of their effects.

Not least among the roadblocks on the way to a more effective and more useful treatment of bias is the question: "bias referred to what?" If there is a real finite population to be studied, exactly what aspect ought to concern us? The arithmetic mean? The median? These questions are far from easy, and their answer seem more likely to be field-dependent than universal.

Even if fluctuations are negligibly small, the bias of a specific method of measurement can only be defined in terms of a *better* method of measurement. A question of bias adjustment that, in the United States, has generated considerable controversy and litigation, is the question of adjusting census counts for underreporting. Here the better method, against which counts in the hundreds of millions are to be compared, has to be a sample survey conducted with greater care than the bulk census.

About the only bias-in-calculation correction that we know how to make in rather general circumstances is bias associated with finiteness of sample size. To correct generally means making calculations for at least two different sample sizes, ordinarily through some sort of jackknife.

7. Randomization: The Challenge of Reliable Stochastics

Over the last half-dozen decades, we have seen a succession of steps to strengthen the faith we have placed in our inferences. Student's t and the F-tests of the classical analysis of variance were originally "derived" within the narrow, over-utopian Gaussian framework – and they appeared to work properly. The application of the conventional distribution to s^2/σ^2 – a χ^2 – was equally justified under the same narrow hypotheses. It proved not to work even tolerably, and was tacitly abandoned. This difference was explained, in large measure, by the work of Pitman and Welch on randomization analysis. Their work showed that, for randomized block designs and Latin squares, at least, the t and F-distributions were good approximations not only to overall but even *conditional* on the values that were randomized. We should draw at the very least four conclusions from these results:

1) Some Gaussian-assumption results work, and others do not.
2) Reasons, both for working and for non-working, can be found.
3) Reasons for working, though applicable to sampling from a wide variety of distributions, depend upon the randomization behavior following from conditioning on unassigned collections of values, not on distribution theory for such collections.
4) Randomization can be trustworthy – but distribution theory need not be.

Full recognition of what had been learned was unacceptable 50 years ago, because

1) Most other statistical "results" were based on Gaussian theory, and did not seem easy to support by any simple application of randomization theory.

2) Computation was expensive.

3) Much data came from non-experimental sources, so that its analysis could only receive the weaker blessing based on distribution theory.

From the recognized weakness sprang first the non-parametric approach and then the robust approach. These attempted to do what could be done within the framework of distribution theory, and did, so far as we can see today, much of what can be done for either non-experimental data or data from non-randomized experiments.

The most secure probability-based inferences come in experiments where randomization is an explicit part of both the data collection and the analysis (Tukey, 1990). Such re-randomization analysis are probability-wise almost completely trustworthy, because their probability structure depends only upon what we, ourselves, have done. An essential in such an analysis, if we want a "tight" answer, is to plan to put on record in advance of data collection just what set of re-randomizations are to be used in the analysis, and how it has been ensured that the actual assignment was randomly selected from the set.

TOOLS AND TOOL KITS

8. Automatic Robustness

Robustness offers us tools that are resistant to outliers and yet reasonably efficient. There is a ladder of techniques available that goes from simple to involved. For the location problem we have

1) the median, a simple estimator that is highly resistant;

2) Huber's M-estimator, which is resistant, and asymptotically efficient in a neighborhood of the Gaussian model;

3) the biweight M-estimator which is resistant and efficient in broader circumstances;

4) configural polysampling estimators that are resistant and efficient for the biggest spectrum of tail heavyness.

With the growth of computer-based data analyses, it becomes increasingly difficult for statisticians to control and even influence the quality of data analysis practiced by users. This implies that we must make sure that only good procedures are being released. Today, good means robust/resistant against outliers and other untypical idiosyncrasies of a data set. For regression and ANOVA models such resistant procedures exist. The same is true for other multivariate methods. It seems to us that procedures that give an automatic protection are underused in favor of worse methods.

9. Resampling

Below the platinum standard of re-randomization, but far above the cadmium standard of inadequate analysis, comes the use of resampling techniques for assessing significance and confidence. These require, as a foundation, two essentials:

1) A well-chosen, sharply-pointed, danger-limited summary of the outcome (a description of how this summary is to be created, not just a numerical result).

2) Agreement upon which division of the data into pieces comes close enough to reflecting the underlying variablity whose impact on the summay we want to assess.

If the carefully selected estimate is linear – is a linear function of the results for the different pieces – and if either a single comparison is at issue or Bonferroni allowance for multiplicity is satisfactory, then the answer is easy:

- first summarize over pieces
- then use Student's t.

If the need to assess something less simple, or the need to make the assessment robust, leads us to a non-linear summary, we would still like to find our way back to our familiar t-procedure. Two kinds of *jackknives* are available for our use:

- (more nearly classical) leave-out-one jackknife, and

• multihalver jackknives.

The first of these is based upon r pieces and upon the value of the summary y_{all} for all pieces combined and upon the values $y_{(i)}$, one per piece, of the summary when that one piece is left out. Had the combinations actually been linear, so that no jackknife was needed, then the pseudovalues

$$y_{*i} = r y_{all} - (r-1) y_{(i)}$$

would have been the summary values for the individual pieces, which we could then insert directly into a t-procedure. Heuristically, we can regard the pseudovalue calculation as nearly linearizing our problem, and happily go ahead with applying Student's t to the pseudovalues.

For harder problems, especially those with order-statistic-like summaries, we can use the multihalver jackknife, in which data is many times divided into halves, with each half summarized separately, each pair of halves contributing, first to two accumulations and then to the means of each of these accumulations:

1) An accumulation of ((result for one half) - (result for another))^2 each value of which estimates 4 times the variance of the final result.
2) An accumulation of (result for all)-(result for one half)/2 - (result for other half)/2 each value of which estimates the correction for bias to be applied to the result for all.

It is important to make the various splittings into halves nearly (formally) orthogonal, preferably in a balanced way. In some problems, such as weather-related data extending over one or more years, it may be desirable or essential to maintain the balance of the seasons. Thus the pieces of leave-out-one-half might each involve one half of each month.

A variety of resampling techniques, mainly based on sampling *with* replacement from the available data, are usually referred to as *bootstrap techniques*. Beside sampling with replacement, they typically involve recreating samples – or more complex structures – of the same size as the whole data, from which they thus must differ by both omission and repetitions. Bootstraping has a simpler asymptotic theory than the jackknifing. Computational effort in bootstraping is typically greater, and all the considerations involved in making the resampling reflect the proper error term apply to the bootstrap.

10. The Orderly Tool Kit

Many useful calculations involving order statistics make use of one or more of the following:

1) The median of the ith order statistic of n observations is close to $F^{-1}(3i - 1/(3n+1))$, where F is the cumulative distribution function.
2) The gaps – differences between adjacent order statistics – are approximately exponentially distributed.
3) The gaps are approximately independent.

We can use the same basic idea to get hold of the distribution of particular order statistics in whatever detail we may need. The distribution of the $i|n^{th}$ order statistic from the unit rectangular distribution (uniform on [0,1]) is an incomplete Beta-distribution with parameter values n-i+1 and i. (See Pearson and Hartley, 1970, Table 16). If we apply F^{-1} to these values, we get the corresponding %-points for order statistics from F.

If, for instance, we wish to combine spread information from a number of gaps the natural construction is to rescale each gap to (approximately) the same average values. If we then sum k rescaled gaps (often called leaps), the approximate distribution of the sum will be proportional to chi-square on 2k degrees of freedom.

The orderly tool-kit both gives indications about how to combine information from various parts of the sample and offers approximate distributions for certain interesting statistics.

11. Re-expression

Among the most dangerous heresies is the assumed nobility of raw values – the belief that because values are in terms of a scale on a measuring instrument, or are in terms commonly used in the field in question, it follows that they are in the terms in which they should be analyzed. It is worth much effort and thought to learn what expressions are likely to be most useful for a particular data set, for data from a moderately specific kind, or even, for many kinds of data.

Why do we choose one expression over another? To make both analysis and understanding easier and stronger. How do we do this? By making bahavior simpler. What kinds of behavior can often be made at least somewhat simpler? Above all, additive impact on the response of changes in each of two or more circumstances. (Done by re-expression of response.) Next, simple dependence of the response on the values of one or more factors. (Done by re-expressing the factors.) If not in conflict with the above, constancy of variance, and, not for its own sake, but as a useful indicator of good performance in one or more of the ways above, rough symmetry of distribution.

Exploratory techniques based on particular parametric families of transformations together with diagnostic plots (Hoaglin, Mosteller & Tukey, 1983) have been used with success. Recently, computer-based methods that re-express explanatory factors and the response in a regression analysis have been introduced (Breiman & Friedman, 1985). The idea is to find a transformation that maximizes an estimate of the multiple correlation between the transformed response and the transformed factors. Such a method tries to linearize the functional dependence of the response on the factors. Better fits in a regression situation can also be obtained by direct search among larger sets of possible surfaces. Projection pursuit regression is such a technique. Other nonparametric regression methods that give good descriptions of a large class of surfaces are available. Nonparametric descriptions are, however, more complex than what can be obtained by regression methods together with re-expressions, and they contain less specific information.

In order to re-express the response, one can use variance stabilizing transformations. And, as with the multiple correlation, it is possible, to base such a choice on the available data, i.e., to estimate local variances and to stabilize them.

The ability to visualize an automatic choice of re-expressions for various variables is a valuable tool in an interactive setting, where the user keeps the ultimate control over which transformation to use.

12. Borrowing Strength

The use of parallel data to strengthen what we learn from what is clearly directly relevant is – and should be – widely practiced. We need to be more conscious of doing it, however. If ANOVA is focused on comparisons as we believe, then what is directly relevant to the variability of a specific comparison, say of version A3 with version A7, come from data involving those two versions. As an important innovation, the classical ANOVA allowed data for all versions of factor A to contribute to the various mean squares, thus borrowing strength. We need to be keenly aware of borrowing strength, not only for variability, but for distributional shape and even for values themselves.

13. Models

If we are to be good data analysts we must not behave like applied probabilists! Our aim must be to avoid a need for careful stochastic modelling whenever the target of the analysis is itself not stochastic. A careful probability model is the hallmark of a good applied probabilist – reliance on such a model is the hallmark of a reckless data analyst.

When functional and stochastic models are given equal importance, or the stochastic is emphasized, we are clearly supporting confirmation. As we put more and more emphasis on the functional model, we are clearly confirming our support more and more to exploration.

14. Encyclopedias

If data analysis is to grow, there will have to be both more and more effective collaboration among workers in the field than the present pattern of scattered papers in the scientific literature is likely to provide. Starting an encyclopedic account is not easy – keeping it rolling, even with a plan of nearly continuous updating, segment by segment, is probably harder, but how else can the proper progress be made?

We need a relatively current encyclopedia of data analysis and graphical displays, which we must plan to prepare and to update on a rolling basis.

MAJOR KINDS OF ANALYSES

In these sections we select a few points to illustrate points of view and selected directions of growth. We do not intend to give a complete account.

15. Regression

All of us are familiar with "slope and intercept" as the descriptor of a fitted line. All of us need to come to realize why this is often a bad description, why we ought to focus on "slope and centercept" as a better description. This corresponds to replacing

$$a + bx$$

by

$$(a + bx_0) + b(x - x_0) = \tilde{a} + b(x - x_0),$$

where x_0 is a convenient value near the center of the data.

A bad fit leaves us very doubtful that what was fit is capable of describing what is really going on. A good fit need *not* mean that we have done well in describing what is *really* going on. We may not even use the variables required for such a description. We ought to expect that there will be a variety of good fits – often involving different variables, surely involving variables in different ways. Techniques for finding "all regressions", given a y and p x's were pioneered by Daniel & Wood (1980). This class of techniques is useful in helping to pick out a single regression we might like, as well as for more general purposes. We should keep in mind, that any ordinary amount of data will not contain enough information to choose the "right" subset. What we can hope for – and often make use of – is a relatively close fit to the data at hand. What we are not at all likely to get is insight into which x's really matter. Best-of-all-subset techniques should therefore not be over-interpreted.

The ideas behind "all regressions" can be used in other ways. If $b_{i,\lambda}$ denotes the fitted value of the coefficient of x_i when the set λ is used as carriers, and s_i^2 is the variance of x_i, one measure of the effect of x_i in our data is $b_{i,\lambda}s_i$. We could maybe look at a few (4 or 5) of these:

- one where $b_{i,\lambda+} = \max_\lambda b_{i,\lambda}$
- one where $b_{i,\lambda-} = \min_\lambda b_{i,\lambda}$
- one where $b_{i(i)}$ is for fitting only x_i
- one where $b_{i(F)}$ is for fitting all x_i
- one where $b_{i(S)}$ is for fitting λ_S, the selected subset.

Rules for selecting, such as Mallow's C_p, Allen's PRESS and the ratio MEAN SQUARE/DEGREES OF FREEDOM of residuals, all behave similarly.

Various classes of location estimators – among them L-estimators, M-estimators, R-estimators, and S-estimators – have been generalized to the regression case. As was to be expected, the location case is not representative of the typical regression data which does not contain replicated observations, and new problems in robustness have arisen because of this.

The possibility of leverage points in a design has led to the development of M-estimators – the other classes could be used to a similar effect – with a tuning constant that depends on the position of the observation in the factor space. Observations at outlying points are more readily downweighted than observations at points inside the main cluster. This is notably the outcome of an (asymptotic) variance optimization under the constraint of a bounded influence function. These second-generation robust methods suffer from a severe loss of efficiency for some designs, but the fit one obtains can be useful in identifying outliers.

The breakdown point of bounded-influence M-estimators of regression is not necessarily large ($\geq 1/4$, say) and the search for procedures with breakdown point close to 50% – outlier-repellant methods – led to estimators that do not belong to the M-class. We now have several proposals for high-breakdown methods, including repeated medians, least median of squares and appropriate S-estimators. (See Rousseeuw & Leroy, 1987, for more discussion.) The computation of such estimators is difficult, typically based on a search. Like the bounded-influence estimators, the high-breakdown methods are very inefficient for some designs. But it is clear that they are **the** methods of choice for identifying outliers. How well the least-squares method preceeded by outlier rejection based on a high-breakdown fit will perform is unclear. It would in any case be better to follow the outlier rejection by an application of iteratively reweighted least squares with a redescending weight function.

In comparison to the estimation problem, relatively moderate efforts have also gone into research of resistance against non-constant variances, bias due to a wrong functional model and robust designs, robust tests and robust confidence intervals.

In the near future – if robust methods want to gain widespread use – we need a basis for comparison among the many methods that are available, in order to discard the less important procedures. What we need most, are agreed upon situations in order to simulate and learn about the small sample behavior of the competing proposals. Simulations are necessary because first-order asymptotics will not discriminate between procedures and because asymptotic variances are **not** the criteria of interest.

We think that the study of regression methods that depend on the design has not gone far enough. We should combine an analysis of the design matrix with a subsequent regression analysis. For example, if the design points in the factor space are clustered, it may be interesting and illuminating to regress within the clusters first and combine later.

An interesting question involves the study of estimators that are "as efficient as possible for a few different error distributions" (O'Brien, 1986). These can be computed with the help of numerical integration routines. It is not clear whether such estimators can be approximated reasonably well by design-dependent M-estimators.

The above comments apply equally well to the problems of testing and interval estimation. Many procedures have been proposed and we need to sift through them. Robust versions of model selection procedures, causal analysis etc. can be built, once estimation and testing is well understood.

We do not have in place a systematic procedure for allowing for changes in variablity from place to place or response level to response level.

16. ANOVA

The basis of ANOVA is the presentation of each data value as a sum of parts. All the parts of a given kind, arranged in the same way as the original data, make up an overlay. Thus, the entire data is presented as the sum of the overlays. Most overlays, e.g. column effects, are repetitive, so that we only need to write down the basic element, which we can call a box.

For a classical, least-squares ANOVA, what we add and subtract will be means of subsets. More generally, in a robust ANOVA, what we add and subtract will be some other summary of the same subsets, and iteration may be required.

17. Fourier

We have known for some time that the most naive analyses of spectra can be dangerously misleading. Recent work by Thomson and his collaborators is quite persuasive that we should routinely use a combination of (1) segmentation in time, (2) use of Slepian-sequence-based multiple windows before Fourier transformation, and (3) jackknife estimation of variability (Thomson & Chave, 1990). We badly need a software package, with an accompanying description directed to the user, rather than to the mathematician, so that these techniques can receive much wider use, giving broad experience a chance to identify them as "well tested".

Wherever rigid translation of the location of our data points, most often in time or space, is a natural operation, Fourier methods of data analysis are likely to be important and effective (Tukey, 1985, Vol. II).

18. Smoothing

We define smoothing as a way to avoid distraction. We use it more as a kind of summarization, rather than as a way to bring individual values closer to underlying or long-run values. In particular, we regard smoothing any scatter plot as likely to help us understand better what that plct was trying to tell us, ofter rather ineffectively.

As a consequence, we do not expect smoothing to be restricted to only a few techniques – nor to even a few kinds of techniques – among which choice is directed by both our purposes and the data's behavior. There can be no single "correct" smoother.

Classical smoothing techniques, like running means, suffer from two major ills:

- exotic values, even if isolated, affect the resulting appearance very substantially, by conversion to "tents",
- hills are eroded downward and valleys are sedimented upward (most classical smoothers leave straight-line behavior unchanged, but distort quadratic behavior).

Curing "tent-making" requires the use of smoothers that are non-linear in the sense that

$$\text{smooth } \{y_t + z_t\} \neq \text{smooth } \{y_t\} + \text{smooth}\{z_t\}.$$

For, if we are not to respond at all to isolated exotic values, superposability would imply not responding to any sequence, since any sequence can be presented as a superposition of isolated exotic values. Robust smoothers can be quite effective in this connection. Curing erosion and sedimentation is most simply done by re-roughing, where we add to the smooth a smoothed version of the rough (the residuals from the first smooth).

Median smoothers of odd length – alone, repeated, or re-roughed – preserves "edges", breaks in the sequence. Median smoothers of even length do not. Sometimes it is important to preserve edges, at other times it is important to round them off. Sometimes it is desirable to copy through end values of the sequence, at others it is important to do something more complicated. Sometimes it is important to smooth lightly, at others it is important to smooth vigorously. And so on, diversity of procedures is clearly essential.

19. Clustering

Data analysis of a single data set rarely reveals deep truths with any certainty, but the appearances it does reveal are often very useful. A sample mean is not the same as the population average, but is often close to it and can center a confidence limit for it. A data-guided re-expression of a collection of data values need not be the best expression for those values, but using it tends to help a lot. The clusters into which a set of data points usually described in two or more dimensional space, are divided by some clustering procedure need not be exactly the same as some intrinsic underlying subdivision of the data, but they may still be very helpful for further study and analysis.

Indeed, such clusters may be very useful when there are no intrinsic underlying units. Consider a long, thin island that is inhabited at a near uniform density. For school districts, post office areas, and many other purposes, it is very convenient to segment the long island. Clustering is for convenience – its results are likely to be more convenient if there are natural underlying groups – but deteminining clusters is no ground for asserting that such groups are real.

When clusters are more obvious, certain properties of distances of pairs of points among and between clusters are quite obvious. Selection of some such property as a criterion, and use of that criterion to define clusters among the data points was a natural initial approach. In very clear situations, as we would expect, such procedures work well. They do not seem, however, to do nearly as well as the human eye in, say, two dimensions.

There seems to be no way to set up a clustering problem without specifying, directly or indirectly, a measure of distance between points. In many circumstances, different definitions will lead to different apparent clusters. We may think this good or bad, but have, it would seem, to accept it as inevitable.

Often, however, although classical techniques are likely to involve comparing a distance in one part of the data set with another in a far away part, it suffices to compare distances somewhat locally. A simple way to enforce this is to replace the numerical distances between two points by their orderly distance, namely by the pair of integers that tell us which nearest neighbor each is of the other. This is probably not always a good idea, but seems usually to be one.

If we try to learn to use orderly distances effectively by starting with a simple two-dimensional test bed (Hansen & Tukey, 1990), we learn how to do much better than classical procedures. In doing this we learn that it does not seem to be enought to do our cluster generation either only once or in only one way. Rather, we are led to go through a clustering one way, save part of the results, and return to use a different clustering technique.

What about higher dimensional cases? If there are many clusters to consider we have as yet little to offer. If we find only a few clusters, say 3, in our first pass at the data, however, there seems to be something useful to do. We can pass a two-dimensional place through the centers (somehow defined) of the three clusters and then change all the distances, preserving the part in the plane and replacing the part perpendicular to the plane.

20. Summary

We have discussed the importance that the computer has for data analysis. It allows us to perform quite complex computations quickly and interactively. As we have stressed, both numerical and graphical output have their importance, but at this time, graphics needs more promotion.

Three central issues of data analysis that we have commented on are multiplicity, bias and re-randomization. Any data analyst must understand their implications.

The second half of the paper deals with data analysis techniques. We first introduced broadly useful ideas, and then turned our attention to specific kinds of analyses.

21. References

Breiman, L. and Friedman, J.H. (1985): "Estimating Optimal Transformations for Multiple Regression and Correlation", *J. Am. Statist. Assoc.*, **80**, 580–598.

Daniel, C. and Wood, F.S. (1980): *Fitting Equations to Data*, Wiley, New York.

Hansen, K. and Tukey, J.W. (1990): "Tuning a Major Part of Clustering Algorithm", submitted to *International Statistical Review*.

Hoaglin, D.C., Mosteller, F. and Tukey, J.W. (1983): *Understanding Robust and Exploratory Data Analysis*, Wiley, New York.

O'Brien, F.L. (1984): *Polyefficient and Polyeffective Simple Linear Regression Estimators and the Absolute Polyefficiency of the Biweight Regression Estimator*, Ph.D Thesis, Princeton University.

Pearson, E.S. and Hartley, H.O. (1970): *Biometrika Tables for Statisticians, Vol. I*, Cambridge University Press, Cambridge.

Rousseeuw, P.J. and Leroy, A. (1987): *Robust Regression and Outlier Detection*, Wiley, New York.

Thomson, D.J. and Chave, A.D. (1990): "Jackknifed Error Estimates for Spectra, Coherences, and Transfer Functions", submitted to *Advances in Spectrum Estimation*.

Tukey, J.W. (1962): "The Future of Data Analysis", *Ann. Math. Statist.*, **33**, 1-67.

Tukey, J.W. (1985): The Collected Works of John W. Tukey, Vol. II, *Time Series: 1965-1984*, (D.R. Brillinger, ed.), Wadsworth & Brooks/Cole, Monterey, Cal.

Tukey, J.W. (1986): The Collected Works of John W. Tukey, Vol. IV, *Philosophy and Principles of Data Analysis: 1965-1986*, (L.V. Jones, ed.), Wadsworth & Brooks/Cole, Monterey, Cal.

Tukey, J.W. (1988): The Collected Works of John W. Tukey, Vol. V, *Graphics: 1965-1985*, (W.S. Cleveland, ed.), Wadsworth & Brooks/Cole, Monterey, Cal.

Tukey, J.W. (1990): "Platinum Standard", Technical Report, Department of Statistics, Princeton University.

FACTOR ANALYSIS

LEAST SQUARES CRITERION FOR ASYMMETRIC DEPENDENCE MODELS IN THREE-WAY CONTINGENCY TABLES

Luigi D'Ambra
Dipartimento di Scienze Sociali, 1st. Univ. Orient. Napoli (Italy)

SUMMARY

This paper deals with logbilinear and logtrilinear models proposed by Goodman fot three–way contingency tables. The proposed approach allows to study the dependence or association structure between two o three qualitative variables. The parameters of the models are estimated by means of the least squares methods.

Keywords: Three-way contingency tables, Non symmetrical correspondence analysis, Log-bilinear and log-trilinear models.

1. INTRODUCTION

In the last years a particular attention has been dedicated to the study of the qualitative data.

Two different approaches, to the analysis of contingency tables, appeared in literature; the first is model–based with log-linear and logbilinears analysis characterized by models whose parameters are generally estimated by maximum likelihood method; the second is an exploratory one, as is the case of symmetrical correspondence analysis (Benzécri, 1973) and a non symmetrical one (D'Ambra & Lauro, 1988) whose parameters are estimated by least squares method.

The model–based approach lies on the hypothesis that data are generated by a probabilistic model; assuming that model is true, estimations are made and the estimates are compared with the observed frequencies to evaluate how the model fits to data.

In this way it is possible to make inferences about the population on the basis of sample data.

The author is very grateful to Prof. Yves Escoufier, University of Montpellier, for his helpful comments and suggestions. This paper has been supported with grants from MPI 60% (L. D'Ambra).

On opposite, the correspondence analysis is used in a descriptive context, without assuming any underlying distribution or model-only a decomposition of the data is obtained to study the structure of the data. The primary interest is in the presentation of the structure of the data in the optimal way. In other words, the observed data table is approximated, in a least squares sense, by a lower rank table.

Recently, many contributions appeared in literature with the purpose of pointing out the possible links that occur between approaches (Lauro & De Carli, 1982; Van Der Heijden & Meijerink, 1988; Van Der Heijden & De Leeuw, 1985; Escoufier, 1988: Lauro & Siciliano, 1988).

Among the Authors who compared both approaches, it must be cited Goodman (1986), who proposed association models in order to look for a link with canonical correlation models (i.e. correspondence analysis), likewise Escoufier (1986), who considered logbilinear models related to a two-way contingency table, in order to show, in an explorative context, a least squares estimation procedure for calculating the parameters.

Having as start-point the different models proposed by Goodman (1986), for the analysis of a three-way contingency table, models which are useful to describe the dependence or association structure between qualitative variables, our goal is to estimate the parameters of the models by mean of the least squares method, and also to determine the tables, derived from the models, on which to apply the generalized singular values decomposition (GSVD).

2. GENERALIZED SINGULAR VALUES DECOMPOSITION

Consider two real vector spaces E and F, of dimensions p and n respectively.

Let (E, Q) be a real Euclidean space with a bilinear symmetrical positive definite form given by the matrix Q.

Let (F, D) be a real Euclidean space with a bilinear symmetrical positive definite form given by the matrix D.

Let X be the matrix associated to a linear application from E^* into F, with E^* dual space of E.

The generalized singular values decomposition (GSVD) of the matrix X consists in the research of a matrix U whose columns are D-orthonormals, of a matrix V whose columns are Q-orthonormals and of a diagonal matrix Λ positive and such that

$$X = U \Lambda V' = \Sigma_\alpha^r \lambda_\alpha u_\alpha v_\alpha' \tag{1}$$

where r is the rank of $X, r = \min(n, p)$.

The columns of U and V are respectively the basis of the representation of the columns and rows of the matrix X.

If the first $h(h < r)$ terms of (1) are considered, the matrix $X^h = \Sigma_\alpha^h \lambda_\alpha u_\alpha v_\alpha'$ represents the best approximation, in the least squares sense, of the matrix X that minimizes

$$\text{tr } D \left(X - X^h\right) Q \left(X - X^h\right)' \tag{2}$$

between all the matrices X^h of rank h (Greenacre, 1984).

3. NON SYMMETRICAL CORRESPONDENCE ANALYSIS AND SINGULAR VALUES DECOMPOSITION

Let **IJK** be three qualitative variables with $I\,J\,K$ categories respectively. Consider the three-way contingency table P_{IJK} which general term P_{ijk} indicates the relative frequency of the cell i,j,k; furthermore, let be

$$\begin{aligned}
&\Sigma_j\, P_{ijk} = P_{i.k} \ ;\ \Sigma_k\, P_{ijk} = P_{ij.} \ ;\ \Sigma_i\, P_{ijk} = P_{.jk}\\
&\Sigma_i \Sigma_j\, P_{ijk} = P_{..k} \ ;\ \Sigma_i \Sigma_k\, P_{ijk} = P_{.j.} \ ;\\
&\Sigma_j \Sigma_k\, P_{ijk} = P_{i..} \ ;\ \Sigma_i \Sigma_j \Sigma_k\, P_{ijk} = 1
\end{aligned}$$

From the three-way contingency it is possible to derive three different multiple tables $P_{I(JK)}$ $P_{J(IK)}$ $P_{K(IJ)}$ of dimensions $(I, J\times K)$ $(J, I\times K)$ $(K, I\times J)$, each of them crosses a qualitative variable with a composite variable given by the cartesian product of the categories of the two others qualitative variables.

It is supposed that the variable $\mathbf{I}$ plays a different role in respect to the variables $\mathbf{J}$ and $\mathbf{K}$, so that the $P_{I(JK)}$ table is considered for the GSVD. Let $P_{I(JK)}$ be the multiple contingency table and let D_I be the diagonal matrix including the marginal frequencies $P_{i..}\,(i = 1\ldots I)$ and D_{jk} be the diagonal matrix with general term $P_{.jk}$ for $jk = 1\ldots JK$.

Let $Y_{I(JK)} = \left(P_{I(JK)} - D_I 1_I 1'_{JK} D_{JK}\right) D_{JK}^{-1}$, where $1'_I\,(1'_{JK})$ is the row vector with $I(JK)$ unitary elements.

The non symmetrical correspondence analysis (Lauro & D'Ambra, 1984; D'Ambra & Lauro, 1988) of the matrix $P_{I(JK)}$, in the GSVD sense, consists in the research of the rank h matrix $Y^h_{I(JK)}$ of dimension $(I\times JK)$ that minimizes

$$\operatorname{tr}\left(Y_{I(JK)} - Y^h_{I(JK)}\right) D_{JK} \left(Y_{I(JK)} - Y^h_{I(JK)}\right)' \tag{3}$$

with $h \le s = \min(I-1, JK-1)$ where s is the rank of $Y_{I(JK)}$, or the singular values decomposition of the matrix $Y_{I(JK)}\, D_{JK}^{1/2}$.

As consequence of the decomposition we get

$$Y^h_{i(jk)} = \Sigma^h_\alpha\, \lambda_\alpha\, u_{i\alpha}\, v_{jk\alpha}$$

furthermore

$$\Sigma_i \Sigma_{jk} \left(Y_{i(jk)} - Y^h_{i(jk)}\right)^2 P_{.jk} = \Sigma^s_{\alpha=h+1}\, \lambda^2_\alpha$$

where $Y_{i(jk)}$ is general term of matrix $Y_{I(JK)}$

$$Y_{I(JK)} D_{JK} Y'_{I(JK)} u_\alpha = \lambda^2_\alpha u_\alpha \ ;\ Y'_{I(JK)} Y_{I(JK)} D_{JK}\, v_\alpha = \lambda^2_\alpha v_\alpha$$

with $u'_\alpha u_\alpha = 1$, $v'_\alpha D_{JK}\, v_\alpha = 1$, and $\lambda_1 \ge \lambda_2 \ge \ldots \ge \lambda_h$.

4. DEPENDENCE MODELS AND ASSOCIATIVE MODELS FOR TWO QUALITATIVE VARIABLES

The model–based approach, for a two- or a multi-way table analysis, is based on the hypothesis that data are generated by a probabilistic model characterized by parameters that must be estimated through observations.

The associative models introduced by Goodman (1986) are useful to analyze both interdependence and dependence between qualitative variables, in terms of principals effects and/or interaction effects.

For a two-way contingency table P_{IJ}, aiming at studying the association between two variables, Goodman (1986) considers the following logbilinear model

$$P_{ij} = \alpha_i \alpha_j \exp\left[\Sigma_\alpha \phi_\alpha u_{i\alpha} v_{j\alpha}\right] \tag{4}$$

for $\alpha = \min(I-1, J-1)$ where α_i, α_j are positive parameters, while $u_{i\alpha}$ $v_{j\alpha}$ are the coordinates such that $1'_I D_I u_I = 1'_J D_J v_J = 0$, the parameter ϕ_α is a measure of the "intrinsic" association between the qualitative variables.

Furthermore the parameter ϕ_α is susceptible of an interesting interpretation in entropy frame, it may be interpreted as a measure of uncertainty in the information–theoretic sense (Gilula Krieger Ritov 1988).

The estimation of the parameters of the model is obtained by the maximum likelihood method; since the model remains invariant when $u_{i\alpha}$ and $v_{\alpha j}$ are replaced by similar functions, in order to identify the parameters let us consider the following conditions

$$\Sigma_i u_{i\alpha} P_{i.} = \Sigma_j v_{j\alpha} P_{.j} = 0 \qquad \text{, for } \alpha = 1, \ldots \min\left[I-1,\ J-1\right]$$

The introduction of weighting system is merely necessary for the comparison with the correspondence analysis.

The parameters of the associative model for $\alpha = 1$ are estimated by Goodman (1979) by the maximum likelihood method. A process generalization, for any α based on Newton algorithm was recently proposed by Choulakian (1988). A different process to parameters estimate of the previous model was proposed by Escoufier (1986); in effect, taking the logarithms of (4)

$$\log(P_{ij}) = \log\alpha_i + \log\alpha_j + \Sigma_\alpha \phi_\alpha u_{i\alpha} v_{j\alpha} \quad ,$$

taking into account the constraints $\Sigma_i P_{i.} u_{i\alpha} = \Sigma_j P_{.j} v_{j\alpha} = 0$ we get

$$\Sigma_i (\log P_{ij}) P_{i.} = \Sigma_i (\log\alpha_i) P_{i.} + \log\alpha_j$$

$$\Sigma_j (\log P_{ij}) P_{.j} = \Sigma_j (\log\alpha_j) P_{.j} + \log\alpha_i \ .$$

Having as conditions $\Sigma_i (\log\alpha_i) P_{i..} = 0$ we get:

$$\log(\alpha_j) = \Sigma_i (\log P_{ij}) P_{i.} \ ;$$

$$\log(\alpha_i) = \Sigma_j (\log P_{ij}) P_{.j} - \Sigma_i \Sigma_j (\log P_{ij}) P_{.j} P_{i.}$$

Let us consider, then, the matrix Y_{IJ} with general term

$$Y_{ij} = \log P_{ij} - \log \alpha_i - \log \alpha_j$$

that verifies the property of being centered by rows and columns

$$\Sigma_j Y_{ij} P_{.j} = \Sigma_i Y_{ij} P_{i.} = 0$$

To estimate the quantity $\phi_\alpha u_{i\alpha} v_{j\alpha}$ we carry out the singular values the composition of the matrix $D_I^{1/2} Y_{IJ} D_J^{1/2}$, that can be considered the correspondence analysis of the table having general term

$$P_{i.} P_{.j} (Y_{ij} + 1)$$

and margins $P_{i.}$ and $P_{.j}$.

In case of the antecedence of a qualitative variable in respect to another, it seems more logic to analyze the dependence of a variable in respect to another.

To this end, it is possible to consider for a two-way contingency table an asymmetric logbilinear model (Escoufier, 1988) of the kind:

$$P_{ij}/P_{.j} = \alpha_i \exp\left[\Sigma_\alpha \phi_\alpha v_{i\alpha} v_{j\alpha}\right] \tag{5}$$

In this case, too, the estimation of the parameters of the model is carried out by the maximum likelihood method, imposing some orthonormality conditions on the parameters. As regards of the estimate of the parameters by least squares method, let us consider the logarithm of (5)

$$\log(P_{ij}/P_{.j}) = \log \alpha_i + \Sigma_\alpha \phi_\alpha v_{i\alpha} v_{j\alpha} \tag{6}$$

and taking into account the constraint $\Sigma_j P_{.j} v_{j\alpha} = 0$ we get the estimate of $\log \alpha_i = \Sigma_j (\log P_{ij}) P_{.j}$; then, we consider the matrix M with general term

$$M_{ij} = (\log(P_{ij}/P_{.j})) - \Sigma_j (\log(P_{ij}/P_{.j}))\, P_{.j}$$

To estimate the parameters $\phi_\alpha v_{i\alpha} v_{j\alpha}$, we carry out the GSVD of the matrix M with weighting matrix D_j , whose general term is $P_{.j}$.

It must be pointed out that this approach represents a generalization of the principal components analysis based on composition data as proposed by Aitchinson (1983). Actually, this author, starting from different hypothesis suggested a principal components analysis of the matrix of general term $(\log(P_{ij}/P_{.j})) - \Sigma_j (\log(P_{ij}/P_{.j}))$, not considering the different weighting system of each category of the variable $\mathbf{J}$.

5. DESCRIPTION OF ASYMMETRIC MODELS FOR A THREE–WAY CONTINGENCY TABLE

In the last paragraph, a two-way contingency table was considered to study either the association or the dependence of two qualitative variables.

In this paragraph, starting from the models proposed by Goodman (1986), for a three-way contingency table and considering Escoufier approach, we intend to analyze the dependence structure between the variable $\mathbf{I}$ and the variables $\mathbf{J}$ and $\mathbf{K}$ using GSVD.

In the first column of the table 1, the principal asymmetric models are shown, which describe the dependence of a qualitative variable $\mathbf{I}$ with respect to the other two qualitative variables $\mathbf{J}$ e $\mathbf{K}$. On first, among the three-way contingency tables models we had distinguished, in bilinear and trilinear models.

The first models describe the dependence link, considering either a multiple two-way table that crosses the $\mathbf{I}$ dependence variable with the $\mathbf{J} \times \mathbf{K}$ composite variable obtained by the product of the categories of the two explanatory variables, or a two-way table obtained by adding respect to an index.

Dependence can be explained by three or two factors according to the multiplicative effect is supposed to be composed either of a simple factor and a composite one, the latter related to the composite variable (three factor models), or of two simple factors (two factor models).

The second class of models (trilinear models) describes the dependence link considering a three-way table and directly, in multiplicative terms, the effect of three simple factors.

Furthermor, we divided the three-factor logbilinear models in conditional models and multiple models; the first of them analyze the dependence between two variables, given the level of the third variable, (each strata analysis), the second ones analyze the strata set of a contingency table.

6. ASYMMETRIC DEPENDENCE MODELS

6.1. THREE-FACTOR BILINEAR MODELS

The model (a) (see: tab.1)

$$P_{ijk}/P_{.jk} = \alpha_{ik} \exp\left[\Sigma_\alpha \phi_{k\alpha} v_{ik\alpha} \omega_{jk\alpha}\right] \tag{7}$$

starts from the contingency table $P_{I(JK)}$ of dimension $I \times JK$.

Taking the logarithms of (7) and imposing the condition $\Sigma_j \omega_{jk\alpha} P_{.jk}/P_{..k} = 0$ for $k = 1..K$ we get the estimate of the parameter

$$\log \alpha_{ik} = \Sigma_j \left(\log(P_{ijk}/P_{.jk})\right) P_{.jk}/P_{..k} \qquad \text{for } k = 1 \ldots K$$

To the $\phi_{k\alpha} v_{ik\alpha} \omega_{jk\alpha}$ estimate, let us consider the matrix Z_k $(k = 1 \ldots K)$ of general term

$$(\log(P_{ijk}/P_{.jk})) - \Sigma_j \left(\log(P_{ijk}/P_{.jk})\right) P_{.jk}/P_{..k} \quad \text{for } k = 1 \ldots K$$

that verifies the property of being centered in respect to the weighting system $P_{.jk}/P_{..k}$ for $k = 1 \ldots K$.

Thus, the following function is minimized:

$$\Sigma_i \Sigma_j \left(Z_{ij(k)} - \Sigma_\alpha \phi_{k\alpha} v_{ik\alpha} \omega_{jk\alpha}\right)^2 P_{.jk}/P_{..k} \text{ for } k = 1 \ldots K$$

and then, the parameters $\phi_{k\alpha} v_{ik\alpha} \omega_{jk\alpha}$ are, for (3), the result of GSVD of the matrix Z_k with weighting system equal to $P_{.jk}/P_{..k}$ for $k = 1 \ldots K$.

The characteristic of this analysis is to consider each strata of the contingency table. From a practical point of view, in case it should be made a comparison between that tables, it is necessary to relate each frequency of the kth table to its total, to eliminate the different weight of each table.

A more general approach for studying dependence is described by models $b1$, $b2$, $b3$. These models too, as the previous, start from the contingency table that crosses the variable $\mathbf{I}$ with the composite variable $\mathbf{J} \times \mathbf{K}$ describing, then, the dependence of two variables in three-factor terms on a two-way contingency table.

We shall examine the model $b3$, for its majority related to the other two; the principal results of the $b1$ and $b2$ models are reported into Table 1.

Let us consider now the model

$$P_{ijk}/P_{.jk} = \alpha_{ik} \alpha_{ij} \exp \left[\Sigma_\alpha \phi_\alpha v_{i\alpha} \omega_{jk\alpha}\right]$$

or, taking the logarithms, we get

$$\log(P_{ijk}/P_{.jk}) = \log \alpha_{ik} + \log \alpha_{ij} + \Sigma_\alpha \phi_\alpha v_{i\alpha} \omega_{jk\alpha}$$

with the constraints $\Sigma_k \omega_{jk\alpha} P_{..k} = \Sigma_j \omega_{jk\alpha} P_{.j.} = 0$

Considering the previous constraints we get

$$\log \alpha_{ik} = \Sigma_j \left(\log(P_{ijk}/P_{.jk})\right) P_{.j.}$$

$$\log \alpha_{ij} = \Sigma_k \left(\log(P_{ijk}/P_{.jk})\right) P_{..k} - \Sigma_j \Sigma_k \log(P_{ijk}/P_{.jk}) P_{.j.} P_{..k}$$

thus, to estimate the bilinear part, we consider the matrix Z_1 of general term

$$Z_{i(jk)} = \log(P_{ijk}/P_{.jk}) - \log \alpha_{ij} - \log \alpha_{ik}$$

and the following function is minimized

$$\Sigma_i \Sigma_j \Sigma_k \left(Z_{i(jk)} - \Sigma_\alpha \phi_\alpha v_{i\alpha} \omega_{jk\alpha}\right)^2 P_{.j.} P_{..k}$$

or in matrix terms:

$$\text{tr } (Z_1 - \gamma) D_J \otimes D_K (Z_1 - \gamma)'$$

with $\gamma = \Sigma \phi_\alpha v_\alpha \omega'_\alpha$ and then the parameters constitute the result of GSVD of the matrix Z_1 with weighting system given by $(P_{.j.} P_{..k})$.

6.2. TWO-FACTOR BILINEAR MODELS

Another class of bilinear models is that relative to the study of dependence of a variable **I** with respect to the **J** and **K** variables, by means two factors, supposing the interaction with the third factor is null.

In particular, the models $c1$, $c2$ and $c3$ that, as the previous models, start from a three-way contingency table, they yet suppose that the data structure can be specified by parameters that reflect the dependence between two qualitative variables, considering then a two-way table obtained by adding in respect to an index.

We shall consider into details the model $c1$ (that establishes in the bilinear part an extension of the models $c2$, $c3$). Let us consider now the following two-factor bilinear model:

$$\log\left[P_{ijk}/(P_{.j.}P_{..k})\right] = \log\alpha_i + \Sigma_\alpha\left[\phi_{1\alpha}v_{i\alpha}v_{j\alpha} + \phi_{2\alpha}\omega_{i\alpha}\omega_{k\alpha}\right]$$

with constraints

$$\Sigma_j v_{j\alpha}P_{.j.} = \Sigma_k\omega_{k\alpha}P_{..k} = 0 \tag{8}$$

In effects, this model supposes that the data structure can be explained in multiplicative terms by an combination of simple factors.

To estimate the parameters, on first, let us observe that the model can be decomposed in two parts; in effects, by adding respect to j and k and taking into account (8) we get:

$$\Sigma_j \log\left[P_{ijk}/(P_{.j}P_{..k})\right]P_{.j.} = \log\alpha_i + \Sigma_\alpha\phi_{2\alpha}\omega_{i\alpha}\omega_{k\alpha}$$

$$\Sigma_k \log\left[P_{ijk}/(P_{.j}P_{..k})\right]P_{..k} = \log\alpha_i + \Sigma_\alpha\phi_{1\alpha}v_{i\alpha}v_{j\alpha}$$

moreover, considering (8) again, we obtain:

$$\log\alpha_i = \Sigma_j\Sigma_k \log\left[P_{ijk}/(P_{.j.}P_{..k})\right]P_{.j.}P_{..k}$$

The parameters $\phi_{2\alpha}\omega_{i\alpha}\omega_{k\alpha}$, $\phi_{1\alpha}v_{i\alpha}v_{j\alpha}$ constitute respectively the result of GSVD of the matrices Z_{IK}, Z_{IJ}, respectively of dimension $(I \times K), (I \times J)$, whose general terms are respectively:

$$Z_{ik} = \Sigma_j \log\left[P_{ijk}/(P_{.j.}P_{..k})\right]P_{.j.} - \log\alpha_i$$

$$Z_{ij} = \Sigma_k \log\left[P_{ijk}/(P_{.j.}P_{..k})\right]P_{..k} - \log\alpha_i$$

with weighting matrix D_K, D_J, (diagonal matrices of general terms $P_{..k}, P_{.j.}$).

6.3. TRILINEAR MODELS

The previous models supposed that the dependence structure of can be explained either considering the effect of a simple factor and a composite one, or the effect of two simple factors.

A class of models more complex than previous, and that request a different speech, is represented by trilinear models included in Table 1.

We shall just analyze the model $d1$, referring to Table 1 for the principal results of the model $d2$.

Let us consider the following model

$$P_{ijk}/P_{.j.}P_{..k} = \alpha_{ij}\alpha_{ik}\exp\Sigma_\alpha\phi_\alpha\eta_{i\alpha}\delta_{j\alpha}\gamma_{k\alpha}$$

or

$$\log[P_{ijk}/(P_{.j.}P_{..k})] = \log\alpha_{ij} + \log\alpha_{ik} + \Sigma_\alpha\phi_\alpha\eta_{i\alpha}\delta_{j\alpha}\gamma_{k\alpha} \tag{9}$$

with constraints

$$\Sigma_j\delta_{j\alpha}P_{.j.} = \Sigma_k\gamma_{k\alpha}P_{..k} = 0$$

that describes the dependence in multiplicative terms of the $\mathbf{I}$ variable with respect to the $\mathbf{J}$ and $\mathbf{K}$ variables, and consists in a dependence model in which the three simple factors effect comes directly in multiplicative terms.

Developing (9) and taking into account the constraints we get

$$\log\alpha_{ik} = \Sigma_j\log(P_{ijk}/P_{.j.}P_{..k})P_{.j.}$$

$$\log\alpha_{ij} = \Sigma_k\log(P_{ijk}/P_{.j.}P_{..k})P_{..k} - \Sigma_j\Sigma_k(\log P_{ijk}/P_{.j.}P_{..k})P_{.j.}P_{..k}$$

Thus, let us consider the matrix Z_{IJK} with general term

$$Z_{ijk} = [\log P_{ijk}/(P_{.j.}P_{..k})] - \log\alpha_{ik} - \log\alpha_{ij}$$

that verifies the following conditions

$$\Sigma_j Z_{ijk}P_{.j.} = \Sigma_k Z_{ijk}P_{..k} = \Sigma_j\Sigma_k Z_{ijk}P_{..k}P_{.j.} = 0$$

The estimate of the parameters $\phi_\alpha\eta_{i\alpha}\delta_{j\alpha}\gamma_{k\alpha}$ is performed minimizing the function

$$\Sigma_i\Sigma_j\Sigma_k(Z_{ijk} - \pi_{ijk})^2 P_{.k.}P_{..j} \tag{10}$$

with

$$\pi_{ijk} = \Sigma_\alpha\phi_\alpha\eta_{i\alpha}\delta_{j\alpha}\gamma_{k\alpha}$$

The (10) can be also written:

$$\Sigma_i\Sigma_j\Sigma_k(Z_{i(jk)} - \pi_{i(jk)})^2 P_{.j.}P_{..k}$$

where $Z_{i(jk)}$ is the general term (i, jk) of the two-way matrix $Z_{I(JK)}$ obtained by juxtaposition of the k strata of matrix Z_{IJK} .

As first step we perform the GSVD of the matrix $Z_{I(JK)}$, with weighting matrix $D_J \otimes D_K$ that enables to obtain the best approximation in the least squares sense of the matrix $\pi_{I(JK)}$

$$\hat{\pi}_{i(jk)} = \Sigma_\alpha \Phi_\alpha \tau_{i\alpha} \psi_{(jk)\alpha}$$

Then we consider the matrix $\psi_{(JK)\alpha}$ of dimension $J \times K$.
The matrix $\psi_{(JK)\alpha}$ verifies the following conditions:

$$\Sigma_j \psi_{(jk)\alpha} P_{.j.} = \Sigma_k \psi_{(jk)\alpha} P_{..k} = 0 \quad \text{for } \alpha = 1 \dots \min(J-1\,,\, K-1)$$

To estimate the coordinates of the categories of the $\mathbf{J}$ and $\mathbf{K}$ variables we perform GSVD of the matrix $\psi_{(JK)\alpha}$ so that

$$\psi_{(jk)\alpha} = \Sigma_\beta \psi_{\alpha\beta} \mu_{j\alpha\beta} \varsigma_{k\alpha\beta}$$

The parameters $\mu_{j\alpha\beta}\ \varsigma_{k\alpha\beta}$ verify the same conditions of constraints, in fact

$$\Sigma_j \mu_{j\alpha\beta} P_{.j.} = \Sigma_k \varsigma_{k\alpha\beta} P_{..k} = 0$$

thus

$$\begin{aligned} \hat{\pi}_{ijk} &= \Sigma_\alpha \Phi_\alpha \tau_{i\alpha} \left(\Sigma_\beta \psi_{\alpha\beta} \mu_{j\alpha\beta} \varsigma_{k\alpha\beta} \right) \\ &= \Sigma_\alpha \Sigma_\beta \Phi_\alpha \psi_{\alpha\beta} \tau_{i\alpha} \mu_{j\alpha\beta} \varsigma_{k\alpha\beta} = \Sigma_\alpha \hat{\phi}_\alpha \hat{\eta}_{i\alpha} \hat{\delta}_{j\alpha} \hat{\gamma}_{k\alpha} \quad \forall \ \beta \end{aligned}$$

where

$$\hat{\phi}_\alpha = \Phi_\alpha \psi_{\alpha\beta}\,,\ \tau_{i\alpha} = \hat{\eta}_{i\alpha}\,;\ \mu_{j\alpha\beta} = \hat{\delta}_{j\alpha}\,,\ \varsigma_{k\alpha\beta} = \hat{\gamma}_{k\alpha}$$

and which are estimates of the parameters $\phi_\alpha \eta_{i\alpha} \delta_{j\alpha} \gamma_{k\alpha}$.

Till now, we considered the bilinear models and the trilinear models separately, but it is possible to suppose a class of models more general than the previous one, that includes both bilinear and trilinear terms, as i.e.

$$\log(P_{ijk}/P_{.j.}P_{..k}) = \log\alpha_i + \Sigma_\alpha \left[\phi_{1\alpha} v_{i\alpha} v_{j\alpha} + \phi_{2\alpha} \omega_{i\alpha} \omega_{k\alpha} + \phi_{3\alpha} \eta_{i\alpha} \delta_{j\alpha} \gamma_{k\alpha} \right]$$

This model aims at the description of the dependence structure of the $\mathbf{I}$ variable with respect to the $\mathbf{J}$ and $\mathbf{K}$ variables, and such a structure is studied in two phases: the former consists into estimating the bilinear term, in a least squares sense, minimizing the functions:

$$\Sigma_i \Sigma_k \left[\Sigma_j \log(P_{ijk}/P_{.j.}P_{..k}) P_{.j.} - \log\alpha_i - \Pi_{ik} \right]^2 P_{..k}$$

$$\Sigma_i \Sigma_j \left[\Sigma_k \log(P_{ijk}/P_{.j.}P_{..k}) P_{..k} - \log\alpha_i - \Pi_{ij} \right]^2 P_{.j.}$$

where

$$\log\alpha_i = \Sigma_j \Sigma_k \log\left(P_{ijk}/\left(P_{.j.}P_{..k}\right)\right) P_{.j.}P_{..k}$$

with

$$\Pi_{ij} = \Sigma_\alpha \phi_{1\alpha} v_{i\alpha} v_{j\alpha} \; ; \; \Pi_{ik} = \Sigma_\alpha \phi_{2\alpha} \omega_{i\alpha} \omega_{k\alpha}$$

Then, we consider

$$T_{ijk} = \left[\log(P_{ijk}/P_{.j.}P_{..k}) - \log \alpha_i - \hat{\Pi}_{ij} - \hat{\Pi}_{ik}\right]$$

where $\hat{\Pi}_{ij}$ and $\hat{\Pi}_{ik}$ are respectively an estimation of Π_{ij} and Π_{ik} and we proceed to estimate, in a least squares terms, the trilinear part minimizing the following function:

$$\Sigma_i \Sigma_j \Sigma_k (T_{ijk} - \sigma_{ijk})^2 P_{.j.} P_{..k}$$

with

$$\sigma_{ijk} = \Sigma_\alpha \phi_{3\alpha} \eta_{i\alpha} \delta_{j\alpha} \gamma_{k\alpha} \quad .$$

A different approach to the parameters estimation, that considers all three way at the same time, symmetrically, is the solution proposed by Kroonenberg (1983). It uses algorithms Alternative Least Square kind and it does not attain to a univocal solution, since it depends on the number of the components linked to each mode.

Our solution, two stages based, has the advantage of being easily interpreted and of taking into account the non symmetric role of the qualitative variables.

7. ASSOCIATION STRUCTURE ANALYSIS OF THREE–WAY CONTINGENCY TABLES

In this paragraph we shall consider some models that start from the three-way contingency table and suppose, instead of the dependence structure as described above, the association among the three qualitative variables.

Among the various models proposed by Goodman (1986) we shall just consider two of them, in particular a bilinear model in which the three variables association structure is described by the multiplicative effect, given by the simple factors combinations (two-factor logbilinear model on a two-way contingency table) and a trilinear model (three-factor model on a three-way contingency table).

Let us consider the following bilinear model:

$$P_{ijk} = \alpha_i \alpha_j \alpha_k \exp \Sigma_\alpha (\phi_{1\alpha} v_{i\alpha} v_{j\alpha} + \phi_{2\alpha} \omega_{i\alpha} \omega_{k\alpha} + \phi_{3\alpha} \nu_{j\alpha} \nu_{k\alpha})$$

or

$$\begin{aligned} \log(P_{ijk}) &= \log \alpha_i + \log \alpha_j + \log \alpha_k \\ &\quad + \Sigma_\alpha (\phi_{1\alpha} v_{i\alpha} v_{j\alpha} + \phi_{2\alpha} \omega_{i\alpha} \omega_{k\alpha} + \phi_{3\alpha} \nu_{j\alpha} \nu_{k\alpha}) \end{aligned} \tag{11}$$

with constraints

$$\begin{aligned} \Sigma_i v_{i\alpha} P_{i..} &= \Sigma_j v_{j\alpha} P_{.j.} = \Sigma_i \omega_{i\alpha} P_{i..} = \Sigma_k \omega_{k\alpha} P_{..k} \\ &= \Sigma_j \nu_{j\alpha} P_{.j.} = \Sigma_k \nu_{k\alpha} P_{..k} = 0 \end{aligned} \tag{12}$$

Developing (11) and taking into account (12) we get:

$$\text{a) } \Sigma_i (\log P_{ijk}) P_{i..} = \Sigma_i (\log \alpha_i) P_{i..} + \log \alpha_j + \log \alpha_k + \Sigma_\alpha \phi_{3\alpha} \nu_{j\alpha} \nu_{k\alpha}$$

$$\text{b) } \Sigma_j (\log P_{ijk}) P_{.j.} = \log \alpha_i + \Sigma_j (\log \alpha_j) P_{.j.} + \log \alpha_k + \Sigma_\alpha \phi_{2\alpha} \omega_{i\alpha} \omega_{k\alpha}$$

$$\text{c) } \Sigma_k (\log P_{ijk}) P_{..k} = \log \alpha_i + \log \alpha_j + \Sigma_k (\log \alpha_k) P_{..k} + \Sigma_\alpha \phi_{1\alpha} v_{i\alpha} v_{j\alpha}$$

From the first equation, imposing the conditions:

$$\Sigma_i (\log \alpha_i) P_{i..} = \Sigma_k (\log \alpha_k) P_{..k} = 0$$

we get

$$\log \alpha_j = \Sigma_i \Sigma_k \log(P_{ijk}) P_{i..} P_{..k}$$

$$\log \alpha_k = \Sigma_i \Sigma_j \log(P_{ijk}) P_{.j.} P_{i..} - \Sigma_i \Sigma_j \Sigma_k \log(P_{ijk}) P_{..k} P_{.j.} P_{i..}$$

Furthermore, imposing the condition $\Sigma_j (\log \alpha_j) P_{.j.} = 0$ we estimate the parameters of the equation b) and c).

After estimated the parameters $\log \alpha_i$, $\log \alpha_j$, $\log \alpha_k$ we consider the matrices Z_{JK} , Z_{IK} , Z_{IJ} , respectively of dimension $(J \times K), (I \times K), (I \times J)$, whose general terms are:

$$\text{a) } Z_{jk} = \Sigma_i (\log P_{ijk}) P_{i..} - \log \alpha_j - \log \alpha_k$$

$$\text{b) } Z_{ik} = \Sigma_j (\log P_{ijk}) P_{.j.} - \log \alpha_i - \log \alpha_k$$

$$\text{c) } Z_{ij} = \Sigma_k (\log P_{ijk}) P_{..k} - \log \alpha_i - \log \alpha_j$$

The estimation of the parameters $\phi_{1\alpha}\, v_{i\alpha}\, v_{j\alpha'}$, $\phi_{2\alpha}\, \omega_{i\alpha}\, \omega_{k\alpha}$, $\phi_{3\alpha}\, \nu_{j\alpha}\, \nu_{k\alpha}$ are respectively the result of the decomposition in singular values of the following matrices:

$$D_I^{1/2} Z_{IJ} D_J^{1/2} \,,\, D_I^{1/2} Z_{IK} D_K^{1/2} \,,\, D_J^{1/2} Z_{JK} D_K^{1/2} \;.$$

A particular attention must be paid on the following trilinear model:

$$P_{ijk} = \alpha_{ij} \alpha_{ik} \alpha_{jk} \exp \left[\Sigma_\alpha \Phi_\alpha \eta_{i\alpha} \gamma_{j\alpha} \nu_{k\alpha} \right]$$

or

$$\log P_{ijk} = \log \alpha_{ij} + \log \alpha_{ik} + \log \alpha_{jk} + \Sigma_\alpha \Phi_\alpha \eta_{i\alpha} \gamma_{j\alpha} \nu_{k\alpha} \tag{13}$$

with constraints

$$\Sigma_i \eta_{i\alpha} P_{i..} = \Sigma_j \gamma_{j\alpha} P_{.j.} = \Sigma_k \nu_{k\alpha} P_{..k} = 0 \tag{14}$$

that describes the association, in multiplicative terms, of the three single factors.

Developing (13) and taking into account the constraints, we get

$$\log \alpha_{jk} = \Sigma_i \log(P_{ijk}) P_{i..}$$

$$\log \alpha_{ik} = \Sigma_j \log(P_{ijk}) P_{.j.} - \Sigma_i \Sigma_j \log(P_{ijk}) P_{.j.} P_{i..}$$

$$\log \alpha_{ij} = \Sigma_k \log(P_{ijk}) P_{..k} - \Sigma_j \Sigma_k \log(P_{ijk}) P_{..k} P_{.j.} + \\ + \Sigma_i \Sigma_j \Sigma_k \log(P_{ijk}) P_{i..} P_{..k} P_{.j.} - \Sigma_i \Sigma_k \log(P_{ijk}) P_{i..} P_{..k}$$

Let us consider the matrix S_{IJK}, whose general terms is

$$S_{ijk} = \log P_{ijk} - \log \alpha_{ij} - \log \alpha_{ik} - \log \alpha_{jk}$$

that verifies the following conditions

$$\Sigma_i S_{ijk} P_{i..} = \Sigma_j S_{ijk} P_{.j.} = \Sigma_k S_{ijk} P_{..k} = 0$$

The estimate of the trilinear term consists in minimizing the function

$$\Sigma_i \Sigma_j \Sigma_k (S_{ijk} - \Gamma_{ijk})^2 P_{.j.} P_{..k} P_{i..}$$

with $\Gamma_{ijk} = \Sigma_\alpha \Phi_\alpha \eta_{i\alpha} \gamma_{j\alpha} \nu_{k\alpha}$

As first step we consider the matrix $S_{I(JK)'}$ obtained by juxtaposition of the k strata of S_{IJK} and we minimize the function

$$\Sigma_i \Sigma_j \Sigma_k (S_{i(jk)} - \Gamma_{i(jk)})^2 P_{.j.} P_{..k} P_{i..}$$

or the decomposition in singular value of the matrix $D_I^{1/2} S_{I(JK)} (D_J \otimes D_K)^{1/2}$, where $(D_J \otimes D_K)$ has as general term $P_{.j.} \times P_{..k}$ $(j = 1 \ldots J\,,\, k = 1 \ldots K)$, that permits to have an estimate of $\Gamma_{i(jk)}$, in other words

$$\hat{\Gamma}_{i(jk)} = \Sigma_\alpha \partial_\alpha v_{i\alpha} \nu_{jk\alpha}$$

where the parameters verify the following conditions:

$$\Sigma v_{i\alpha} P_{i..} = \Sigma_j \nu_{jk\alpha} P_{.j.} = \Sigma_k \nu_{jk\alpha} P_{..k} = 0$$

Then, we diagonalize the matrix $\nu_{(JK}$ which coordinates have the identical properties of (14) , and then can constitute an estimation of the parameters of the trilinear part.

Likewise the non symmetrical way, a model including both a bilinear and a trilinear terms can be considered.

Table 1 - Asymmetric dependence models for three-way contingency tables.

Asymmetric Dependence Model		Model	Table	Weights	Centering
Bilinears	3 Factors	**Conditional**			
		a) $P_{ijk}/P_{.jk} = \alpha_{ik} \exp \Sigma_\alpha \phi_{k\alpha} v_{ik\alpha} \omega_{jk\alpha}$	$\log(P_{ijk}/P_{.jk})$ $k = 1 \ldots K$	$P_{.jk}/P_{..k}$ $k = 1 \ldots K$	$\Sigma_j \log(P_{ijk}/P_{.jk}) P_{.jk}/P_{..k}$ $k = 1 \ldots K$
		Multiple			
		b1) $P_{ijk}/P_{.jk} = \alpha_i \exp \Sigma_\alpha \phi_\alpha v_{i\alpha} \omega_{jk\alpha}$	$\log(P_{ijk}/P_{.jk})$	$P_{.jk}$	$\Sigma_{j,k} \log(P_{ijk}/P_{.jk}) P_{.jk}$
		b2) $P_{ijk}/P_{.jk} = \alpha_{ik} \exp \Sigma_\alpha \phi_\alpha v_{i\alpha} \omega_{jk\alpha}$	$\log(P_{ijk}/P_{.jk})$	$P_{.j.}$	$\Sigma_j \log(P_{ijk}/P_{.jk}) P_{.j.}$
		b3) $P_{ijk}/P_{.jk} = \alpha_{ik}\alpha_{ij} \exp \Sigma_\alpha \phi_\alpha v_{i\alpha} \omega_{jk\alpha}$	$\log(P_{ijk}/P_{.jk})$	$P_{.j.}P_{..k}$	$\left[\Sigma_j \log(P_{ijk}/P_{.jk}) P_{.j.} - [\Sigma_k \log(P_{ijk}/P_{.jk}) P_{..k} - \Sigma_j \Sigma_k \log(P_{ijk}/P_{.jk}) P_{.j.} P_{..k}]\right]$
	2 Factor	c1) $P_{ijk}/P_{.j.}P_{..k} = \alpha_i \exp \Sigma_\alpha [\phi_{1\alpha} v_{i\alpha} v_{j\alpha} + \phi_{2\alpha} \omega_{i\alpha} \omega_{k\alpha}]$	$\Sigma_j \log(P_{ijk}/P_{.j.}P_{..k}) P_{.j.}$	$P_{.j.}$	$\Sigma_k \Sigma_j \log(P_{ijk}/P_{.j.}P_{..k}) P_{.j.}P_{..k}$
			$\Sigma_k \log(P_{ijk}/P_{.j.}P_{..k}) P_{..k}$	$P_{..k}$	$\Sigma_k \Sigma_j \log(P_{ijk}/P_{.j.}P_{..k}) P_{.j.}P_{..k}$
		c2) $P_{ijk}/P_{.j.}P_{..k} = \alpha_i \alpha_{ik} \exp \Sigma_\alpha [\phi_\alpha \mu_{i\alpha} \mu_{j\alpha}]$	$\Sigma_k \log(P_{ijk}/P_{.j.}P_{..k}) P_{..k}$	$P_{..k}$	$\Sigma_k \Sigma_j \log(P_{ijk}/P_{.j.}P_{..k}) P_{.j.}P_{..k}$
		c3) $P_{ijk}/P_{.j.}P_{..k} = \alpha_i \alpha_{ij} \exp \Sigma_\alpha [\phi_\alpha \mu_{i\alpha} \mu_{k\alpha}]$	$\Sigma_j \log(P_{ijk}/P_{.j.}P_{..k}) P_{.j.}$	$P_{.j.}$	$\Sigma_k \Sigma_j \log(P_{ijk}/P_{.j.}P_{..k}) P_{.j.}P_{..k}$
Trilinears		d1) $P_{ijk}/P_{.j.}P_{..k} = \alpha_{ij}\alpha_{ik} \exp \Sigma_\alpha [\phi_\alpha \eta_{i\alpha} \xi_{j\alpha} \gamma_{k\alpha}]$	$\log(P_{ijk}/P_{.j.}P_{..k})$	$P_{.j.}P_{..k}$	$\left[\Sigma_j \log(P_{ijk}/P_{.j.}P_{..k}) P_{.j.} - [\Sigma_k \log(P_{ijk}/P_{.j.}P_{..k}) P_{..k} - \Sigma_j \Sigma_k \log(P_{ijk}/P_{.j.}P_{..k}) P_{.j.}P_{..k}]\right]$
		d2) $P_{ijk}/P_{.j.}P_{..k} = \alpha_i \exp \Sigma_\alpha [\phi_\alpha \eta_{i\alpha} v_{j\alpha} v_{k\alpha}]$	$\log(P_{ijk}/P_{.j.}P_{..k})$	$P_{.j.}P_{..k}$	$\Sigma_j \Sigma_k \log(P_{ijk}/P_{.j.}P_{..k}) P_{.j.}P_{..k}$

REFERENCES

AITCHINSON J., "*Principal component analysis of compositional data*" Biometrika, 1983, n. 70

BENZÈCRI J.P., "*L'Analyse des Donneés – Tome 2 : L'Analyse des Correspondances*" Dunod, Paris, 1973 .

CHOULAKIAN V., "*Exploratory analysis of contingency tables by loglinear formulation and generalization of correspondence analysis*" Psychometrika, vol. 53 n. 2, 1988

D'AMBRA L., LAURO N., "*Non symmetrical analysis of three-way contingency tables*" Roma, MULTIWAY, 27–30 marzo 1988

ESCOUFIER Y., JUNCA S., "*Least-squares approximation of frequencies or their logarithms*" Int.Stat.Rev., 54 n. 3, 1986

ESCOUFIER Y., "*Beyond correspondence analysis. Classification and Related Methods of Data Analysis*", 1988

GILULA Z., KRIEGER A.M., RITOV Y., "*Ordinal association in contingency tables: some interpretative aspects*" J.Am.Stat.Ass., June 1988, vol. 83

GOODMAN L., "*Simple models for the analysis of association in cross-classifications having ordered categories*" J.Am.Stat.Ass., vol. 74, 1979

GOODMAN L., "*Some useful extensions of the usual correspondence analysis approach and the usual loglinear models approach in the analysis of contingency tables*" Int. Stat. Rew., 54, n. 3, 1986

GREENACRE M.J., "*Theory and applications of correspondence analysis*" Academic Press, 1984

KROONENBERG P.M., "*Three-mode principal component analysis*" Leiden, DSWO, 1983

LAURO N., DE CARLI A., "*Correspondence analysis and loglinear models in multi-way contingency tables study: Some remarks on experimental data*" Metron 40, 221-234, 1982

LAURO N., D'AMBRA L., "*L'analyse non symmétrique des correspondances*" in Data Analysis ana Informatics III, Diday E. et al. eds, 1983

LAURO N., SICILIANO R., "*Correspondence analysis and modelling for contingency tables: symmetrical and non symmetrical approaches*" Vienna, Third International Workshop on Statistical Modelling, July 1988

RAO R., "*The use and interpretation of principal component analysis in applied research*" Shankhya A, 26, 1965

van der HEIJDEN P.G.M., MEIJERINK F., "*Generalized correspondence analysis of multi-way contingency tables and multi-way (super)indicators matrices*"

Roma Multiway – 88, 27 – 30 Marzo, 1988

van der HEIJDEN P.G.M., de LEEUW J. "*Correspondence analysis used complementary to loglinear analysis*" Psychometrika, 50, 429-447. 1985

A previous version of this paper was published on Unite' de Biometrie, Montpellier Technical Report n. 8802 December 1988.

ANALYZING ASYMMETRY OF TWO-WAVE TWO-VARIABLE PANEL DATA WITH GENERALIZED CORRESPONDENCE ANALYSIS AND LOGLINEAR MODELS

Patrick J.F. Groenen and Peter G.M. van der Heijden
Department of Data Theory and Department of Psychometrics, University of Leiden, Wassenaarseweg 52
P.O. Box 9555, 2300 RB Leiden, (Netherlands)

Abstract

In this paper it is our objective to give a graphical display of the asymmetry of a square contingency table formed by two variables measured at two time points. This is done by making combined use of generalized correspondence analysis and loglinear symmetry models. The residuals of loglinear symmetry models have various forms of asymmetry in it. These residuals are submitted to a skewsymmetric form of generalized correspondence analysis. In this way a graphical display of asymmetry is obtained.

1. Introduction

This paper discusses the analysis of so-called two-wave two-variable data, i.e. two variables measured at two time points, with correspondence analysis and loglinear analysis, both techniques for analyzing categorical data.
An example of a two-wave two-variable table is taken from Plewis (1985) (see table 1). In this example mental health (with categories good, moderate and poor) and loneliness (with categories absent and present) are measured at two occasions. In fact we have a four-way table, and table 1 shows this table in two-way form: the state of persons at time 1 is coded in the rows, and the state of persons at time 2 is coded in the columns. Many questions can be answered from these data, for example, in what way do persons change generally, does loneliness influence mental health or the other way around, and so on. In this paper we will concentrate on *asymmetry* in these data, i.e. we concentrate on the fact that the number of persons changing from joint state X at time 1 to joint state Y at time 2 is not the same as the number of persons changing from joint state Y to joint state X.

Table 1. Mental health (A_1 and A_2) and loneliness (B_1 and B_2) interactively coded at two time points with marginal frequencies.

			A_2						
			good		moderate		poor		
			B_2		B_2		B_2		
label	A_1	B_1	abs	pres	abs	pres	abs	pres	margin
R11	good	abs	34	3	13	1	2	0	53
R12	good	pres	7	1	2	1	1	3	15
R21	mod.	abs	13	3	12	4	6	2	40
R22	mod.	pres	4	3	2	2	1	1	13
R31	poor	abs	7	2	6	0	11	7	33
R32	poor	pres	1	2	3	9	8	12	35
margin			66	14	38	17	29	25	

We will analyze these data using a combined approach of both correspondence analysis and loglinear analysis. Loglinear analysis (LLA) is a technique for analyzing categorical data with which diverse hypotheses about the relation between variables can be verified. Various loglinear models for two-wave two-variable panel data are known from the literature (see Hagenaars, 1986; Plewis, 1985; see Groenen, 1988, for an overview). Correspondence analysis (CA) is another technique for analyzing categorical data. LLA and CA are closely related in the sense that CA can be interpreted as a method that decomposes the residuals from loglinear models (see van der Heijden et al., 1989, for an overview). In this paper we will use the two techniques jointly in the following way: we use loglinear models to obtain different forms of symmetry in the data. Then generalized CA (GCA) is used to analyze the residuals from these models. In this way GCA provides us a graphical representations of the asymmetry. We start now with a short introduction to CA, LLA and GCA, and discuss whether ordinary CA can help us to interpret the asymmetry of two-wave two-variable data.

2. Correspondence analysis and loglinear analysis

Correspondence analysis will be discussed here only briefly. For more details we refer to Benzécri et al. (1973), Gifi (1981) and Greenacre (1984). CA gives a multi-dimensional representation of the dependence of the row and column variable of a two-way contingency table. The representation is formed by assigning scores to row and column categories. Let us define matrix **P** as a two-way contingency table with proportions p_{ij}, where $p_{++} = 1$, $\mathbf{D}_r$ as a diagonal matrix with marginal row proportions p_{i+}, $\mathbf{D}_c$ as a diagonal matrix with marginal proportions p_{+j} and **E** as the matrix with the independent proportions $e_{ij} = p_{i+}p_{+j}$. The decomposition provided by CA is

$$\mathbf{P} = \mathbf{E} + \mathbf{D}_r\mathbf{R}\mathbf{\Lambda}\mathbf{C}'\mathbf{D}_c = \mathbf{D}_r(\mathbf{tt}' + \mathbf{R}\mathbf{\Lambda}\mathbf{C}')\mathbf{D}_c \qquad (1)$$

where **R** are the row scores normalized so that $\mathbf{RD}_r\mathbf{R} = \mathbf{I}$, **C** are the column scores normalized so that $\mathbf{CD}_c\mathbf{C} = \mathbf{I}$, $\mathbf{\Lambda}$ is a diagonal matrix with the singular values and **t** is a vector of ones of appropriate length. So (1) shows that CA can be interpreted as the decomposition of the departure from independence of matrix **P**.
Let us, for example, analyze table 1 with CA. The first and second dimensions of row and column scores are graphically presented in figure 1. In the sequel we will give the singular values and the proportion explained chi square distances. Plots are made with coordinates $\mathbf{R\Lambda}$ and $\mathbf{C\Lambda}$ so that the Euclidean distances are equal to chi-square distances. The first and second dimension have a singular values .591 (73%) and .274 (16%), respectively. We see in figure 1

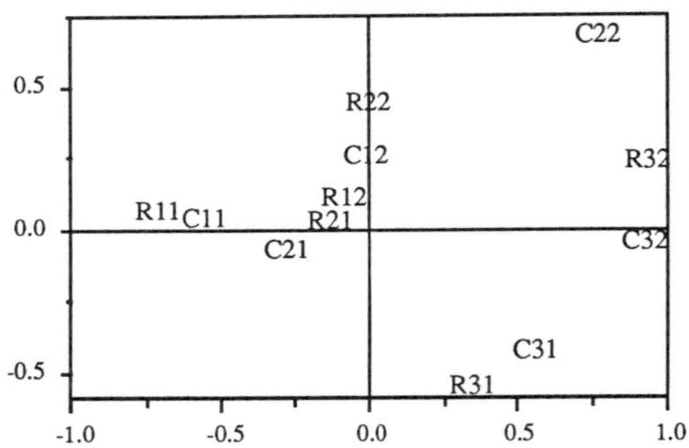

Figure 1. Dimension one and two of CA on table 1. Row points are labeled with R, column points with C.

that having good mental health and not being lonely at t_1 (R11) and t_2 (C11) have a higher proportion than expected. In the same way having poor mental health and being lonely at t_1 (R32) and t_2 (C32) also have a higher proportion than expected. Furthermore, we see that R11 and C11 are opposed to R32 and C32 in the first dimension. This leads us to the most important characteristic of this dimension, being that feeling alright (R11) at t_1 is followed more often than average by feeling alright at t_2 (C11) than by feeling bad (C32) at t_2. Feeling badly at t_1 (R32) is followed more than expected by feeling badly at t_2 (C32) than by feeling alright (C11). The second dimension shows analogously that a moderate mental health and feeling lonely at t_1 (R22) is not changed at t_2 (C22) more than expected and people having a poor mental health without feeling lonely at t_1 (R31) remain the same more than expected at t_2 (C31).

Loglinear analysis is a technique used for analyzing contingency tables. Only a brief explanation of LLA is given so that the reader is able to understand the next sections. For more details we refer to Bishop, Fienberg and Holland (1975). LLA decomposes the logarithm of probabilities in an ANOVA like way. For a two-way table this gives:

$$\log \pi_{ij} = u + u_{1(i)} + u_{2(j)} + u_{12(ij)} \tag{2}$$

In this decomposition there is one parameter for the mean, two sets of parameters for the margins and one set of parameters for the first order interactions. The parameters sum to zero for each index. Equation (2) is generalized to higher-way tables in a straightforward way.
In LLA one usually specifies a model with restrictions on the parameters. For (2) a well-known example is the independence model, for which $u_{12(ij)} = 0$. Once this model is specified, ML-estimates of the expected frequencies are calculated and the differences between the expected and observed frequencies are evaluated with the Pearson's goodness-of-fit statistic χ^2 (= Σ (observed - expected)^2 / expected) or the likelihood-ratio chi-squared test L^2 (= 2Σ (observed) log (observed/expected)). Both statistics are asymptotically chi-square distributed when the model is true and the data stem from a (product-)multinomial or a Poisson distribution. The number of degrees of freedom equals the number of cells minus the number of fitted parameters.
For our purpose it is important to know that ML estimates of loglinear models have the property that observed marginal proportions are equal to expected marginal proportions, for example, for the independence model $p_{i+} = \pi_{i+}$ and $p_{+j} = \pi_{+j}$. For this reason loglinear models are often denoted by the margins that are fitted to the expected proportions, for example, the independence model can be denoted as [1][2].
CA and LLA are closely related in the sense that CA decomposes the departure from the loglinear independence model (for other relations, see Goodman, 1986). In Van der Heijden and de Leeuw (1985) it was shown that this relation also holds for higher-way table. For example, they showed that CA of table 1 decomposes the departure from loglinear model $[A_1B_1][A_2B_2]$, because estimates of expected proportions π_{ijkl} are $\pi_{ijkl} = p_{ij++}p_{++kl}$. These values are also equal to the values in **E**.
Van der Heijden and de Leeuw (1985) also show how a generalization of CA can be used for the decomposition of other loglinear models. The generalization is due to Escofier (1983). The decomposition of generalized correspondence analysis (GCA) is:

$$\mathbf{P} = \mathbf{Q} + \mathbf{S}_r\mathbf{R}\boldsymbol{\Lambda}\mathbf{C}'\mathbf{S}_c \tag{3}$$

where **R**, **C** and $\boldsymbol{\Lambda}$ have the same form as in (1), $\mathbf{S}_r$ and $\mathbf{S}_c$ are diagonal matrices with weights and **P** and **Q** are matrices of the same size. Note that $\mathbf{S}_r$, $\mathbf{S}_c$, **P** and **Q** do not have the same relation as $\mathbf{D}_r$, $\mathbf{D}_c$,·**P** and **E** have in ordinary correspondence analysis (1). In this paper we will have **P** to be the observed proportions (as in (1)) and take for **Q** the estimates of an other model than the independence model.

The study of asymmetry in table 1 with ordinary CA will be very difficult, because the asymmetry is partly reflected by the margins and CA decomposes the departure from the margins. The asymmetry in table 1 that is not due to the margins is not necessarily displayed optimally in the first few dimensions by ordinary CA. Our objective is showing the asymmetry by making use of GCA. What we are going to do is this. We are going to decompose residuals from loglinear symmatry models with GCA. These residuals consist of asymmetry only. Thus we use loglinear models to filter the data.

3. Loglinear models for symmetry in two-wave two-variable data

For ordinary two-wave one-variable data symmetry models are well known (Bishop, Fienberg and Holland, 1975). However, with two-wave two-variable data symmetry models are more complicated. The ordinary symmetry model assumes that the marginal proportions have not changed. Therefore, asymmetry (i.e. the residuals from the symmetry model) may be due to differences between marginal proportions. Duncan (1979) and Hagenaars (1985) have discussed symmetry models for two-wave two-variable data that do take into account differences between corresponding marginal row and column proportions. We can distinguish three kind of differences between the marginal proportions of t_1 and t_2 of table 1:

1. the marginal proportions of A at t_1 can be different from the marginal proportions at t_2,
2. the marginal proportions of B at t_1 can be different from the marginal proportions at t_2,
3. the interaction of A and B at t_1 can be different from the interaction at t_2.

Making use of these three kind of differences four symmetry models can be formulated. The first model is the ordinary symmetry model. The ML-estimates Π of the two-wave two-variable counterpart of the symmetry model can easily be found by computing:

$$\hat{\pi}_{ijkl} = \hat{\pi}_{klij} = (p_{ijkl} + p_{klij})/2 \tag{4}$$

The associated χ^2 and L^2 of the symmetry model have IJ(IJ-1)/2 degrees of freedom. This model states that the probability to go from joint state (i, j) at t_1 to joint state (k, l) at t_2 is the same as the probability to go from (k, l) at t_1 to (i, j) at t_2. The marginal probabilities of rows are equal to those of the columns. To put it in another way, the above mentioned three types of differences are all absent. Model (4) is the symmetry model for two-way tables.
An easy procedure to fit symmetry with computer packages like SPSS^X or BMDP is an four-way extension of the two-way procedure of Bishop et al. (1975). Construct a five-way table of order (I x J x I x J x 2) with the four-way table **P** in the first layer of the fifth variable and the transposed table **P'** in the second layer (see Hagenaars, 1985). Fitting loglinear model $[A_1B_1A_2B_2]$ on this new five-way table will also give the estimates of the symmetry model. Every cell of the original table is estimated twice (namely in the first and in the second layer), so the chi-square and L^2 have to be divided by two. The number of degrees of freedom must be calculated as indicated above.
The next symmetry model is model (4) allowing for A_1A_2 asymmetry. This model is the symmetry model taking taking difference 1 into account. So the marginal proportions of variable B and the relation between A and B are supposed not to change in time. The estimates of this model can be found by fitting on top of model (4):

$$\hat{\pi}_{i+++} = p_{i+++} \quad \text{and} \quad \hat{\pi}_{++k+} = p_{++k+} \tag{5}$$

Now an extra I-1 parameters are needed for fitting the marginal proportions of variable A_1 and A_2 separately. Consequently the associated degrees of freedom decrease to (IJ(IJ-1)/2)-(I-1). If we define our fifth variable in the fitting procedure to be C then fitting $[A_1B_1A_2B_2][A_1C]\ [A_2C]$ in the five-way table yields the estimates of this model.
The third model is a even less restrictive model of the symmetry model. This model relaxes not only the equality restrictions on the marginal proportions of A, but also those of B. The model can be computed by fitting on top of model (5):

$$\hat{\pi}_{+j++} = p_{+j++} \quad \text{and} \quad \hat{\pi}_{+++l} = p_{+++l} \tag{6}$$

Another J-1 parameters are needed to lift the equality constraints on the marginal frequencies of B_1 and B_2. Consequently the associated chi-square test has (IJ(IJ-1)/2)-(I-1)-(J-1) degrees of freedom. The estimates of this model are found by fitting $[A_1B_1A_2B_2][A_1C][A_2C][B_1C]\ [B_2C]$ in the five-way table.
The last model is the quasi-symmetry model for the two-wave two-variable contingency table, which can be found by fitting on top of model (6):

$$\hat{\pi}_{12++} = p_{12++} \quad \text{and} \quad \hat{\pi}_{++34} = p_{++34} \tag{7}$$

This model permits all three kinds of differences mentioned above. Model (7) differs from model (6) in that the interactions of variable A and B are allowed to be different at t_1 and t_2. An additional (I-1)(J-1) parameters are needed for loosening the equality constraints on the marginal proportions of A_1B_1 and A_2B_2. The associated chi-square test has (IJ-1)(IJ-2)/2 degrees of freedom. This model totally reconstructs the margins of table 1. Fitting model (9) comes to the same as fitting the quasi-symmetry model for two-wave one-variable tables (Caussinus, 1965). Model (7) can be estimated by fitting $[A_1B_1A_2B_2][A_1B_1C]\ [A_2B_2C]$ in our five-way table.
The symmetry models above are nested and the difference between the various symmetry models can therefore be tested. One such test is the conditional test between the ordinary

symmetry model (4) and the quasi symmetry model (7) for marginal homogeneity. This test tells us whether the marginal row frequencies of table 1 are different from the marginal column frequencies or not. A conditional test of model (5) and model (7) can tell us whether there are changes in marginal frequencies of B and AB or not. Differences in marginal frequencies of variable A_1 and A_2 are allowed. A conditional test of model (6) and model (7) gives the answer to the question whether the interaction A_1B_1 equals A_2B_2 or not. Table 2 gives the results of a sequence of the four nested models and conditional tests for the example given in table 1. We may conclude that the three types of changes mentioned in the beginning of this section are not significant.

Table 2. Results for model (4) to model (7) and the conditional tests for table 1.

Model	L^2	df	*p*	χ^2
4. Symmetry	19.38	15	>.10	17.24
5. Symmetry and A_1A_2 asymmetry	14.20	13	>.25	12.93
6. Symmetry and A_1A_2 and B_1B_2 asymmetry	13.29	12	>.25	12.05
7. Quasi-symmetry	12.13	10	>.25	11.10
8. Marginal homogeneity (4) - (7)	7.25	5	>.25	
9. (5) - (7)	2.07	3	>.50	
10. (6) - (7)	1.16	2	>.50	

We have discussed some symmetry models that allow for asymmetry due to differences in marginal frequencies. The residuals of these models contain both asymmetry due to changes in marginal frequencies and asymmetry not due to changes in marginal frequencies. Using these four models we can filter the residuals of symmetry for the asymmetry due to changes in marginal frequencies.

4. A skewsymmetric generalization of CA

The residuals of the four symmetry models can be analyzed with GCA. For all models these residuals have the special property that the residuals are skewsymmetric, i.e. a residual $n_{ijkl} = (p_{ijkl} - \hat{\pi}_{ijkl})$ above the diagonal equals $-n_{klij} = -(p_{klij} - \hat{\pi}_{klij})$ below the diagonal. This can be seen from the fit procedure used above, i.e. construct a five-way table **Z** with elements $z_{ijkl1} = p_{ijkl}$ and $z_{ijkl2} = p_{klij}$. In all four models z_{ijkl+} is fitted and the maximum likelihood estimate $\hat{z}_{ijkl+}$ equals $z_{ijkl+} = p_{ijkl} + p_{klij} = \hat{\pi}_{ijkl} + \hat{\pi}_{klij}$. From the right most equation one easily derives that $(p_{ijkl} - \hat{\pi}_{ijkl}) = -(p_{klij} - \hat{\pi}_{klij})$ which proves that the residual matrix **N** is skewsymmetric. Gower (1977; see also Constantine & Gower, 1978) showed that the singular value decomposition of a skewsymmetric matrix **N** has the following form:

$$\mathbf{N} = \mathbf{K \Lambda L}' = \mathbf{K \Lambda J K}' \qquad (8)$$

where $\mathbf{K'K} = \mathbf{I} = \mathbf{L'L}$, $\mathbf{\Lambda}$ is a diagonal matrix with singular values, ordered in pairs $\lambda_1, \lambda_1, \lambda_2, \lambda_2$, and so on. Matrix **J** is a block diagonal matrix with blocks $\begin{pmatrix} 0 & 1 \\ -1 & 0 \end{pmatrix}$. If the number of rows is odd, the last singular value and the last diagonal element of **J** equal 0. A plot of the row scores **K** and column scores **KJ** for paired dimensions shows us that the configuration of column points is merely a rotation of 90° of the configuration of row points. Because the singular values are paired the decomposition is not unique (Constantine and Gower, 1982). This means that the paired dimensions of row and column scores may be rotated. A more simple plot is obtained by showing only the row points and by giving the direction of rotation. This is illustrated in figure 2, where 2 row and 2 column points are plotted. Gower (1977) shows that for figure 2 element n_{12} and $-n_{21}$ are approximated by twice the *area* of the triangle of the origin, R1 and R2. Figure 2 shows that the value n_{12} is positive since R1 and C2 have an angle smaller than 90 degrees, and n_{21} is negative since the angle is larger than 90 degrees. So we can also plot only R1 and R2 and indicate that the direction of rotation is counter-clockwise, showing that value n_{12} is positive. Here we see that rotation of paired dimension does not change the area.

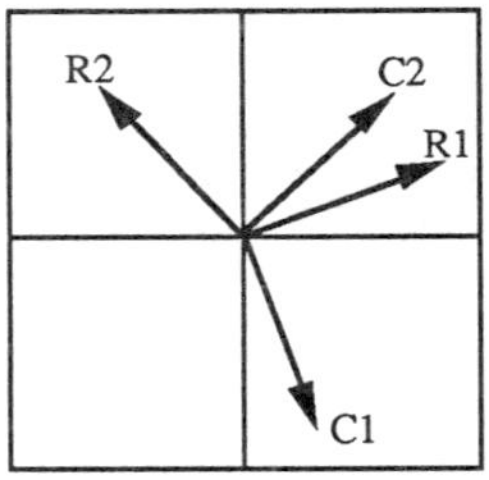

Figure 2. An example of two row points and two column points of a correspondence analysis solution of residuals of a symmetric matrix.

In Van der Heijden (1987) the properties of skewsymmetric matrices above was used to propose a skewsymmetric generalization of CA in the following way:

$$\mathbf{P} = \Pi + \mathbf{DR\Lambda JR'D} \qquad (9)$$

where Π is the matrix with expected proportions of a symmetry model, **R** are the row scores, Λ is a diagonal matrix with singular values and **D** are weights equal to $(p_{ij++} + p_{++kl})/2$. The weights are used to scale the residuals, just as in ordinary CA. The row and column weights are identical in order to retain the skewsymmetric properties of the singular value decomposition above. We are going to use (9) to decompose the residuals from models (4) to (7). For these applications figure 2 can be interpreted as that joint state 1 is followed more by joint state 2 than vice versa.
Models (4), (5) and (6) may not fit because of two types of asymmetry: asymmetry due to differences in the margins and asymmetry not due to the margins. Residuals of model (7) only contain asymmetry not due to the margins. Decomposing the residuals of model (4) with (9) shows symmetry due to differences in marginal proportions of A_1B_1 and A_2B_2. So the margins can differ due to asymmetry of A_1 and A_2, B_1 and B_2, and A_1B_1 and A_2B_2. The residuals of model (5) have asymmetry controlled for changes in marginal proportions of variable A. Asymmetry due to marginal differences of B_1 and B_2 and marginal differences of A_1B_1 and A_2B_2 are still present, but the marginal differences of A_1 and A_2 are not present. During the interpretation of the GCA decomposition one has to be careful not to derive conclusions which refer to the marginal proportions of variable A_1 and A_2, because these effects have been filtered out of the solution. The same can be told for the residuals of model (6). Now the GCA solution shows asymmetry controlled for differences in marginal proportions of A_1 and A_2 and of B_1 and B_2. The solution still shows the marginal differences of A_1B_1 and A_2B_2. Here one should be careful not to interpret the solution in terms of differences in marginal frequencies of variable A and of variable B. The residuals of model (7) only have the asymmetric effects controlled for differences in the interaction of A_1B_1 and A_2B_2, so the GCA solution shows only asymmetry that is not due to marginal differences.

Example

GCA will be performed on residuals of the four models. Taking the c2 values in table 2 seriously would stop us from a residual analysis. Nevertheless, for purpose of illustration we will analyze the residuals with the method above.
GCA on residuals of the first model can be interpreted by looking at the plot of the first two dimensions of the row scores. Figure 3.a displays these first two dimensions of the CA solution with singular values .217, which explain 86% of the inertia.
The points for absence of loneliness (R11, R21 and R31) have small residuals; they are near to the origin and do not contribute much to the solution. Going clockwise from R32 to R22 the figure shows us that if people are lonely, a poor mental health (R22) is followed more by a moderate mental health (R32) than vice versa. This is the most important characteristic of the solution, because the area of the origin, R32 and R22 is the largest area. If people are lonely, a moderate mental health (R22) is more often followed by a good one (R12) than vice versa. There are roughly as many people (who are lonely and) who have a poor mental health at time point one (R32) and a good one at time point two (R12) as people who decline from good mental

health to a poor one, since the area spanned by R32, R12 and the origin is practically zero. This can easily be checked in table 1 by comparing the corresponding frequencies (2 and 3).

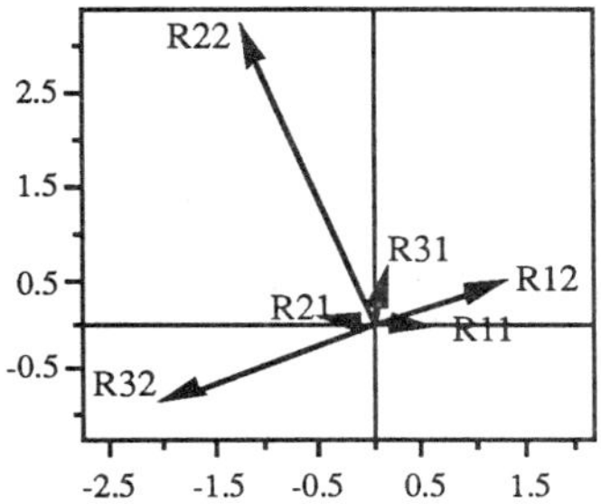

Figure 3.a GCA on residuals of model (4)

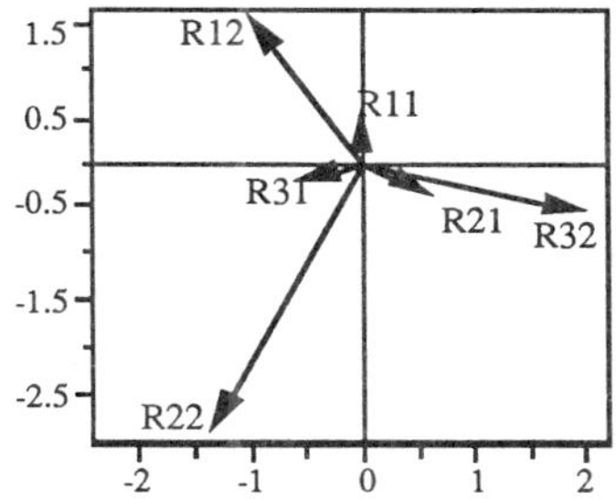

Figure 3.b GCA on residuals of model model (5)

Model (5) is the symmetry model corrected for asymmetry due to the margins of A. The residuals of this model therefore contain asymmetry due to margins of B, due to interaction differences of AB and asymmetry not due to margins. Figure 3.b shows the CA solution on residuals of model (5). In the plot we see that again only the points referring to 'loneliness present' (R12, R22 and R32) contribute much to the CA solution. The solution in two dimensions, with singular values .164, is still very acceptable because 75% of the inertia is shown.

Considering only people who feel lonely, the plot shows us, going clockwise, that good mental health (R12) is followed more often by a bad one (R32) than vice versa. This effect is not very strong because the area spanned by R12, R32 and the origin is small. On the other hand, just as we have seen in figure 3.a, bad mental health (R32) is more followed by moderate mental health (R22) than the other way around and moderate mental health (R22) is more followed by good mental health (R12) than vice versa. This solution compared with the previous one shows us that taking asymmetry due to differences in marginal proportions of mental health into account there is some asymmetry between R12 and R32: for lonely people a good mental health is followed more often by a bad one.

The last two solutions, GCA on residuals of model (6) and (7), are very similar to figure 3.b and does not give us new information.

The residual analysis we propose will be more helpful when dealing with variables that have many categories. The symmetry models will tend not to fit adequately and our skewsymmetric generalization of CA can be of great help for analyzing the residuals.

5. Conclusion and discussion

The decomposition of the residuals of model (4), (5), (6) and (7) is a powerful tool with which one investigates the asymmetry in the data between two time points while suppressing the effects due to specific marginal proportions. Our method can be particularly helpful when dealing with variables with many categories. This method can be used for graphically displaying asymmetry of the data, when there is a need to suppress differences in margins between two time points. When, for example, CA on residuals of the symmetry model is completely dominated by very different marginal frequencies of A_1 and A_2, then it is advised to use the residuals of the second model. Now the solution will be more easy to interpret because the marginal effects of the first variable have been filtered out of the solution.

Models (4), (5), (6) and (7) can also be extended in a straightforward way to two-wave multi-variable contingency tables. Then various symmetry models can be formulated allowing for asymmetry due to differences in main effects (like model (5) and (6)), asymmetry due to differences in first order interactions (like model (7)) and asymmetry due to differences in higher order interactions. For reasons of simplicity we have discussed two-wave two-variable contingency tables only.

We conclude that GCA on the residuals of these symmetry models gives us an interesting graphical display of asymmetry in the data. The interpretation is made in a straightforward way in terms of joint-category X at t_1 which is followed more by joint-category Y at t_2 than the other way around. Therefore residual analysis of loglinear symmetry models using GCA provides an attractive instrument for analyzing two-wave two-variable data.

References

Bishop, Y.M.M., Fienberg, S.E. & Holland, P.W. (1975). Discrete Multivariate Analysis. Cambridge: M.I.T. Press.

Benzécri, J.P. et al. (1973). Analyse des données (2 vols). Paris: Dunod.

Caussinus, H. (1965). Contribution a l'analyse statistique des tableaux de corrélation. Annales de la Faculté des Sciences de l'Université de Toulouse, 29, pp.77-182.

Constantine, A.G., Gower, J.C. (1978). Graphical representation of asymmetric matrices. Applied statistics, 27, pp.297-304.

Constantine, A.G., Gower, J.C. (1982). Models for the analysis of interregional migration. Environment and Planning, 14, pp.477-497.

Duncan, O.D. (1979). Testing key hypotheses in panel analysis. In: K.F. Schuessler (Ed.), Sociological Methodology 1980. San Fransisco: Jossey-Bass, pp.279-289.

Escofier, B., (1983). Analyse de la difference entre deux mesures sur le produit de deux mêmes ensembles. Cahiers de l'Analyse des Données, 8, pp.322 - 333.

Gifi, A., (1981). Non-linear multivariate analysis. Leiden: University of Leiden, Department of Data Theory.

Goodman, L.A. (1986). Some useful extensions of the usual correspondence analysis approach and the usual log-linear models approach in the analysis of contingency tables. International Statistical Review, 54, pp.243-309.

Gower, J.C. (1977). The analysis of asymmetry and orthogonality. In: J. Barra et al. (Ed.), Recent developments in statistics. Amsterdam: North Holland.

Greenacre, M.J. (1984). Theory and applications of correspondence analysis. New York: Academic Press.

Groenen, P.J.F. (1988). Generalized correspondence analysis of two-wave two-variable panel data, with loglinear approaches. Leiden: University of Leiden, Department of Psychometrics and Research Methods (PRM 88-07).

Hagenaars, J.A.P. (1985). Loglineaire analyse van herhaalde surveys; panel-, trend- en cohortonderzoek. Unpublished doctoral, Catholic University of Brabant, Tilburg, The Netherlands.

Hagenaars, J.A. (1986). Symmetry, quasi-symmetry, and marginal homogeneity on the latent level. Social Science Research, 15, pp.241-255.

Heijden, P.G.M. van der (1987). Correspondance analysis of longitudinal categorical data. DSWO Press, Leiden.

Heijden, P.G.M. van der & de Leeuw, J. de (1985). Correspondance analysis used complementary to loglinear analysis. Psychometrika, 50, pp.429-447.

Heijden, P.G.M. van der, Falguerolles, A. de & de Leeuw, J. de (1989, in press). A combined approach to contingency table analysis using correspondence analysis and loglinear analysis. Applied Statistics.

Plewis, I. (1985). Analyzing change. Measurement and explanation using longitudinal data. New York: John Wiley & sons.

DISSYMMETRICAL ASSOCIATION ANALYSES BETWEEN TWO QUALITATIVE VARIABLES

R. Abdesselam. Y. Schektman
AEGIDE, Greco CNRS 59
Toulouse Le Mirail University, 31058 Toulouse Cedex (France)

Key words :

Inner product with relationship effects, Pearson coefficient mean square contingency, Tschuprov coefficient, Goodman-Kruskal tau.

1. GENERALITIES

The methods which analyse associations, generally only take into account symmetrical association. In fact, practically the two sets of observed variables do not often play the same role, some are predictor variables and others predictand variables. In this work, qualitative variables denoted by x and y are respectively predictor and predictand .

Let (x^j ;j=1,p) and (y^k ;k=1,q) be the indicator variables associated to the modalities of these variables and $E_x = IR^p$ $E_y = IR^q$ be the corresponding subspaces in the individual-space $E = E_x \oplus E_y$. The set of individuals is represented in E_x by a configuration of points denoted by N_x : the values x^j_i (j=1,p) of variables (x^j;j=1,p) for the i_{th} individual are coordinates of the corresponding point in N_x. N_y is defined in the same way. For any definition referring to x, a similar definition referring to y exists. So, we shall not give the definition referring to y subsequently. The formal approach for defining general association coefficients, in section 2, is based on inner products with relationship effects [6].

Let $\{u_j(x),U^j(x)$;j=1,p} be a set of principal vectors and components of N_x ,in a geometrical way, the inner product in E is with relationship effect for (E_x,E_y) if :

$$\cos[u_j(x),u_k(y)] = \cos[U^j(x),U^k(y)] \text{ if } ||U(x^j)|| \neq 0 \text{ and } ||U(y^k)|| \neq 0$$
$$= 0 \qquad\qquad \text{else.}$$

The inner products in E_x and E_y being given, it is shown [5] that the inner product M in E is with relationship effect for (E_x,E_y) if :

$$^{t}In_x \; M \; In_y = M_x[(V_xM_x)^{1/2}]^+ \; V_{xy} \; M_y[(V_yM_y)^{1/2}]^+ \tag{1}$$

where,

- $^{t}In_x$ is the transpose of the canonical injection matrix from E_x into E,
- M_x is the inner product matrix in E_x ,
- V_x is the covariance matrix of variables $\{x^j\}$,
- V_{xy} is the covariance matrix between variables $\{x^j\}$ and $\{y^k\}$,
- $[(V_xM_x)^{1/2}]^+$ is the Moore-Penrose generalized inverse of $(V_xM_x)^{1/2}$ weighted by M_x .

2. RELATIONSHIP ASSOCIATION COEFFICIENTS

We use the results of this section for describing a new formal framework for classical methods, like correspondence analysis [2], and also for new ones, like dissymmetrical correspondence analysis [4,7],which are extended. It is based on the definition of general relationship association coefficients between two sets of variables. We propose two complementary approachs for defining them.

2.1. Formal relationship approach

The inner product M in E, is supposed to be with relationship effect for (E_x,E_y). Note that the inner products in E_x and E_y depend upon the problem, so we shall specify them when they are needed .

DEFINITION 1: Symmetrical (SRAC) and Dissymmetrical (DRAC) Relationship Association Coefficients are defined by

$$SRAC[x,y \;/M_x=V_x^+,M_y=V_y^+] = I_x(N_y) \;/\; [I_x(N_x) \; I_y(N_y)]^{1/2}$$

$$DRAC[x\to y \;/M_x,M_y] = I_x(N_y)/I_y(N_y)$$

where,

- V_x^+ is the Moore-Penrose generalized inverse of V_x ,
- $I_x(N_y)$ is the inertia of the orthogonal projection of N_y into E_x relative to its centre of gravity.

For a qualitative variable x, it is shown [5] that V_x^+ can be replaced by the chi-square distance matrix χ^2_x .

PROPERTY 1: $I_x(N_y) = \emptyset^2(x,y)$ if $M_y = \chi^2_y$

$= I_y(N_y)\ \tau(x\text{->}y)$ if $M_y = I_q$

where,

- $\emptyset^2(x,y)$ is the Pearson coefficient of mean-square contingency,
- $\tau(x\text{->}y)$ is the Goodman-Kruskal tau,
- I_q is the identity matrix .

proof:

As $I_x(N_y) = tr(P_x\ I_{ny}\ V_y\ {}^tI_{ny}\ M)$ and $P_x = I_{nx}\ M_x^+\ {}^tI_{nx}\ M$ (2)
using (1), one has,

$$I_x(N_y) = tr(I_{nx}\ M_x^+\ M_x\ [(V_xM_x)^{1/2}]^+\ V_{xy}\ M_y[(V_yM_y)^{1/2}]^+\ V_y\ {}^tI_{ny}\ M)$$

$$= tr(V_{xy}\ M_y\ V_{yx}\ V_x^+)$$

which leads to $I_x(N_y) = \emptyset^2(x,y)$ for $M_y = \chi^2_y$

and to $I_x(N_y) = \sum_{k=1}^{q} ||\ Q_x(y^k)||^2 = \tau(x\text{->}y)\ I_y(N_y)$ for $M_y = I_q$,

where,

> $Q_x(y^k)$ is the orthogonal projection of the variable vector y^k on the subspace spanned by the vectors $\{x^j;j=1,p\}$, associated to the variable x^j, assumed to have zero means.

One can deduce immediately the following property

PROPERTY 2: For two qualitative variables x and y

$SRAC[x,y\ /M_x = \chi^2_x, M_y = \chi^2_y]$ is the Tschuprov coefficient

$DRAC[x\text{->}y\ /M_x\ , M_y = I_q\ ]$ is the Goodman-Kruskal tau .

2.2 Empirical algebraic approach

x and y being qualitative variables, for measuring dissymmetrical association (x->y), one can reasonably propose the general formula

$$F(x\text{->}y) = \sum_{j=1}^{p} n_{j.}/n \sum_{k=1}^{q} m_k\ [n_{jk}/n_{j.} - n_{.k}/n]^2 \quad (3)$$

where $n_{jk} = \sum_{i=1}^{n} x_i^j\ y_i^k$ and $n_{j.} = \sum_{k=1}^{q} n_{jk}$.

Let x_j^* be the point in IR^q whose coordinates are $\{n_{jk}/n_{j.}\ ;k=1,q\}$

and let set $N_x^* = \{x_j^*;j=1,p\}$, $G_x^* = \sum_{j=1}^{p}(n_{j.}/n)\ x_j^*$ (the centre of gravity of N_x^*) and $M_x^* = Diag[m_k\ ]$.

Clearly one has $F(x\text{->}y) = \sum_{j=1}^{p} n_{j.}/n\ \ d^2(x_j^*, G_x^*)$

i.e. $F(x\to y)$ is equal to the inertia of N_x^* relative to G_x^* with respect to M_x^*.

Starting from (3) and replacing $\{m_k ; k=1,q\}$ by the convenient values, it is easy to show [1] the following property.

PROPERTY 3:

- If $m_k = n/n_{.k}$ for all k then

  $F(x\to y)$ is the Pearson coefficient of mean-square contingency,

- if $m_k = n^2 / \sum_{k=1}^{q} n_{.k}(n-n_{.k})$ for all k then

  $F(x\to y)$ is the Goodman-Kruskal tau.

Obviously, with (3) we can define more general dissymmetrical coefficients. For that, let us define before, in a geometrical way, two new coefficients. These definitions are justified by the classical geometrical definition of the Goodman-Kruskal tau.

One has
$$\tau(x\to y) = \sum_{k=1}^{q} \frac{\|y^k\|^2}{\sum_{l=1}^{q}\|y^l\|^2}\,\frac{\|Q_x(y^k)\|^2}{\|y^k\|^2}$$

i.e. the weighted sum of the proportions of $\|y^k\|^2$ (k=1,q) linearly explained by the variables $\{x^j\}$.

DEFINITION 2:

- Equally weighted tau, denoted by τ_{ew}, is defined by

$$\tau_{ew}(x\to y) = 1/q \sum_{k=1}^{q} \frac{\|Q_x(y^k)\|^2}{\|y^k\|^2} \tag{4}$$

- Weighted tau, denoted by τ_w, is defined by

$$\tau_w(x\to y) = \sum_{k=1}^{q} \frac{n_{.k}\,\|y^k\|^2}{\sum_{l=1}^{q} n_{.l}\,\|y^l\|^2}\,\frac{\|Q_x(y^k)\|^2}{\|y^k\|^2} \tag{5}$$

For $\{y^k\}$ and $\{x^j\}$ assumed to be centered, let us take the inner product matrix in the variable-space IR^n, denoted by $D_v = 1/n\ I_n$, where n is the number of individuals.

As $\|y^k\|^2 = n_{.k}(n-n_{.k})/n^2$ and $\|Q_x(y^k)\|^2 = \sum_{k=1}^{q} n_{j.}/n\ [n_{jk}/n_{j.} - n_{.k}/n]^2$

starting from (4) and (5), we show the following property.

PROPERTY 4:

- $F(x \to y) = \tau_{ew}(x \to y)$ if $m_k = \dfrac{n^2}{q\, n_{.k}\,(n - n_{.k})}$ for all k

- $F(x \to y) = \tau_v(x \to y)$ if $m_k = \dfrac{n^2\, n_{.k}}{\sum_{l=1}^{q} n_{.l}^2\,(n - n_{.l})}$ for all k .

Let us give now a useful property as far as symmetrical associations are concerned. We give it for two qualitative variables, but it can be extended to mixed variables.

PROPERTY 5: $\{x^j\}$ being centered, if M is with relationship effect for (E_x, E_y) with $M_y = \chi^2_y$, then coordinates in E_x of $P_x(y_i)$, y_i is being such that $y_i^k = 1$, are equal to $\{(n_{jk}/n_{.k} - n_{j.}/n)\ ; j=1,p\}$.

proof:

As the matrix expression of (2) is given by $P_x = \left[\begin{array}{c|c} I_p & V_{xy}\,\chi^2_y \\ \hline 0 & 0 \end{array}\right]$,

one has $\quad {}^t I_{nx}\, P_x(y_i) = V_{xy}\, \chi^2_y\, {}^t I_{ny}\, e_k(y) \quad$ (6)

where $\quad [{}^t I_{ny}\, e_k(y)]_l = \delta_{lk} \quad$ if $y_i^k = 1$.

As $[V_{xy}]_{jk} = n_{jk}/n - n_{j.}n_{.k}/n^2$ and $[\chi^2_y]_{jk} = \delta_{jk}\, n/n_{.k}$

the expected result is obtained by developping (6).

3. ANALYSIS OF ASSOCIATIONS IN PRINCIPAL ELEMENTS

3.1 The method

The classical correspondence analysis [2] is the most popular example of this approach. The association coefficient analysed is the (symmetrical) Pearson mean square contingency. It consists of the two following Principal Components Analysis (PCA)

$$\text{PCA}\ \{\ N^*_x\ ;\ M^*_x = \text{Diag}[n/n_{.k}]\ ;\ \text{Diag}[n_{j.}/n]\ \} \quad (7)$$

and

$$\text{PCA}\ \{\ N^*_y\ ;\ M^*_y = \text{Diag}[n/n_{j.}]\ ;\ \text{Diag}[n_{.k}/n]\ \}\ . \quad (8)$$

These analysis correspond to the empirical approach described in section 2.2 .

Expressed in the formal approach (section 2.1), these two previous PCA are equivalent to the two following PCA (property 5)

$$\text{PCA}\ \{\ P_y(N_x)\ ;\ M\ ;\ D_v\ \} \quad \text{and} \quad \text{PCA}\ \{\ P_x(N_y)\ ;\ M\ ;\ D_v\ \}\ .$$

In the same way, for analysing dissymmetrical association, $(x \to y)$ for example, one must do now the two following PCA (property 5 is now false).

$$\text{PCA } \{ N_x^* \ ; \ M_x^* \ ; \ \text{Diag}[n_{j.}/n] \ \} \tag{9}$$

and

$$\text{PCA } \{ P_x(N_y) \ ; \ M \ ; \ D_v \ \} \tag{10}$$

where M_x^* and M is depending now upon the dissymmetrical coefficient which is analysed.

So, for the weighted tau, M_x^* will be defined using the property 4 and M will be defined using the following result [1].

PROPERTY 6:

$$\tau_v \ (x \rightarrow y) = \text{DRAC}[x \rightarrow y \ / M_x, \ M_y = \text{Diag}[n_{.k}/n] \ ].$$

3.2 Illustrative example (Part I)

In figure 1, we give the illustrative data used for applying the methods described above.

	x^1	x^2	x^3	x^4	
y^1	50	50	1	1	102
y^2	51	1	50	2	104
y^3	1	350	2	650	1003
y^4	2	2	500	500	1004
	104	403	553	1153	2213

a) Frequencies

	x^1	x^2	x^3	x^4	
y^1	.481	.125	.002	.001	.046
y^2	.490	.002	.090	.002	.047
y^3	.010	.868	.004	.564	.453
y^4	.019	.005	.904	.433	.454

b) Probabilities $[y^k/x^j]$

Figure 1: Illustrative data

$\{x^j\}$ (resp.$\{y^k\}$) are indicator variables associated to the modalities of the predictor x (resp.predictand y).

The scatterplots of the modalities $\{x_j^*\}$ and $\{y_k\}$, on the first principal plane, carry out by the classical correspondence analysis, [PCA (7) and (8)], are given in figure 2.

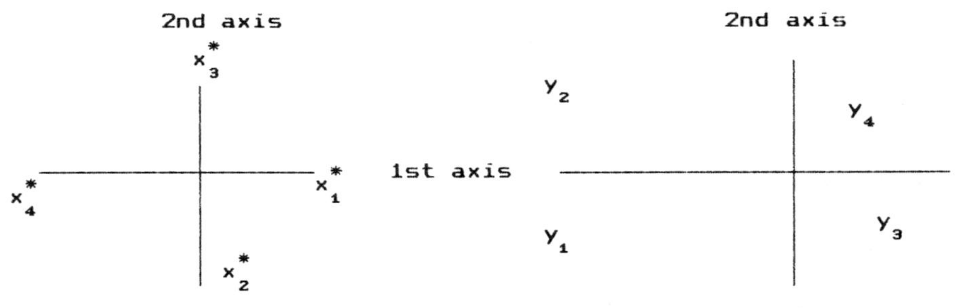

Figure 2: Results of classical correspondence analysis $[\emptyset^2(x,y)]$

The similar results but for the analysis of the dissymmetrical association coefficient $\tau(x\to y)$ [Goodman - Kruskal], by doing the two PCA (9 and 10), are given in figure 3 respectively a) and b).

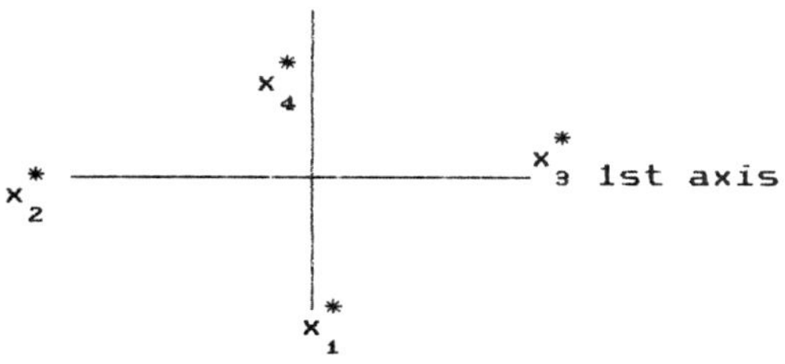

a) PCA $\{N^*_x;M^*_x\}$

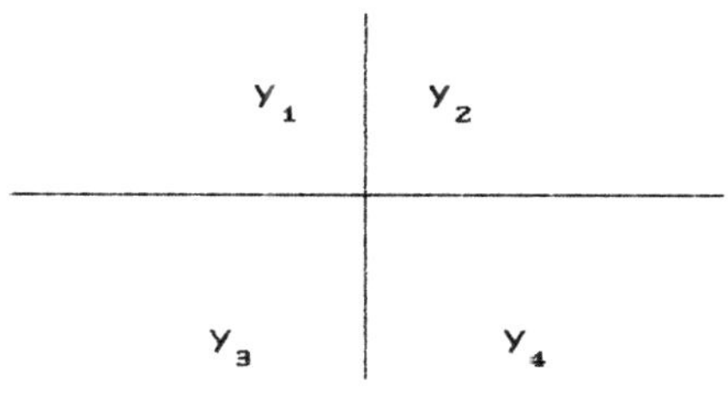

b) PCA $\{P_x(N_y);M_y=I_4\}$

Figure 3: Results of a dissymmetrical analysis [$\tau(x\to y)$]

Scatterplots of figure 3 are approximatively obtained, from those of figure 2, by a rotation of 90 degrees. As it can be shown in figure 1-b, the probabilities in row y^3 and y^4 are more different of the marginal probabilities than those of row $y^1$and y^2. So, y^3 and y^4 being better forcasted than y^1 and y^2, the scatterplot in figure 3-b carried out by the analysis of the $\tau(x\to y)$ is better than the correspondence one carried out by the analysis of the $\emptyset^2(x,y)$ [figure 2].

3.3 Possible improvements of the method

One can provide a simultaneous graphical representation of $\{x_j\}$ and $\{y_k\}$ on the same principal plane, by locating in figure 3-b, each x_j at the gravity centre of the points $\{y_k\}$ with respect to a well chosen system of weights.

One can provide also an "optimal" simultaneous graphical representation. One way is to search inner products in E_x and E_y, in the context of the formal approach [figure 3-b for $\{y_k\}$], which optimize criteria given by the user. Practically these criteria will be functions of the probabilities $P[y^k/x^j]$ and the distances between (x_j,y_k). More over, we suggest to search inner products (with relationship effect) where M_y (resp. M_x) is built as follow

$$\begin{cases} M_y(\alpha) = (1-\alpha)\ \chi^2_y + \alpha\ I_q \\ \alpha \in [0,1] \end{cases} \quad \text{and} \quad \begin{cases} M_y(\alpha) = (1-\alpha)\ I_q + \alpha\ V_y \\ \alpha \in [0,1] \end{cases}$$

3.4 Illustrative example (Part II)

For the data described in 3.2, figure 4 gives the values of the distances between all pairs (x_j,y_k) for the analysis of the $\emptyset^2(x,y)$ and for the analysis of the dissymmetrical association coefficient corresponding to the maximum of the KENDALL rang concordence coefficient between the two samples $\{\max[d(x_i,y_m)]-d(x_j,y_k)\ ;\text{for all } j,k\}$ and $\{P[y^k/x^j]\ ;\text{for all } j,k\}$.

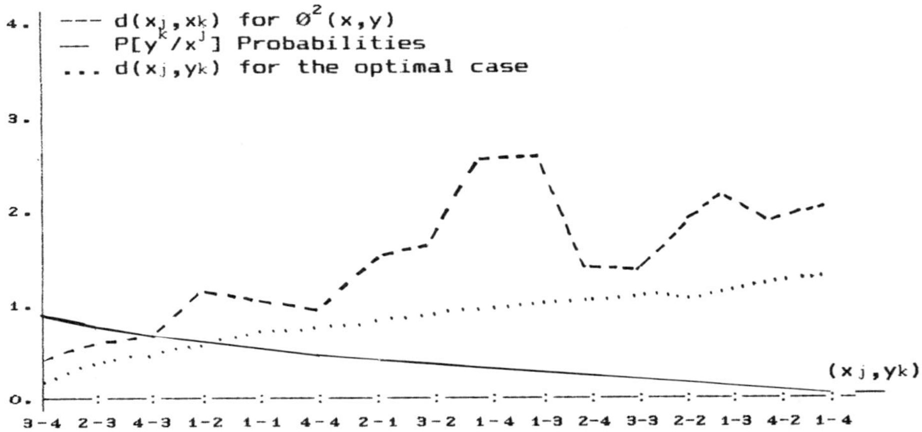

Figure 4: Distances

Note that, in figure 4, the pairs (x_j,y_k) are ordered in the decreasing order according to the $\{P[y^k/x^j]\}$ values. Obviously, a graphical representation will be reasonable if $d(x_j,y_k)$ is small when $P[y^k/x^j]$ is great.
So, we observe, in figure 4, that the results are better when we analyse a dissymmetrical association coefficient by chosing inner product in E in an optimal way, than when we do classical correspondence analysis.

REFERENCES

[1] R.Abdesselam (1988) : "Contribution a l'analyse des associations dissymetriques" These 3me cycle, Universite Paul Sabatier (U.P.S - Toulouse).

[2] J.-P.BENZECRI (1976) :l'Analyse des Donnees, Tome 2 "l'analyse des correspondances" Dunod 2me edition.

[3] C.Fabre (1986) : "Contribution a la protection des methodes relationnelles" These 3me cycle, Universite Paul Sabatier (U.P.S - Toulouse).

[4] N.Lauro,L.D'Ambra (1983) : "Analyse non symetrique des correspondances" Data analysis and informatics. E.Diday and all. Editors, North-Holland Publishing Company 433-446.

[5] Y.Schektman (1984) : "Metriques et Analyse de donnees. Contribution et Synthese" deuxieme colloque Franco-Sovietique. Statistique et Analyse des Donnees CNET Paris.

[6] Y.Schektman (1983) : "A general euclidean approach for mesuring and describing associations between several set of variables" In recent developments in clustering and data analysis. Proceeding of the Japanese-French scientific seminar. Ed. C.Hayashi and all., academic Press,Inc., Tokyo, 37-48.

[7] YVONNE M.M.BISHOP, STEPHEN E.FIENBERG, and P.W.HOLLAND "Discrete multivariate analysis; theory and practice" the MIT press Cambridge, Masschusetts and London, England.

CONTRACTIONS FOR THREE-WAY TABLES

R.G. Flöres Jr
UFRJ and CIDE, Rio de Janeiro (Brasil)

1. INTRODUCTION

The exploratory analysis of three-way contingency tables relies many times on special two-way tables built upon the original array. Among these, the so-called Burt's table plays a most important role, as it allows the simultaneous analysis of all the categories involved in the three-way classification. Nevertheless, beyond keeping only information related to the two-way marginals, its special features have been frequently criticized.

In this paper we develop a generalization of the Burt's table idea which partially overcomes some of its shortcomings. The generalization can be linked to a geometrical representation for the global analysis of three-way tables, enhancing its meaning. The application of some simple, approximate formulas can provide a further insight on the transformations to be used.

2. THE CASE WITH THE STANDARD BURT'S TABLE.

Consider a three-way set of counts n_{ijt}, where $i \in I$, $j \in J$ and $t \in K$, all finite sets with cardinals n, m, and k, respectively. Empirical distributions (relative frequencies) will be denoted by f_{ijt}, f_{jt}, f_i, f^i_{jt}, etc, those with a superscript meaning a conditional distribution, the conditioning set corresponding to the symbol at the top. The total number of counts is $N = \Sigma\, n_{ijt}$.

The standard Burt's table is a symmetric two-way table on $(I+J+K)x(I+J+K)$, whose general element $x_{aa'}$, for a and a' belonging to different marginal sets, say $(a,a')=(i,j)$, is $x_{aa'}= x_{ij}= \sum_t n_{ijt}$; and for a and a' belonging to the same marginal set, say I, is $x_{aa'}= x_{ii}= \sum_{j,t} n_{ijt}$, if $a=a'$, and $x_{aa'}=\emptyset$, if $a\neq a'$.

This table is usually analysed through correspondence analysis (CA), Lebart, Morineau et Tabard(1977), its chi-square being equal to:

$$X^2 = N (n+m+k-3) + 2 (X^2_{IxJ} + X^2_{IxK} + X^2_{JxK}) \qquad (1)$$

The first term in (1) stands for the fact that the submatrices corresponding to the same marginal set are diagonal matrices with the absolute marginal frequencies in the diagonal. The second term is exactly the pooling of the chi-squares of the three possible two-way marginal tables.

The inertia of the corresponding cloud in R^{n+m+k} is:

$$I = (1/9N)\ X^2 = (1/9)\ (n+m+k-3) + (2/9N)\ (X^2_{IxJ} + X^2_{IxK} + X^2_{JxK}) \qquad (2)$$

In spite of having two zero eigenvalues, Cazès(1977a, 1977b), Benzécri(1977), the spectral decomposition of this inertia tensor usually presents a rather uniform set of values, decreasing very slowly. Since Benzécri(1979)'s remarks, other authors also considered this same point, de Leeuw(198Ø), Greenacre(1984, 1988a,b).

As the mean of the square root of the eigenvalues is, in this case, 1/3, Benzécri(1979) suggests to concentrate the analysis on those eigenvalues such that $\lambda_\alpha^{1/2} > 1/3$, and to consider as the explained inertia for axis α the value $\rho(\alpha) = (\lambda_\alpha^{1/2} - 1/3)^2$.

This amounts to analyse, restricted to the subset of direct factors, a modified Burt's table where $x_{aa'}=\emptyset$, whenever a and a' belong to the same set. Actually, de Leeuw(198Ø) has showed that the quantity $9N \sum_{\alpha} (\lambda_{\alpha}^{1/2} - 1/3)^2$ corresponds to the pooled chi-square of the six off-diagonal tables, being then equal to the second term in (1). This value, nevertheless, is not the chi-square of the modified table with zeroed matrices in the diagonal:

$$2N + 2/3 (X^2_{IxJ} + X^2_{IxK} + X^2_{JxK}) \quad (3)$$

All this poses a certain objection to the proposed corrections. Moreover, as Kendall and Stuart(1967) point out, behind the search of scoring systems with maximum correlation for the margins of a two-way table, lies an attempt to produce a bivariate normal distribution which would fit somewhat the table probabilities. From this viewpoint, to suppress completely a Burt's table diagonal terms may be a doubtful, if not too strong, outliers' rejection.

3. GENERALIZED BURT'S TABLES.

Instead of focusing on the diagonal terms, we look at other operators to generate the table values. In the standard Burt's table this happens to be the addition operator. In the general case, let Φ denote a family of symmetric functions $\{\Phi_n : R^n \rightarrow R,\ n \geq 2\}$, where each Φ_n performs the same operation, their only difference being the number of variables in their domain. Given a three dimensional array of counts on IxJxK, we define a generalized Burt's table or a Φ-Burt's table as a two-way table $\{x_{aa'}\}$ on (I+J+K)x(I+J+K) such that:

i) for a and a' belonging to different marginal sets, say (a,a') = (i,j): $\quad x_{aa'} = x_{ij} = \Phi_K(n_{ij1}, n_{ij2}, \ldots, n_{ijK})$

ii) for a and a' belonging to the same marginal set, say I: $x_{aa'} = x_{ii} = \Phi_{mk}(vec\{n_{ijt}, j \in J, t \in K\})$, if a=a' (the symbol vec standing for the vectorization of the corresponding array) and $x_{aa'} = \emptyset$ otherwise.

It is worth emphasizing that any symmetric operator can be chosen, the idea being that a suitable one should either attenuate the "diagonal effect" or extract particular features of the three-way structure. Of primary interest are the cases where the operator described by the family Φ is some kind of mean (arithmetic, geometric, etc), or a robust location or dispersion measure.

For keeping the properties of CA, the weighing system should be that one generated by the new table. In the medians case, for instance, the weight assigned to element $i \in I$, say, will be:

$$\sum_t \operatorname*{med}_j n_{ijt} + \sum_j \operatorname*{med}_t n_{ijt} + \operatorname*{med}_{jt} n_{ijt} \qquad (4)$$

divided by the total counts in the table. As the summation operator does not commute with the medians (med) operator, the final expression is not very informative. In some cases, however, these marginals will not be very different from those of the standard table.

4.ANALYTICAL MOTIVATION -INERTIA TRANSFORMATIONS.

Consider a set of n points in R^q and their corresponding inertia tensor I. If M is the qxq diagonal matrix with the weights p_i, $i=1,\ldots,n$, and Q is tne metric tensor, calling X the qxn array formed by placing columnwise the points, we know that $I= XMX'Q$. Given a general transformation $\Phi_{q,1}:R^q \text{--->} R^1$, $1\leq q$, how will the inertia be transformed?

For $1=q$ and $\Phi_{q,1}$ a linear transformation, well known results provide a solution. In the general case, supposing $\Phi_{q,1}$ differentiable, a first order Taylor expansion around the gravity centre x_{gc} gives, for a general element x_i of the cloud (writing, from now on in this section, $\Phi_{q,1}$ without the subscripts):

$$\Phi(x_i) = \Phi(x_{gc}) + D\Phi(x_{gc}).(x_i-x_{gc}) \qquad (5)$$

where $D\Phi$ stands for the derivative of Φ.

Supposing that the weighing system is the same and that the metric is the euclidean one (Q=identity), the new inertia can be approximated by:

$$I* = \Phi(x_{gc})\ 1'M\ 1\ \Phi(x_{gc})' + D\Phi(x_{gc})\ X_c\ M\ X_c'\ D\Phi(x_{gc})' \qquad (6)$$

where 1 is a 1x1 vector of ones and X_c stands for the matrix of the centred data in the original space. As in a first order approximation:

$$\Sigma p_i \Phi(x_i) = \Phi(x_{gc}) + D\Phi(x_{gc})\ \Sigma p_i(x_i - x_{gc}) = \Phi(x_{gc}) \qquad (7)$$

formula (6) gives the new inertia as a function of a standard transformation of the original centred inertia plus the part due to the displacement of the gravity centre.

Formula (6), although approximate, may be of some help for examining the behaviour of certain transformations. To see this we draw on a special geometric view of the Burt's table.

5. CONTRACTIONS OF SUPERMATRICES' CLOUDS.

In Flōres(1988b) the Burt's table is viewed as arising from a combination of contractions of all the possible clouds produced by the three supermatrices from a given three-way table. According to this, given say a supermatrix on Ix(JxK), the cloud of the n points in R^{mk} can be contracted to a cloud either in R^m or in R^k, depending on how the respective homomorphisms are chosen.

The ideas in the previous sections can be translated to this representation, by substituting the general functions Φ for the homomorphism. To illustrate, we consider the median operator on supermatrix Ix(JxK). Putting, in the notation of the previous section, q=mk and l=k or m, point $(f^i{}_{jt}) \in R^{mk}$ can be contracted to, respectively, points:

$$\Phi_{mk,k}(f^i{}_{jt})=(\operatorname{med}_j(f^i{}_{jt})) \quad \text{and} \quad \Phi_{mk,m}(f^i{}_{jt})=(\operatorname{med}_t(f^i{}_{jt})) \qquad (8)$$

Completion of the third space, i.e. R^n, can be made by multiplying the canonical vector corresponding to $i \in I$ by $\operatorname{med}_{jt} f^i{}_{jt}$, and , for analysing the combined cloud, Flôres(1988b), a proper decision has to be made on the metric and weighing system.

Returning to supermatrix Ix(JxK), in the standard case, $\Phi_{mk,k}$ is a linear transformation whose matrix is the horizontal juxtaposition of k mxm identity matrices and formula (6) is valid exactly. But, if $\Phi_{mk,k}$ is the contraction through medians described in (8) - supposing, for simplicity, m odd and all coordinates of the gravity centre unequal - $D\Phi(x_{gc})$ is a matrix that, for each row, will present only one 1, what allows an immediate comparison of this contracted inertia with that for the linear homomorphism.

Another interesting case is when contraction is achieved through a geometric means operator. In this instance, reminding that, for the supermatrix cloud at stake $x_{gc}= (f_{jt})$, the entry (i, t) corresponding to $D\Phi(x_{gc})\ X_c$ in formula (6) applied to a contraction to R^k will be:

$$\prod_j f_{jt}{}^{1/m} \left(1/m \sum_j (f_{ijt}/f_i f_{jt}) - 1 \right) \qquad (9)$$

what means that the usefulness of the departing supermatrix - as a means of studying the interaction of set I with the (JxK) combination, Flôres(1988a) - has been somewhat preserved. The greater this interaction the higher will the inertia of the geometric contraction be with respect to that of the standard one.

7.EXAMPLES AND CONCLUSIONS.

The first example refers to a 4x5x12 matrix with monthly unemployment data, in a 0-100 basis, for the main four Brazilian metropolitan areas, during the years 83 to 87, IBGE(1988). As the table values did not vary too much, one might guess that a medians contraction would not give very different results from a standard contraction. Nevertheless, comparison of the results from a standard

EXHIBIT 1: Inertia decomposition.

	Burt's table	Benzecri's correction	medians' table
total inertia	2.0064	0.0013	0.5246
% in axes 1 to 2	12.69%	84.33%	72.55%
% in axes 1 to 4	24.37%	97.76%	78.45%

Burt's table and the median's table, in Exhibit 1, shows how the medians table quickly decomposes its inertia, while the standard table presents its usually slow decrease. Of course, use of Benzécri's idea improves the situation. Of the standard table total inertia, 99.7% is due to the diagonal effect (which amounts to 2 in this case), leaving almost no room for the contribution of the supermatrices contractions. Even so, its small value of 0.0064 is greater than the corrected Benzécri's inertia. As for the median, the total inertia of the three contractions amounts to 0.0175, a value to be compared with the inertia of the generalized table.

The second example is related to a table from Waelbroeck and Kol(1987), on exports in manufactures from Japan, US and the EC, to five different world regions and according to four SITC product groups. Exhibit 2 displays the inertiae of each standard and geometric two-way contraction. Most of the information of the original array is due to the rather different trade patterns from US, the EC and Japan.

EXHIBIT 2: Trade flow data: geometric means versus standard contractions. Inertia values.

	geometric	standard
"countries" x world regions	0.3347	0.3025
"countries" x SITC groups	0.0387	0.0336
world regions x SITC groups	0.0168	0.0193
pooled inertia	0.3902	0.3554

The geometric contraction is more convenient for preserving this higher interaction, as shown in the first two lines of Exhibit 2. The resulting pooled inertia is around 10% higher than the standard one.

The above examples show how the different summarizations of the original data can either be more robust or, due to nonlinear effects, preserve some special information contained in the three-way interactions. In general, they also tend to improve on the diagonal effect. Notwithstanding, a general prescription can not be stated and different contractions should be tried in a given situation.

REFERENCES

Benzécri, J.-P.(1977). Sur l'analyse des tableaux binaires associes a une correspondance multiple. Les Cahiers de l'Analyse des Données, vol. II, n 1.

Benzécri, J.-P.(1979). Sur le calcul des taux d'inertie dans l'analyse d'un questionnaire, addendum et erratum a [bin. mult.]. Les Cahiers de l'Analyse des Données, vol. IV, n 3.

Cazès, P.(1977a). Probleme sur l'analyse des questionnaires. Les Cahiers de l'Analyse des Données, vol. II, n 1.

Cazès, P.(1977b). Etude de quelques propriétés extrémales des facteurs issus d'un sous-tableau d'un tableau de Burt. Les Cahiers de l'Analyse des Données, vol. II, n 2.

de Leeuw, J.(1980). Canonical Analysis of Categorical Data. DSWO Press, Leiden.

Flôres, R. G.(1988a). Classical approaches to cubic tables. Cahiers du CERO, vol. 30, 1-2. Bruxelles.

Flôres, R. G.(1988b). A simple, interactive method for three-way tables. MULTIWAY 88, to appear in the North-Holland proceedings.

Greenacre, M.(1984). Theory and Applications of Correspondence Analysis. Academic Press, London.

Greenacre, M.(1988a). The limitations of multiple correspondence analysis. MULTIWAY 88, to appear in the North-Holland proceedings.

Greenacre, M.(1988b). Correspondence analysis of multivariate categorical data by weighted least-squares. Biometrika, vol. 75, n 3.

IBGE(1988). Anuário Estatistico do Brasil - 1987. IBGE, Rio.

Kendall, M. G. and A. Stuart(1967). The Advanced Theory of Statistics, vol II: Inference and Relationship. 2nd ed. Griffin, London.

Lebart, L., Morineau, A. et N. Tabard(1977). Techniques de la Description Statistique. Bordas, Paris.

Waelbroeck, J. and J. Kol(1987). Export opportunities for the South in the evolving pattern of world trade. CEPS Papers n 33, Center for European Policy Studies, Brussels.

MULTIPLE CORRESPONDENCE ANALYSIS AND NEIGHBORING RELATION

Brigitte Escofier
IRISA Campus de Beaulieu 35042 Rennes cedex
IUT B.P. 1104 56008 Vannes Cedex (France)

Abstract
We propose here two methods which allow to take into account a neighbouring relation of individuals in Multiple Correspondence Analysis (M.C.A.) and to study the relations between the variables and the neighbouring properties. The first one eliminates local variations in order to analyse more easily the general trends and the second one analyses these local variations. We represent the neighbouring relation by means of weighted graph. We define an easily respected condition for the graph in such a way that the application of the two methods needs only a classical Correspondence Analysis (C.A.) program. If the graph is induced by a partition, the first method analyse the barycenters, so it is a "between analysis". The second one is the Multiple Conditional Analysis (cf.[9]) .

1 INTRODUCTION

M.C.A. analyses nominal variables defined on a set of individuals. Data sets whose individuals are related by a neighbouring relation are frequent. For instance, in analysing images where individuals correspond to small areas with texture described by nominal variables. We often wish to introduce such a contiguity or geographical structure. Sometimes to decrease the importance of local variations, (maybe partially random) and study general trends on which the areas of a same region group together. Sometimes on the contrary to study local variations and group areas whose differences with theirs neighbours are identical. Sometimes to analyse the relations between geographical structure and the nominal variables.

The objectives are the same for numerical and nominal variables. So the two methods we have proposed (cf.[2] et [3]) in case of numerical variables (and Principal Component Analysis) can be transposed to M.C.A. with some technical modifications and a condition for the graph.
The principle of "smooth analysis" is to replace in the cloud of individual (cf. [10] and [12]) each individual by the barycenter of its neighbours. The principle of "local differences analysis" is to represent an individual by its difference with this barycenter.

2 CONTIGUITY GRAPH

We represent neighbourhood by means of a graph. We denote I as the set of individuals and G as the square matrix of the graph : $g_{ii'}$ is nonnegative and is called the weight of i' in the neighbourhood of i ($g_{ii'}$=0 if i' is not a neighbour). We denote $g_{i.} = \sum_{i'} g_{ii'}$ and $g_{.i}=\sum_{i'} g_{i'i}$.
We shall see that smooth analysis and local differences analysis are easier to apply and that their results are more satisying if $g_{i.}=g_{.i} = g_{i'.}=g_{.i'}$ for all i and i'. All graphs do not satisfy this condition. For instance a symmetrical graph ($g_{ii'} = g_{i'i}$) not weighted ($g_{ii'}$ = 0 ou 1) in which some individuals have many neighbours and others have very few neighbours. In order to $g_{i.}$ equals 1 we can divide $g_{ii'}$ by $g_{i.}$ but the new graph is not symmetrical and the condition is not yet satisfied.

Bistochastic graph

This condition is easy to obtain for a symmetrical graph by modifying the weight of i in its own neighbourhood which appears in the diagonal of G. Then, we suppose without loss of generality that $g_{i.} = g_{.i} = 1$. The matrix G is then a bistochastic matrix. In this case we name the graph a bistochastic graph.

For instance, if we study areas of an image, most of them have 8 neighbours, but the corners have only 3. In order to introduce as neighbourhood all the adjacent areas by means of a bistochastic graph, we can fix for instance $g_{ii'}=0.1$ if i and i' are adjacent, $g_{ii}=0.2$ for a central area and $g_{ii}=0.7$ for a corner area. This choice is satisfying : first, all the areas have the same weight for their neighbours, second a corner keeps an important weight in his own neighbourhood which is poor whereas a central area with a rich neighbourhood has a weaker weight. Notice that the loss about i by smoothing ($g_{i.}-g_{ii}$) is exactly balanced by its influence on the others ($\sum_{i'\neq i} g_{i'i} = g_{.i}- g_{ii}$).

Another example is given by a graph induced by a partition of the set of individual into different groups. The neighbourhood of an individual i is the set of the elements of its group. The weight $g_{ii'}$ is the inverse of the number of the elements in the group. The next two paragraphs give and recall results concerning this extreme case. We generalize them after.

3 PARTITION GRAPH : BETWEEN ANALYSIS

In smooth analysis each individual is replaced by the weighted barycenter of its neighbours calculated with the weight $g_{ii'}$.
In the case of partition, all the elements of the same group are replaced by the same point. So the cloud of individuals (cf.[10] and [12]) is the cloud of the B barycenters of the groups. The mass of each one is equal to the number of the group's elements. So this analysis is a "between-analysis" since the dispersion is exactly the between dispersion.

In order to obtain results of M.C.A., a simple Correspondence Analysis (C.A.) algorithm may be applied to the indicator matrix Z (I rows and J columns). In order to obtain results of between analysis, a C.A. algorithm may be applied either to a matrix T (B rows and J columns) or to a matrix L (I rows and J columns).

The first one T is obtained by summing the rows of Z belonging the same group. As a matter of fact the profiles of the sum of several rows is the barycenter of the profiles of these rows. The mass receveided in C.A. which is the row total is the sum of the masses of the rows. Besides the column totals of T are equals to the column totals of Z, so the chi-squared metric induced in the two C.A. are exactly the same. Notice that the rows of Z can be added as supplementary elements in order to project them onto "between-axes".

In the second one L, the row i is he weighted average of the Z'rows of its group. If i and i' belongs to the same group, their profiles are identical. Notice that the row totals and the columns totals of Z and L are equals, so the chi-squared metrics induced on the clouds of individuals (and on the clouds of columns) are identical. The profiles of the columns of L are the projections of those of Z onto the subspace spanned by the indicators of each group.

4 PARTITION GRAPH : MULTIPLE CONDITIONAL ANALYSIS

The problem is complementary to the precedent one. we are interested in analysing the differences between the Z's profiles row and the barycenter of their group. We have already studied this problem (cf [9]) and introduced a specific technique called Multiple Conditional Analysis which have many interesting properties similar to those of M.C.A. In order to obtain results, C.A. algorithm may be applied either to an IxJ matrix or an square JxJ matrix.
The first one is R=Z-L+P , where p_{ij} is the product of the total row i and the total column j divided by the overall total.

The second one is a symmetrical matrix deduced from Burt's contingency table.

5 SMOOTH MULTIPLE CORRESPONDENCE ANALYSIS

Smooth Multiple Analysis is a generalization of between analysis considered as correspondence analysis of the matrix L.

5.1 Smooth indicator matrix

Let us consider the IxJ matrix L defined, as in between analysis :

$$l_{ij}=\sum_{i'} (g_{ii'}\ k_{i'j} / g_{i.})$$

Smooth Mutiple Correspondence Analysis is a C.A. of L. We study its properties.

For each nominal variable instead of 1, as in Z, indicating a specific category with 0 elsewhere, there is a set of nonnegative values that add up to 1. It is called "fuzzy coding" (cf. [10]). These can even be considered probabilities that the individual lies in the respective categories.
The row totals of L are equal to those of Z :

$$l_{i.}= \sum_j l_{ij} = \sum_j \sum_{i'} (g_{ii'}\ k_{i'j} / g_{i.}) = k_{i.}$$

But the columns totals of L are not ever equal to those of Z :

$$l_{.j}= \sum_i l_{ij} = \sum_i \sum_{i'} (g_{ii'}\ k_{i'j} / g_{i.}) = \sum_{i'} (g_{.i'} / g_{i.}) k_{i'j}$$

That'is what they are if the graph is bistochastic. We assume this to be true in the next paragraphs. The other case is studied in paragraph 7.

5.2 The cloud of individuals

A L row profile i is the weighted barycenter of the Z row profiles of its neighbours. The row totals of L being equal to those of Z, each i received, as in M.C.A., the same mass. The column totals of L being equal to those of Z, the chi-sqared metrics are the same. So the clouds of individuals is deduced from those of M.C.A. by replacing each i by the barycenter of its neighbourhood.

Notice that the L's rows contain information about the *repartition of it neighbourhood* of i between the different categories of each nominal variable. The information is richer than in the case of numerical variables where smoothed matrix rows are only the average of the neighbourhood.

As in between-analysis it is useful to add the rows of Z, as supplementary elements, in order to obtain projections of initial individuals onto the axes. These axes are less dependent on local variations than those of M.C.A. These projections are often the main results and can be used in a classification algorithm.

These projections are useful too, to compare positions of initial and "smoothed" individuals in order to study the relations between each one with its neighbourhood : stable individuals look like their neighbours whereas instable individuals don't.
Notice that it is often difficult to study each individual when they are too numerous. The next paragraph indicate how to study stability of groups of individuals.

5.3 The cloud of modalities

The columns of Z are not exactly logical indicators, l_{ij} is not 0 or 1 but only a number between 1 and 0 and can be considered probability. We called them "fuzzy indicators".

Many results of C.A. of L can be interpreted as those of M.C.A. with respect to this notion of fuzzy indicators. As in M.C.A. the barycenter of each nominal variable's modalities is the barycenter of all the modalities (the origin of the axes). The barycentric relationships between row points and column points are analogous to those of M.C.A. The coordinates of projections of a modality onto axes are not proportional to thoses of the barycenter of the group (as in M.C.A.) but only to thoses of a *weighted barycenter* calculated with the weight $g_{ii'}$.

The columns of Z can be added as supplementary elements. These projections are useful for several reasons. First in order to study the position of nonweighted barycenter of group of smoothed individuals. Second in order to obtain projections of Z's columns onto axes less dependent of local variations. Third, in order to study relationships between graph and nominal variables : when the "fuzzy indicator" (L's column) and the logical indicator (Z's column) are located very near one another, the elements belonging to the group defined by the logical indicator are neighbours. Then the indicator is highly linked to the graph. On the contrary when they are far away from one another, the elements belonging to the group are generally not neighbours and the indicator is not linked to the graph.
Notice that these projections are not interesting in the case of a partition graph. In effect, in this case the L's columns are projections of Z's columns onto the subspace spanned by L's columns. The axes belong to this subspace and the projections of Z's columns onto these axes are identical to those of L's columns.

5.4 Smooth indicator matrix as supplementary elements

In order to study relationships between nominal variables and graph, a C.A. of matrix Z with columns of L as supplementary elements instead of the inverse is also interesting.

6 MULTIPLE CORRESPONDENCE ANALYSIS OF LOCAL DIFFERENCES

The problem studied is complementary to the precedent one. Instead of eliminate local variations (in order to analyse general trends), we are now interested in analysing these local variations. In this paragraph, we suppose than $g_{i.} = g_{.i'}$ for all i and i' (case of bistochastic graph), the other cases are studied in paragraph 8.

6.1 Correspondence analysis of residuals

In order to analyse the differences between characteristics of each individual (rows of Z) ond those of the barycenter of their neighbours (rows of L) we will use a generalization of C.A. (cf.[8]). This generalization can be used to analyse the residuals of a model (cf. for instance ([8], [7]ch. 8 sec. 5.2 and ‹[21]).

Let R be the matrix :

$$R = Z - L + P$$

where p_{ij} is the product of the total row i (of both Z and L) and the total column j (of both Z and L) divided by the overall total.

This matrix is not a contingency table, r_{ij} may be negative. However C.A. computer program may be applied to R. It is not a C.A., but only a mean to compute results of the generalization of C.A. We recall here some of their properties.

Generally Z is a contingency table and L is the model whose total rows and total columns are equal to those of Z. The difference between row (resp. column) profiles of R and the average row (resp. column) profile are exactly the difference between row (resp. column) profiles of Z and L. The totals row and total columns are those of Z and L. So the two chi-squared metrics are those of C.A. of Z and L. There are weighted barycentric relationships between row and column projections onto the axes which have not the same form as in C.A. because r_{ij} may be negative. We can deduce that an individual is locaated on the side of columns j with $l_{ij} > k_{ij}$, and at the opposite of j with $l_{ij} < k_{ij}$.

6.2 The cloud of individuals

An individual is very far from the origin if it is very different from their neighbours The coordinate of i is with respect to axis j is :

$$r_{ij}/ r_{i.} = k_{ij}/k_{i.} - r_{ij}/ r_{i.} = (k_{ij} - l_{ij}) / k_{i.}$$

This coordinate is nonnegative if the individual i lies in category j wheras it is negative if i don't ly in category j and some of its neighbours ly in it. Two rows appears close together in the space only if they ly in the same category and if the repartitions of their neighbours in the different categories are similar : this technique allows to analyse individuals with their context. As in M.C.A. all the individuals have the same mass.

6.3 The cloud of categories

A modality is represented by the difference between its Z profile and its L profile. As in M.C.A. the barycenter of each nominal variable's modalities is the origin of the axes.

The modalities located far from the origin are those which group individuals which are not neighbours : they are not correlated with the graph. On the contrary the modalities which are located near the origin group neighbouring individuals and are highly correlated with the graph. This analysis point at the indicators which are not correlated with the graph. In order to study the relationships between nominal variables and graph, this analysis is very efficient and the interpretation of its results are easier than in the previous one. In the two previous one (C.A. of L with Z as supplementary and C.A. of Z with L as supplementary) the two projections of each modality must be compared and the axes don't fit their differences.
Two modalities far from the origin are close together if the individuals which ly in them are the same and if the neighbours of these ly in same others modalities.

6.4 Relationships between categories and individuals

Projection of a modality j onto an axis is located near the individuals i which ly in it and whose neighbours don't ly in it. It is on the opposite side of individuals which don't ly in it and whose neigbours ly in it.

6.5 Indicator matrice as supplementary element

The rows and the columns of Z can be added as supplementary elements. The projections of the columns give us the position of group's barycenters.

7 GENERAL SMOOTH MULTIPLE CORRESPONDENCE ANALYSIS

If the graph isn't bistochastic a very important property is lost : the total columns of L are not equal to those of Z. The consequences are :

- the chi-sqared metric induced in C.A. of L isn't the M.C.A. metric : the cloud of individuals is disturbed and the points are not exactly the barycenters of Z'ones.

- the masses of individuals are not the same.

Theorically, it's a problem but freqently the totals are not quite different and in this case it's of little importance.

Notice that there is no problem in adding L as supplementary element in C.A of Z.

8 GENERAL M.C.A. OF LOCAL DIFFERENCES

Theorically, it is possible to analyse differences between Z's row profiles and L's row profiles even if their total columns are not equal. (cf.[8]) But a specific computer program is needed and many properties are lost. The cloud of individuals is not centered and the column profiles are not exactly the differences between Z and L's ones. If L total columns are not very different from Z'ones this can be neglected in the interpretation, in the others cases the results are not very interesting.

9 COMPARISON WITH OTHER TECHNIQUES

The nearest technique is "local analysis" which is due to Lebart and Aluja (cf. [1] and [11]°. The computer program is set up in Spadn library (cf.[18]). The neighbouring relation is represented by a nonweighted graph. As factorial analysis of local differences, the local analysis aims to study local variations and eliminate the effect of geographical positions of individual. The comparison between these two techniques is made in [2] in the of numerical variables. The case of nominal variables is analogous. Here, we recall only that local analysis may be considered as an analysis of edges instead of individuals. Local inertia is computed by suppressing pairs of individuals which are not related by the graph. Local analysis is generalized to a weighted graph (cf. [15]) and may be considered as a factorial analysis of the initial matrix data with a special metric (cf.[4]). The two methods may be defined in the same way with others special metrics.

We can also compare these techniques to those which consider projections of the cloud of variables onto subspaces. Theses analyses based on the ideas due to Rao (Principal Component analysis of Instrumental variables, (cf.[16]) are studied by Sabatier in his thesis (cf.[17]). These subspaces are either spanned by variables of another data table or orthogonal to them in order to analyse the part of the variability due (or not due) to the others variables. In smooth anlysis and in analysis of local differences we have a graph instead of instrumental variables. In both of them, generally the cloud of variables is not a projection of the cloud of the indicator variables but in the case of partition graph.

We can also compare them to Canonical Correspondence Analysis (C.C.A.) due to Ter Braak (cf. [19] and [20]), studied and generalized to nominal variables by Lebreton and alls (cf.[11] and [12]). In C.C.A the "other data table" is a contingency table. As in smooth analysis a cloud of weighted barycenter of the row profiles is considered , the weights are given by the contingency table instead of the graph and the metric is the discriminant analysis metric (cf.[5] and [6]).

10 CONCLUSION

Many differents types of neigbouring relations may be introduced by means of bistochastic graph. It is very easy to apply smooth multiple correspondence analysis and multiple correspondence analysis of local differences because they need only classical C.A. program. Their results may be interpreted as those of C.A. and may be the first step before a classification algorithm.
We apply these techniques in order to analyse images by grouping small areas characterised by numerical texture parameters recoded into categories. Smooth analysis groups neighbouring areas whereas analysis of local differences groups contours.

REFERENCES

[1] Aluja T. et Lebart L. (1985) Factorial Analysis upon a graph. Bulletin technique du centre de statistique et informatique appliqué. Vol. 3 pp 4-34

[2] Benali H. et Escofier B. (1988) Smooth factorial analysis and factorial analysis of local differences. Proc. Inter. Conf. on multiway matrices Rome 26-29 mars 1988 North Holland Amsterdam.

[3] Benali H. et Escofier B. Analyse factorielle lissée et analyse des différences locales A paraitre Revue de statistique appliquée.

[4] Carlier A. (1985) Applications de l'analyse factorielle de l'évolution et de l'analyse intra-période. Statistique et analyse des données Vol. 10 pp 27-53

[5] Chessel D. Lebreton J.D. Yoccoz N. (1987) Propriétés de l'analyse canonique des correspondances : Une illustration en hydrobiologie. Revue de Statistique appliqquée XXXV(4) pp.55-72

[6] Doledec S. et Chessel D. (1987) Description d'un plan d'observation complet par projection des variables. Acta oecologica Oecol. Gen. Vol.8 n°3 pp.403-426

[7] Escofier B. Pages J. (1988) Analyses Factorielles simples et multiples. Objectifs, Méthodes et Interprétation Dunod Paris

[8] Escofier B. (1985) Analyse factorielle en référence à un modèle. Application au traitement des tableaux d'échanges. Revue de statistique appliquée1985 n°2.

[9] EscofierB.(1987)Analyse des correspondances multiples conditionnelle Proc. Inter. Symp. Data Analysis and Informatics Versailles. pp.13-22 North Holland Amsterdam

[10] Greenacre M. , Hastie T. (1987) The geometric interpretation of Correspondence Analysis Journal of the American Statistical Association 82, 437-447

[11] Lebart L. (1984) Correspondance analysis of graph structure. Bulletin technique du centre de statistique et informatique appliqué. Vol.2 pp. 5-19

[12] Lebart L., Morineau A, Warwick (1984) Multivariate descriptive statistical analysis Wiley , New York

[13] Lebreton J.D. Chessel D. Prodon R. Yoccoz N. (1988) L'analyse des relations espèces-milieu par l'analyse canonique des correspondances. Variables de milieu quantitavives. Acta oecologica Oecol. Gen. Vol.9 n°1pp.53-67

[14] Lebreton J.D. , Chessel D. ,Richardot-Coulet M. ,Yoccoz N. (1988) L'analyse des relations espèces-milieu par l'analyse canonique des correspondances. Variables de milieu quantitavives. Acta oecologica Oecol. Gen. Vol.9 n°2 pp.137-151[15] Le Foll Y. (1982) Pondérations des distances en analyse factorielle. Statistique et analyse des données n°7,1, pp.13-31

[16] Rao C.R. 1964 The use and interprétation of Principal Component Analysis in applied research. Sankhya A 26,4 p.329-358

[17] Sabatier R. 1987 Méthodes factorielles en analyse des données. Approximation et prise en compte de variables concomitantes. Thèse. Université de Montpellier

[18] Spadn manuel de référence Cisia 1989 Paris

[19] Ter Braak C.J.F. (1986) Canonical correspondence analysis : a new eigenvector technique for multivariate direct gradient analysis. Ecology, 67(5), pp.1167-1179

[20] Ter Braak C.J.F. (1987) The analysis of vegetation-environnment relationships by canonical correspondence analysis. Vegetatio, 69, pp.69-77

[21] Van der Heijden P.G.M. , De Leeuw J. (1985) Correspondence analysis used complementary to loglinear analysis Psychometrica 50 pp.429-447

CLUSTERING

BILINEAR CLUSTERING ALGORITHMS

Mirkin, B.G.
CEMI, Krasikova, 32, Moscow, 117418 (USSR)

1. PREFACE: BILINEAR MODEL AND CLUSTERING

A new class of clustering algorithms is discussed. These are based on the following bilinear model.

The model represents the data matrix $Y = (y_{ik})$ (here $i = 1, \ldots, N$ are the units and $k = 1, \ldots, n$ are the variables) by the cluster structure, which is characterized by some set of the pairs (f_t, a_t) $(t = 1, \ldots, m)$, where f_t is the N-dimensional vector of the values f_{it} which are equal to 0 or 1 (usual clusters) or are between 0 and 1 (fuzzy clusters , with the additional requirement that the equalities $f_{i1}+\ldots+f_{im}=1$ hold for each i), and a_t is the n-dimensional vector with real a_{tk}, which is the cluster "typical" (or "standard") point.

The model consists of the problem of minimizing the residuals e_{ik} of the equalities

$$y_{ik} = a_{1k}*f_{i1} + \ldots + a_{mk}*f_{im} + e_{ik} \tag{1}$$

by unknown a_{tk}, f_{it}. The residuals are given by the term

$$\sum e_{ik}^2 \text{, or } \sum |e_{ik}| \text{, or } \max_{i,k} |e_{ik}| \tag{2}$$

The model (1) is bilinear by unknoun a_t and f_t, so we shall refere to the algorithms for fitting the model as the bilinear algorithms. To fit the model we use special principle of consecutive exhaustion of the data (CONEX), which is a version of the "simple structure principle" in factor analysis. The principle requires to find the solutions (a_t, f_t) in consecutive way by "the residual data" maximizing the "explained share" of the data scatter on each step. The model and the principle lead us to the clustering algorithms, which have the following peculiarities:

(i) there exists an association between (a) the choice of approximation criterion and method of fitting and (b) preliminary transformation of the data and choice of similarity measure between units and/or variables;

(ii) the algorithms are effective for mixed variable scales;

(iii) the parametes of the algorothms may be determined automatically;

(iv) the decomposition of the scatter of the data leads to the set of interpretative characteristics.

2. CONEX METHOD

Applied to the model (1) with the initial data matrix Y, the consecutive exhaustion (CONEX) method has the following stages.

0. Let us define $m=1$, $Y_m = Y$.

1. Resolve the minimisation problem for the criterion

$$\sum_{i,k} (y^m_{ik} - a_k f_i)^2 \qquad (3)$$

with respect to admissible values of a_k and f_i (in the least sum of squares case; the other two forms of the criterion in (2) will be considered later). Here y^m_{ik} are the elements of the residual matrix Y_m at step m.

Let us denote the solution (it may not be optimal but suboptimal only by f_i) by $a_m = (a_{mk})$ and $f_m = (f_{im})$.

2. The new residual matrix

$$Y_{m+1} = Y_m - f_m a_m^T$$

with the elements

$$y^{m+1}_{ik} = y^m_{ik} - f_i a_k \qquad (4)$$

is calculated.

The following equality is true:

$$\sum y^2_{ik} = \sum_{n \le m} l_n + \sum (y^{m+1}_{ik})^2 \qquad (5)$$

where

$$l_n = \sum_k a_{nk}^2 \sum f_{in}^2$$

So we may use as stopping rules

$$\sum_{n \le m} l_n / \sum y^2_{ik} > 1 - d' \quad \text{or} \quad l_m / \sum y^2_{ik} < d''$$

/with previously fixed values d' and d" of the negligible share of the initial data scatter. This implies the following step.

3. If one (or both) of the stopping rules is true we finish the procedure. In opposite case we add 1 to m and return to stage 1 for finding the next pair (f_m, a_m).

The CONEX method is a realisation of usual principal component analysis procedure for more complex situations. The decomposition (5) , which is analogous to the spectral decomposition of the principal component analysis with l_m as singular values of the matrix Y, holds always : the vectors a_m and/or f_m may be not ortho- gonal to each other, moreover some of them may coinside.

3. PRINCIPAL CLUSTER ANALYSIS.

In the cluster analysis problem the values a_k are arbitrary, and f_i are boolean, 0 or 1. The CONEX method for this case has been termed the Principal Cluster Analysis Method (Mirkin, 1987).

Let us denote the cluster corresponding to the vector $f = (f_i)$ by $S = \{i : f_i = 1\}$. For given f the optimal a has the following form:

$$a_k = \sum_{i \in S} y_{ik} / \#S , \qquad (6)$$

where #S is the cardinal of the set S.

So the standard point $a = (a_k)$ for cluster S is its center of gravity ("real type"). It is not difficult to prove that the problem of minimization (3) is equivalent to the problem of maximizing the contribution l of the cluster S in the data scatter, or cluster variance

$$g(S) = \sum_{i,j \in S} b_{ij} \ / \ \#S = \#S \ b(S) \qquad (7)$$

where b_{ij} the is (i,j)-th entry of the proximity matrix $B=Y_m Y_m^T$, and b(S) is an average proximity within S which equals, evidently, the scalar square of the standard vector $a = (a_k)$ (6):

$$b(S) = \sum_{i,j \in S} b_{ij}/(\#S)^2 = a^T a \qquad (8)$$

The following statement may be proved (Mirkin, 1987).

THEOREM 1. The general m-th step of the principal cluster analysis method consists of searching the set of units S_m which maximizes contribution (7) of S in the scatter of data ; the standard vector a_m at the step m equals gravity center of S_m. The contribution of the variable k to the cluster variance $g(S_m)$ equals $a^2_{mk} \#S_m$ (according to (7) and (8)).

The problem of maximizing (7) is NP-complete in the set of all subsets S. So we consider a so-called principal clusters algorithm (Mirkin, 1987), based on the value

$$w_i = b_{ii} + 2 * \#S * b(i,S) \qquad (9)$$

where

$$b(i,S) = \sum_{j \in S} b_{ij}/\#S$$

Algorithm ADDY-SQ

On the initial step for S=0 we define $w_i = b_{ii}$ according to (9); so the furthest data point to the origin y_i^m gives the first unit to be placed in S , because $b_{ii} = (y^m_i, y^m_i)$.

If

$$w_i > \#S * b(S) = g(S) \qquad (10)$$

then the unit i is added to S, and the general step is reiterated for recalculated values #S, b(S) and b(i,S) with S+i substituted for S . The unit i is not added to S if (10) would not be true.In this case the set S is announced as the principal cluster S_m with the interpretative parameters a_{mk}, g(S) and $a^2_{mk} \#S/g(S)$ (see THEOREM 1). The new (m+1)-th principal cluster is found from the new residual data

$$y^{m+1}_{ik} = y^m_{ik} - a_{mk} f_{im}$$

using the same algorithm according to the CONEX method.

NOTE . In general, the clusters S_m may overlap. But the procedure ADDY-SQ may be arranged to find nonoverlapping clusters. It is sufficient to add to S only the units i which up to this point are not included in any preceding cluster S_n (for n < m); that is we have to maximize w_i (9) for the nonincluded i only. But in this case it is necessary to use another stopping rule. We have to stop the clustering process when the set of nonincluded units would be empty. A similar modification covers the case when there are other a priori requirements for interrelating individual units .

THEOREM 2. Each principal cluster S is a strict cluster,

that is $b(i,S) < b(S)/2$ for each $i \notin S$.

Thus though the principal cluster analysis method is a method for fitting the parameters of the approximation model, it reveals some properties which are quite natural and reasonable in the algorithmic framework. In particular, the principal cluster analysis suboptimal algorithm is close to some wellknown heuristical clustering algorithms: B-coefficient method of K.Holzinger and H.Harman (1941), algorithm "Specter" of E.Braverman and A. Dorophеiuk (see Dorophеiuk, (1966) and also E.Braverman and I. Mouchnik (1983)), average neighbour method of R.Socal and C. Michener (1958).

4. QUALITATIVE VARIABLES CASE

Let us consider the least square sum criterion for the model with nonoverlapping clusters for situation when all variables are qualitative. As usual, a qualitative grade s is represented by a boolean column in the data matrix, with entry i equal to 1 when unit i has grade s, otherwise equal to 0.

Let us denote by p_s the proportion of ones in the column corresponding to the grade s. To center the column is to subtract p_s from each entry. There are several ways of normalizing the column:

(i) no normalizing

Using (6) it is not difficult to demonstrate that the loading of the cluster S_m for the grade s in this case is given by

$$a_{ms} = p_{ms} / p_m - p_s \tag{11}$$

where p_{ms} is the proportion of units which belong to S_m and have grade s.

The value of a_{ms} (14) is the change in the probability of s when the complete set of units is changed by the cluster S_m. It may be considered as a natural index of association between the "grade" S_m and the grade s (for rather large values p_s) (see Mirkin, 1985).

The proportion of the grade s in the contribution of S to the scatter of the initial data is determined (by THEOREM 1) by

$$\#I * p_m * a_{ms}^2 = \#I * (p_{ms} - p_m p_s)^2 / p_m \tag{12}$$

So the total contribution of the nominal variable X with grades s to all clusters of the partition $S = \{ S_m \}$ equals

$$D(X/S) = \#I * \sum (p_{ms} - p_m p_s)^2 / p_m = \#I * [\sum p_{ms}^2 / p_m - \sum p_s^2]$$

This value is well-known in data analysis as the decrement in the average number of misclassifications in the proportional projection of grades s when the cluster containing the projected unit is known (see, for example, Goodman and Kruskal, 1954). The relative value of this measure is known as Wallis' coefficient of association between classifications S and X:

$$w(X/S) = D(X/S) / (1 - \sum_s p_s^2) \tag{13}$$

So in the case where qualitative data are not normalized the principal cluster analysis criterion is equivalent to the criterion of the maximization of the sum of association coefficients (12) maximization; the separate term (s,m) of the coefficient equals the contribution (11) of the grade s into the cluster Smand the standard ; point a_{ms} is determined by the changes (10) in the probabilities.

(ii) Normalizing the boolean columns s by quantities p_s with subsequent (or, equivalently, with preliminary) centering these columns.

In this case the loadings equal

$$a_{ms} = p_s * (p_{ms}/p_m - p_s)/p_s = p_s * FI(s,m)$$

where

$$FI(s,m) = p_{ms}/p_m p_s - 1$$

is the relative increment of the probability of s conditional on m compared with the unconditional probability p_s. The values FI(s,m) are rather adequate measures of association between m and s for situations with small p_s (Mirkin, 1985).

The contribution of s to S_m equals

$$\#I*p_m*a^2_{ms} = \#I*(p_{ms} - p_m p_s)^2/p_m p_s = \#I*FI^2(s,m)*p_m p_s \qquad (14)$$

and so the total contribution of the variable X with grades s into the partition S= $\{S_m\}$ is in fact Pearson's coefficient

$$X^2(X,S) = \#I*\sum_{m,s}(p_{ms} - p_m p_s)^2/p_m p_s. \qquad (15)$$

It is interesting to note the formula from (Mirkin, 1985)

$$X^2(X,S) = \sum p_{ms}*FI(s,m)$$

which demonstrates the sense of the coefficient in terms of the relative changes FI(s,m) of the probabilitiies p_s when the conditions m become available .

We see, that in this case the principal cluster analysis criterion is equivalent to the criterion of maximizating the sum of the Pearson's coefficients (19) of association between initial qualitative variables X's and the partition S which has to be constructed.

Thus we have estabilished an equivalence between the usual cluster constructions and "statistical" criteria for "concordant" partitions construction. The author discussed the questions in more details in his book in Russian (Mirkin, 1985). Some of these criteria are also considered in the paper (Saporta, 1988).

5. NONEUCLIDEAN METRIC CASES

We consider now the possibility of using the CONEX clustering method when the stage 1 uses the city-block metric criterion

$$F = \sum_{i,k} |y^m_{ik} - f_i a_k| \qquad (16)$$

THEOREM 3. The cluster S is optimal for the criterion (16) iff it maximizes the value

$$g_1(S) = \sum_{i \in S} (2*b(i,a) - |a|) \qquad (17)$$

where

$$b(i,a) = \sum_{k \in K(i,a)} \min(|y^m_{ik}|, |a_k|), \qquad (18)$$

$$K(i,a) = \{k : y^m_{ik} a_k > 0\}$$

$$|a| = \sum_k |a_k| .$$

The components a_k of the standard vector a are the medians of the variables k in the cluster S.

The value $g_1(S)$ is the contribution of S to the city-block scatter of initial data y_{ik} ,and its decompositions (17), (18) allow us to characterize the contributions to S both for the separate units i (which are equal to $b(i,a)-|a|$) and for the separate variables k (which are equal to

$$[\mathrm{sgn}(y^m_{ik})+\mathrm{sgn}(a_k)]*\min(|y^m_{ik}|, |a_k|)$$

The THEOREM 3 yields also some algorithms for forming suboptimal S (beginning from S=0) adding the units step by step to S . One of the algorithms is considered below.

Algorithm "ADDI-CB"

We begin from the most distant unit (in terms of city-block metric) i which maximizes $|y^m_i| = \sum y^m_{ik}|$ and define $a = y^m_i$, $S = \{i\}$. Then we select two units j, j' which are closest to a by the measure (22) and add these to S, if $g_1(S)$ would increase. Then we have to recalculate the vector a so that it remains the median for the new S , and repeat the process until the $g_1(S)$ stops rising.

The algorithm ,evidently, finds a suboptimal cluster S and so (17) implies for each $i \notin S$

$$b(i,a) < b(a,a)/2$$

which is analogous to the strict cluster concept in Euclidean space .

Let us consider now the third criterion for stage 1 of the CONEX method:

$$FA = \max_{i,k} |y^m_{ik} - f_i a_k| \qquad (19)$$

which now expresses requirement of mutual diminishing the residuals e_{ik} in model (1). The problem of minimizing (19) is equivalent to the following mathematical programming problem:

$$\begin{array}{lr} \min u & \\ -u \le y^m_{ik} \le u & (i \notin S) \\ a_k - u \le y^m_{ik} \le a_k + u & (i \in S) \\ u \ge 0 & \end{array} \qquad (20)$$

It is evident from (20) that the optimal a is the center of the cube $\{y : a - u \le y \le a+u\}$ with the lengths of all edges equal to 2u , where u is that minimal value for which all points of S are contained in the cube . For fixed S the value a_k is determined as the mid-point of the interval between maximal and minimal values y^m_{ik} for i S .

Let us consider a suboptimal addition algorithm for the criterion.

Algorithm "ADDI-CUBE"

We begin with $u_0 = \max_{i,k} |y^m_{ik}| = |y^m_{i_0 k}|$ and assign the unit i_0 to S. For the general step we consider

$$u(S) = \max_{i \notin S} \max_k |y^m_{ik}| = |y^m_{i(S)k}|$$

and put i into S, each time recalculating the standard vector a as the vector of the mid-range values between the maximal and the minimal y^m_{ik} for $i \in S$. Computations are completed when the maximal deviation

$$\max_k \left[\max_{i \in S} y^m_{ik} - \min_{i \in S} y^m_{ik} \right]$$

becomes greater than $2*u(S)$.

It is clear that the algorithm ADDI-CUBE in the CONEX method is appropriate for a special "comet-like" structure of the data, when the cloud of points in the variable space may break up into a "nucleus" and "tail" .The tail forms the first cluster, then the nucleus gives a new tail and so on.

REFERENCES

Braverman, E.M., and Mouchnick, I.B.(1983) ,Structural Methods of Empirical Data Treatment, Moscow ,Nauka Publishers (in Russian).

Goodman ,L.A.,and Kruskal, W.H.(1954), "Measures of Association for Cross Classifications", Journ.Amer.Stat.Assoc., 49, 723-764.

Dorophеiuk, A.A. (1966), "Algorithms for Pattern Recognition without Teachers, Based on the Potential Functions Method" , Automation and Remote Control, 10, 78-87 (in Russian).

Holzinger, K.J., and Harman, H.H.(1941), Factor Analysis, Chicago: Univ.Ch.Press.

Mirkin, B.G. (1985), Groupings in Socio-Economical Researchs, Moscow :Finansy i Statistika Publishers (in Russian).

Mirkin, B.G.(1987)," Method of Principal Cluster Analysis", Automation and Remote Control, 10, 131-142 (in Russian).

Saporta, G.(1988), "About Maximal Association Criteria in Linear Analysis and in Cluster Analysis", in Classification and Related Methods of Data Analysis, ed. H.H.Bock, Amsterdam: Elsevier , 541-550.

Shepard, R.N.,and Arabie, P.(1979), "Additive Clustering: Representation of Similarities as Combinations of Overlapping Properties", Psych.Review, 86,87-123.

Socal, R., and Michener ,C.D.(1958) ,"A Statistical Method for Evaluating Relationship", Univ. Kansas Sci. Bull., 318, 1409-1438.

CLASSIFICATION OF CONCEPTS DESCRIBED BY TAXONOMIC PREORDONNANCE VARIABLES WITH MULTIPLE CHOICE. APPLICATION TO THE STRUCTURATION OF A SPECIES SET OF PHLEBOTOMINE.

I.C. Lerman (IRISA - Rennes) and PH. Peter (IRESTE - Nantes)

INTRODUCTION

The "descriptive variable" notion can be formalized with much more generality and flexibility in Data Classification than in Factorial Analysis. In this latter field a given variable has to be represented geometrically with respect to the whole set of the described objects. But in Data Classification the descriptive variable can have very synthetic expression of combinatorial and logical nature.

The underlied classification method is hierarchical and called "Likelihood Linkage Analysis". It has been set up and developed by I.C. Lerman & Collaborators (8). This method is based on a combinatorial and statistical approach and it leads to "significant" classes formation on both sets : the set of objects or concepts and the set of descriptive variables, and that, for all types of variables. Each variable is represented in the method, as a relation on the set of objects. We proceed to the interpretation of the results in a dual way by situating each class of objects with respect to the different variable classes and vice versa.

The purpose of this paper is to introduce a very new type of "descriptive variable" -or "descriptor" which is of combinatorial and logical nature and which has got a very synthetic and rich descriptive structure. This descriptor enables us to formalize the expert knowledge concerning the biological descriptions of phlébotomine sandflies of French Guiana [J. Lebbe, J.P. Dedet & R. Vignes (Institut Pasteur de la Guyane Française)(4)]. Descriptions are very complex. Each species is a class of specimens and its description must represent not only a prototype, but all possible variations in the species. Thus, the description by a qualitative variable of a given species, requires -the most often- a subset of possible values. Finally, we assume ranking similarity function on the modality set of each variable.

We will distinguish formally two description levels. The first concerns an elementary object x for which we only know that it belongs to a set $\mathcal{O}_C$ of "examples" of a concept C. The second level concerns the deduced description of the concept C. The problem is the to structure a set $\mathcal{C}$ of concepts. In our application x is a specimen and C is a species.

II. QUALITATIVE VARIABLE WITH UNIQUE CHOICE AND QUALITATIVE VARIABLE WITH MULTIPLE CHOICE

Let E be the set of objects or examples, where each object is a representative element of a concept C ; n=card(E). Let $J=\{1,2,...,j,...,m\}$ denote the coding of the modality set of a qualitative descriptive variable v. The structure of which the modality set is provided, is not taken into account in this paragraph. In classical data analysis v has to be defined on the whole set E ; but we consider here that v is a "descriptor" of E which an "unique choice", if there exists a non empty subset D of E, such that v is a mapping of D into J ; in other words $\mathcal{P}(E)$ denoting the set of subsets of E,

$$(\exists D \in [\mathcal{P}(E)-\{\emptyset\}]), (\forall x \in D) \Rightarrow v(x) \in J. \quad (1)$$

In fact by difference with the classical data classification, the "value" v(x) on the representation object x of a concept C, interprets the expert knowledge on C.

The descriptive variable v is "with multiple choice" on E, if there exists a non empty subset D of E, such that v is a mapping of D on a set of logical expressions on J. More precisely, each logical expression can be defined by a disjunction -eventually positively weighted by a probability distribution- of conjunctions, where each conjunction concerns a subset of J (i.e. a subset of modalities of J).

Thus, to a given object x of D, we can associate a sequence of subsets :

$$(J_1^x, J_2^x, \ldots, J_e^x, \ldots, J_k^x) \quad (2)$$

of the set J, provided by a probability distribution :

$$(p_1^x, p_2^x, \ldots, p_e^x, \ldots, p_k^x), \quad (3)$$

where the object x may possess the subset J_e^x of the modalities of the variable v, with the probability (or relative frequency) p_e^x $1 \leq e \leq k$. Notice that x may possess J_1^x or J_2^x or ... or J_k^x, where the "or" defines a strict disjunction. On the other hand, the different subsets J_e^x $(1 \leq e \leq k)$ are not necessarily disjoint.

We denote

$$(\exists D \in [\mathcal{P}(E) - \{\emptyset\}])(\forall x \in D),$$

$$v(x) = (\bigwedge\{j/j \in J_1^x\}, p_1^x) \vee (\bigwedge\{j/j \in J_2^x\}, p_2^x)$$

$$\ldots \vee (\bigwedge\{j/j \in J_k^x\}, p_k^x). \quad (4)$$

We may now extend the variable or descriptor definition on the set $\mathcal{C}$ of the concepts C. Relative to (4) the "value" of v on the concept C that x represents is in fact a conjunction of the following form :

$$v(C) = (\bigwedge\{j/j \in J_1^C\}, p_1^C) \wedge (\bigwedge\{j/j \in J_2^C\}, p_2^C) \wedge$$

$$\ldots \wedge (\bigwedge\{j/j \in J_k^C\}, p_k^C), \quad (5)$$

where $(J_l^C, p^C) = (J_l^x, p^x)$ for $1 \leq l \leq k$, since the "value" J_1^C occurs with the relative frequency p_1^C and the "value" J_2^C occurs with the relative frequency $p_2^C, \ldots$

III. TOTAL PREORDONNANCE STRUCTURE ON THE MODALITY SET OF A QUALITATIVE VARIABLE. REPRESENTATION

A "preordonnance" qualitative variable is a qualitative variable of which the modality set J is provided by a total preordonnance which is a total preorder on the set of unordered (or ordered) modality couples. In the case of interest it concerns here the following set :

$$J^{\{2\}} = \{(j,h)/1 \leq j < h \leq m\} \quad (1)$$

(cf.[10] & [16]). This total preorder is established directly by the expert or by means of an Artificial Intelligence program taking into account the expert knowledge and a theory of the concerned field. The total preorder is established from the highest couples to the lowest ones, according to an ordinal comparison of the similarities on $J^{\{2\}}$ {i.e.$(j,h)\succ(j',h')$ [resp.$(j,h)\sim(j',h')$] if j and h are strictly more [resp. equally similar than j' and h'].

This last total preorder is coded by means of a "ranking function" which leads to the table

$$\{r_{jh}/(j,h)\in J^{\{2\}}\}, \quad (2)$$

where the rank r_{jh} is computed with the following formula :

$$r_{jh}=l_1+l_2+\ldots+l_{(p-1)}+\frac{1}{2}(l_p+1), \quad (3)$$

where l_q denotes the qth class cardinal of the total preorder on $J^{\{2\}}$ and where (j,h) belongs to the pth class, according to an increasing ordering of the preorder classes.

Example. Let consider the 33rd variable of phlebotomine sandflies description (cf.[4]), "Aspect of individual duct". The modalities of this variable are :

1- Smooth non-sclerotized
2- Smooth sclerotized
3- Transversely striated or annulated
4- With small prominent tubercles.

By going from the most similar couples to those, the least similar, the expert has given the following preordonnance on J={1,2,3,4} :

$$11\sim22\sim33\sim44>12\sim13\sim23>14\sim24\sim34, \quad (4)$$

where jh represents the pair {j,h}, $1\leq j\leq h\leq 4$.

Then -by ordering lexicographically the couples (j,h)- the table (2) becomes :

$$\{8.5,5,5,2,8.5,5,2,8.5,2,8.5\} \quad (5)$$

IV. TAXONOMIC VARIABLE ORGANIZING A SET OF QUALITATIVE VARIABLES. REPRESENTATION

Such variable that we denote ω consists of a sequence of sequences of qualitative variables of the following form :

$$\omega=\{v^{(1)};v_1^{(2)},\ldots,v_{k_2}^{(2)};\ldots;v_1^{(p)},\ldots,v_{k_p}^{(p)};\ldots;v_1^{(q)},\ldots,v_{kq}^{(q)}\}. \quad (1)$$

The first sequence is necessarily reduced to one element : the variable $v^{(1)}$ which is associated to the root of the tree figuring the taxonomic variable. On the one hand, each variable $v_{j_p}^{(p)}$ $(1\leq j_p\leq k_p)$ is the "daughter" variable of at least one variable $v_h^{(p-1)}$ $(1\leq h\leq k_{p-1})$ and on the other hand, when two variables $v_j^{(p)}$ and $v_{j'}^{(p)}$ have the same "mother" variable $v_h^{(p-1)}$, they are respectively defined on two distinct subsets where each subset is characterized by one modality of $v_h^{(p-1)}$.

Example. Let us consider the variables 1,18,19 and 20 of [4]; where v^1 is the "Sex", v^2 is the "Number of style spines", v^3 is the "Distribution of insertion of 4 style spines" and v^4 is the "Distribution of insertion of 5 style spines".

We obtain the following taxonomic structure :

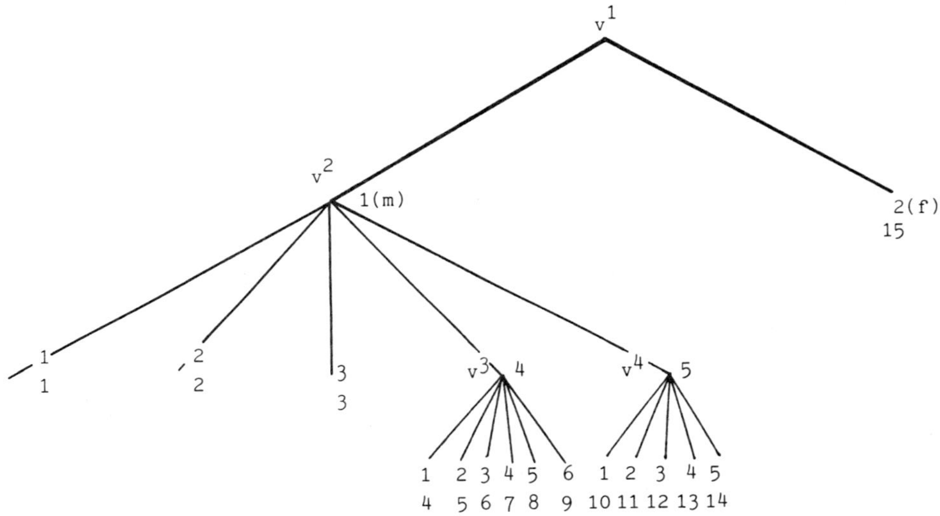

Figure 1

We represent the variable defined by this structure by means of what we call an "ultrametric preordonnance" (cf.[7] & [10]) on the set of the taxonomy leaves (the cardinality of this set is 15 in the above example). According to this total preorder on the set of unordered leaf couples, the higher is the rank of a given couple, the lower is the first node -for a decreasing construction of the taxonomic tree from the root to the leaves- which underlies the two leaves. Thus, in the above example, the pair {6,8} has the same rank as the pair {10,12}. This last is greater than that one of { 7,12}, which is equal to the {2,3} rank, and so on...

Let L denote the set of the leaves, a ranking function r coding the ultrametric preordonnance is characterized by the following property :

$$(\forall \{x,y,z\} \in P_3(L)),\ r(x,z) \geq \min[r(x,y),(r(y,z)]$$

We adopt as in the general case (cf.§III), the notion of "mean rank" which can be determined by a formula from the "type" of the taxonomy (cf. [14]). We must precise that the preordonnance class at which is assigned the highest rank is composed of the couples (x,x), x$\in$ L.

By this way, the taxonomic variable appears as a particular case of the preordonnance variable.

V. TAXONOMIC PREORDONNANCE VARIABLE. REPRESENTATION

Relative to the above defined structure of a taxonomic variable ω, we further assume that the set $M^{(p)}_{j_p}$ of the modalities of a qualitative variable $v^{(p)}_{jp}$, is provided by a total preordonnance expressing -in a ordinal way- the modality resemblances ($1 \leq j_p \leq k_p$, $1 \leq p \leq q$). On the other hand, we suppose that these k_p preordonnances can be extended in an unique total preorder on the set :

$$\bigcup\{P_2(M^{(p)}_{j_p})/1 \leq j_p \leq k_p\}, \quad (1)$$

according to an ordinal similarity on the modality pairs, where $P_2(X)$ denotes the unordered object pairs of X.

In these conditions, we have to define a total preordonnance on the set of the leaves of the taxonomy, or -equivalently- on the set of complete chains, going from the root to the leaves. This preordonnance must take into account both the preordonnance defined in the previous paragraph IV, and those that we have just introduced.

Such preordonnance is built step by step, in a decreasing way with respect to -perceived or deduced- resemblance between terminal modalities represented by the leaves of the taxonomy. Then, the first class of the preordonnance concerns the pairs {x,x}, where x is a leaf modality ; x belongs to {1,2,...,15} in the above paragraph IV example. <u>The general principle consists of refining the ultrametric preordonnance associated to the taxonomy by means of the preordonnances, locally defined</u>. Each one of these last concerns the set of the node pairs separated at a given taxonomy level and joined at the consecutive level. Thus, in the above example, if A={4,5,6,7,8,9} and B={10,11,12,13,14}, the continuation of the preordonnance will concern $[P_2(A) \cup P_2(B)]$, where $P_2(A)$ [resp.$P_2(B)$] denotes the unordered element pairs of A (resp. B). Now, if C denotes the set of the leaves {1,2,...,14}, the continuation of the preordonnance will concern $\{P_2(C)-[P_2(A) \cup P_2(B)]\}$.

The finally obtained structure is a total preordonnance on the set of the taxonomy leaves, that we code by means of a ranking function [see formula (3) §III].

VI. PREORDONNANCE TAXONOMIC VARIABLE WITH MULTIPLE CHOICE

The description structure of the variable that we consider here is identical to that one defined in the preceding paragraph V ; but now, the "value" of a given variable $v^{(p)}_{jp}$ ($1 \leq p \leq q, 1 \leq j_p \leq k_p$) on a given concept is defined by a <u>weighted disjunction of conjunctions on the set $J = M^{(p)}_{jp}$</u> of its modalities ; that is to say, an expression as that (4) of paragraph II above.

By referring to this logical expression, description [1] can be formalized by the particular case where J^x_e includes only one element, and where $p^x_e=1/k$, $1 \leq e \leq k$. Besides, description [4] can be expressed by the case where the sequence of subsets $(J^x_1, J^x_2, \ldots, J^x_k)$ is reduced to one non empty subset J^x strictly included in J.

The "value" of such variable on an object representing a given concept will finally correspond to a weighted disjunction of conjunctions. Each conjunction concerns a subset of terminal leaves, or -equivalently- of complete chains, joining the root to the terminal leaves. The weight assigned to a conjunction of complete chains will be computed multiplicatively, with respect to their respective compositions.

Example : Let us imagine that -for a given concept x- we have for the descriptive variables introduced in the paragraph IV example :

$$v^1(x)=1$$
$$v^2(x)=(1\&2,0.4)\vee(2\&3\&4,0.2)\vee(4\&5,0.4)$$
$$v^3(x)=(1\&2,0.4)\vee(2\&3,0.6)$$
$$v^4(x)=(2\&3\&4,0.8)\vee(3\&5,0.2),$$

where & denotes a conjunction.

A given path is represented by the sequence of the modalities of the different variables organized according to the taxonomy. The value of the "taxonomic preordonnance variable with multiple choice" ω, is :

$$\begin{aligned}\omega(x)=&(11\&12,0.4)\vee(12\&13\&141\&142,0.08)\vee\\&(12\&13\&142\&143,0.12)\vee(141\&142\&152\&153\&154,0.128)\vee\\&(141\&142\&153\&155,0.032)\vee(142\&143\&152\&153\&154,0.192)\vee\\&(142\&143\&153\&155,0.048);\end{aligned}$$

or, by using the coding of the taxonomy leaves with the integers 1 to 15 (cf. Figure 1),

$$\begin{aligned}\omega(x)=&(1\&2,0.4)\vee(2\&3\&4\&5,0.08)\vee(2\&3\&5\&6,0.12)\vee\\&(4\&5\&11\&12\&13,0.128)\vee(4\&5\&12\&14,0.032)\vee\\&(5\&6\&11\&12\&13,0.192)\vee(5\&6\&12\&14,0.048)\end{aligned}$$

We can check that the weighting sum is equal to unity. This point -as we may show- is general.

Now, the "value" of v on the concept C represented by x, can be put in the following form :

$$\begin{aligned}\omega(C)=&(1\&2,0.4)\wedge(2\&3\&4\&5,0.08)\wedge(2\&3\&5\&6,0.12)\wedge\\&(4\&5\&11\&12\&13,0.128)\wedge(4\&5\&12\&14,0.032)\wedge\\&(5\&6\&11\&12\&13,0.192)\wedge(5\&6\&12\&14,0.048)\end{aligned}$$

where $\wedge$ is another notation for a conjunction.

VII. ROUGH SIMILARITY INDEX BETWEEN TWO OBJECTS (resp. CONCEPTS) DESCRIBED BY A TAXONOMIC PREORDONNANCE VARIABLE WITH MULTIPLE CHOICE.

Let us recall the case of a preordonnance variable with unique choice (cf. §III,[10] & [16]). To classify the object set, we have to consider ordinal similarity as symmetrical notion coded by the ranking function introduced in paragraph III cf. (2) & (3) §III).

Let x and y be two given objects. If ω denotes the preordonnance variable with unique choice and if $\omega(x)$ [resp. $\omega(y)$] is the value of ω on x [resp. y] [i.e. the modality of ω possessed by x (resp. y)], then the rough similarity index between x and y, is

$$s(x,y)=r[\omega(x),\omega(y)] \quad (1)$$

Now, let us consider the case where ω is a preordonnance variable with multiple choice. The value of the variable ω on the object x (resp. y) has the form (4) of the above paragraph II.

According to the notations of this last paragraph, let us denote by

$$\{(J_e^x,p_e^x)/1\leq e\leq k\} \qquad (2)$$

the components of the weighted disjunction corresponding to $\omega(x)$.

On the other hand

$$\{(J_f^y,p_f^y)1\leq f\leq l\} \qquad (3)$$

are the components of the weighted disjunction corresponding to $\omega(y)$.

The estimation of the similarity between x and y requires matching coefficient between (J_e^x,p_e^x) and (J_f^y,p_f^y), $1\leq e\leq k$, $1\leq f\leq l$.

Let us introduce, for comparing J_e^x and J_f^y :

$$(\forall j\in J_e^x),r(j)=\max\{r(j,h')/h'\in J_f^y\} \qquad (4)$$

$$(\forall h\in J_f^y),r(h)=\max\{r(j',h)/j'\in J_e^x\} \qquad (5)$$

Then, the contribution that we consider of (J_e^x,J_f^y) to the rough similarity index between x and y, is

$$s_{ef}(x,y)=\mathcal{M}\{\{r(j)/j\in J_e^x\},\{r(h)/h\in J_f^y\}\}, \qquad (6)$$

where $\mathcal{M}$ represents mean operation over $[\mathrm{card}(J_e^x)+\mathrm{card}(J_f^y)]$ integers.

The integration of the whole following set of weighted couples of J subsets [cf.(7) below] -to evaluate global rough index s(x,y)- is done by a mean operation $\mathcal{M}^p$ [cf. (8) below], with respect to the distribution (9) :

$$\{[(J_e^x,p_e^x),(J_f^y,p_f^y)/1\leq e\leq k,\ 1\leq f\leq l\}, \qquad (7)$$

$$s(x,y)=\mathcal{M}^p\{s_{ef}(x,y)/1\leq e\leq k,\ 1\leq f\leq l\}, \qquad (8)$$

$$\{p_e^x \times p_f^y/1\leq e\leq k,\ 1\leq f\leq l\} \qquad (9)$$

Let us now define the rough similarity index $s(C^x,C^y)$ between the two concepts C^x and C^y that x and y represent respectively. $\omega(C^x)$ [resp. $\omega(C^y)$] can be expressed by (5) of paragraph II, where $\{(J_d^C,p_d^C)/1\leq d\leq k\}$ is replaced by $\{(J_e^{C^x},p_e^{C^x})/1\leq e\leq k\}$ [resp. $\{(J_f^{C^y},p_f^{C^y})/1\leq f\leq l\}$]

Taking into account the above expression (6) -which matches J_e^x and since $\omega(C^x)$ [resp. $\omega(C^y)$] is defined by a conjunction, we have to determine for each e ($1\leqslant e\leqslant k$) [resp. f ($1\leqslant f\leqslant l$) the best matching of J_e^{Cx} with a J_f^{Cy} (resp. of J_f^{Cy} with a J_e^{Cx}). Namely, we define

$$s_{ef(e)}(x,y)=\max\{s_{ef'}(x,y)/1\leqslant f'\leqslant l\}, \qquad (10)$$

and

$$s_{e(f)f}(x,y)=\max\{s_{e'f}(x,y)/1\leqslant e'\leqslant k\}. \qquad (11)$$

The rough similarity index that we adopt between the two concepts C^x and C^y can be written as follow :

$$s(C^x,C^y)=\mathcal{M}_0^p\{\{s_{ef(e)}(x,y)/1\leqslant e\leqslant k\}, \{s_{e(f)f}(x,y)/1\leqslant f\leqslant \ \}\} \qquad (12)$$

where $\mathcal{M}_0^p$ denotes mean operation with respect to the following distribution

$$\{\{p_e^x p_{f(e)}^y/1\leqslant e\leqslant k\}, \{p_{e(f)}^x p_f^y/1\leqslant f\leqslant l\}\}. \qquad (13)$$

VIII. SIMILARITY STRUCTURE BETWEEN OBJECTS (CONCEPTS) COMPATIBLE WITH LIKELIHOOD LINKAGE CRITERION IN CASE OF SEVERAL VARIABLES.

We have studied this similarity structure in [10]. Let E denote here a set of objects or a set of concepts and let $\mathcal{W}$ be a set of descriptive variables of any type. Relative to a given variable $\omega(\omega\in\mathcal{W})$, we have built the rough index $s_\omega(x,y)$ between two elements x and y of E, in case where ω is a preordonnance variable with multiple choice and in case where E is a set of objects [cf. (8) above] or a set of concepts [cf. (12) above].

In our approach we normalize the contribution $s_\omega(x,y)$ with respect to the set of unordered element pairs of E : $F=P_2(E)$. This normalization is statistical ; namely, we define :

$$S_\omega(x,y)=\frac{[s_\omega(x,y)-moy_e(s)]}{\sqrt{var_e(s)}} \qquad (1)$$

where $moy_e(s)$ and $var_e(s)$ are respectively the empirical mean and variance of $s_\omega(x,y)$ over $F=P_2(E)$.

To evaluate the similarity relative to the whole set $\mathcal{W}$ of the variables, we begin by considering

$$S(x,y)=\frac{1}{\sqrt{card(\mathcal{W})}}\sum\{S_\omega(x,y)/\omega\in\mathcal{W}\} \qquad (2)$$

which is statistically standardized over $F=P_2(E)$. We obtain :

$$S_\nu(x,y)=\frac{[S(x,y)-moy_e(S)]}{\sqrt{var_e(S)}}, \qquad (3)$$

where $moy_e(S)$ and $var_e(S)$ are respectively the empirical mean and variance of S(x,y over $F=P_2(E)$.

At this level, we consider a hypothesis of independence or no relation for which we associate to the set of observed variables $\mathcal{V}$, a set $\mathcal{V}^*$ of independent random variables, respectively of the same types as those of $\mathcal{V}$. In these conditions the computed similarity index (3), $S^*_\nu(x,y)$, based on $\mathcal{V}^*$, is a random variable. In case where the number of variables, card($\mathcal{V}$) is not too small, it can be shown that $S^*_\nu(x,y)$ is normally N(0,1) distributed. Then the similarity coefficient that we adopt finally takes the following form

$$P(x,y)=\Phi[S_\nu(x,y)] \qquad (4)$$

where Φ is the cumulativedistribution function of the standardized normal variable.

This similarity index is referring to a probability scale. The similarity table is then

$$\{P(x,y)/\{x,y\}\in F=P_2(E)\}. \qquad (5)$$

The agglomerative building of hierarchical classification tree on E, is based in our method on the "likelihood of the maximal linkage" criterion that we have introduced in 1970 [cf. (6),(8)]. This criterion has got clear statistical basis with respect to statistical independence hypothesis between the two disjoint subsets C and D (defining two classes) of E, to be compared at a given level of the tree construction. Its expression is the following :

$$P(C,D)=[\max\{P(x,y)/(x,y)\in C\times D\}]^{c\times d} \qquad (6)$$

where c=card(C) and d=card(D).

Computational reasons lead us to use the strictly increasing function on the interval [0,1] :

$$f(\xi)=-\mathrm{Log}[-\mathrm{Log}(\xi)] \qquad (7)$$

and technically to adopt the equivalent criterion

$$Q(C,D)=-\mathrm{Log}\{-\mathrm{Log}[P(C,D)]\} \qquad (8)$$

In order to build the classification tree, we have recently established reactualization formulas in case of <u>multiple aggregations</u> at a given level for the most classical hierarchical classification criteria ("Single Linkage", "Complete Linkage", "Average Linkage", "Ward Criterion" and "Likelihood Maximal Linkage Criterion") (cf.[11]).

We note in passing that the "Likelihood Maximal Linkage Criterion" has got the "Reductibility" property [cf.(8)] introduced by Bruynooghe in 1977. This property makes possible algorithm implementation for classification of large data set [cf.(1)].

A decisive stage of the "Likelihood Linkage Analysis" (L.L.A.) classification method consists of the recognition of the "Significant" levels and nodes of hierarchical classification tree, that we reduce to the levels where "significant" nodes appear [cf.(7),(8) & (9)]. This is done by means of an association criterion between the emerged partition at a given level anda suitable structure of the information concerning the resemblances between the elements of the set E to be classified. This criterion that we have assessed from both points of view, formal and statistical, can be based on the preordonnance on E associated to the Similarity function (on E) or directly on the Similarity function itself [cf.(8) & (9)]. The recognition of the most "significant" levels and the most "significant" nodes is based on the behaviour of the distribution of the mentionned criterion, on the increasing sequence of the levels of the classification tree. A "significant" level indicates natural partition to consider with a given degree of synthesis. The significant nodes indicate completion stages of the different classes appearing in the tree.

We show in the following figure 2 a branch of the hierarchical classification tree that we have obtained on a set of phlebotomine sandflies species that we have described by means of taxonomic preordonnance variables with multiple choice. In this structure, the significant nodes are represented by the darker and thicker lines.

Figure 2

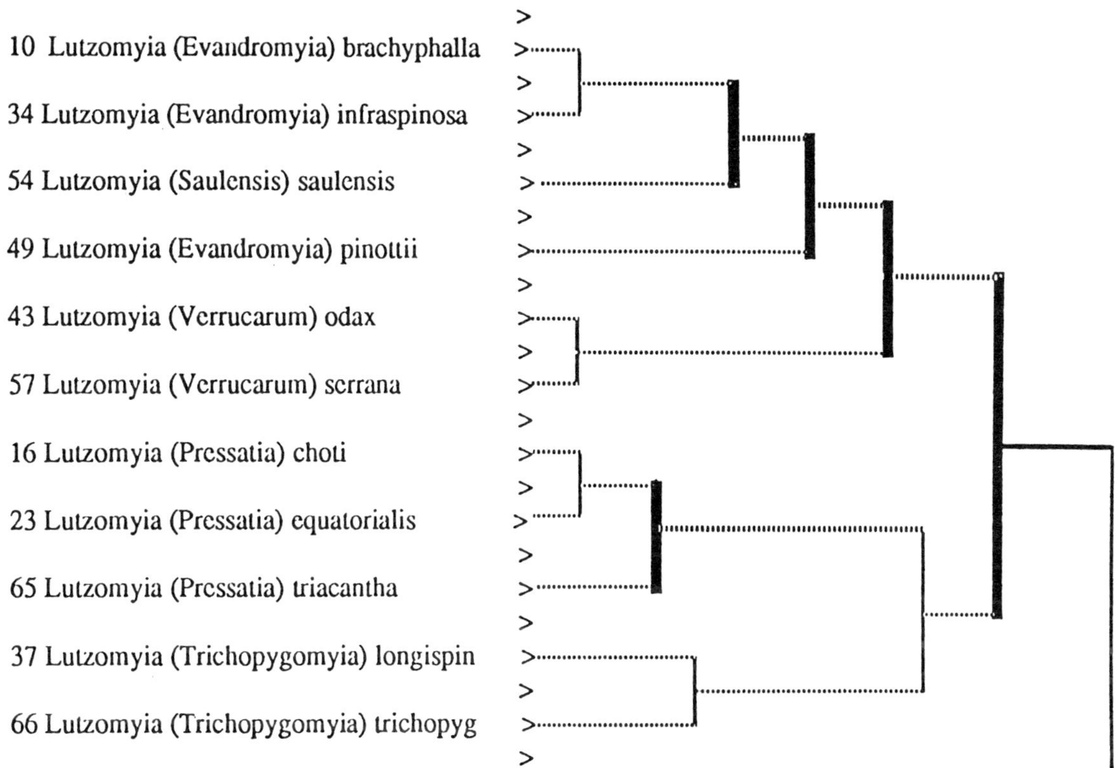

10 Lutzomyia (Evandromyia) brachyphalla
34 Lutzomyia (Evandromyia) infraspinosa
54 Lutzomyia (Saulensis) saulensis
49 Lutzomyia (Evandromyia) pinottii
43 Lutzomyia (Verrucarum) odax
57 Lutzomyia (Verrucarum) serrana
16 Lutzomyia (Pressatia) choti
23 Lutzomyia (Pressatia) equatorialis
65 Lutzomyia (Pressatia) triacantha
37 Lutzomyia (Trichopygomyia) longispin
66 Lutzomyia (Trichopygomyia) trichopyg

REFERENCES

[1] BRUYNOOGHE M. (1989) ; "Nouveaux algorithmes en classification automatique applicables aux très grands ensembles de données, rencontrés en traitement d'images et en reconnaissance des formes", Thèse de Doctorat d'Etat, Univ. de Paris VI, 23 Janv. 1989.

[2] DECAESTECKER C. (1988) ; "Conceptual clustering with hierarchical and conjunctive variables", Rapport technique I.A. -03-88, CADEPS, Univ. Libre de Bruxelles.

[3] DIDAY E. (1987) ; "Introduction à l'approche symbolique en analyse des données", Journées "Symbolique-Numérique", Université Paris 9 Dauphine, 8-9 déc. 87.

[4] LEBBE J., DEDET J.P. & VIGNES R. (1987) ; "Identification assistée par ordinateur des phlébotomes de la Guyane Française", Institut Pasteur de la Guyane Française, Version 1.02 - 10/07/1987.

[5] LECLERC B. & CUCUMEL G. (1987) ; "Consensus en Classification : une revue bibliographique", Rev. Math. & Sc. Humaines 100, p. 109-118.

[6] LERMAN I.C. (1970a) ; "Sur l'analyse des données préalable à une classification automatique ; proposition d'une nouvelle mesure de similarité", Rev. Math. & Sc. Humaines n° 32, p. 5-15.

[7] LERMAN I.C. (1970b) ; "Les bases de la classification automatique", Gauthier Villars "collection Programmation", Paris.

[8] LERMAN I.C. (1981) ; "Classification et analyse ordinale des données", Dunod, Paris.

[9] LERMAN I.C. (1983) ; "Sur la signification des classes issues d'une classification automatique" in Numerical Taxonomy, NATO ASI series vol. G1, edited by J. Felsenstein, Springer Verlag (1983), p. 179-198.

[10] LERMAN I.C. (1987) ; "Construction d'un indice de similarité entre objets décrits par des variables d'un type quelconque. Application au problème du consensus en classification". Rev. de Stat. Appl., XXXV (2), 39-60.

[11] LERMAN I.C. (1988) ; "Formules de réactualisation en cas d'agrégations multiples", Rapport Interne IRISA n° 409. A paraître dans RAIRO-R.O. (Vol. 23, n° 2, 1989).

[12] LERMAN I.C., HARDOUIN M. et CHANTREL T. (1980) ; "Analyse de la situation relative entre deux classifications floues", in Data Analysis and Informatics, E. Diday et al. (eds), North Holland (1980).

[13] LERMAN I.C., NICOLAS J. & PETER Ph. (1988) ; "Classification Conceptuelle: une approche dans la combinaison de méthodes numériques et symboliques", Deuxièmes Journées "Symbolique-Numérique" pour l'Apprentissage de Connaissances à partir de Données, Orsay, 12-13 Décembre 1988.

[14] LERMAN I.C. & PETER Ph. (1988) ; "Classification en présence de variables préordonnances taxonomiques à choix multiple. Application à la structuration des phlébotomes de la Guyane Française", Public. Int. n° 426 IRISA, Rennes, Septembre 1988.

[15] MICHALSKI R.S. (1980) ; "Pattern recognition as rule-guided inductive inference", IEEE Transactions on pattern analysis and machine intelligence, vol. PAMI-2, n° 4, July 1980.

[16] PETER Ph. (1987) ; "Méthodes de classification hiérarchiques et problèmes de structuration et de recherche d'informations assistées par ordinateur", Thèse de l'Univ. de Rennes I, 6 mars 1987.

A METHOD FOR ANALYSING MULTIDIMENSIONAL EXPERIMENTAL DATA

O. Cadet, Y. Schektman
AEGIDE, Greco-CNRS 59, Université Toulouse le Mirail
31058 Toulouse Cedex (France)

Key words : MANOVA, Inner product with relationship effects, symmetrical, dissymmetrical association coefficients.

I GENERALITIES.

Let (y^k ;k=1,q) be q observed variables and (x^j ;j=1,p) be the indicator variables associated to the p levels of an experimental factor. Usually in particular in MANOVA, analysis of the factor effects on the variables (y^k) is based on statistics which are functions of the canonical correlations between the two sets of variables (x^j) and (y^k) .So, in the analyses, the observed variables and the factor play symmetrical roles .Note that this choice is not necessarily the best in all cases. So, we propose a method [2] which takes into account both symmetrical and dissymmetrical points of view. The method described here, uses general association coefficients [7] based on inner products with relationship effects. Let $E = R^{p+q} = E_x \oplus E_y$ be the individual-space where $E_x = R^p$ (resp. $E_y = R^q$) is the subspace associated to the variables (x^j;j=1,p) [resp. (y^k;k=1,q)]. In each subspace , an individual is represented by a point (vector): so, the coordinates of the ith individual point in E_y denoted by y_i, are (y^k_i; k =1,q), where y^k_i is the value of y^k for the ith individual . In E_y , the configuration of points (y_i) is denoted by N_y. N_x is defined in the same way. Subsequently, for avoiding redundancy,we shall give only definitions referring to y.

Let [$c_k(y)$, $C^k(y)$; k=1,q] be a set of principal vectors and components of N_y . Inner products in E_x being given, the inner product with relationship effect for (E_x, E_y) in E is defined geometrically by

$$\cos(c_j(x),c_k(y)) = \cos(C^j(x),C^k(y)) \quad \text{if} \quad \|C^j(x)\| \neq o \text{ and } \|C^k(y)\| \neq o$$
$$= o \qquad\qquad \text{else}$$

According to the partition of the variables in two sets and to the corresponding partition of E in two subspaces, the covariance matrix between all variables, denoted by V, and the inner product matrix in E, denoted by M, may be written

$$V = \begin{bmatrix} V_x & V_{xy} \\ V_{yx} & V_y \end{bmatrix} \qquad M = \begin{bmatrix} M_x & M_{xy} \\ M_{yx} & M_y \end{bmatrix}$$

where - V_{xy} is the covariance matrix between the two sets of variables $\{x^j\}$ and $\{y^k\}$,

- M_y is the inner product matrix in E_y.

A general algebraic expression of an inner product M with relationship effect [7] may be expressed like that:

$$M_{xy} = M_x\ [(V_x\ M_x)^{1/2}]^+\ V_{xy}\ M_y\ [(V_y\ M_y)^{1/2}]^+ \qquad (1)$$

where $[(V_yM_y)^{1/2}]^+$ is the Moore-Penrose generalized inverse of $(V_yM_y)^{1/2}$(weighted by M_y) .

Subsequently , M_x and M_y will depend upon the problem to be solved.

The general symmetrical (SRAC) or dissymmetrical (DRAC) relationship association coefficients are defined [9] by

$$\text{SRAC}\ [x,y\ /\ M_x=\chi^2_x,\ M_y=V_y^+\ ] = \frac{I[P_x(N_y)]}{(\ I[N_x]\ \ I[N_y]\ )^{1/2}} \qquad (2)$$

$$\text{DRAC}\ [x\text{->}y\ /\ M_x\ ,\ M_y\ ] = \frac{I[P_x(N_y)]}{I[N_y]} \qquad (3)$$

where

- $I[N_y]$ is the inertia of N_y relative to its centre of gravity,

- $P_x(N_y)$ is the orthogonal projection of the points of N_y onto E_x.

If $\{x^j\}$ and $\{y^k\}$ are defined as previously, then SRAC[x,y] is the generalized correlation ratio divided by $(I[N_x]\ I[N_y])^{1/2}$. But, if the variables $\{y^k\}$ are the indicator variables of the modalities of a qualitative variable, then SRAC[x,y] is the Tschuprov coefficient and DRAC[x->y /$M_y=I_q$] is the Goodman-Kruskal tau [4]. Finally , if $\{x^j\}$ and $\{y^k\}$ are quantitative variables, DRAC[x->y /$M_y=I_q$] is the Stewart-Love coefficient [10]. Some elements for demonstrating these results are given in section 2.

The method proposed consists in two complementary steps described in the two following sections.

II DESCRIPTIVE ANALYSIS OF THE FACTOR EFFECTS

II-1 The method

In the definitions (2) and (3) given in section 1, the numerators are the inertia of the configuration of points $P_x(N_y)$ relative to different inner products. Concerning SRAC we have $M_y = V_y^+$, but M_y must be made more explicit the DRAC expression. For analysing the effects of a factor, it is reasonable to analyse the corresponding association coefficient between $\{x^j\}$ and $\{y^k\}$ i.e. here to analyse $I[P_x(N_y)]$. This naturally leads us to analyse the scattering of the points corresponding to $P_x(N_y)$. To achieve this, a statistical tool which appears to be well fitted , is the Principal Component Analysis (P.C.A.).

Note that this approach is used in the classical method of Correspondence Analysis [1] for describing the mean square contingency coefficient $\Phi^2(x,y)$. Fortunately, it is easy to show that

$$\Phi^2(x,y) = I[P_x(N_y)] \qquad (4)$$

This result comes from the general following property.

Property 1: If the inner product in the individual-space is with relationship effect for (E_x,E_y) then

$$I[P_x(N_y)] = \text{trace}\,[\, V_{xy}\, M_y\, V_{yx}\, V_x^+ \,] \qquad (5)$$

proof

We have $I\,[P_x(N_y)] = \text{trace}\,[\, P_x\, In_y\, V_y\, {}^t In_y\, M \,]$

where ${}^t In_y$ is the transpose of the injection matrix from E_y into E.

As $P_x = In_x\, M_x^+\, {}^t In_x\, M$ it follows that

$$I[P_x(N_y)] = \text{trace}\,[\, M_x^+\, M_{xy}\, V_y\, M_{yx}].$$

Finally, the result is obtained by replacing M_{xy} by its expression given in (1).

If we replace M_y by V_y^+ in $I[P_x(N_y)]$, we recognize in (5) a classical expression of the square canonical correlation sum of the two sets of variables $\{x^j\}$ and $\{y^k\}$. So (4) is established.

Consequently, as far as the previous approach is concerned for analysing the association, we shall use P.C.A, of $P_x(N_y)$ relative to an inner product M with relationship effect for (E_x, E_y):

- inner product M_y will be chosen according to what seems appropriate for example

- $M_y = V_y^+$ for analysing a symmetrical association,

- $M_y = I_q$ for analysing a dissymmetrical association corresponding to a classical point of view (cf. remarks in section 1),

- as $I[P_x(N_y)]$ does not depend upon M_x, M_x will be chosen equal to the chi-square distance matrix (χ_x^2) for simplifying calculi,

- as in our problem, we are concerned with experimental data, obviously each individual will have the same weight.

Let w_i be the weight of the ith individual,

G(j) be the centre of gravity of set of the individual points $\{P_x(y_i) \;/\; x_i^j = 1\}$

The following property links our formal approach with classical Factorial Discriminant Analysis. In the considered problem, all the individuals have the same weight and the following property is demonstrated in that case.

Property 2 : $\{y^k\}$ being centered, if M is with relationship effect for (E_x, E_y), with $M_x = \chi_x^2$ and $w_i = 1/n$ for all i, then $P_y(x_i)$ is locate at the centre of gravity of the set of the individual points $\{P_x(y_i) \;/\; x_i^j = 1\}$.

Proof

The matrix expression of $P_y = In_y \, M_y^+ \, {}^t In_y \, M$ is

$$P_y = \begin{bmatrix} 0 & 0 \\ V_x \, \chi_x^2 & I_q \end{bmatrix}$$

For x_i such that $x_i^j = 1$, we have

$$ {}^t In_y \, P_y(x_i) = V_{yx} \, {}^t In_x \, e_j(x) \qquad (6)$$

where $[{}^t In_x \, e_j(x)]_l = \delta_{lj}$

$$[\chi_x^2]_{jk} = \delta_{jk} \; n \,/ n_{j.} \quad \text{with } n_{j.} = \sum_{\langle i / x_i^j = 1 \rangle} x_i^j \quad \text{and} \quad n = \sum_j n_{j.}$$

V_{yx} is the covariance matrix between $\{y^k\}$ and $\{x^j\}$.

As $[V_{yx}]_{lm} = \sum_{\langle i / x_i^m = 1 \rangle} \frac{1}{n} \, y_i^l$, (6) may be written :

$$[\,{}^{t}\mathrm{In}_y\ P_y(x_i)\,]_k = \sum_{\langle i/x_i^j=1\rangle} \frac{n}{n_{j.}}\ y_i^k$$

which is the kth coordinate of G(j) in E_y.

For analysing a dissymmetrical association, the problem which arises now, is to define inner product M_y "better" than in the classical case where $M_y = I_q$. We suggest a simple formula

$$M_y(\alpha) = [\ \alpha\ V_y + (\ 1 - \alpha\ )\ I_q\ ]^{+} \qquad \alpha \in [o,1]$$

where $M_y(\alpha)$ depends upon only one parameter α .

Note that $M_y(0) = V_y^{+}$ corresponds to the symmetrical case and $M_y(1) = I_q$ to the dissymmetrical case.

For choosing $M_y(\alpha)$ it is reasonable to use the two following complementary criteria.

First criterium : $M_y(\alpha)$ is such that C1/C2 be minimum

where

$$C1 = \sum_{j=1}^{p} \sum_{\langle i/x_i^j=1\rangle} w_i\ \|\ P_x(y_i) - G(j)\ \|^2$$

and

$$C2 = \sum_{j=1}^{p} W(j)\ \|\ G(j) - G\ \|^2$$

$$\text{with}\quad W(j) = \sum_{\langle i/x_i^j=1\rangle} w_i \qquad \text{and}\quad G = \sum_{j=1}^{p} W(j)\ G(j)$$

Second criterium : $M_y(\alpha)$ is such that DRAC [x->y /M_x,$M_y(\alpha)$] be maximum .

The choice of $M_y(\alpha)$ is the result of a compromise between the two preceding criteria and some more information such as the variation of the correlations between $\{y^k\}$ and the principal components.

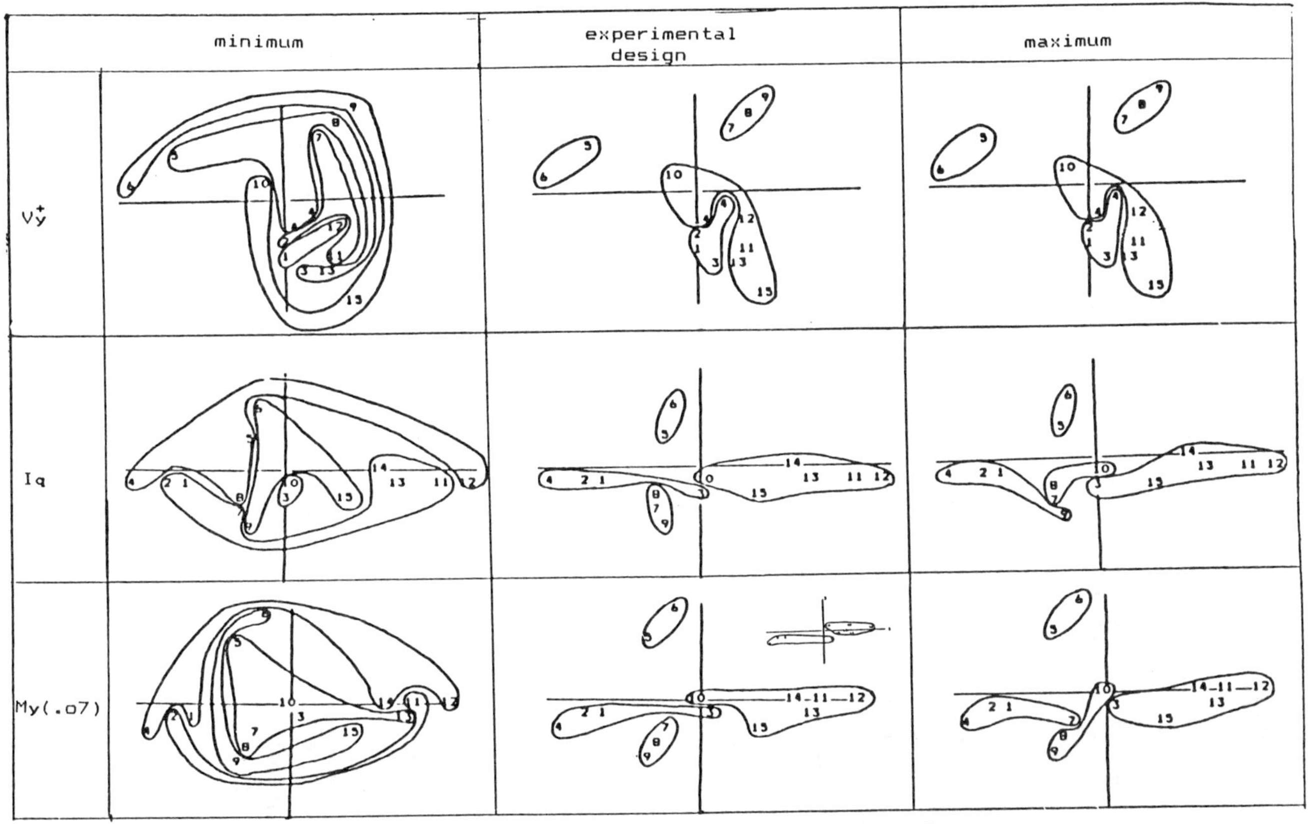

Figure 2: scatterplots matrix with distance choices in rows and valuation in column.

II-2 Example

The data are given in figure 1

f	1				2		3			4					
i	1	2	3	4	5	6	7	8	9	10	11	12	13	14	15
y_1	0	.2	.4	.4	-5	-7.2	2	3	4	.2	.2	.4	.6	0	.8
y_2	.25	.1	1.5	-.05	-7	-8.4	.5	1	1.5	.1	2	2	2.5	1	3
y_3	13	14	0	17	2.5	1.5	3.5	4	4.5	0	-15	-20	-10	-10	-5

Figure 1: illustrative data
(f = factor levels, i = individual, y_1,y_2,y_3 = observed variables)

In the column "experimental design" of figure 2, we give three scatterplots of individuals: the first two ones are concerned with symmetrical and classical dissymmetrical associations and the third one with the new choice $M_y=M_y(.07)$. One can see easily improvements the individuals treated with the same factor level being underlined the such configurations are better separated for $M_y=M_y(.07)$ than for $M_y^+ = V_y$ or $M_y = I_q$. Note that the clusters corresponding to the first and forth factor level are better separated in the case $M_y(.07)$ by the third principal axe than in the case I_q, because in this last case, the third principal moment is quite equal to zero.

III VALUATION OF THE FACTOR EFFECTS

III-1 Principal clusters.

In this section, we describe the Principal Clusters (PC) [5,8] used in the method proposed in III-2 for valuing the factor effects.
A qualitative variable z, associated with an unknown partition of individuals, defines PC of the variables $\{y^k\}$ if

$$DRAC\,[z\to y] = \max_{(S)} DRAC[s\to y]$$

$$SRAC\,[z,y] = \max_{(S)} SRAC[s,y]$$

where s is any qualitative variable having the same modality number as z.
Note that a dual definition of PC in the variable space is given in [6].

III-2 The method

For valuing the factor effect, we suggest to compare the partitions of individuals, corresponding to the experimental design, with two new partitions corresponding to the maximum (resp. minimum) of the DRAC or SRAC. Obviously, the two new partitions will be in agreed with the experimental design according to the size of the p classes. Note that the maximum (resp. minimum) of the DRAC corresponds to the maximum (resp.

minimum) effect of the factor with respect to the observed data. Consequently, we adapt the PC algorithm to this problem.

III-3 Example

The data are the same as in II-2. In figure 2, second column corresponds to the experimental design and the two others respectively to the minimum and the maximum of the SRAC (first row) and of the DRAC (2th and 3th rows). Obviously, without more information, the factor x seems really to have effects on the $\{y^k\}$. Note that the observation is easier for $M_y = M_y(.07)$ than for $M_y = I_q$.

---

[1] J.P. BENZECRI (1976):Analyse des données, tome 2 "L'analyse des Correspondances" Dunod 2ème édition.

[2] O. CADET (1988): Contribution à l'analyse de données multidimensionnelles planifiées. Thèse 3ème cycle, U.P.S. Toulouse.

[3] C. FABRE (1986): Contribution à la protection des méthodes relationnelles . Thèse 3ème cycle, U.P.S. Toulouse.

[4] L.A. GOODMAN, W.H. KRUSKAL (1954) Measures of association for cross classifications. J.A.S.A., 49, 732,764.

[5] A. IBRAHIM (1982): Quelques applications des analyses en composantes sous contraintes et des métriques à effets relationnels en analyse de données . Thèse 3ème cycle, Toulouse.

[6] B,G MIRKINE (1987): Méthode des clusters principaux. Actes du IIIème Colloque Franco-Soviétique, 15-18.

[7] Y. SCHEKTMAN (1978): Contribution à la mesure en facteurs dans les sciences expérimentales et à la mise en oeuvre des calculs numériques. Thèse d'Etat, U.P.S.Toulouse.

[8] Y. SCHEKTMAN ,A. IBRAHIM (1988) Evolving Principal Clusters Theory and application to management monitoring in data, ExpertKnowledge and Decisions (W.Gaul,M.Schader,eds.).Berlin: Springer-Verlag, 291-298.

[9] Y. SCHEKTMAN (1988) A general Euclidiean Approach for Measuring and Describing Associations between Several Sets of Variables. In Recent Developments in Clustering and Data Analysis (E.Diday,C.Hayashi, M.Jambu,NOhusumi,eds.),Tokyo:Academic Pres., Inc. 37-48.

[10] D,STEWART, W,LOVE (1968) A general canonical correlation index Psychol. Bull. vol. 70.

CLUSTERING MODEL AND METRIC WITH CONTINUOUS DATA

Gérard GOVAERT

Université de Metz, Ile du Saulcy 57045 Metz
INRIA-Lorraine, BP 239 54506 Vandoeuvre les Nancy Cedex

Clustering methods often come down to the optimization of a numeric criterion defined from a metric. It is possible to show that this problem is often equivalent to the estimation of a probabilistic model by a clustering approach. For instance, we know that the inertia criterion corresponds to the hypothesis of a population arising from a Gaussian mixture.
Here, we study the links between these two approaches when the data are quantitative. First, we show that a probabilistic criterion can always be considered as a metric criterion and then we establish that, under some conditions, the converse is true. Finally, these results are applied to the quadratic and to the city-block metrics. With this approach, we can explain the differences between the adaptive distances method and the Gaussian mixture problem, and we can show that the criterion defined with the city block metric is associated to a bilateral exponential distributions mixture.

Keywords : Clustering, mixture model, quadratic metric, L_1 metric.

INTRODUCTION

Clustering methods often come down to the optimization of a numeric criterion defined from a metric. In particular, all the dynamical clustering algorithms (Diday and Simon 1976) follow this scheme. In this case, the choice of the criterion and of the metric is the first and the most important problem to solve.

When it is possible to find a probabilistic model of mixture such that the estimation of the parameters of the model by the clustering approach (Scott et Symons 1971, Schroeder 1976, Celeux 1988) leads to the optimization of a numeric criterion, we obtain a new interpretation of the criterion and the metric which can justify or reject their choice. For instance, Celeux shows that the inertia criterion corresponds to the hypothesis of a population arising from a mixture of gaussian probability density functions which all have the same covariance matrix a.I where a is a real and I the identity matrix. In the same way, we can show (Govaert 1989) that the criterion optimized by the MNDBIN algorithm defined for binary data corresponds to a multivariate Bernoulli mixture.

Here, we develop this approach for continuous data. Fisrt, we define two kinds of clustering criteria and we study what conditions are necessary to obtain equivalent criteria (section 1 and 2). Section 3 is concerned with the links between gaussian distributions and quadratic metrics : for instance, the differences between the adaptive distance method (Govaert 1975, Diday et Govaert 1977, Celeux 1988) and the Gaussian mixture problem (Schroeder 1976). Lastly, a criteria defined with the city block metric is studied and we show that it corresponds to a bilateral exponential distributions mixture.

1. DEFINITION OF THE TWO KINDS OF CRITERIA

Throughout this work, the data will be given by an (n,p) array which contains the values measured on n objects by p **quantitative** variables. Two kinds of criteria are studied : **metric criteria** which use a dissimilarity measure and **probabilistic criteria** which use the concept of probabilistic mixture model. First, we define these criteria.

1.1. METRIC CRITERION

1.1.1. Définition

In this approach, the data are considered as a set Ω of n objects of $\mathbf{R}^p$. Each class of a partition is represented by a "kernel" . The set L of kernels will depend on the criterion and is not defined more precisely for the moment. Lastly, a function $D : \mathbf{R}^p \text{x } L \rightarrow \mathbf{R}^+$ is used to measure the dissimilarity between an object of $\mathbf{R}^p$ and a kernel.

The problem is to find the partition $P=(P_1,...,P_K)$ of Ω and the K-tuple $(\lambda_1,...,\lambda_K)$ of kernels (one for each class) which minimize the criterion

$$\sum_{k=1}^{K} \sum_{x \in P_k} D(x,\lambda_k).$$

This criterion, which depends on the dissimilarity measure D, will be called **metric criterion** and denoted $\mathbf{CP(R^p,L,D)}$.

Remark : The dynamic cluster method (Diday et al 1980), which defines a sequence of couples partitions-kernels such that the criterion is decreasing, gives a solution to this problem. This method uses the following functions :

- a representation function g defined by $g(P) = g(P_1,...P_K) = (\lambda_1,...\lambda_K)$ where λ_k is the object of L minimizing $\sum_{x \in P_k} D(x,.)$
- an affectation function h defined by $h(\lambda) = h(\lambda_1,...\lambda_K) = (P_1,...P_K)$ where $P_k = \{x \in \Omega \mid D(x,\lambda_k) \leq D(x,\lambda_m)$ with $k<m$ if there is equality$\}$.

Under some hypothesis, this algorithm converges and gives a local optimum of the criterion.

1.1.2. Equivalence of metric criteria

Définition : *Two metric criteria are said to be equivalent if and only if they are defined on the same sets $\mathbf{R}^p$ and L and if there is a non-decreasing bijection φ on $\mathbf{R}^p$ satisfying*

$$CM(\mathbf{R}^p,L,D_1) = \varphi \circ CM(\mathbf{R}^p L,D_2)$$

where D_1 and D_2 are the measures of dissimilarity associated to the two criteria.

Remark : All the optimun solutions corresponding to equivalent criteria are identical. Moreover with equivalent criteria, local optimum algorithms, as the dynamic cluster method, will give the same results .

It is easy to show that if we replace D by a increasing linear function of D, we obtain an equivalent metric criterion :

Proposition 1: *$\forall \alpha \in \mathbf{R}^+$ et $\beta \in \mathbf{R}$, the criteria $CM(\mathbf{R}^p,L,D)$ and $CM(\mathbf{R}^p,L,\alpha D+\beta)$ are equivalent.*

1.1.3. Constraint on the kernels

With the same criterion, we can modify the problem by adding a constraint to the k-tuple of kernels $(\lambda_1,...,\lambda_K)$. For instance, if the kernel is a couple (a,b), we can force the first member of the couple to be the same for all the kernels : $\lambda = ((a,b_1),...,(a,b_K))$.

1.2. PROBABILISTIC CRITERION

1.2.1. Definition

The data are now considered as a n-sample Ω of a $\mathbf{R}^p$ random variable whose probability density function (p.d.f.) are $f(x) = \sum_{k=1}^{K} p_k\, f(x,\lambda_k)$ with $\forall k=1,K$ $p_k \in]0,1[$, $\sum_{k=1}^{K} p_k=1$ and where $f(.,\lambda)$ belongs to a familly $\mathbf{F}$ of p.d.f which depend on the parameter λ and p_k is the probability of the distribution $f(.,\lambda_k)$. The set of parameters is denoted L.
The problem is the estimation of the number K of composants of the mixture and of the parameters $\{p_k,\lambda_k \,/\, k=1,K\}$. In the clustering approach (Scott et Symons 1971, Schroeder 1976, Celeux 1988), this estimation problem is replaced by the following problem :

Find a partition $P=(P_1,...P_K)$, where K is assumed to be known, such that each class can be considered as a sample with a p.d.f. $f(.,\lambda_k)$.

Hence we try to maximize the **maximum likelihood criterion**

$$W(P,\lambda) = \sum_{k=1}^{K} \text{Log } L(P_k,\lambda_k)$$

where λ is the p-tuple $(\lambda_1,...\lambda_k)$ and $L(P_k,\lambda_k)$ is the likelihood of the subsample P_k for the p.d.f $f(.,\lambda_k)$.
This criterion which depends on the familly $\mathbf{F}$ of p.d.f. defined on $\mathbf{R}^p$ will be called the **probabilistic criterion** and denoted $\mathbf{CP(R^p,F)}$.

Remark : As for the metric criterion, this criterion can be maximized by the dynamic cluster method with the following functions :

- a representation function g defined by $g(P) = g(P_1,...P_K) = (\lambda_1,...\lambda_K)$ where λ_k is the **maximum likelihood estimator** of the parameter of the p.d.f. associated to the sub-sample P_k.
- an affectation function h defined by $h(\lambda) = h(\lambda_1,...\lambda_K) = (P_1,...P_K)$ where $P_k=\{x\in \Omega \,/\, f(x,\lambda_k)\geq f(x,\lambda_m)$ with $k<m$ if there is equality$\}$

With some hypotheses, it can be shown that this algorithm converges. Then, we obtain a partition P and an estimation of the parameters λ_k. The proportions p_k of the mixture are given by the frequencies of the classes P_k.

1.2.2. Constraints on the parameters

We have seen previously that is was possible to modify the problem by forcing constraints on the kernels without changing the metric criterion. In the same way, with the probabilistic criterion, we can modify the problem with constraints on the parameters of the p.d.f. For instance, if the familly $\mathbf{F}$ is the set of gaussian distribution on $\mathbf{R}^p$, we can force all the covariance matrices of the gaussian distributions associated to each class to be the same.

2. LINKS BETWEEN THE TWO KINDS OF CRITERIA

Here we show that a metric criterion can be associated to any probabilistic criterion and we extend the notion of equivalent criterion, defined for the metric criterion, to the set of metric and probabilistic criteria. Then, we try to answer the reciprocical question : is any metric criterion associated to a probabilistic criterion ? This property is not generally true, but we find a condition on the metric criterion to obtain this property. Lastly, we show than a less strong condition is sufficient for a metric criterion to be equivalent, and not associated, to a probabilistic criterion.

2.1. METRIC CRITERION ASSOCIATED TO A PROBABILISTIC CRITERION

We can easily show the following proposition :

Proposition 2 : *If L is the set of parameters of the family F and D is defined by* $D(x,\lambda) = -\, Log\, f(x,\lambda)\ \forall x \in \boldsymbol{R}^p,\ \forall \lambda \in L$, *we have* $CP(\boldsymbol{R}^p,\boldsymbol{F}) = -\, CM(\boldsymbol{R}^p,L,D)$.

This metric criterion, defined from a probabilistic criterion, will be called the **associated metric criterion.** Thus, the maximization of a probabilistic criterion is equivalent to the minimization of the associated metric criterion and probabilistic criteria can be considered as metric criteria.

2.2. EQUIVALENCE OF PROBABILISTIC AND METRIC CRITERIA

The definition of equivalent metric criteria can be extended :

Definitions : *Two probabilistic criteria are equivalent if their associated metric criteria are equivalent. A probabilistic criterion* CP_1 *and a metric criterion* CM_2 *are equivalent if the metric criterion* CM_1 *associated to* CP_1 *is equivalent to the metric criterion* CM_2.

2.3. CONDITIONS OF THE EXISTENCE OF A METRIC CRITERION ASSOCIATED TO A PROBABILISTIC CRITERION

We have seen previously that a metric criterion can be associated to any probabilistic criterion. Now, we study the converse property. This property is not generally true, but the following proposition can be established :

Proposition 3 : *A metric criterion* $CM(\boldsymbol{R}^p,L,D)$ *is associated to a probabilistic criterion iff* $\forall \lambda \in L$ *the function* $x \rightarrow e^{-D(x,\lambda)}$ *is continuous and satisfies* $\int_{\boldsymbol{R}^p} e^{-D(x,\lambda)}\, dx = 1$.

2.4. PROBABILISTIC CRITERION EQUIVALENT TO A METRIC CRITERION

The following proposition can be easily shown :

Proposition 4 : *Let* $CM(\boldsymbol{R}^p,L,D)$ *be a metric criterion, if there exists a real* $r>1$ *such that* $s = \int_{\boldsymbol{R}^p} r^{-D(x,\lambda)}\, dx$ *is independant of* λ, *then the probabilistic criterion* $CP(\boldsymbol{R}^p,\boldsymbol{F})$ *where* $\boldsymbol{F}$ *is defined by the p.d.f.* $f(x,\lambda) = 1/s\ r^{-D(x,\lambda)}$ *is an equivalent criterion.*

3. QUADRATIC METRICS ANG GAUSSIAN DISTRIBUTIONS

In this section we study several criteria, all of which use quadratic metrics. We show that, with some conditions, each criterion can be associated or equivalent to a gaussian probabilistic criterion. Obviously, we obtain well–known results, but moreover, this approach permits us to go further and, for instance, to explain the differences between the distance adaptive method (Govaert 1975 et 1977) and the maximum likelihood estimate of a mixture of multivariate normal distributions (Schroeder 1976). In this section, the set Ω to cluster is assumed to be always included in $\mathbf{R}^p$.

3.1. FIXED QUADRATIC METRIC

Here the kernel is assumed to be an element of $\mathbf{R}^p$, i.e. $L=\mathbf{R}^p$, and the function D is defined with a fixed positive definite symetric matrix M

$$\forall x \text{ et } \lambda_\kappa \in \mathbf{R}^p \quad D(x,\lambda_\kappa) = \alpha.{}^t(x-\lambda_\kappa).M.(x-\lambda_\kappa) + \beta \qquad \text{where } \alpha\in \mathbf{R}^+ \text{ and } \beta\in \mathbf{R}.$$

The criterion is equivalent for any values α and β (proposition 1). Thus, we only study the simplest case which corresponds to the function D :

$$\forall \ x \text{ and } \lambda \in \mathbf{R}^p \quad D(x,\lambda_\kappa) = {}^t(x-\lambda_\kappa).M.(x-\lambda_\kappa)$$

It is easily seen that $\int_{\mathbf{R}^p} e^{-D(x,\lambda)} = \Pi^{p/2}.(\det M)^{-1/2}$. Hence, we can deduce from the proposition 4 that this metric criterion is equivalent to the probabilistic criterion whose family **F** is defined by

$$\forall \ x \text{ et } \lambda_\kappa \in \mathbf{R}^p \quad f(x,\lambda_\kappa) = \Pi^{-p/2}.(\det M)^{1/2}.e^{-{}^t(x-\lambda_\kappa).M.(x-\lambda_\kappa)}$$

which corresponds to a gaussian distribution with parameters λ_κ and $2.M^{-1}$.
If M is the identity matrix, we prove that the minimization of the inertia criterion corresponds to a mixture of spheric gaussian distributions with the same covariance matrix.

3.2. VARIABLE QUADRATIC METRICS DEPENDING ON EACH CLASS

Here, we assume that the metric is not fixed and moreover depends on each class. We consider two cases :

3.2.1. First case

Now, the kernel are couples (a,M) where a is an element of $\mathbf{R}^p$ and M a positive definite symetric matrix. We have $D(x,(a_k,M_k)) = {}^t(x-a_k).M_k.(x-a_k)$. The metric criterion is associated to a probabilistic criterion if the function D satisfies (proposition 3) :

$$\int_{\mathbf{R}^p} e^{-D(x,(a_k,M_k))}.dx = 1 \quad \Rightarrow \quad \pi^{p/2}.(\det M_k)^{-1/2} = 1 \quad \Rightarrow \quad \det M_k = \pi^p$$

Hence we obtain the adaptive distances method (Govaert 1975, Celeux 1988) which provides a justification of this method and especially of the constraint on the metrics (det M = constant) which had been arbitrarily chosen : this method corresponds to a mixture of gaussian distributions with covariance matrices of fixed determinant.

3.2.2. Second case

Here, we assume that $D(x,(a_k,M_k,\alpha_\kappa)) = \alpha_k + {}^t(x-a_k).M_k.(x-a_k)$. The metric criterion is associated to a probabilistic criterion if the function D satisfy :

$$\int_{\mathbf{R}^p} e^{-D(x,(a_k,M_k,\alpha_k))}.dx = 1 \quad \Rightarrow \quad \alpha_k = \frac{p}{2}.\mathrm{Log}\ \pi - \frac{1}{2}.\ \mathrm{Log}\ (\det M_k)$$

Hence α_k is a function of M_k and we obtain the same kernels obtained in the previous case. We have

$$D(x,(a_k,M_k)) = \frac{p}{2}\ \mathrm{Log}\ \pi - \frac{1}{2}\mathrm{Log}\ \det (M_k)) + {}^t(x-a_k).M_k.(g-a_k).$$

Thus, with the use of the constant α_k, the constraint on the matrices M_k vanishes. The probabilistic criterion corresponding to the metric criterion is defined by the p.d.f. :

$$f(x,(a_k,\Gamma_k)) = \frac{1}{(2\Pi)^{p/2}.(\det \Gamma_K)^{1/2}} e^{-\frac{1}{2}{}^t(x-g_k).\Gamma_k^{-1}.(x-g_k)} \quad \text{where } \Gamma_k = 1/2 \,.\, M_k^{-1}.$$

This situation corresponds to the most general situation of gaussian distributions. Then it is known (likelihood maximum estimation Schroeder 1976) that the a_k are the centers of gravity of the classes and that the Γ_k are the $V_k = W_k / \text{Card } P_k$ with $W_k = \sum_{x \in P_k} (x-g_k).{}^t(x-g_k)$. We obtain the same metric :

$$D(x,a_k) = \frac{p}{2} \text{ Log } \pi - \frac{1}{2}\text{Log det} \left(\frac{1}{2}.V_k^{-1}\right) + \frac{1}{2}\ {}^t(x-g_k).V_k^{-1}.(g-g_k)$$

The comparison of the results obtained in this section explains the difference between the distance adaptive method and the maximum likelihood estimation of a mixture of multivariate normal distributions. The former is a particular case of the latter : the determinant of all the matrices must be constant. Thus the adaptive distances method corresponds also to a gaussian mixture model.

3.3. ONE QUADRATIC METRIC

We assume now that the metric, still not fixed, is the same for all the classes. Since the matrix M does not depend on the class k, the kernels will be denoted (a_k,M). As in the previous section, two cases are examined.

3.3.1. First case

$$D(x,(a_k,M)) = {}^t(x-a_k).M.(x-a_k)$$

This criterion is associated to a probabilistic criterion if this function D satisfies :

$$\int_{R^p} e^{-D(x,(a_k,M))}.dx = 1 \quad \Rightarrow \quad \pi^{p/2}.(\det M)^{-1/2} = 1 \quad \Rightarrow \quad \det M = \pi^p$$

Hence, we obtain the second version of adaptive metric method which uses only one metric (Diday, Govaert et Lemoine 1978) with a similar constraint on the matrix. Then the p.d.f. is

$$f(x,a_k) = e^{-1/2\ {}^t(x-g_k).\Gamma^{-1}.(x-g_k)} \quad \text{where } \Gamma = 1/2\ M^{-1}$$

This p.d.f. corresponds to a gaussian distribution whose covariance matrix is Γ since $(2\pi)^{p/2}.(\det \Gamma)^{1/2} = 1$.

3.3.2. Second case

We assume now that $D(x,(a_k,M,\alpha)) = \alpha + {}^t(x-a_k).M.(x-a_k)$. This criterion is associated to a probabilistic criterion if this function D satisfies :

$$\int_{R^p} e^{-D(x,(a_k,M,\alpha))}.dx = 1 \quad \Rightarrow \quad e^{-\alpha}.\pi^{p/2}.(\det M)^{-1/2} = 1 \quad \Rightarrow \quad \alpha = \frac{p}{2} \text{ Log } \pi - \frac{1}{2}\text{Log det} (M))$$

Hence α depends on M and we obtain the same kernels obtained in the previous case, but with the following function D :

$$D(x,(a_k,M)) = \frac{p}{2} \text{ Log } \pi - \frac{1}{2}\text{Log det} (M)) + {}^t(x-a_k).M.(g-a_k)$$

Thus, here again with the use of the constant α, the constraint on the matrices M vanishes. The probabilistic criterion corresponding to the metric criterion is defined by the p.d.f. :

$$f(x,(a_k,\Gamma)) = \frac{1}{(2\Pi)^{p/2}.(\det \Gamma)^{1/2}} e^{-1/2\ {}^t(x-a_k).\Gamma^{-1}.(x-a_k)} \quad \text{where } \Gamma = 1/2\ M^{-1}.$$

This situation corresponds to a mixture model of gaussian distributions with the same covariance matrix (Celeux 1988). We can easily show that the two last cases are equivalent : the use of the additive term α has no effect , which was not true in the previous section with the terms α_k.

4. L_1 METRIC

We can replace in the inertia criterion the quadratic metric by the city-bloc metric. It is easily shown that all the convergence properties are preserved and that the center of gravity is replaced by the median.
Here again we show that this approach corresponds to a probabilistic mixture model and we use this model to obtain weighted L_1 metrics which have been already proposed in (Govaert 1975) but without references to a probalistic model..

4.1. FIXED METRIC

The kernels are assumed to be element of $\mathbf{R}^p$, i.e. $L=\mathbf{R}^p$, and the function D defined with fixed real positive values α^j is $D(x,\lambda_k) = \sum_{j=1}^{p} \alpha^j \mid x^j-\lambda_k^j \mid$. This criterion is associated to a probabilistic criterion if the function D satisfies :

$$\int_{\mathbf{R}^p} e^{-D(x,\lambda_k)}.dx = 1 \quad \Rightarrow \quad \int_{\mathbf{R}^p} \exp(-\sum_{j=1}^{p} \alpha^j. \mid x^j - \lambda_k^j \mid).dx = 1 \quad \Rightarrow \quad \prod_{j=1}^{p} \alpha^j = 2^p$$

Hence, if this condition is true, the metric criterion defined via the function

$$D(x,\lambda_k)=\sum_{j=1}^{p} \alpha^j. \mid x^j-\lambda_k^j \mid$$

is associated to a probabilistic criterion whose p.d.f. is

$$f(x,\lambda_k) = e^{-D(x,\lambda_k)} = \prod_{j=1}^{p} \exp(-\alpha^j. \mid x^j - \lambda_k^j \mid) = \prod_{j=1}^{p} \frac{\alpha^j}{2} \exp(-\alpha^j \mid x^j - \lambda_k^j \mid) \qquad (\text{car} \prod_{j=1}^{p} \alpha^j = 2^p).$$

Hence, in this model, it is assumed that, within each composant, the p variables are mutually independant and are distributed according to the **bilateral exponential distributions** $L(\lambda_k^j,\alpha^j)$.

When the α^j does not satisfy the constraints, we can divide the function D by $1/2.(\prod_{j=1}^{p} \alpha^j)^{1/p}$ and obtain equivalent metric criteria (proposition 1). Hence, all the criteria defined with the function $D(x,\lambda_k) = \sum_{j=1}^{p} \alpha^j \mid x^j - m_k^j \mid$ are equivalent to probabilistic criteria based on bilateral exponential distributions. For instance, we can study the simplest case $D(x,\lambda_k) = \sum_{j=1}^{p} \mid x^j - m_k^j \mid$. Here we have $\prod_{j=1}^{p} \alpha^j = 1$. So we use the following equivalent metric criterion $D(x,\lambda_k)=1/2.\sum_{j=1}^{p} \mid x^j-m_k^j \mid$ since $1/2.(\prod_{j=1}^{p} \alpha^j)^{1/p} = 1/2$.
This criterion is the metric criterion associated to the probabilistic criterion whose p.d.f. is

$$f(x,\lambda_k) = \prod_{j=1}^{p} 1/2 . \exp(- \mid x^j - \lambda_k^j \mid) .$$

Hence, with the L_1-metric criterion, each class is assumed to be a sample of a product of p bilateral exponential distributions $L(\lambda_k^j,1)$.

4.2. VARIABLE L_1 METRICS DEPENDING ON EACH CLASS

We assume now that the metric is not fixed and moreover depends on each class :

$$D(x,(a_k,\alpha_k,\beta_k)) = \sum_{j=1}^{p} \alpha_k^j \mid x^j - a_k^j \mid) + \beta_k$$

This criterion is associated to a probabilistic criterion if we have $\int_{\mathbf{R}^p} e^{-D(x,(a_k,\alpha_k,\beta_k)))}.dx = 1$

which gives $\beta_k = p.\text{Log}2 - \sum_{j=1}^{p} \text{Log}\,\alpha_k^j$.

Hence $D(x,(a_k,\alpha_k)) = p.\text{Log}2 - \sum_{j=1}^{p} \text{Log}\,\alpha_k^j + \sum_{j=1}^{p} \alpha_k^j \mid x^j - m_k^j \mid$

To obtain the weighted L_1 metric, we must have $\sum_j \text{Log}\beta_k^j = p \text{ Log } 2 \Rightarrow \prod_j \beta_k^j = 2^p$.

Hence we obtain the L_1-adaptive distances method (Govaert 1975) with exactly the same constraint which provides an a posteriori justification.

CONCLUSION

We saw that the comparison of metric criterion and probabilistic criterion have provided a new approach of several clustering algorithms and justified constraints which had often been necessary only for technical reasons of optimization. Also, this approach gives a way to define new criteria and, moreover, this comparison explains the interest and the flexibility of the dynamical cluster method whose main idea was the use of a kernel associated to each cluster. This kernel corresponds merely to the parameters of the p.d.f. associated to each class in the probabilistic criterion. It remains to be studied if the other criteria used in the dynamical cluster method are also equivalent to probabilistic criteria.

References

Celeux, G. , Diday, E. , Govaert, G., Lechevallier, Y. and Ralambondrainy, H., *Classification automatique des données : Environnement statistique et informatique*. Dunod 1989.

Diday, E. and Collaborateurs (1980), *Optimisation en Classification Automatique*. INRIA. Rocquencourt .

Diday, E. and Govaert, G. (1977), Classification avec distances adaptatives. *RAIRO* **11**, n°4, 329–349.

Diday, E., Govaert, G. and Lemoine, Y. (1978), A new kind of representation in clustering. *Proocedings of the 4th IJCPR*. Kyoto.

Celeux, G. (1988), Classification et modèles. *R.S.A.* **36** n°4, 43-58.

Govaert, G. (1975), *Classification Adaptative*. Thesis Université Paris 6.

Govaert, G. (1983), *Classification croisée*. Thesis Université Paris 6.

Govaert, G. (1989), Classification binaire et modèles, *R.S.A.* **37** (to appear).

Schroeder, A. (1976), Analyse d'un mélange de distribution de probabilité de même type. *RSA* **24**, n°1, 39-62

Scott, A.J. and Symons, M.J. (1971), Clustering methods based on likelihood ratio criteria. *Biometrics* **27**, 387-397.

SPACE-DISTORTING PROPERTIES IN AGGLOMERATIVE HIERARCHICAL CLUSTERING ALGORITHMS AND A SIMPLIFIED METHOD FOR COMBINATORIAL METHOD

Noboru Ohsumi
Department of Statistical Methodology
The Institute of Statistical Mathematics
4-6-7, Minami-Azabu, Minato-ku
Tokyo (Japan)
Nagatomo Nakamura
Department of Construction
Nihon University Junior College
7-24-1, Narashinodai, Funabashi-shi
Chiba (Japan)

1. Introduction

The objective of this paper is to examine the relationship between the properties of parameters which express agglomerative hierarchical clustering algorithms (AHC algorithms) and "space distortion" resulting from updated distances, and to generalize the Lance and Williams(1967) formula (combinatorial method). The AHC algorithms are based on the following formula proposed by Lance and Williams :

$$d(C_i \cup C_j, C_k) = \alpha_i\, d(C_i, C_k) + \alpha_j\, d(C_j, C_k) + \beta\, d(C_i, C_j) + \gamma\, |d(C_i, C_k) - d(C_j, C_k)| . \qquad (1)$$

Using this formula, the dissimilarities, or distances, between a newly-merged cluster $C_i \cup C_j$ and the remaining other clusters C_k are updated and a set of parameters, $\theta \equiv \{\alpha_i, \alpha_j, \beta, \gamma\}$, are used to characterize the clustering methods and these parameters determine the linkage process.

With this formula, Lance and Williams introduced the concepts of "space distortion" (which may be space-conserving, space-contracting, or space-dilating) and the "monotonicity" of updated distances. One of the conditions for monotonicity are given by

$$\alpha_i + \alpha_j + \beta \geq 1. \tag{2}$$

However, the AHC algorithms may produce reversal of the resulting tree structure. Thus, Milligan (1979) and Batagelj (1981) have presented following necessary and sufficient conditions for suppressing such reversals.

$$\begin{aligned} \gamma &\geq -\min\{\alpha_i , \alpha_j\}, \\ \alpha_i + \alpha_j &\geq 0. \end{aligned} \tag{3}$$

At the same time, the concept of space distortion, as discussed by Lance and Williams, is intuitive. In addition, DuBien and Warde (1979) have derived, under some assumptions, a more sophisticated, theoretical concept of space distortion among distances obtained at different cluster merging levels. Unfortunately, these studies have concentrated on characterization of only a sub-family of the AHC algorithms, namely, the sub-family characterized by the (β,γ) space defined in formula (1).

2. Conditions for Space Distortion

Thus, this paper proposes several extensions to the concept of space distortion which will increase the sophistication of the AHC algorithms. Let us first assume the following condition concerning the distances among three clusters (C_i, C_j, and C_k):

$$d(C_i, C_j) < d(C_i, C_k) < d(C_j, C_k). \tag{4}$$

In addition, let

$$\Delta^m = \{ d(C_i \cup C_j , C_k) \text{ at } \theta \mid d(C_i, C_j) < d(C_i, C_k) < d(C_j, C_k)\} \tag{5}$$

be a set of all distances obtained from the result of an agglomeration at the m-th step in a clustering process. In this case, the conditions causing space distortion are defined as follows:

1) Space conservation:
$$d(C_i, C_k) < d(C_i \cup C_j , C_k) < d(C_j, C_k) ; \quad d(C_i \cup C_j , C_k) \in \Delta^m,$$

2) Space dilation:

$$d(C_j, C_k) \leq d(C_i \cup C_j, C_k) ; \; d(C_i \cup C_j, C_k) \in \Delta^m,$$

3) Space contraction:

$$d(C_i, C_k) \geq d(C_i \cup C_j, C_k) ; \; d(C_i \cup C_j, C_k) \in \Delta^m.$$

The paper examines the properties of space distortion occurring in various clustering methods and the results are summarized (see Table 2). Most clustering methods are based on general agglomerative algorithms using the $(\alpha_i, \alpha_j, \beta, \gamma)$ parameter space (i.e., formula (1)). However, examination of these strategies under the assumptions above clarifies the relationship between space distortion and the parameter space occupied by the particular strategy. Figure 1 shows the region occupied by various methods in the parameter space $(\alpha_i, \alpha_j, \beta)$, as defined below.

1) Region in which space conservation occurs:

$$\{ (\alpha_i, \alpha_j, \beta) \mid \alpha_i + \alpha_j + \beta = 1, \; 0 < \alpha_i , \alpha_j < 1, \; \beta = 0 \}.$$

2) Region in which space conservation or space dilation occurs:

$$\{ (\alpha_i, \alpha_j, \beta) \mid \alpha_i + \alpha_j + \beta > 1, \; 0 < \alpha_i , \alpha_j < 1, \; -1 < \beta < 1 \}$$
$$\cup \{ (\alpha_i, \alpha_j, \beta) \mid \alpha_i + \alpha_j + \beta = 1, \; 0 < \alpha_i , \alpha_j < 1, \; -1 < \beta < 0 \}.$$

3) Region in which space conservation or space contraction occurs:

$$\{ (\alpha_i, \alpha_j, \beta) \mid \alpha_i + \alpha_j + \beta < 1, \; 0 < \alpha_i , \alpha_j < 1, \; -1 < \beta < 1 \}$$
$$\cup \{ (\alpha_i, \alpha_j, \beta) \mid \alpha_i + \alpha_j + \beta = 1, \; 0 < \alpha_i , \alpha_j < 1, \; 0 < \beta < 1 \}.$$

3. Simplification of the Lance and Williams formula

This paper proves that the single linkage and complete linkage methods are characterized as special cases of the flexible method by simplifying the Lance and Williams formula (1). First, the following two parameters are defined for updating the distances to usage two clusters:

$$\delta = d(C_j,C_k) - d(C_i,C_k) > 0, \quad \varepsilon = d(C_i,C_k) - d(C_i,C_j) > 0. \qquad (6)$$

The proof involves substituting the values $\delta/(\delta+2\varepsilon)$ for parameter β to derive the single linkage method from the flexible method in formula (1).

$$d(C_i \cup C_j, C_k) = \{ d(C_i, C_k) + d(C_j, C_k) \} (1-\beta)/2 + \beta \, d(C_i, C_j)$$
$$= \{ d(C_i, C_k) + d(C_j, C_k) \} \varepsilon / (\delta + 2\varepsilon) + \delta \, d(C_i, C_j) /(\delta+2\varepsilon).$$

By substituting $d(C_j,C_k)= \delta + d(C_i,C_k)$, and $d(C_i,C_j)= - \varepsilon + d(C_i,C_k)$ for the above expression, we can obtain the following relation.

$$\begin{aligned} d(C_i \cup C_j , C_k) &= \varepsilon \{ 2d(C_i, C_k) + \delta \} /(\delta+2\varepsilon) + \delta \{ d(C_i,C_k) - \varepsilon \} /(\delta+2\varepsilon) \\ &= \{ 2\varepsilon d(C_i, C_k) + \varepsilon\delta + \delta d(C_i,C_k) - \varepsilon\delta \} /(\delta+2\varepsilon) \\ &= d(C_i, C_k). \end{aligned}$$

♦

Similarly, substituting the value $-\delta/(\delta+2\varepsilon)$ for parameter β, we can obtain the complete linkage method. Thus, specifying parameter γ can be eliminated, and the formula (1) is simplified as following expression:

$$d(C_i \cup C_j , C_k) = \alpha_i\, d(C_i, C_k) + \alpha_j\, d(C_j, C_k) + \beta\, d(C_i, C_j).$$

Moreover, the condition of monotonicity which contains parameter γ is unnecessary to be considered. Thus, simplification of the Lance and Williams formula is completed.

REFERENCES

[1]Batagelj, V. (1981): Note on ultrametric hierarchical clustering algorithms. *Psychometrika*, **46**, 351-352.

[2]DuBien, J.L. and Warde, W.D. (1979): A mathematical comparison of the members of an infinite family of agglomerative clustering algorithms. *The Canadian Journal of Statistics*, **7**, 1, 29-38.

[3]Lance, G.N. and Williams, W.T. (1967): A general theory of classificatory sorting strategies, I. Hierarchical systems. *The Computer Journal*, **9**, 373-380.

[4]Milligan, G.W. (1979): Ultrametric hierarchical clustering algorithms. *Psychometrika*, **44**, 343-346.

[5]Nakamura, N. and Ohsumi, N. (1988): Space-distorting properties in agglomerative hierarchical clustering and a new simplified algorithm. *The Japan and Korea 5th conference.*

[6]Ohsumi, N. and Nakamura, N. (1988): Some relations among space-distorting properties in agglomerative hierarchical clustering algorithms (in Japanese). *Proceedings of the Annual Meeting of the Japanese Applied Statistics Society 1988*, 16-20.

[7]Ohsumi, N. and Nakamura, N. (1988): On a hierarchical classification method with the adjustable parameters (in Japanese). *Proceedings of the 2nd Annual Meeting of the Japanese Society of Computational Statistics*, 29-32.

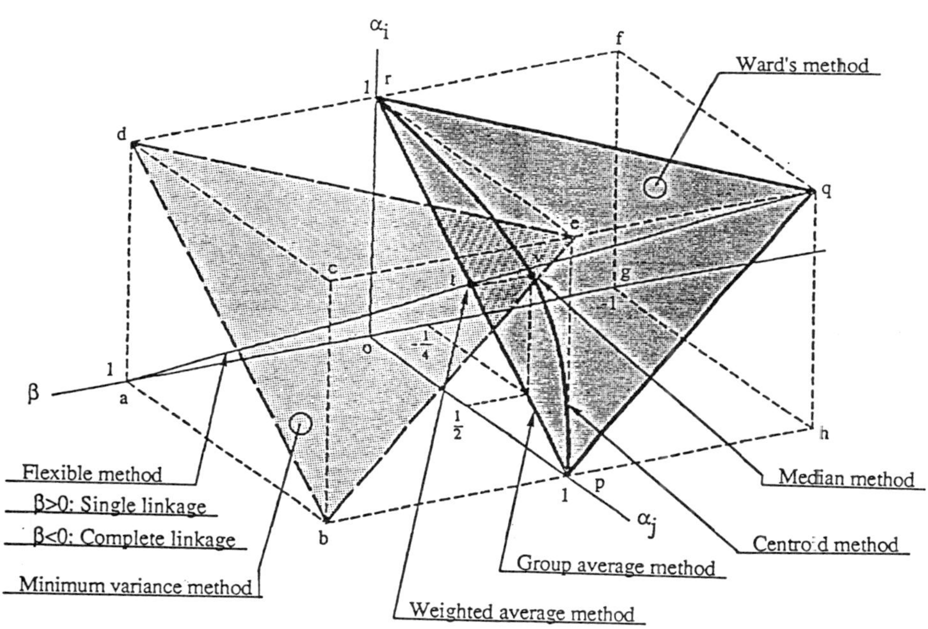
Ward's method
Flexible method
β>0: Single linkage
β<0: Complete linkage
Minimum variance method
Weighted average method
Group average method
Centroid method
Median method

FIGURE 1

TABLE 1. Hierarchical Clustering Algorithms

Methods	α_i	α_j	β	γ	$\alpha_i+\alpha_j+\beta$
Single linkage	1 / 2	1 / 2	0	- 1 / 2	1
Complete linkage	1 / 2	1 / 2	0	1 / 2	1
Group average	$n_i/(n_i+n_j)$	$n_j/(n_i+n_j)$	0	0	1
Weighted average	1 / 2	1 / 2	0	0	1
Ward's method	$(n_i+n_k)/n_t$	$(n_j+n_k)/n_t$	$-n_k/n_t$	0	1
Centroid method	$n_i/(n_i+n_j)$	$n_j/(n_i+n_j)$	$-\alpha_i*\alpha_j$	0	$1+\beta$
Median method	1 / 2	1 / 2	- 1 / 4	0	3/4
Flexible method	$(1-\beta) / 2$	$(1-\beta) / 2$	$\beta < 1$	0	1

Note: $n_t = n_i + n_j + n_k$ and n_i is Cluster size of C_i.

TABLE 2. Space Distortion Conditions

Methods	Contracting	Conserving	Dilating
Single linkage	Yes	No	No
Complete linkage	No	No	Yes
Group average	No	Yes	No
Weighted average	No	Yes	No
Ward's method	No	Yes [for $\varepsilon / \delta < n_i / n_k$]	Yes [for $\varepsilon / \delta \geq n_i / n_k$]
Centroid method	Yes [for $\delta(n_i+n_j)/n_i \leq d_{ij}$]	Yes [for $\delta(n_i+n_j)/n_i > d_{ij}$]	No
Median method	Yes [for $2\delta \leq d_{ij}$]	Yes [for $2\delta > d_{ij}$]	No
Flexible method	Yes [for $\delta / (\delta+2\varepsilon) \leq \beta$]	Yes [for $-\delta / (\delta+2\varepsilon) < \beta < \delta / (\delta+2\varepsilon)$]	Yes [for $-\delta / (\delta+2\varepsilon) \geq \beta$]
Flexible method (Lance-Williams)	Yes [for $\beta > 0$]	Yes [for $\beta = 0$]	Yes [for $\beta < 0$]

Note: n_i is cluster size of C_i, $d_{ij} = d(C_i,C_j)$, $d(C_i,C_j) < d(C_i,C_k) < d(C_j,C_k)$,

$\delta = d(C_j,C_k) - d(C_i,C_k)$, $\varepsilon = d(C_i,C_k) - d(C_i,C_j)$.

IMPROVEMENTS IN NHMEAN METHOD

F. COSTA NICOLAU - M. Paula BRITO
Faculty of Sciences, University of Lisbon
Center of Statistics and Applications
(I.N.I.C.)

ABSTRACT

We recall the non hierarchical clustering method NHMEAN, the first of this type to be included in the probabilistic family of VL clustering methods. The NHMEAN comparison function arises from adapting the objective function Γ_{AVM} of the AVM hierarchical clustering method to a non hierarchical procedure.

Let E be the set to be submitted to the method, card E = n. Let us suppose that all the information available is contained in the similarity matrix $\{ \gamma_{xy} , x, y \in E\}$ which is considered , under the independence hypothesis, as a reference, to be an i.i.d. sample of the uniform distribution in $[0 , 1]$.

Starting with k seed-points, chosen by one of several possible strategies, the remaining elements $x \in E$ are assigned, at each step of the iterative procedure, to the cluster E_i which corresponds to the maximum of $\Gamma_{AVM} (\{x\} , E_i)$. Here $E_i , \dots , E_k$ are the present kernels of the partition clusters. The procedure is iterated until stabilization is achieved. The method is adaptative as concerns the number k of clusters, allowing both the merging of clusters considered to be too close, and the creation of a new cluster around an element which is too "far" from all existing clusters. The processes of merging and refinement are based on probabilistic criteria consistent with the whole method.

In the large sample case, a learning process is adopted at the initialization stage, based on a sub-sample, chosen at random in E. The dependence shown by most non hierarchical methods on the initialization stage led us to try, along with traditional random techniques, a new guided (non random) strategy - "wise-choice" - which seems to yield good results (speed of convergence and quality of the resulting partition).

KEY-WORDS - AVM Method, VL Similarity, Clustering, Partition, Cluster, Validity of the Mean Link

1. INTRODUCTION

It is well known that, conceptually, most non-hierarchical clustering methods arise from a geometric framework.

The NHMEAN method, dealt with in this paper (cf. F. Nicolau (1985,1986) and Paula Brito (1988)), which is adaptative as concerns the number k of clusters of the partition to be obtained, is essentially a probabilistic method. Its probabilistic character stems both from the nature of the similarity measure and of the allocation function as well as the criteria of merging and refinement of the partition. However, it is distinct from the class of mixture-based methods, since there is no underlying parametric model.

Let us further specify that the NHMEAN method belongs to the VL family of clustering methods (it is in fact the first non-hierarchical method in this family) : VL similarities between units (variables or individuals) of the set to be clustered are supposed to be computed beforehand, and the allocation and merging criteria are linked to the distribution function of the mean of unit uniform random variables.

Results obtained so far on several data sets together with comparison with other methods allow us to conclude that it is a very quick procedure whose performance is at least as good - as concerns quality of the results - as other classic methods. Four factors account for this fact :

- the choice of "wise-choice" strategy to determine the k initial seed-points;
- the use of an estimation-learning procedure to determine the k initial seed-points, together with "wise-choice" strategy, in the case where the set to be clustered is large (n = card E > 50);
- the choice of an effective allocation function which is the adaptation of the AVM hierarchical clustering method objective function to a non-hierarchical method;
- the use of merging and refinement options, based on criteria which are consistent with the whole method.

On the other hand, the present version of the algorithm does not allow the demonstration of its convergence by local optimization of a criterion closely related to the probabilistic nature of the allocation function.

2. VL SIMILARITY AND RESEMBLANCE IN NHMEAN

Let E, card E = n, be the set to be clustered. We measure the resemblance between pairs $\{x_i,x_j\}$ of E by means of a statistics $\gamma_{x_i x_j}$ which takes its values in [0 , 1]. Generally, $\gamma_{x_i x_j}$ will be defined such that it can be considered as following an unit uniform distribution of probability, under a reference hypothesis of independence, adapted to the problem. There are several ways of making this choice, depending on the nature of E and of its descriptors (cf. Lerman (1981), Bacelar-Nicolau (1981), Nicolau (1983)). This is known as VL resemblance (V stands for Validity and L for Link).

To define the comparison/allocation function it is assumed that all the information needed to compare two clusters A and B such that $A \cap B = \emptyset$, is contained in the table

$$[\gamma_{ab} \ , (a,b) \in A \times B] \quad (1)$$

taken as an iid sample of the unit uniform distribution, under the reference hypothesis (Nicolau (1983)). This is a simple working hypothesis, well adapted to the problem of defining a probabilistic measure of similarity between clusters. Based on this hypothesis, several hierarchical methods which are now well known (AVL, AVM, ...), have been deduced and new research trends have been developed.

The AVL method, which is based on the distribution function of the maximum, was first introduced by Lerman (1970), its first properties and, particularly, the initial computing algorithm and program have been proposed by Bacelar-Nicolau (1972) and then improved by several other authors. The comparison function between clusters of the AVL method is

$$\Gamma_{AVL}(A,B) = \gamma_{AB}^{\alpha\beta}$$

where $\gamma_{AB} = \max\ \{\gamma_{ab} : (a,b) \in A \times B\}$, $\alpha = \text{card } A$, $\beta = \text{card } B$.

This function has been taken to evaluate the proximity of an element x to a cluster, in view of a non-hierarchical clustering method. In fact, to assign an element to a cluster may well be considered as the union of two clusters, one of which is a singleton. We then obtain :

$$\Gamma_{AVL} = \Gamma_{AVL}(A,\{x\}) = \gamma_{Ax}^{\alpha}$$

where $\gamma_{Ax} = \max\ \{\gamma_{ax} : a \in A\}$, $\alpha = \text{card } A$.

The results thus obtained were not satisfactory.

The good behavior of the AVM method, which is mainly due to its comparison function, led us to take it as the measure of proximity of an element to a cluster, in a non-hierarchical clustering algorithm.

In the AVM method, which is based on the distribution function of the mean, the comparison function between clusters A and B is:

$$\Gamma_{AVM} = \frac{1}{(\alpha\beta)!} \sum_{0 \le j \le \alpha\beta Z_{AB}} (-1)^j \binom{\alpha\beta}{j} (\alpha\beta Z_{AB} - j)^{\alpha\beta} \qquad (2)$$

where $Z_{AB} = \sum\ [\gamma_{ab}, (a,b) \in A \times B] \ / \ \alpha\beta$, $\alpha = \text{card } A$, $\beta = \text{card } B$.

Hence, the objective function of NHMEAN method is

$$\Gamma_{x,A} = \Gamma_{AVM}(x,A) = \frac{1}{\alpha!} \sum_{0 \le j \le \alpha\bar{\gamma}_{A,x}} (-1)^j \binom{\alpha}{j} (\alpha\bar{\gamma}_{A,x} - j)^{\alpha} \qquad (3)$$

where, of course,

$$\bar{\gamma}_{A,x} = \sum\ [\gamma_{ax}, a \in A] \ / \ \alpha , \quad \alpha = \text{card } A$$

Notice that in practice the computing of $\Gamma_{A,x}$ is simplified by the use of symmetry properties of the density function of the mean and the quick convergence of the distribution of this statistics to the normal distribution (cf. F. Nicolau (1981--a), Paula Brito (1988)).

So $\Gamma_{A,x}$ is simply $\Gamma_{Ax} = \text{Prob}(M_\alpha \le \bar{\gamma}_{Ax})$, that is, the value of the d.f. of the random variable M_α, sample mean of α unit uniform random variables, computed for the observed value $\bar{\gamma}_{Ax}$ of the mean link of x to the elements of the already existing cluster A. Formula (3) governs the assignment of x to the cluster E_i, i = 1,..., k, of the partition present at a given stage, whose value $\Gamma_{E_i,x}$ is maximum; that is, x will be assigned to the cluster E_i whose value $\Gamma_{E_i,x}$ is most unlikely to be exceeded under the reference hypothesis.

Notice that this rule is not the same as assigning on the largest value $\overline{\gamma_{A\,x}}$. In fact, it can be shown that, if $\alpha < \beta$, then

$$\text{Prob}(M_\alpha \leq m) < \text{Prob}(M_\beta \leq m) , \text{ if } m > 1/2$$
$$\text{Prob}(M_\alpha \leq m) > \text{Prob}(M_\beta \leq m) , \text{ if } m < 1/2$$

where M_α and M_β are sample means of α and β i.i.d. unit uniform random variables respectively. Hence x may be assigned to cluster B rather than to cluster A, with $\overline{\gamma_{A\,x}} > \overline{\gamma_{B\,x}}$, $\overline{\gamma_{A\,x}} > 1/2$ providing that these values are not too different and that β = card B is sufficiently larger than α = card A .

3. INITIALIZATION STRATEGIES. "WISE-CHOICE" METHOD.

Several techniques to choose the k initial seed-points, used by most k-partition methods, may be found in the literature (cf. for example, Anderberg (1973), Lerman (1981), F. Nicolau (1981-b), Villoing (1983)). These techniques differ in their random nature (random vs non-random choice, sometimes subjective, but mainly guided) and in their set or non-set based formulation .

Among others, we may cite :

a) the simple choice method where the initial seed points are the k first elements of the sorted sequence of elements of E;

b) the systematic choice method, where we take as seed points the elements whose indices are given by the arithmetic sequence with first element i - i taken at random between 1 and m, m = n/k, integer division - and common difference m, that is : i, i+m, i+2m, ... , i+(k-1)m;

c) the random choice method, where the k seed points are taken at random between x_1 et x_n ;

d) choice of the k elements which are simultaneously most distant from one another;

e) the "attraction poles" initialization method (Leredde & Lerman, cf. Lerman (1981));

f) the "wise-choice" method, which we introduce, and which is described below.

All these methods are implemented in the program that executes NHMEAN.

Let us now go back to method f) which we call "wise-choice". This technique is based on the idea of taking as initial seed points elements of E which are likely to be located near the central regions of the existing clusters.The seed points will then be centered in dense neighborhoods and distant from one another.

Let $\{ s(x_i,x_j) \; ; \; x_i , x_j \in E \}$ be the set of similarities, computed beforehand. We start by determining the L-neighborhoods $V_L(x_i)$, $\forall\, x_i \in E$, L fixed integer :

$$V_L(x_i) = \{ x_j : x_j \in E \text{ is one of the L elements closest to } x_i \}$$

Obviously, $x_i \in V_L(x_i)$, and hence in $V_L(x_i)$ there will just be L-1 elements different from x_i.

Let

$$\bar{s}_L(x_i) = \frac{1}{L-1} \sum_{x_j \in V(x_i) \setminus \{x_i\}} s(x_i,x_j) \quad , \; \forall\, x_i \in E$$

The first seed point is the element of E, which we shall henceforth note c_1, such that

$$\bar{s}_L(c_1) = \max_{x \in E} \bar{s}_L(x)$$

The other seed-points, c_j , j=2, . . . , k are successively chosen so as to meet the condition :

$$c_j : \min_{1 \le l \le j-1} \frac{\bar{s}(c_j)}{s(c_j, c_l)} =$$

$$= \max_{x \in E \setminus \{c_1, \dots, c_{l-1}\}} \min_{1 \le l \le j-1} \frac{\bar{s}(x)}{s(x, c_l)}$$

c_j is hence, among the remaining elements of E , the most distant from all seed points already determined and located in a dense neighborhood.

Notice that the use of blind methods, that is, methods that are non-guided or random, is essential in the testing of a method to evaluate its interest and convergence properties.

4. MERGING AND REFINEMENT. ADJUSTING THE NUMBER OF CLUSTERS.

Most partition methods are designed to yield a partition in k clusters, k being fixed at the start and invariable until the end of the procedure.

The NHMEAN method we propose is adaptative as concerns the number of clusters, that is, it includes procedures that allow the control of the fit of the current value of k during the execution of the basic algorithm.

4.1 Merging Clusters

At the end of each iteration, that is, at the end of each complete passage over E, we check whether some clusters have become too close.

Let P = { $E_1, \dots, E_k$ } be the current partition. For each pair { E_i, E_j }, we compute the value :

$$\bar{\gamma}_{E_i E_j} = \frac{1}{\alpha_i \alpha_j} \sum [\gamma_{ab}, (a,b) \in E_i \times E_j]$$

which is considered a value of the random variable $M_{\alpha_i \alpha_j}$, α_i = card E_i , α_j = card E_j , sample mean of $\alpha_i \alpha_j$ i.i.d. unit uniform random variables (the usual reference hypothesis in the VL similarity based methods) .

The d.f. value corresponding to the observed value $\bar{\gamma}_{E_i E_j}$ is:

$$\Gamma_{AVM}(E_i, E_j) = \frac{1}{(\alpha_i \alpha_j)!} \sum_{0 \le h \le \alpha_i \alpha_j \bar{\gamma}_{E_i, E_j}} (-1)^h \binom{\alpha_i \alpha_j}{h} (\alpha_i \alpha_j \bar{\gamma}_{E_i, E_j} - h)^{\alpha_i \alpha_j}$$

which we shall take as the criterion-value to evaluate the proximity of clusters { E_i , E_j }. Merging E_i and E_j is decided, that is, the current partition P is replaced by $(P \setminus \{ E_i, E_j \}) \cup \{ E_i \cup E_j \}$ if $\Gamma_{AVM}(E_i, E_j) > 0.95$. This corresponds to taking the decision of merging on a 95% confidence test on the proximity of the two clusters.

This procedure could be further generalized by fixing a significance level α (confidence level 1 - α) and deciding on merging $\{ E_i, E_j \}$ if $\Gamma_{AVM}(E_i, E_j) > 1 - \alpha$.

4.2 Refinement

On the other hand, if at a given stage h of an iteration an element $x \in E$ is such that

$$\max_{1 \le i \le k} \Gamma(x, E_i^h) < 0.5$$

where $\{ E_1^h, \ldots, E_k^h \}$ is the set of the cluster kernels present at this stage, x is considered to be too distant from all kernels and a new kernel is created, the k+1th kernel, by replacing k by k+1 and presenting to the next stage, h+1, the kernels $\{ E_1^{h+1}, \ldots, E_k^{h+1}, E_{k+1}^{h+1} \}$ where

$$E_i^{h+1} = E_i^h , \quad i = 1, \ldots, k$$
$$E_{k+1}^{h+1} = \{x\}$$

5. THE BASIC ALGORITHM OF THE NHMEAN METHOD

Let E be the set to be clustered, $E = \{ x_1, \ldots, x_n \}$, card E = n.

In the algorithm used at the present stage of research, all VL-similarity values $\gamma_E = \{ \gamma\, x_i x_j , 1 \le i \le j \le n, x_i, x_j \in E \}$ should be known at the start. Also, the desired number of clusters of the final partition, k, should be fixed.

The algorithm may be summarized as follows :

Stage 0. Table γ_E is either computed by the program itself or obtained from an external file, in the case where it has already been computed, for instance, during the construction of a classification tree by a VL method.

The initial number of clusters k is only the value desired by the user, since it will possibly be changed during the execution of NHMEAN by the merging and refinement procedures. The maximum number of transfers, T (in general T=0) and the maximum number of iterations I, should also be fixed.

Stage 1. Allocating procedure. The elements of E are assigned, one at a time, to the clusters ("set-based attraction" objective function, refinement procedure, updating).

a) In choosing the seed points, two situations may arise :

1. Case f the 1st iteration, h=1.

We choose the k initial seed-points by one of the strategies presented in paragraph 4. We advise the use of at least one random method together with the "wise-choice" option.

2. Case of the h th iteration, h>1.

In this case the k seed-points will be the elements of each cluster that maximize the mean link to the remaining elements of the corresponding cluster. This solution consists in taking as seed-points, in each cluster, the element whose mean link to the cluster is least likely to be exceeded (this criterion is consistent with the whole NHMEAN method).

b) Table initialization : cardinal table, sorted sequence of elements of E, assignment table, table of the sums of similarities of all non-assigned elements (at the start, all but the seed points) to each cluster.

Stage 2. Each element x_{k+i} , $i = 1, \ldots , n-k$, is assigned to one of the cluster kernels already formed, according to the criterion :

$$\max \{ \Gamma (x_{k+i}, E^{i}_{j,h}) / j = 1 ,\ldots, k \}$$

where $\{ E^{i}_{1,h} , E^{i}_{2,h} , \ldots , E^{i}_{k,h} \}$ is the set of disjoint kernels present at the start of step i of iteration h. Γ is the measure described in paragraph 2. Assignment is done if the criterion value is above 1/2, otherwise the refinement procedure described in paragraph 4 takes place .

Stage 3. Merging test, as described in paragraph 4. Should merging take place, all control tables must be updated.

Stage 4. Where we decide to stop the procedure or to proceed to the next iteration, h+1.

Stopping is decided if:

- either $h = I$, where I = maximum number of iterations allowed (defined at stage 0);

- or stabilization is achieved. The process is considered to have stabilized if the number t of transfers of elements between clusters that have taken place between the final partition of iteration h-1 and the final partition of iteration h is such that $t \leq T$, T having been fixed at stage 0.

If it is decided to proceed, the present partition must be stored in place of the previous one, before updating h $(h \rightarrow h+1)$ and going back to stage 1.

CONCLUSION

We have presented a new non-hierarchical clustering method.

A clear disadvantage of the NHMEAN method is the non-existence of an algorithmic convergence proof, which is due to its probabilistic nature. The method remains, nevertheless, fairly consistent throughout its different probabilistic-based stages : assignment, merging/refinement, choice of the seed points after the 1st iteration. On the other hand, two notable aspects stand out regarding the method's practical applications :

- the speed at which stabilization is achieved, especially when coupled with the"wise-choice" strategy;

- the obtained results, which are at least comparable to those obtained by other classic methods.

As concerns further research, the two following lines may be taken up :

1) Applying the merging procedure at each step of each iteration, together with the refinement procedure;

2) Coupling the method, after stabilization, with an exchange procedure aiming at optimizing the partition by taking an objective function based on the likelihood of the sum of the links of each element to the cluster it belongs to.

REFERENCES

Anderberg, M.R. (1973) - Cluster Analysis for Applications Academic Press

Bacelar-Nicolau, H. (1972) - Analyse d'un Algorithme de Classification Automatique - Thèse de 3ème cycle, Univ. Paris VI

Bacelar-Nicolau, H. (1981) - Contribuição ao Estudo dos Coeficientes de Comparação em Análise Classificatória - PhD Thesis , Fac. of Sciences , Univ. of Lisbon

Brito, M. Paula (1988) - Métodos Não Hierárquicos de Classificação - O Método NHMEAN - Master Dissertation - Fac. of Sciences , Univ. of Lisbon

Kendall & Stuart (1976) - The Advanced Theory of Statistics - Vol.3 , 3ª ed

Lerman, I.C. (1970) - Sur l'Analyse des Données Préalable à une Classification Automatique - Proposition d'une Nouvelle Mesure de Similarité - in Mathématiques et Sciences Humaines (M.S.H.), Vol. 32, pag. 5-15, Paris

Lerman, I.C. ; Leredde, H. (1977) - La Méthode des Pôles d'Attraction - in Actes du Colloque "Journées Analyse des Données et Informatique", I.R.I.A., Versailles, Sept. 77

Lerman, I.C. (1981) - Classification et Analyse Ordinale des Données - Dunod, Paris

Nicolau, F. (1981-a) - Critérios de Análise Classificatória Hierárquica Baseados na Função Distribuição - PhD Thesis, Fac. of Sciences, Univ. of Lisbon

Nicolau, F. (1981-b) - Métodos Não-Hierárquicos de Classificação - O Método das K-Médias - Complementary PhD Thesis - Fac. of Sciences, Univ. of Lisbon

Nicolau, F. (1983) - Cluster Analysis and Distribution Function - in Meth. of Op. Research, 45, pag. 431-433

Nicolau, F. (1985) - Analysis of a Non-Hierarchic Clustering Method Based on VL-Similarity - Center of Statistics and Applications / I.N.I.C., Note Nº18/85

Nicolau, F. (1986) - Non-Hierarchic Clustering by NHMEAN Method - in Proc. of COMPSTAT-1986, Short Communication, Rome, 1986

Villoing, P. (1983) - Méthodes de Classification Automatique Non Hiérarchique (Clustering) - Centre de Calcul Scientifique de l'Armement (France), Note d'Étude LA Nº 56

VALIDITY

USING THE BOOTSTRAP AND THE RV COEFFICIENT IN THE MULTIVARIATE CONTEXT

Susan Holmes
Unité de Biométrie
ENSA.M-INRA-USTL
9, Place Pierre Viala
34060 Montpellier (France)

Key words : Bootstrap, Pivot, Confidence Interval, Confidence Region, Principal Components, RV-coefficient

Abstract : When boostrapping in the multivariate context several difficulties arise, one of these is the need to find an appropriate vector space where the different bootstraps can be studied. One solution to this problem is provided through the analysis of a cube of data as in Conjoint Analysis. This method uses the concepts of characteristic operators and their inner product which will be defined in the first part of this paper. Another possibility is to bootstrap a uni-dimensionsal statistic, namely the RV coefficient, this provides a way of choosing the number of components to be retained in PCA, without the indecision of graphical procedures and the distributional assumptions in the usual parametric procedures.

Résumé : Cet article présente quelques développements autour de l'application du principe du bootstrap (ou méthode à la Cyrano) dans le cadre multi-dimensionnel en soulignant leur application à la construction de régions de confiance, dans les cas paramétrique simple et non-paramétrique.

L'Analyse en Composantes Principales est prise comme exemple type de procédure à laquelle peut s'appliquer la construction d'intervalles de confiance pour les valeurs propres ainsi que de cônes de confiance pour les vecteurs propres, enfin une application inédite au coefficient de corrélation RV montre l'utilité de la méthode dans le cas de statistiques dont on ignore la distribution asymptotique.

AMS : 62E25,62G15,62H20,62H25.

---

*visiting Stanford University for the Spring Quarter

1 Bootstrap: Motivations and Notations

Given a probability distribution F on $\mathbf{R}^k$, using a sample of random variables drawn from F $\mathcal{X}_n = (X_1, \ldots, X_n)$ allows the construction of the empirical cumulative distribution function defined as

$$F_n(x) = \frac{1}{n}\sum_{i=1}^{n} 1_{(-\infty,x]}(X_i).$$

Glivenko-Cantelli's theorems ensures us that F_n is a consistent estimate of F in the following sense:

$$\sup_x |F_n(x) - F(x)| \rightarrow_{p.s.} 0 \text{ quand } n \rightarrow \infty$$

The idea behind the bootstrap is that if one wants to know the cumulative distribution function $J_n(., F)$ of a random variable $T_n(\mathcal{X}_n, F)$ depending both on the sample and on its distribution, then one may consider the approximation provided by the empirical cumulative distribution function $J_n(., F_n)$ of $T_n(\mathcal{X}_n^*, F_n)$ where $\mathcal{X}_n^*$ is a random sample of observations which are independent and identically distributed according to F_n. Using Monte Carlo methods one can draw with replacement n observations from $\mathcal{X}_n$, thus constructing B such resamples provides B values of $T_n(\mathcal{X}_n^*, F_n)$, thus providing a way of estimating the cumulative distribution function $J_n(., F_n)$ only containing a simulation error. This function is then used as an estimate for the cumulative distribution function of $T_n(\mathcal{X}_n, F)$.

The theoretical justification of such a procedure has to be provided in each particular case and usually depends on the Triangular Array Convergence Property cited below, and the verification of the following **Theorem** (BERAN 1984 [1]).

Triangular Array Convergence : The class $\mathcal{F}$ obeys the triangular array convergence property for the distance d defined on $\mathcal{F}$ if for every sequence of c.d.f.'s $\{F_n, F_n \in \mathcal{F}\}$ such that $\lim_{n\to\infty} d(F_n, F) = 0$, we have the weak convergence of $J_n(., F_n)$ towards $J(., F)$ (which must not depend on F).

Theorem : Under the condition that the family $\mathcal{F}$, obeys the **Triangular Array Convergence Property** for a distance d and that the sequence $\{\hat{F}_n\}$ verifies $d(\hat{F}_n, F) \xrightarrow{p}_{n\to\infty} 0$, for $\hat{F}_n \in \mathcal{F}$ then

$$\sup_x |J_n(x, \hat{F}_n) - J_n(x, F)| \xrightarrow{p}_{n\to\infty} 0$$

The advantage of this resampling procedure is that it allows construction of confidence regions even in a distribution-free context the first articles published on the subject are due to EFRON 1979 [10], 1981 [11].

A more theoretical justifiaction of the procedure as developped in sections 2 and 4 can find their source in work done by P. HALL [15], 1987 [16], and R. BERAN 1987 [2]. Such refinements as smoothing, usage of the Bayesian paradigm, importance sampling have also enhanced the method in recent years. (Some references may be found at the end of this presentation).

We are going to present an example of how such a procedure can be used in Principal Component Analysis. However in this case the application is not a straightforward one because once B new arrays have been drawn from the original one, there remains the problem of choosing an *optimal* space in which to compare replicate arrays. This motivates the use of conjoint analysis as explained in section 5.

2 Confidence Regions

Construction of confidence intervals and regions is always dependent on existence of a function of the parameter of interest and the sample called the **pivot**, whose distribution is supposed to be independent of the parameter.

2.1 Percentile Method for Confidence Intervals

In order to define confidence intervals for a one-dimensional parameter $\theta = \vartheta(F)$, EFRON, 1981, [11] supposes that one could replace the unknown distribution function $G(x, F) = P_F(\hat{\theta} < x)$ by the bootstrap approximation $G(x, F_n)$ however recent theoretical work has shown that it is better to use bootstrap approximations to the distributions $H(x, F)$ and $K(x, F)$ of the quantities $n^{\frac{1}{2}}(\hat{\theta} - \theta)/\sigma$ and $n^{\frac{1}{2}}(\hat{\theta} - \theta)/\hat{\sigma}$ (where $n^{-1}\sigma^2$ is the asymptotic variance of $\hat{\theta}$).
Under the assumption that θ is a smooth function of the mean μ, one can show that bootstrap critical points constructed for confidence regions having a nominal probability of α are first order correct when G is used and second order correct when the studentized bootstrap is used, i.e. when built using $K(x, F)$. This encourages use of the studentized root whenever a satisfactory estimate of σ is available. However when this is not the case, iteration of the bootstrap mechanism as suggested by BERAN, 1987, [2] and HALL & MARTIN, 1988, [18], provides a way of improving poor coverage accuracy.

2.2 Confidence Regions

Generalization of the above to the multidimensional case is quite natural through use as a root of the quantity $n^{\frac{1}{2}}\hat{S}_n^{-\frac{1}{2}}(\hat{\theta} - \theta)$, however confidence regions are not uniquely defined as confidence intervals are, one possibility, as suggested by HALL, 1987, [16], is to choose the region C that has minimum content subject to $P(\theta \in C) = \alpha$, this is equivalent to imposing that C should be such that all parameter values inside C have higher likelihood than those outside. In order to do this it will be necessary to provide an estimate $\hat{f}$ for the density

function.
Other ways of addressing this problem is to study the p-dimensional parameter as defined in polar coordinates, thus providing confidence cones for its direction.
We will see how such an approach can be useful in section 6.3.

2.3 Real Functions of a covariance-matrix

BERAN & SRIVASTAVA 1985, [5] bring down the multivariate problem to a one-dimensional one through use of the root $T_n = u(n^{\frac{1}{2}}[g(S_n) - g(\Sigma_F)])$ where

- g is a function defined from $\mathbf{R}^{p(p+1)/2}$ $uvec(\Sigma_F)$ in $\mathbf{R}^k$
- u is a real-valued function defined $\mathbf{R}^k$ that is not constant on any set other than those having zero measure.

They show that if F has finite fourth moments and if g is continuously differentiable, then justification of the use of the bootstrap in construction of confidence regions for $g(\Sigma_F)$ is provided by the convergence of T_n^*'s distribution towards that of a transformed multi-normal distribution.

A confidence interval that has an asymptotic level of $1 - \alpha$ is provided by

$$D_{n,g}(\alpha) = \{g(\Sigma_F) : u(n^{\frac{1}{2}}[g(S_n) - g(\Sigma_F)]) \leq c_{n,g}(\alpha, \hat{F}_n)\}$$

where $c_{n,g}(\alpha, \hat{F}_n)\}$ is an upper α quantile of the bootstrap distribution of T_n^*.
The following examples use these results and are of practical interest :

1. $g(\Sigma_F) = \rho_{F,ij}$, confidence intervals can be provided for the correlation coefficient using the root $T_n = |r_{n,ij} - \rho_{F,ij}|$

2. In order to construct simultaneous confidence intervals for simple eigenvalues $\lambda_1 \geq \lambda_2 \geq \ldots \lambda_p \geq 0$ BERAN & SRIVASTAVA 1985, [5] use the root

   $$T_n = \max_{1 \leq i \leq p} |\log(\lambda_i(S_n)) - \log(\lambda_i(\Sigma_F)|$$

   The confidence region thus obtained by resampling is

   $$\{\lambda : \lambda_1 \geq \lambda_2 \geq \ldots \lambda_p \geq 0 \text{ et } \forall i, 1 \leq i \leq p, \lambda_i(S_n)A_n^{-1} \leq \lambda_i \leq \lambda_i(S_n)A_n\}$$

   with $A_n = \exp(\hat{c}_{n,g}(\alpha))$, where $\hat{c}_{n,g}(\alpha)$ is an upper α-quantile of the bootstrap distribution of the root T_n.

   Justification of the use of the bootstrap in this case is provided by the fact that for any semi-definite matrix S that has simple eigenvalues the function that associates these eigenvalues to the matrix is continuously differentiable. (see KATO, [22]).

3. If F is such that the eigenvalues of Σ_F are all simple and if we denote by $\gamma_i(S)$ the i th normed eigenvector of S. Then BERAN & SRIVASTAVA use the root

$$T_n = 1 - |< \gamma_i(S_n), \gamma_i(\Sigma_F) >|$$

in order to construct confidence cones for the eigenvectors with fixed coverage.

Remark : An interesting application of this result is the possibility of choosing in the confidence cone of a given level (for instance 95%) the vector that is the easiest to interpret

BERAN & SRIVASTAVA [5] give a good example of such an approximation.

3 Motivations and Notations for the RV-coefficient

Suppose that the data to be studied is contained in a matrix $X_{n \times p}$ of p measurements made on n observations, to which we can associate a positive semi-definite matrix $Q_{p \times p}$ that defines distances between observations :

$$d^2(x_{i.}, x_{j.}) = (x_{i.} - x_{j.})'Q(x_{i.} - x_{j.})$$

and we will also associate weights p_i to the observations these will be the elements of a diagonal matrix $D_{n \times n}$ whose trace is usually one.
Unless otherwise stated X is supposed to be centred with respect to D , i.e. $X'D1_n = 0$ where $1_n = (1, 1....1)'$, then the variance-covariance matrix of the sample is $X'DX$.
In the subsequent sections we will always consider X together with the metrics defining distances in the observation space R^p and the variable space R^n, so we will always be analyzing triples (X, Q, D).
Our objectives are

- 1. to simplify the triple (X, Q, D)
- 2. to compare several triples $\{(X_k, Q_k, D_k)_{,k=1..K}\}$

This will involve associating a characteristic operator to each triple, for instance when we have several arrays of different measurements made on the same observations the characteristic operator will be

$$O_k = X_k Q_k X_k' D \text{ so that } O_k \in R^{n \times n}$$

if the different studies are relative to the same variables then

$$O_k = X_k' D_k X_k Q \text{ and } O_k \in R^{p \times p}$$

Comparison of orperators is done through the use of vectoriel covariance as defined by ESCOUFIER & ROBERT, 1973 [24] :

$$CovV(O_1, O_2) = Tr(O_1 O_2)$$

and the vectoriel correlation defined for two symmetric operators by :

$$RV(O_1, O_2) = \frac{Tr(O_1 O_2)}{[Tr(O_1^2) Tr(O_2^2)]^{\frac{1}{2}}}$$

In fact when X is an array with only one variable and the triple associated is $(X, 1, \frac{1}{n} I_n)$, if one wants to compare it to another 1-variable array $(Y, 1, \frac{1}{n} I_n)$ using the RV coefficient between the two charateristic operators $\frac{1}{n} X'X$ and $\frac{1}{n} Y'Y$, which is precisely ρ^2 the square of the ordinary correlation coefficient.

Properties: When comparing two triples of measurements (X_1, Q_1, D) and (X_2, Q_2, D) relative to the same observations, the RV-coefficient has properties similar to that of the square of the correlation coefficient i.e.

1. $0 \leq RV(O_1, O_2) \leq 1$
2. $RV(O_1, O_2) = 1$ iff $O_2 = kO_1$, for some k in R,
3. $RV(O_1, O_2) = 0$ iff the two groups of variables are uncorrelated i.e. $X_1' D X_2 = 0$.

4 Principal Component Analysis

This is the solution to the first problem cited in the objectives, that is the simplification of (X, Q, D).

4.1 Classical approach

This is through the generalized singular value decomposition.

$$X = USV' \text{ with } \begin{cases} V'QV = I_r \\ U'DU = I_r \\ S \text{ diagonal} \end{cases}$$

One can construct the best M-ranked approximation to X through

$$X^{(M)} = US^{(M)}V' \text{ where } S^{(M)} = \begin{pmatrix} s_1 & 0 & & & & \\ 0 & \ddots & 0 & & & \\ & 0 & s_M & 0 & & \\ & & 0 & 0 & 0 & \\ & & & 0 & \ddots & 0 \\ & & & & 0 & 0 & 0 \end{pmatrix}$$

This is equivalent to the eigenanalysis of the characteristic operator associated to the triple.

4.2 PCA seen through the RV-coefficient

Find a triple $(Y_{n\times M}, I, D)$, with $M < p$ as good an approximation of rank M as possible i.e. that maximises

$$RV(O_X, O_Y) = RV(XQX'D, YY'D) = \frac{Tr(XQX'DYY'D)}{\sqrt{Tr(XQX'D)^2 Tr(YY'D)^2}}$$

This maximum is attained for Y being composed of the first M eigenvectors of $XQX'D$ such that $Y'DY = \Lambda^{(M)}$ and the value of this maximal RV-coefficient is

$$RV(O_X, O_Y) = \sqrt{\frac{\sum_{i=1}^{M} \lambda_i^2}{\sum_{i=1}^{p} \lambda_i^2}}$$

4.3 Correspondence Analysis

This is a special case of the above procedure when th data is a contingency table N and the data analysed is $X = \frac{1}{\nu} N$ where $\nu = \sum_{i=1}^{n} \sum_{j=1}^{p} n_{ij}$ is the grand total.
If r denotes the vector of row sums ($r = \frac{1}{\nu} N.1_p$), and c the vector of column sums ($c = \frac{1}{\nu} N'.1_n$), then independance is equivalent to $X = rc'$.
Correspondence analysis is equivalent to the PCA of (X, D_c^{-1}, D_r^{-1}), where D_v^{-1} denotes the diagonal matrix whose elements are those of the inverses of the components of the vector v. This means that the methods developped hereafter in the case of PCA can be applied in just the same way to Correspondence Analysis.

5 Analysis of a Data Cube (Conjoint Analysis)

Here we will try to simplify a situation where K different triples $\{(X_k, Q_k, D_k), k = 1..K\}$ are to be compared, where the arrays X_k must have one dimension in common, i.e. measurements on the same observations or the same variables measured on different groups of observations. The technique used here has been developped in Escoufier 1987, [13], Glaçon ,1981, [14] and Lavit, 1988, [23]. In the following we will consider for instance that the variables remain the same throughout the K studies, i.e. $Q_k = Q$ (which is the case when the different triples are bootstrap replicates of the original one).
Then three different problems will be addressed and solved, namely representing distances between triples, summarizing the K triples and then comparing the observations and variables as *seen* through the different triples.

5.1 Distances between the arrays : Interstructure

Each triple is characterized by its associated operator : $O_k = X_k' D_k X_k Q$, distance between two operators can be measured through their vectoriel covariance:

$$\delta_{kl} = CovV(O_k, O_l) = Tr(O_k O_l)$$

these can be represented by a symmetric matrix $\Delta = (\delta_{kl})_{\{k=1..p,l=1..p\}}$.
Diagonalisation of Δ will provide an optimal planar representation of the arrays, as in an odinary multidimensional scaling problem. Two equivalent triples will appear as collinear. If one wants equivalent triples to have identical projections then as in PCA one can use the reduced distances i.e. diagonalize the matrix composed of the elements

$$\rho_{kl} = RV(O_k, O_l) = \frac{\delta_{kl}}{\sqrt{\delta_{kk}\delta_{ll}}}$$

5.2 A summary of the K triples : the compromise

In principal components the objects compared are variables and the inner product is their covariance, here the objects are the operators and their inner product will be their CovV or their RV-coefficient. The first eigenvector u of the diagonalised matrix allows the construction of a linear combinataion

$$\tilde{O} = \sum_{k=1}^{K} u_k O_k$$

which has the largest norm and which maximises the sum of squares of the inner products with the initial operators.
Remark : It can be shown that all the elements of u have the same sign, so they can be chosen positive, thus $\tilde{O}$ is a semi-definite positive operator. In fact $\tilde{O}$ is the operator related to the triple $(\mathcal{X}, Q, \mathcal{D})$ where $\mathcal{X} = (X_1|X_2|..|X_K)$ and

$$\mathcal{D} = \begin{pmatrix} \sqrt{u_1}D_1 & & \\ & \ddots & \\ & & \sqrt{u_K}D_K \end{pmatrix}$$

Diagonalization of $\tilde{O}$ will provide a representation of the p variables, in which proximity of two points is to be interpreted as an average likeness of these variables.

5.3 Comparing the evolution of the variables and the observations through the different triples

Projection onto the first principal planes obtained by using the first two eigenvectors of $\tilde{O}$ provides a means of comparing the variables as they are *seen* by the different triples, each variable will thus have K different projections, forming the variable's trajectory.

The same can be done for the observations, those that are not present in a triple can be represented all the same through projections, as can supplementary observations in ordinary PCA.

6 PCA and the bootstrap

6.1 Confidence Regions for the observations' representations

As is customary in Boostrapping, from the original array X resamples X_b^* are drawn, for b varying from 1 to B, the number of bootstrap replicates needed. At the same time a new diagonal weight matrix is created through associating the weights to the observations resampled and then normalizing, so that D_b^* has unit trace.
The conjoint analysis explained above is then applied to the cube of B replicates $\{(X_b^*, Q, D_b^*)_{,b=1..B}\}$.
The number M of components to be retained should already have been estimated through either cross-validation (cf HOLMES-JUNCA, 1985, [20]), or a graphical *cut-point* test, then the K different M-ranked approximations can be compared through conjoint analysis.
The representation in the first principal plane of the compromise shows for instance K different projections of each of the observations, the spread of each of these clusters will indicate the stability of the representation for each observation.
Another indicator is provided by constructing the convex hull of the projections for each observation.
Empirical confidence regions can be obtained from these clusters of B projections by constructing the successive convex hulls envelopping the points.
One could also use P. Hall's likelihood confidence region approach by doing a two dimensional density estimate for each cloud and taking the confidence region to be defined by an iso-contour for the estimated density $\hat{f}$.(cf. HALL 1987 [16])
This procedure provides *smoother* confidence regions without sacrificing indications of first-order departure from normality as is the case when one uses ellipses.

6.1.1 Confidence Intervals for the RV-Coefficient

The quality of a PCA can be measured, as has been already explained by the coefficient

$$RV_n^{(M)} = RV(O_n^{(M)}, O_n) = \frac{\sum_{i=1}^{M} \lambda_i(O_n)^2}{\sum_{i=1}^{p} \lambda_i(O_n)^2}$$

One could suppose that $RV_n^{(M)}$ estimates some unknown coefficient $\rho V^{(M)}$ that measures the quality of the approximation of $O^{(M)}$ to O. The most frequent case is of course when the population (unknown) covariance matrix is Σ_F and $O = \Sigma_F Q$. When all the eigenvalues of O are simple ones, $\rho V^{(M)}$ is a continuously differentiable function of the covariance matrix, and following BERAN & SRIVASTAVA,1985, [5] we can use the quantity $T_n = |RV_n^{(M)} - \rho V^{(M)}|$ as the root to be bootstrapped in the usual way so to obtain c_n^* such that the confidence interval :

$$D_n(\alpha) = \left\{\rho V^{(M)} : |RV_n^{(M)} - \rho V^{(M)}| \leq c_n^*\right\}$$

has an asymptotic coverage of (1-α).

Remark : Other roots based on the RV coefficient can provide other bounds for the quality of a PCA's approximation are:

$$T_n = \max_{1 \leq M \leq p-1} |RV_n^{(M)} - \rho V^{(M)}|$$

Through analogy with the correlation coefficient case (HALL, MARTIN & SCHUCANY, 1989,[19]) one knows in advance that a variance stabilization transformation will give better results suggesting the roots:

$$T_n = |\text{arctanh}(RV_n^{(M)}) - \text{arctanh}(\rho V_n^{(M)})|$$

$$\text{or } T_n = \left|\log \frac{1 + RV_n^{(M)}}{1 - RV_n^{(M)}} - \log \frac{1 + \rho V^{(M)}}{1 - \rho V^{(M)}}\right|$$

6.2 Directional Data

6.2.1 Theory

This section details results due to DUCHARME & AL., 1985, on the construction of confidence cones for the mean direction defined as

$$\theta(F) = \frac{\mu_F}{||\mu_F||}$$

whereμ_F denotes the mean of a random p-lengthed vector X that belongs to the p-dimensionned sphere S_p. DUCHARME & AL. construct confidence cones based on the *root*

$$T_n(F_n) = n\,(1- < \theta(F), \theta(F_n) >)$$

Indeed if $J_n(F_n) = \mathcal{L}(T_n(F_n))$ denotes T_n's distribution, they show that $J_n(\hat{F}_n)$ converges to a distribution that does not depend on the direction of $\mu(F)$. In order to do this DUCHARME & AL. define a function g on $\mathbf{R}^{p*}$ by

$$g(x) = \frac{< \mu(F), x >}{||\mu(F)||||x||}$$

Denote by H_F g's Hessian at μ_F, B_F the matrix of normed eigenvectors of H_F.
Then the following theorem justifies the use of the bootstrap through application of the

triangular array principle:

Theorem: Denoting by J(F) the distribution of the random vector

$$\frac{1}{2||\mu(F)||^2}\sum_{i=2}^{p} Z_i^2 \text{ where } \mathcal{L}(Z_1 \dots Z_p) = \mathcal{N}(0, B'\Sigma_F B)$$

then one has the following weak convergence:

$$J_n(\hat{F}_n) \overset{p.s.}{\Longrightarrow} J(F)$$

This theorem implies almost sure convergence of the bootstrap critical point $c(\alpha, \hat{F}_n)$ towards the true critical point $c(\alpha, F)$, and thus validity of the resampling procedure.

The set

$$C_B = \left\{v \in S_p | < v, \theta(\hat{F}_n) > \geq 1 - (2n)^{-1} c(\alpha, \hat{F}_n)\right\}$$

is a confidence region whose approximate probabilty coverage is $1 - \alpha$ for any cumulative distribution function F on the sphere such that $\mu(F) \neq 0$.

6.3 Confidence Cones for the Characteristic Operator

If the function g is defined as the function that associates to a definite positive symmetric matrix S, its vectoriel correlation with regards to O , $g(S) = RV(O, S)$, then g is a continuously differentiable function. Application of the preceding section's results to the roo:

$$T_n = n(1 - RV(O, O_n))$$

leads to confidence cones for the unknown caracteristic operator O of the study without having to make any other distributional assumptions except the existence of fourth order moments.

Moreover these confidence cones also provide a way of choosing the number M of components to retain in PCA. One can choose the first integer M for which $O_n^{(M)}$ belongs to the confidence cone that has a preassigned nominal level $\alpha = 0.8$, for instance.

References

[1] BERAN R., 1984, *"Bootstrap methods in Statistics"*, Jbder. d. Dt. Math. Verein., 86, pp. 14-30.

[2] BERAN R., 1987, *"Pre-pivoting to reduce level error of confidence sets"*, Biometrika, vol.74 , pp. 457-468.

[3] BERAN R., 1988, *"Balanced Simultaneous Confidence Sets"*, J.A.S.A., vol.83, pp.679-686.

[4] BERAN R. AND MILLAR P.W., 1986, *"Confidence Sets for a multivariate distribution"*, 14, pp.431-443.

[5] BERAN R. AND SRIVASTAVA M.S., 1985, *"Bootstrap sets and confidence regions for functions of a covariance matrix "*,Ann. Stat. 13, pp.95-115.

[6] BUNKE O. AND DROGE B., 1984, *"Bootstrap and Cross-Validation estimates of prediction error for linear regression models"*, Ann. Stat., vol. 12, pp.1400-1424.

[7] DIACONIS P. AND EFRON B., 1983, *"Computer-Intensive Methods in Statistics"*, Scientific American, vol.248, pp.116-130.

[8] DICICCIO T. AND ROMANO J.P., 1988, *"A Review of Bootstrap Confidence Intervals"*, J.R.S.S. B, 50, pp. 338-354.

[9] DUCHARME G.R., JHUN M., ROMANO J.P., TRUONG K.N.,1985, *"Bootstrap confidence cones for directional data"*, Biometrika, vol. 72, pp. 637-645.

[10] EFRON B., 1979, *"Bootstrap Methods : Another Look at the Jackknife"*, Ann. Stat., 7, pp. 1-26.

[11] EFRON B., 1982, *"The Jackknife, the Bootstrap and Other Resampling Plans"*, SIAM, Philadelphia.

[12] EFRON B., 1987, *"Better Bootstrap Confidence Intervals"*, J.A.S.A., vol. 82, pp. 171-185.

[13] ESCOUFIER Y., 1987, *"Three-Mode Data Analysis : The statis method"*,in Methods for Multidimensional Analysis, Lecture Notes from the European Courses in Statistics.

[14] GLAÇON F.,1981, *" Analyse Conjointe de Plusieurs Matrices, Comparaison de Différentes Méthodes"*, Thèse 3ème cycle, Grenoble.

[15] HALL P.,1986, *"On the Bootstrap and Confidence Intervals"*, Ann. Stat., vol. 14, pp.1431-1452.

[16] HALL P.,1987, *"On the Bootstrap and Likelihood-Based Confidence Regions"*, Biometrika, vol. 74, pp.481-493.

[17] HALL P.,1988, *"Theoretical Comparison of Bootstrap Confidence Intervals"*, Ann. Stat., 16, pp. 927-953.

[18] HALL P. AND MARTIN M.A.,1988, *"On Bootstrap Resampling and Iteration"*, Biometrika, 75, pp.661-671.

[19] HALL P.,MARTIN M.,SCHUCANY W.,1989, *"Better Nonparametric Bootstrap Confidence Intervals for the corrėlation coefficient"*, (to appear)

[20] HOLMES-JUNCA S.,1985, *"Outils Informatiques pour l'Evaluation des Resultats en Analyse des Donnees"*, These 3eme cycle, USTL, Montpellier.

[21] JOHNS V., 1988, *"Importance Sampling for bootstrap confidence intervals"*, J.A.S.A., vol. 83, pp.709-714.

[22] KATO T., 1966, *"Perturbation Theory for Linear Operators"*,Springer, N.Y.

[23] LAVIT C., 1988, *"Analyse Conjointe de Tableaux Quantitatifs"*, Ed. Masson.

[24] ROBERT P. AND ESCOUFIER Y., 1976, *"A Unifying Tool for Linear Multivariate Methods: the RV-coefficient"*, Applied Statistics, vol.25, pp. 257-265.

ON THE DETECTION OF MULTIVARIATE DATA OUTLIERS AND REGRESSION OUTLIERS

Aziz Lazraq[1] and Robert Cléroux[2]

(1) Ecole Nationale de l'Industrie Minérale, Rue Elghafiki, B.P. 753, Agdal, Rabat, Maroc.

(2) Département d'informatique et de recherche opérationnelle, Université de Montréal, C.P. 6128, Succ. A, Montréal, Québec, Canada, H3C 3J7

In this paper we introduce new methods for detecting multivariate outliers. The problems of outlier detection in a multivariate sample and in the multivariate regression model are considered separately. Informal exploratory techniques are proposed and approximate tests are developed. Some examples are given in order to compare these techniques with previous ones. They are easy to use and we believe they should prove useful in practice.

1. INTRODUCTION

Detecting the presence of outliers is a challenging problem. The outliers may be important in themselves (in detecting pollution events, for example) or because they may contaminate subsequent analyses.

It is hard to define exactly what a multivariate outlier really is. An observation may be identified as an outlier by a given method and not by another, it may be an outlier in a given model and not in another and it may be an outlier in a p-dimensional space and not in a given subspace. Projections may thus not lead to the detection of multivariate outliers.

Much has been written on the problem of outlier detection. A complete bibliography has been given in Cléroux, Helbling and Ranger (1986) and it will not be reproduced here.

These authors have proposed some methods for detecting outliers in a multivariate sample (data outliers) and in the multivariate linear regression model (model outliers) using the influence functions of two multivariate association measures: the RV coefficient introduced by Escoufier (1973) and the RV_{reg} coefficient introduced by Robert and Escoufier (1976).

In this paper a similar approach is taken using two other multivariate association measures: a coefficient introduced by Lingoes and Schonemann (1974) and the redundancy index of Stewart and Love (1968). In Section 2 we recall the Lingoes and Schonemann coefficient. Its influence function is obtained in Section 3 and it is used for multivariate data outlier detection. In Section 4 we are concerned with the detection of outliers in the multivariate linear regression model. The Stewart and Love coefficient is recalled and its influence function is obtained. Approximate tests for outling observation are given in Section 5 and a conclusion follows in Section 6.

The examples considered in this paper are taken from Cléroux, Helbling and Ranger (1986) in order to allow a comparison between the methods introduced there and those developed in the present paper.

2. LINGOES AND SCHONEMANN'S COEFFICIENT OF VECTOR CORRELATION

Let $X^{(1)} : p \times 1$ and $X^{(2)} : q \times 1$ be two random vectors with $E(X^{(i)}) = \mu^{(i)}$ and $\Sigma_{ij} = E(X^{(i)} - \mu^{(i)})(X^{(j)} - \mu^{(j)})'$ for $i = 1,2$ and $j = 1,2$. Write

$$X = \begin{bmatrix} X^{(1)} \\ X^{(2)} \end{bmatrix} , \ \mu = \begin{bmatrix} \mu^{(1)} \\ \mu^{(2)} \end{bmatrix} , \ \Sigma = \begin{bmatrix} \Sigma_{11} & \Sigma_{12} \\ \Sigma_{21} & \Sigma_{22} \end{bmatrix}$$

and assume Σ is positive definite. Suppose μ and Σ are unknown and consider a random sample $X_\alpha = \begin{bmatrix} X_\alpha^{(1)} \\ X_\alpha^{(2)} \end{bmatrix}$, $\alpha = 1, 2, \ldots, n$ from X . Let $\bar{X}^{(i)} = \frac{1}{n}\sum_{\alpha=1}^{n} X_\alpha^{(i)}$, $S_{ij} = \frac{1}{n-1}\sum_{\alpha=1}^{n}(X_\alpha^{(i)} - \bar{X}^{(i)})(X_\alpha^{(j)} - \bar{X}^{(j)})'$ for $i, j = 1, 2$ and write $\bar{X} = \begin{bmatrix} \bar{X}^{(1)} \\ \bar{X}^{(2)} \end{bmatrix}$ and $S = \begin{bmatrix} S_{11} & S_{12} \\ S_{21} & S_{22} \end{bmatrix}$.

Lingoes and Schonemann (1974) introduced the following sample correlation coefficient between the two vectors $X^{(1)}$ and $X^{(2)}$:

$$RLS = RLS(X^{(1)}, X^{(2)}) = \frac{tr(S_{12}S_{21})^{\frac{1}{2}}}{(trS_{11}trS_{22})^{\frac{1}{2}}} \tag{2.1}$$

where $tr(.)$ is the trace operator. The algebraic properties of RLS are as follows:

(i) if $p = q = 1$ then $RLS = r$, the simple correlation coeficient

(ii) $0 \leq RLS \leq 1$ and $RLS = 0$ if and only if $S_{12} = 0$

(iii) if A is a $p \times p$ matrix such that $A'A = kI$ where k is a scalar and I is the identity matrix, then $RLS(AX^{(1)}, X^{(2)}) = RLS(X^{(1)}, X^{(2)})$

(iv) let $X^{(1)} : p \times 1$ and $X^{(i)} : q \times 1, i = 2, 3, \ldots, m$ and let $a_i, i = 2, 3, \ldots, m$ be scalars. Let S_{ij} be the sample covariance matrix between $X^{(i)}$ and $X^{(j)}$ for $i, j = 1, 2, \ldots, m$. Then the quantity $RLS(X^{(1)}, \sum_{i=z}^{m} a_i X^{(i)})$ reduces to R, the multiple correlation coefficient, when maximized with respect to the a_i's, when $p = q = 1$.

Moreover by analogy to Robert and Escoufier (1976), a distance between two data matrices can be defined in terms of RLS. Consider the following data matrices

$$Y_1 = (X_1^{(1)} - \overline{X}^{(1)}, X_2^{(1)} - \overline{X}^{(1)}, \ldots, X_n^{(1)} - \overline{X}^{(1)}) : p \times n$$

$$Y_2 = (X_1^{(2)} - \overline{X}^{(2)}, X_2^{(2)} - \overline{X}^{(2)}, \ldots, X_n^{(2)} - \overline{X}^{(2)}) : q \times n .$$

If $q < p$, say, then we can add a column of $p - q$ zeros to each $X_i^{(2)} - \overline{X}^{(2)}$.

For any square matrix E, $||E|| = \sqrt{trE'E}$ is a norm and a distance between Y_1 and Y_2 is

$$\text{dist}(Y_1, Y_2) = \left\| \frac{Y_1}{\sqrt{trY_1'Y_1}} - \frac{Y_2}{\sqrt{trY_2'Y_2}} \right\| . \tag{2.2}$$

It can be shown that (see Seber (1984, pp. 253-256)), restricting ourselves to orthogonal matrices $T : p \times p$ we have

$$\min_T \text{dist}(Y_1, TY_2) = \sqrt{2}\sqrt{1 - RLS(X^{(1)}, X^{(2)})} . \tag{2.3}$$

Thus $RLS = 1$ if and only if there exists a procrustes rotation T that matches the two configurations of n points completely. The RLS coefficient is a correlation coefficient in that sense and a data point which will be influential on RLS will at the same time be influential on the minimum distance between Y_1 and TY_2. Let

$$\rho LS = \frac{tr(\Sigma_{12}\Sigma_{21})^{\frac{1}{2}}}{(tr\Sigma_{11}tr\Sigma_{22})^{\frac{1}{2}}} \tag{2.4}$$

be the population analogue of RLS.

3. INFLUENCE FUNCTION OF ρLS

Let F be a distribution function and $\tilde{F} = (1 - \varepsilon)F + \varepsilon\delta_x$ be a perturbation of F by δ_x, the distribution function which assigns unit probability to the point x. Let $\theta = T(F)$ be any parameter expressed as a functional of the distribution function F and let $\tilde{\theta} = T(\tilde{F})$. Hampel (1974) defines the theoretical influence function $I(x; \theta)$ at x as

$$I(x; \theta) = \lim_{\varepsilon \to 0} \left[\frac{\tilde{\theta} - \theta}{\varepsilon} \right] . \tag{3.1}$$

The sample influence function is defined in Devlin, Gnanadesikan and Kettenring (1975) by

$$I\_(x_i; \hat{\theta}) = (n - 1)(\hat{\theta} - \hat{\theta}_{-i}), \; i = 1, 2, \ldots, n \tag{3.2}$$

where $\hat{\theta}$ is an estimator of θ based on n observations $x_1, x_2, \ldots, x_n$ and $\hat{\theta}_{-i}$ is an estimator of the same form as $\hat{\theta}$ but determined without the ith observation x_i. The sample influence function is an approximation of the empirical influence function obtained from (3.1) by replacing F by F_n, the empirical distribution function, and ε by $\frac{-1}{n-1}$ (Mallows, 1976).

The theoretical influence function can be used to determine the influence of the point x on the parameter θ and the sample influence function is used to determine the influence of x_i on the estimator $\hat{\theta}$. In this section the theoretical influence function of RLS is obtained and extended to the sample case. It will be used together with the sample influence function to detect multivariate outliers in the context of vector correlation in a way similar to that used in Devlin, Gnanadesikan and Kettenring (1975) in the context of bivariate correlation or in Cléroux, Helbling and Ranger (1986) in another context of multivariate correlation.

Let $X = \begin{bmatrix} X^{(1)} \\ X^{(2)} \end{bmatrix}$ be a random vector having the same distribution as the population from which we are sampling and let x be the vector of values taken by X.

The theoretical influence function of the vector x on the parameter ρLS is

$$I(x;\rho LS) = \lim_{\varepsilon \to 0} \frac{(\rho \tilde{L}S - \rho LS)}{\varepsilon}. \tag{3.3}$$

Since

$$\lim_{\varepsilon \to 0} \frac{\rho \tilde{L}S^2 - \rho LS^2}{\varepsilon} = 2\rho LS \lim_{\varepsilon \to 0} \frac{\rho \tilde{L}S - \rho LS}{\varepsilon}$$
$$= 2\rho LS\ I(x;\ \rho LS) \tag{3.4}$$

the left-hand member of (3.4) will be considered first and only terms of order ε need be retained. We have

$$\Sigma_{ij} = E(X^{(i)} - \mu^{(i)})(X^{(j)} - \mu^{(j)})' \quad \text{for } i, j = 1, 2,$$

$$\rho LS^2 = \frac{[tr(\Sigma_{12}\Sigma_{21})^{\frac{1}{2}}]^2}{tr\,\Sigma_{11} tr\,\Sigma_{22}}, \quad \rho \tilde{L}S^2 = \frac{[tr(\tilde{\Sigma}_{12}\tilde{\Sigma}_{21})^{\frac{1}{2}}]^2}{tr\,\tilde{\Sigma}_{11} tr\,\tilde{\Sigma}_{22}}$$

and

$$\frac{\rho \tilde{L}S^2 - \rho LS}{\varepsilon} = \frac{\frac{1}{\varepsilon}\,[(tr(\tilde{\Sigma}_{12}\tilde{\Sigma}_{21})^{\frac{1}{2}})^2 tr\,\Sigma_{11} tr\,\Sigma_{22} - tr\,\tilde{\Sigma}_{11} tr\,\tilde{\Sigma}_{22}(tr(\Sigma_{12}\Sigma_{21})^{\frac{1}{2}})^2]}{tr\,\tilde{\Sigma}_{11} tr\,\tilde{\Sigma}_{22} tr\,\Sigma_{11} tr\,\Sigma_{22}} \tag{3.5}$$

For $j = 1, 2$, we can write $\Sigma_{ij} = \int (x^{(i)} - \mu^{(i)})(x^{(j)} - \mu^{(j)})' dF$ and it is seen that $\tilde{\Sigma}_{ij} = (1-\varepsilon)\Sigma_{ij} + \varepsilon z^{(i)} z^{(j)\prime}$ where $z^{(i)} = x^{(i)} - \mu^{(i)}$ and $z^{(j)} = x^{(j)} - \mu^{(j)}$. Using the relation

$$tr(A + \varepsilon B)^{\frac{1}{2}} = tr(A^{\frac{1}{2}}) + \frac{\varepsilon}{2} tr(A^{-\frac{1}{2}} B) + 0(\varepsilon^2) \tag{3.6}$$

where A and B are symmetric matrices with A positive definite, and following an approach similar to that of Cléroux, Helbing and Ranger (1986) we finally obtain

$$I(x;\ \rho LS) = \frac{\rho LS}{2}\left[\frac{2z^{(1)\prime}(\Sigma_{12}\Sigma_{21})^{-\frac{1}{2}}\Sigma_{12} z^{(2)}}{tr(\Sigma_{12}\Sigma_{21})^{\frac{1}{2}}} - \frac{z^{(1)\prime} z^{(1)}}{tr\,\Sigma_{11}} - \frac{z^{(2)\prime} z^{(2)}}{tr\,\Sigma_{22}}\right] \tag{3.7}$$

We can easily show that $E[I(X;\ \rho LS)] = 0$ and if X has the multivariate $N(\mu, \Sigma)$ distribution, then $\mathrm{Var}[I(X;\ \rho LS)] = \delta_1^2$ where

$$\delta_1^2 = \rho LS^2 \left[\frac{tr\,\Sigma_{11}^2}{2(tr\,\Sigma_{11})^2} + \frac{tr\,\Sigma_{22}^2}{2(tr\,\Sigma_{22})^2} + \frac{tr\,\Sigma_{12}\Sigma_{21} + trH'\Sigma_{11}H\Sigma_{22}}{(tr(\Sigma_{12}\Sigma_{21})^{\frac{1}{2}})^2} + \frac{tr\,\Sigma_{12}\Sigma_{21}}{tr\,\Sigma_{11} tr\,\Sigma_{22}}\right.$$
$$\left. - \frac{2tr\,\Sigma_{11}(\Sigma_{12}\Sigma_{21})^{\frac{1}{2}}}{tr\,\Sigma_{11} tr(\Sigma_{12}\Sigma_{21})^{\frac{1}{2}}} - \frac{2tr\,\Sigma_{21} H \Sigma_{22}}{tr\,\Sigma_{22} tr(\Sigma_{12}\Sigma_{21})^{\frac{1}{2}}}\right] \tag{3.8}$$

where $H = (\Sigma_{12}\Sigma_{21})^{-\frac{1}{2}}\Sigma_{12}$.

When $p = q = 1$, formula (3.7) reduces to

$$I(x;\ \rho) = \bar{y}_1\bar{y}_2 - \frac{\rho}{2}(\bar{y}_1^2 + \bar{y}_2^2) \tag{3.9}$$

where $\bar{y}_1 = \dfrac{x_1 - \mu_1}{\sigma_1}$ and $\bar{y}_2 = \dfrac{x_2 - \mu_2}{\sigma_2}$, and where σ_1 and σ_2 are the standard deviations of X_1 and X_2 respectively. From formula (3.4) it is seen that the term in brackets is the influence function of ρ as used in Devlin, Gnanadesikan and Kettenring (1975) and Gnanadesikan (1977).

In practice, one would replace $\mu^{(i)}$ and $\sum_{ij}$ by $\overline{X}^{(i)}$ and S_{ij} in (3.7) to obtain $\hat{I}(x; \rho LS)$ and then compute the influence of each data point to pinpoint possible outliers. The sample influence function

$$I_{-}(x_i; RLS) = (n - 1)(RLS - RLS_{-i}),\ i = 1, 2, \ldots n \tag{3.10}$$

can be used similarly to detect possible outliers.

Example 3.1

Consider the data (Anderson (1935) and Fisher (1936) on 50 Iris virginica plants. They consist of vectors of size 4 where the components are $X_1(X_2)$ = sepal length (width) in cm, $X_3(X_4)$ = petal length (width) in cm. The data can be found in Kendall, Stuart and Ord (1983, p. 375). Let $Y^{(1)'} = (\ln(10X_1), \ln(10X_2))$ and $Y^{(2)} = (ln(10X_3), \ln(10X_4))$ and suppose we are interested in the correlation between subvectors $Y^{(1)}$ and $Y^{(2)}$ in the sense of Lingoes and Schonemann. Figure 3.1 shows the graph of $I_{-}(x_i; RLS)$ given by (3.10) as a function of i. Here RLS and RLS_{-i} are the sample vector correlation coefficients between $Y^{(1)}$ and $Y^{(2)}$. Figure 3.1 also shows an horizontal line at $2\hat{\delta}_1$, where $\hat{\delta}_1$ is obtained from (3.8) replacing ρLS and $\sum_{ij}$ by RLS and S_{ij}, $i, j = 1, 2$. The horizontal line at $-2\hat{\delta}_1$, is the x axis. The vertical scale has been multiplied by 10. We see that the most influential observations are data points 20 and 7 with respective relative influences $\dfrac{100 I_{-}(x_i; RLS)\%}{(n-1)RLS}$ given by 4.33% and 3.24%. Points 19 and 32 must be examined also. Their relative influences are -2.62% and -2.48% respectively. Here the value of RLS is .66 and $\hat{\delta}_1 = .45$.

These data have also been considered by Cléroux, Helbling and Ranger (1986) using the RV vector correlation coefficient introduced by Escoufier (1973) and the most influential observations were found to be data points 7, 30, 20 and 15. Two other methods of detecting multivariate outliers have been applied to the same data for the sake of comparison: a procedure based on the concept of gaps between points (Rohlf (1975)) and a prodecure based on the scatter ratios (Wilks (1963)). In the first method the longuest gaps corresponded to data points 7, 20, 10 and 15 and in the second method, observations 19, 35, 7 and 32 had the smallest scatter ratios. It is seem that the proposed method based on the sample influence function of RLS has some agreement with each of the other three methods but is similar to none of them.

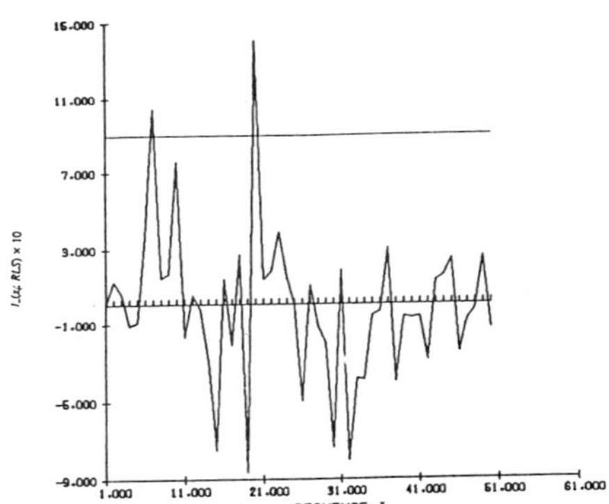

Figure 3.1: Sample influence function of RLS between $(ln(10X_1), \ln(10X_2))$ and $(ln(10X_3), ln(10X_4))$ for Iris virginica data; $\hat{\delta}_1 = .45$.

Example 3.2

Consider the data given in BMDP Statistical Software (1981, p. 38). They consist of 188 vectors of size 8. The variables are age, height, weight, birth pill, cholesterol, albumin, calcium and uric acid. Variable birth pill is omitted and the 7 vectors having missing components are also omitted. Moreover, the authors have modified the value of cholesterol to represent an extreme value.

Let $X^{(1)}$ be the vector whose components are the first three variables (the physical variables) and $X^{(2)}$ be the vector whose components are the last four variables (the blood chemistry variables) and suppose we are interested in the vector correlation between $X^{(1)}$ and $X^{(2)}$.

Here $p = 3$, $q = 4$ and $n = 181$. Figure 3.2 shows the graph of $I_{-}(x_i; RLS)$ as given by (3.10) together with horizontal lines at $\pm 2\hat{\delta}_1$ and $\pm 3\hat{\delta}_1$. The vertical scale has been multiplied by 10. It is seen that data points 154,3 and 46 are possible outliers. Their respective relative influences are -9.92%, -7.52% and -7.15%. We also have $RLS = .19$ and $\hat{\delta}_1 = .64$. The same influential observations have been found in Cléroux, Helbling and Ranger (1986) using the RV coefficient of Escoufier (1973).

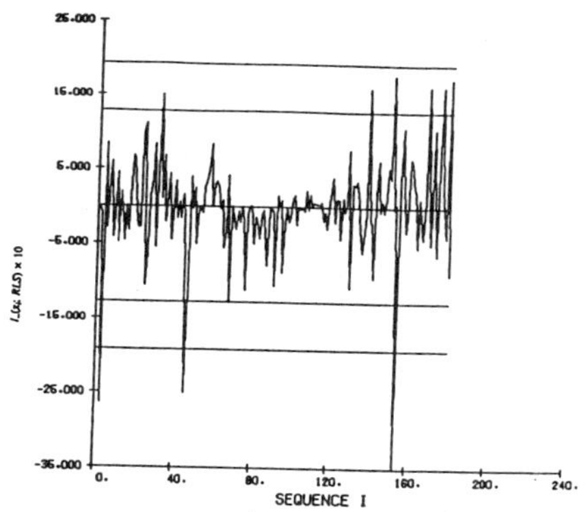

Figure 3.2: Sample influence function of *RLS* between $X^{(1)}$ and $X^{(2)}$ for BMDP data, $\delta_1 = .64$.

4. OUTLIERS IN THE MULTIVARIATE LINEAR REGRESSION MODEL

The problem of multivariate linear regression can be formulated in the following way: find the $q \times p$ matrix M such that $RLS\,(X^{(1)}, M'X^{(2)})$ is maximum subject to $S_{12}M - M'S_{22}M = 0$ (the regression equations). The solution is $M^{*\prime} = S_{12}S_{22}^{-1}$ and $M^{*\prime}X^{(2)}$ are the p linear functions obtained by multivariate linear regression of $X^{(1)}$ on $X^{(2)}$. The maximum value of RLS is such that

$$RI = RLS^2(X^{(1)}, M^{*\prime}X^{(2)}) = \frac{trS_{12}S_{22}^{-1}S_{21}}{trS_{11}} \tag{4.1}$$

and RI is known as the Stewart and Love redundancy index (see Stewart and Love (1968)). When $p = 1$, $RI = R^2$, the squared multiple correlation coefficient.

In this context, an outlying observation can be found by considering the influence function of RI. Here again we start from (3.4) with

$$\rho I = \frac{tr\,\Sigma_{12}\Sigma_{22}^{-1}\Sigma_{21}}{tr\,\Sigma_{11}}\ ,\ \tilde{\rho} I = \frac{tr\,\tilde{\Sigma}_{12}\tilde{\Sigma}_{22}^{-1}\tilde{\Sigma}_{21}}{tr\,\tilde{\Sigma}_{11}}$$

and

$$\frac{\tilde{\rho} I - \rho I}{\varepsilon} = \frac{1}{\varepsilon}\left[\frac{tr\,\tilde{\Sigma}_{12}\tilde{\Sigma}_{22}^{-1}\tilde{\Sigma}_{21}tr\,\Sigma_{11} - tr\,\Sigma_{12}\Sigma_{22}^{-1}\Sigma_{21}tr\,\tilde{\Sigma}_{11}}{tr\,\tilde{\Sigma}_{11}tr\,\Sigma_{11}}\right]. \tag{4.2}$$

We have seen that $\tilde{\Sigma}_{ij} = (1-\varepsilon)\Sigma_{ij} + \varepsilon z^{(i)}z^{(j)\prime}$ where $z^{(i)} = x^{(i)} - \mu^{(i)}$ and $z^{(j)} = x^{(j)} - \mu^{(j)}$. We also have, from Press (1972, p. 23) that

$$\tilde{\Sigma}_{22}^{-1} = \frac{1}{1-\varepsilon}\left[\Sigma_{22}^{-1} - \frac{\varepsilon\Sigma_{22}^{-1}z^{(2)}z^{(2)\prime}\Sigma_{22}^{-1}}{1-\varepsilon+\varepsilon z^{(2)\prime}\Sigma_{22}^{-1}z^{(2)}}\right].$$

Using these results in (4.2), after some algebraic manipulations and taking the limit as $\varepsilon \to 0$ we obtain

$$I(x;\rho I) = \rho I\left[\frac{2z^{(1)\prime}Bz^{(2)}}{tr\,\Sigma_{11}^{*}} - \frac{z^{(1)\prime}z^{(1)}}{tr\,\Sigma_{11}} - \frac{z^{(2)\prime}B'Bz^{(2)}}{tr\,\Sigma_{11}^{*}}\right] \tag{4.3}$$

where $\Sigma_{11}^{*} = \Sigma_{12}\Sigma_{22}^{-1}\Sigma_{21}$ and $B = \Sigma_{12}\Sigma_{22}^{-1}$ is the regression matrix of $X^{(1)}$ on $X^{(2)}$.

It is seen that $E[I(X;\rho I] = 0$ and if X has the multivariate $N(\mu, \Sigma)$ distribution, then $\mathrm{Var}[I(X;\rho I)] = \delta_2^2$ where

$$\delta_2^2 = 2\rho I^2\left[\frac{tr\,\Sigma_{11}^{2}}{(tr\,\Sigma_{11})^2} - \frac{4tr\,(\Sigma_{11}\Sigma_{11}^{*}) - 2tr\,\Sigma_{11}^{*2}}{tr\,\Sigma_{11}^{*}tr\,\Sigma_{11}} + \frac{2tr\,\Sigma_{11}\Sigma_{11}^{*} - tr\,\Sigma_{11}^{*2}}{(tr\,\Sigma_{11}^{*})^2}\right]. \tag{4.4}$$

If $p = 1$, formula (4.3) reduces to

$$I(x; \bar{R}^2) = \frac{1}{\sigma_1^2} [2z_1 B' z^{(2)} - z_1^2 \bar{R}^2 - (B' z^{(2)})^2] \tag{4.5}$$

where $B' = \sigma_{(1)} \Sigma_{22}^{-1}$ is the regression vector of X_1 on $X^{(2)}$, $\sigma_1^2 = \text{var}(X_1)$, $\sigma_{(1)} = \Sigma_{12}$ is a row vector and $\bar{R}$ is the population coefficient of multiple correlation. In Cook and Weisberg (1982, pp. 107-110) and in Critchley (1985) the influence function of $B' = \sigma_{(1)} \Sigma_{22}^{-1}$ is obtained. Moreover, if $p = q = 1$, formula (4.5) reduces to (3.9).

The sample influence function of RI is given by

$$I_{-}(x_i; RI) = (n - 1)(RI - RI_{-i}), \; i = 1, 2, \ldots, n \; . \tag{4.6}$$

Example 4.1

Consider again the BMDP data of Example 3.1. Suppose we are interested in detecting the influential observations in the multivariate linear regression of $X^{(2)}$ on $X^{(1)}$. Figure 4.1 shows the graph of $I_{-}(x_i; RI)$ as given by (4.6) as a function of i together with horizontal lines at $\pm 2\hat{\delta}_2$ and $\pm 3\hat{\delta}_2$ where $\hat{\delta}_2$ is obtained from (4.4) replacing ρI by RI and Σ_{ij} by S_{ij}, $i, j = 1, 2$. The vertical scale has been multiplied by 10. Here $RI = .14$ and $\hat{\delta}_2 = .64$. It is seen that the most influential observations are data points 154, 46, 176 and 181 with respective relative influences -9.87%, -8.49%, 7.72% and -7.55%. Note that points 154 and 46 are possible data outliers and model outliers and that data point 3, which was deliberately modified to represent an extreme value is not an outlier in the regression model. In Cléroux, Helbling and Ranger (1986), using the influence function of the RV_{reg} coefficient introduced in Robert and Escoufier (1976), the same model outliers have been found. This could be explained by the fact that the same solution $M' = S_{12} S_{22}^{-1}$ is used in the maximisation to obtain RI or RV_{reg} .

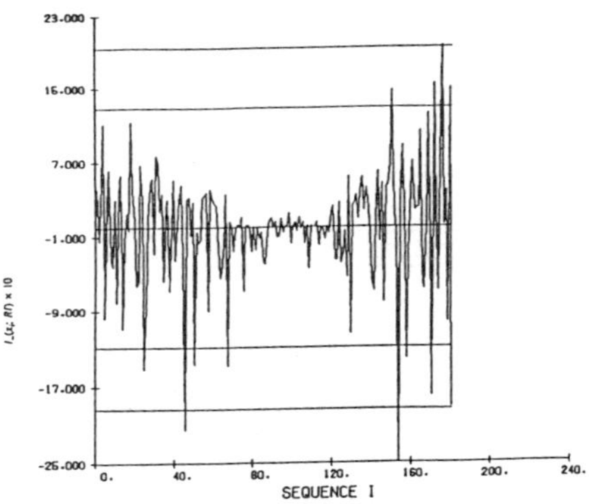

Figure 4.1: Sample influence function of RI between $X^{(1)}$ and $X^{(2)}$ for BMDP data; $\hat{\delta}_2 = .64$.

5. DISTRIBUTION OF I(X; ρLS) and I(X; ρI) AND TESTS FOR OUTLYING OBSERVATIONS

In the preceeding sections the sample influence function was used as a graphical mean of identifying outliers. In this section we obtain the distribution of the theoritical influence function and propose an approximate test for outliers. The procedure is similar to that of Cléroux, Helbling and Ranger (1986).

From (3.7) and (4.3) we can write

$$I(X; \rho LS) = \frac{\rho LS}{2} (Z^{(1)\prime} Z^{(2)\prime}) \begin{bmatrix} \dfrac{-I_p}{tr \Sigma_{11}} & \dfrac{H}{tr(\Sigma_{12}\Sigma_{21})^{\frac{1}{2}}} \\ \dfrac{H'}{tr(\Sigma_{12}\Sigma_{21})^{\frac{1}{2}}} & \dfrac{-I_q}{tr \Sigma_{22}} \end{bmatrix} \begin{bmatrix} Z^{(1)} \\ Z^{(2)} \end{bmatrix} \tag{5.1}$$

and

$$I(X;\rho I) = \rho I \ (Z^{(1)\prime} Z^{(2)\prime}) \begin{bmatrix} \dfrac{-I_p}{tr\sum_{11}} & \dfrac{B}{tr\sum_{11}^{*}} \\ \dfrac{B'}{tr\sum_{11}^{*}} & \dfrac{-B'B}{tr\sum_{11}^{*}} \end{bmatrix} \begin{bmatrix} Z^{(1)} \\ Z^{(2)} \end{bmatrix} \tag{5.2}$$

In both cases the influence function can be written as an indefinite quadratic form $T = Z'AZ$ where Z has mean vector 0 and covariance matrix $\sum$. If Z has the multivariate normal distribution, it is well known (see for example Johnson and Kotz (1970)) that T has the same distribution as $\sum_{i=1}^{p+q} \lambda_i W_i^2$ where the random variables W_i are iid $N(0, 1)$ and where the λ_i's are the eigenvalues of $\sum A$.

In practice the percentiles of this distribution are computed using the Imhof (1961) algorithm which is an exact method, or the approximation suggested by Solomon and Stephens (1977). In the computations below the former has been used. A computer program of the Imhof algorithm can be found in Koerts and Abrahamse (1969).

As mentioned previously, Example 4.2 where $p = 1$ and $RI = RV_{reg} = R^2$ has also been considered in Cléroux, Helbling and Ranger (1986). There the Solomon and Stephens (1977) algorithm was used. The computations of this paper for the same example will provide a comparison of the two algorithms.

The criterion for selecting and testing an extreme observation as a significant outlier is based on the extreme values of the influence function. The distributions of $T_{(1)} = \min_{1 \le i \le n} T_i$ and of $T_{(n)} = \max_{1 \le i \le n} T_i$ are thus needed where $T_i = Z_i'AZ_i$, $i = 1, 2, \ldots, n$ and where the Z_i's and A are given in (5.1) or (5.2). $T_{(1)}$ will usually be negative and $T_{(n)}$ will usually be positive. Since the T_i's are independent it follows that the distribution of $T_{(n)}$ is $G(t) = [F(t)]^n$ and that of $T_{(1)}$ is $H(t) = 1 - [1 - F(t)]^n$, where $F(t)$ is the distribution of T.

An approximate procedure for testing extreme observations as significant outliers is the following:

(i) determine matrix A from (5.1) if data outliers are of interest and from (5.2) if regression outliers are of interest

(ii) compute the eigenvalues of $S\hat{A}$ where $\hat{A}$ is obtained from A by replacing $\sum$ by S and ρLS (or ρI) by RLS (or RI)

(iii) obtain the distribution $F(t)$ of T, $G(t)$ of $T_{(n)}$ and $H(t)$ of $T_{(1)}$

(iv) go to (v) and (vi) if data outliers are of interest or to (vii) and (viii) if, on the other hand, regression outliers are of interest

(v) consider the data point x_i such that $\hat{I}(x_i; \rho LS)$ has the largest positive value. Use $G(t)$ to compute the probability p_1 of exceeding this value. If $p_1 < \alpha$ then x_i can be considered as being an outlier at level α

(vi) consider the data point x_j such that $\hat{I}(x_j; \rho LS)$ has the smallest negative value. Use $H(t)$ to compute the probability p_2 of not exceeding this value. If $p_2 < \alpha$ then x_j can be considered as being an outlier at level α

(vii) consider the data point x_k such that $\hat{I}(x_k; \rho I)$ has the largest positive value. Use $G(t)$ to compute the probability p_3 of exceeding this value. If $p_3 < \alpha$ then x_k can be considered as being an outlier at level α

(viii) consider the data point x_l such that $\hat{I}(x_l; \rho I)$ has the smallest negative value. Use $H(t)$ to compute the probability p_4 of not exceeding this value. If $p_4 < \alpha$ then x_l can be considered as being an outlier at level α.

These tests are approximate tests and nothing is known about their power of outlier detection. Such a study remains to be done.

The tests were applied to the data points of Examples 3.1, 3.2 and 4.1 corresponding to the largest positive values or to the smallest negative values of the respective influence functions. The results are shown in Table 5.1.

	Outliers by graphical method	Largest positive value or smallest negative value of influence function	Probability of exceeding largest value or probability of not exceeding smallest value
Example 3.1 Figure 3.1	20	1.40	$p_1 = .28$
Example 3.2 Figure 3.2	154	-3.48	$p_2 = .09$
Example 4.1 Figure 4.1	176 154	1.95 -2.50	$p_3 = .82$ $p_4 = .49$

Table 5.1: Approximate tests for outlier detection in Examples 3.1, 3.2 and 4.1.

It is seen that observation 154 is an outlier at the 10% level in Example 3.2 and no other data point is a significant outlier. it should be recalled however that the preceeding tests are approximate tests and nothing is known abouth their power. Consequently the graphs of $I_{-}(x_i; RLS)$ and $I_{-}(x_i; RI)$ will prove useful as exploratory techniques for multivariate outlier detection. The same conclusion has been rearched in Cléroux, Helbling and Ranger (1986). Concerning the graphs of $I_{-}(x_i; RV)$ and $I_{-}(x_i; RV_{reg})$.

6. CONCLUSION

Some new methods have been introduced for detecting multivariate data outliers and outliers in the multivariate linear regression model. The approach was similar to that used in Cléroux, Helbling and Ranger (1986). Only the association measures have been changed. The same examples have been used in order to compare the different methods. It was seen in one example that different influential points may be pinpointed depending on the correlation measure used. Since no omnibus outlier detection method exists, the methods proposed in this paper will prove useful in practice.

REFERENCE

1. ANDERSON, E.: The Irises of the Gaspé Peninsula, Bulletin of the American Iris Society, 1935, 59, 2-5.
2. BMDP Statistical Software, 1981, W.J. Dixon, Chief Editor, University of California Press, Berkeley.
3. CLEROUX, R., HELBLING, J.M. and RANGER, N.: Some Methods of Detecting Multivariate Outliers, Computational Statistics Quarterly, 1986, 3, 177-195.
4. COOK, R.D. and WEISBERG, S.: Residuals and Influence in Regression, Chapman and Hall, London, 1982.
5. CRITCHLEY, F.: Influence in Principal Components Analysis, Biometrika, 1985, 72, 627-636.
6. DEVLIN, S.J., GNANADESIKAN, R. and KETTENRING, J.R.: Robust Estimation and Outlier Detection with Correlation Coefficients, Biometrika, 1975, 62, 531-545.
7. ESCOUFIER, Y.: Le traitement des variables vectorielles, Biometrics, 1973, 29, 751-760.
8. FISHER, R.A.: The Use of Multiple Measurements in Taxonomic Problems, Annals of Engenics, 1936, 7, 179-188.
9. GNANADESIKAN, R.: Methods for Statistical Data Analysis of Multivariate Observations, John Wiley and Sons, New York, 1977.
10. HAMPEL, F.R.: The Influence Curve and its Role in Robust Estimation, Journal of the American Statistical Association, 1974, 69, 383-393.
11. IMHOF, P.: Computing the Distribution of Quadratic Forms in Normal Variates, Biometrika, 1961, 48, 419-426.
12. JOHNSON, N.L. and KOTZ, S.: Continuous Univariate Distributions, Vol.2, Houghton Mifflin Company, Boston, 1970.
13. KENDALL, M.G., STUART, A. and ORD, J.K.: The Advanced Theory of Statistics, vol. 3, 1983, Macmillan, New York.
14. KOERTS, J. and ABRAHAMSE, A.P.J.: On the Theory and Application of the General Linear Model, Rotterdam University Press, Rotterdam, 1969.
15. LINGOES, J.C. and SCHONEMANN, P.H.: Alternative Measures of Fit for the Schonemann-Carroll Matrix Fitting Algorithm, Psychometrika, 1974, 39, 423-427.
16. MALLOWS, C.L.: On Some Topics in Robustness, Bell Laboratories Memorandum, 1976.
17. PRESS, S.J.: Applied Multivariate Analysis, Holt, Rinehart and Winston, New York, 1972.
18. ROBERT, P. and ESCOUFIER, Y.: A Unifying Tool for Linear Multivariate Statistical Methods: The RV-Coefficient, Applied Statistics, 1976, 25, 257-265.
19. ROHLF, F.J.: Generalization of the Gap Test for the Detection of Multivariate Outliers, Biometrics, 1975, 31, 93-101.
20. SEBER, G.A.F.: Multivariate Observations , John Wiley and Sons, New York, 1984.
21. SOLOMON, H. and STEPHENS, M.A.: Distribution of a Sum of Weighted Chi-Square Variates, Journal of the American Statistical Association, 1977, 72, 881-885.
22. STEWART, D. and LOVE, W.: A General Canonical Correlation Index, Psychometric Bulletin, 1968, 70, 160-163.
23. WILKS, S.S.: Multivariate Statistical Outliers, Sankhya, 1963, 25, 407-426.

SENSITIVITY ANALYSIS IN LEAST SQUARES FACTOR ANALYSIS

Yutaka Tanaka* and Yoshimasa Odaka**

*Department of Statistics, Okayama University, Tsushima, Okayama 700, Japan
**Information Processing Center, Okayama University of Science, Ridaicho
Okayama 700, Japan

The theoretical influence function for the unique variance matrix is derived for investigating the influence of a small change of data in least squares factor analysis. A numerical example is shown to illustrate the proposed procedure and a brief numerical comparison is made with the cases of principal factor analysis and maximum likelihood factor analysis.

1. Introduction

We consider sensitivity analysis in factor analysis, that is, the influence of a small change of data on the result of factor analysis. Similar problems have been discussed by many authors for other multivariate methods. For example, Campbell[3] considered the influence in discriminant analysis and Benasseni[1], Critchley[4], Pack et al.[9], Tanaka[11] among others did the same for principal component analysis. However, as far as we know, relatively few papers such as Kodake and Tanaka[5], Yamane[16] and Tanaka and Odaka[13-15] discussed the influence in factor analysis. Kodake and Tanaka[5] proposed a method of sensitivity analysis in maximum likelihood factor analysis(MLFA) based on Taylor series expansion. However, their method is too time-consuming for actual application. Yamane[16] tried to derive the influence function $I(\underline{\mathbf{x}}; L)$ of factor loadings L by using the influence functions $I(\underline{\mathbf{x}}; \lambda_s)$ and $I(\underline{\mathbf{x}}; \underline{\mathbf{v}}_s)$ of eigenvalues λ_s and eigenvectors $\underline{\mathbf{v}}_s$ given by Radhakrishnan and Kshirsagar[10] in each cycle of the iterative procedure of principal factor analysis(PFA). His procedure also has some problems in computing cost and accuracy. Tanaka and Odaka[13] proposed a better procedure in PFA. They applied the influence function $I(\underline{\mathbf{x}}; \sum_{s=1}^{q} \lambda_s \underline{\mathbf{v}}_s \underline{\mathbf{v}}_s^T)$ derived by Tanaka[11] instead of $I(\underline{\mathbf{x}}; \lambda_s)$ and $I(\underline{\mathbf{x}}; \underline{\mathbf{v}}_s)$ to the converged solution of the iterative procedure of PFA. Tanaka and Odaka[15] applied the similar technique to MLFA.

In the present paper we propose a method of sensitivity analysis in least squares factor analysis(LSFA). Comparing with the case of PFA and MLFA the situation is a little more complicated. In this case the influence function $I(\underline{\mathbf{x}}; \sum_{s=1}^{q} \lambda_s \underline{\mathbf{u}}_s \underline{\mathbf{u}}_s^T)$ related to a generalized eigenvalue problem $(A - \lambda_s B)\underline{\mathbf{u}}_s = \underline{0}$, which was derived by Tanaka[12], plays an important role.

2. Factor analysis model and least squares estimation

We assume the following factor analysis model with ordinary assumptions.

$$\underline{\mathbf{x}} = \underline{\mu} + L\underline{\mathbf{f}} + \underline{\mathbf{e}}.$$

Then the $p \times p$ variance matrix Σ of the random vector $\underline{\mathbf{x}}$ is given in terms of the $p \times q$ loading matrix L and the $p \times p$ unique variance diagonal matrix Δ.

$$\Sigma = LL^T + \Delta. \tag{1}$$

The fundamental equation (1), which is called the common factor decomposition, is derived under the assumption of orthogonal factors. However, this assumption does not cause any loss of generality.

Many procedures have been proposed to obtain L and Δ from the observed Σ. Among them we consider least squares method in the present paper. Let S be an estimate for Σ based on n observations $\underline{\mathbf{x}}_i, i = 1, \dots, n$. So-called generalized least squares method(GLS) is characterized by the following minimization problem (Browne[2], Okamoto[8]):

$$\text{Minimize} \quad L(S, LL^T + \Delta) \equiv \frac{1}{2} trace[(S - (LL^T + \Delta))W]^2, \tag{2}$$

where the positive definite weight matrix W can be defined in various manners such as i) $W = I(ULS$, unweighted least squares), ii) $W = S^{-1}(WLS$, weighted least squares) and iii) $W = [diag(S)]^{-1}(SLS$, simple least squares). The method corresponding to each definition of W is called by the name in the parenthesis. Since the case of ULS is equivalent to PFA discussed in Tanaka and Odaka[13], we discuss the remaing two cases in the present paper.

The above is the description based on sample. However, as in the case of MLFA, to derive "theoretical influence functions", we consider "population least squares method", which is defined by replacing S by Σ. From (2) with S replaced by Σ we obtain

$$\begin{cases} (\Sigma - LL^T - \Delta)WL = 0, & (3) \\ diag[W(\Sigma - LL^T - \Delta)W] = 0. & (4) \end{cases}$$

Letting

$$R^* = \Sigma - \Delta, \tag{5}$$

equation (3) is rewritten as the following generalized eigenvalue problem :

$$(R^* - \lambda_s W^{-1})\underline{\mathbf{u}}_s = \underline{0}, \quad s = 1, \dots, p, \tag{6}$$

where the eigenvector $\underline{\mathbf{u}}_s$ is normalized so as to satisfy

$$\underline{\mathbf{u}}_s^T W^{-1} \underline{\mathbf{u}}_r = \delta_{sr} \ (\delta_{sr} : \text{Kronecker delta}).$$

On the other hand, equation (4) is transformed into

$$diag[WR^*W - U_1 \Lambda_1 U_1^T] = 0, \tag{7}$$

where $\Lambda_1 = diag(\lambda_1, \cdots, \lambda_q)$ consists of the largest q eigenvalues and $U_1 = (\underline{\mathbf{u}}_1, \cdots, \underline{\mathbf{u}}_q)$ consists of the associated eigenvectors. The loading matrix L is obtained as

$$L = W^{-1} U_1 \Lambda_1^{1/2}. \tag{8}$$

Equations (5), (6), and (7) provide the determinantal equations for the generalized least squares factor analysis.

3. Influence function $I(\underline{\mathbf{x}}; \Delta)$ for the unique variance matrix

We introduce a small perturbation $\Sigma \to \Sigma + \varepsilon \Sigma^{(1)}$ into these determinantal equations and try to evaluate the corresponding changes $R^* \to R^* + \varepsilon R^{*(1)}$ and $\Delta \to \Delta + \varepsilon \Delta^{(1)}$. When we consider the influence functions due to the perturbation $F \to (1-\varepsilon)F + \varepsilon \delta_x$, where F is the cdf of the random vector $\underline{\mathbf{x}}$ and δ_x is the cdf of a unit point mass at a specified $\underline{\mathbf{x}}$, then the influence function $\Sigma^{(1)}$ for the variance matrix Σ is given as

$$\Sigma^{(1)} = (\underline{\mathbf{x}} - \underline{\mu})(\underline{\mathbf{x}} - \underline{\mu})^T - \Sigma. \tag{9}$$

Substitute the perturbed quantities into equations (5), (6) and (7) and compare the coefficients of the first order terms of ε. Here we apply the influence function $I(\underline{\mathbf{x}}; \sum_{s=1}^{q} \lambda_s \underline{\mathbf{u}}_s \underline{\mathbf{u}}_s^T)$ for a function of eigenvalues and eigenvectors of the generalized eigenvalue problem (6), which was derived by Tanaka[12]. Then, in the similar manner as in the cases of PFA and MLFA we can derive the following procedure to obtain the theoretical influence function $\Delta^{(1)} = I(\underline{\mathbf{x}}; \Delta)$. The essential part is a system of linear simultaneous equations with p unknowns in Step 2. Here we denote the $(j,k)^{th}$ elements of $\Sigma^{(1)}$, $R^{*(1)}$, W, etc. simply by $\Sigma_{jk}^{(1)}$, $R_{jk}^{*(1)}$, W_{jk}, etc. and the jth element of $\underline{\mathbf{u}}_s$ by u_{js}.

Step 1. Calculate $R_{jk}^{*(1)}$'s for $j \neq k$ by

$$R_{jk}^{*(1)} = \Sigma_{jk}^{(1)}, \quad j \neq k. \tag{10}$$

Step 2. Solve the following linear simultaneous equations for $R_{jj}^{*(1)}$, $j = 1, \ldots, p$.

$$\sum_{i=1}^{p} a_{li} R_{ii}^{*(1)} = b_l, \quad l = 1, \ldots, p, \tag{11}$$

where

$$a_{li} = W_{li}^2 - \sum_{s=1}^{q} \sum_{r=1}^{q} u_{is} u_{ir} u_{ls} u_{lr} - 2 \sum_{s=1}^{q} \sum_{r=q+1}^{p} \lambda_s (\lambda_s - \lambda_r)^{-1} u_{is} u_{ir} u_{ls} u_{lr},$$

$$\begin{aligned} b_l = & -2 \sum_{j=1}^{p} \sum_{k=1}^{p} W_{lj}^{(1)} W_{lk} R_{jk}^{*} - \sum_{\substack{j=1 \\ j \neq k}}^{p} \sum_{k=1}^{p} W_{lj} W_{lk} \Sigma_{jk}^{(1)} \\ & + \sum_{s=1}^{q} \sum_{r=1}^{q} \{ \sum_{\substack{j=1 \\ j \neq k}}^{p} \sum_{k=1}^{p} \Sigma_{jk}^{(1)} u_{js} u_{kr} - (\lambda_s + \lambda_r) \sum_{j=1}^{p} \sum_{k=1}^{p} (W^{-1})_{jk}^{(1)} u_{js} u_{kr} \} u_{ls} u_{lr} \\ & + 2 \sum_{s=1}^{q} \sum_{r=q+1}^{p} \lambda_s (\lambda_s - \lambda_r)^{-1} \{ \sum_{\substack{j=1 \\ j \neq k}}^{p} \sum_{k=1}^{p} \Sigma_{jk}^{(1)} u_{js} u_{kr} - \lambda_s \sum_{j=1}^{p} \sum_{k=1}^{p} (W^{-1})_{jk}^{(1)} u_{js} u_{kr} \} u_{ls} u_{lr}. \end{aligned}$$

Step 3. Calculate $\Delta_{jj}^{(1)}$'s by

$$\Delta_{jj}^{(1)} = \Sigma_{jj}^{(1)} - R_{jj}^{*(1)}, \quad j = 1, \ldots, p. \tag{12}$$

We used W and W^{-1} in the above formulation. We must replace these by

$$\begin{cases} W = \Sigma^{-1}, W^{-1} = \Sigma, \\ (W^{-1})^{(1)} = \Sigma^{(1)}, W^{(1)} = (\Sigma^{-1})^{(1)} = -\Sigma^{-1} \Sigma^{(1)} \Sigma^{-1}, \end{cases} \tag{13}$$

in WLS and

$$\begin{cases} W = [diag(\Sigma)]^{-1}, W^{-1} = diag(\Sigma), (W^{-1})^{(1)} = diag(\Sigma^{(1)}), \\ W^{(1)} = -[diag(\Sigma)]^{-1}[diag(\Sigma^{(1)})][diag(\Sigma)]^{-1}, \end{cases} \tag{14}$$

in SLS, respectively.

In actual data analysis we must analyze real data obtained from sample. Among possible sample versions for investigating the influence of the ith individual we use the empirical influence curve EIC_i. It is calculated by substituting

$$S = (1/n) \sum_{j=1}^{n} (\underline{\mathbf{x}}_j - \bar{\underline{\mathbf{x}}})(\underline{\mathbf{x}}_j - \bar{\underline{\mathbf{x}}})^T \tag{15}$$

and

$$S^{(1)} = (\underline{\mathbf{x}}_i - \bar{\underline{\mathbf{x}}})(\underline{\mathbf{x}}_i - \bar{\underline{\mathbf{x}}})^T - S \tag{16}$$

into Σ and $\Sigma^{(1)}$ in the above procedure.

It is obvious that the property of scale-invariance holds in WLS and SLS as in the case of MLFA (Tanaka and Odaka[15]).

4. Influence measures

Since the influence function $\Delta^{(1)}$ is a vector-valued function, it is useful to summarize it into a scalar-valued quantity. For this purpose we use three scalar-valued measures : Euclidean norm $\|\widehat{\Delta}^{(1)}\|$, univariately-scaled squared distance D_{us} and multivariately-scaled squared distance D_{ms}. The latter two are defined as :

$$D_{us} = (n-1)^{-2}(EIC)^T[diag(\widehat{V}(\widehat{\Delta}))]^{-1}(EIC)$$

$$D_{ms} = (n-1)^{-2}(EIC)^T[(\widehat{V}(\widehat{\Delta}))]^{-1}(EIC)$$

In the case of WLS

$$D_{us} = \frac{1}{2(n-1)} \sum_{j=1}^{p} \widehat{\xi}_{jj} (\widehat{\Delta}_{jj}^{(1)})^2, \tag{17}$$

$$D_{ms} = \frac{1}{2(n-1)} \sum_{j=1}^{p} \sum_{l=1}^{p} \widehat{\xi}_{jl} \widehat{\Delta}_{jj}^{(1)} \widehat{\Delta}_{ll}^{(1)}, \tag{18}$$

and in the case of SLS

$$D_{us} = \frac{1}{2(n-1)} \sum_{j=1}^{p} \sum_{i,i'=1}^{p} \widehat{\xi}_{ji} \widehat{\xi}_{ji'} \omega_{ii'}^2 (\widehat{\Delta}_{jj}^{(1)})^2, \tag{19}$$

$$D_{ms} = \frac{1}{2(n-1)} \sum_{j=1}^{p} \sum_{l=1}^{p} \sum_{i,i'=1}^{p} \widehat{\xi}_{ji} \widehat{\xi}_{li'} \omega_{ii'}^2 \widehat{\Delta}_{jj}^{(1)} \widehat{\Delta}_{ll}^{(1)}, \tag{20}$$

where $\widehat{\xi}_{jl}$ is the $(j,l)^{th}$ element of $\widehat{\Xi}$ and

$$\widehat{\Xi} = (\widehat{\xi}_{jl}) = (\widehat{\phi}_{jl}^2), \quad \widehat{\Phi} = (\widehat{\phi}_{jl}) = \widehat{\Delta}^{-1} - \widehat{\Delta}^{-1}(\widehat{L}\widehat{\Gamma}^{-1}\widehat{L}^T)\widehat{\Delta}^{-1},$$

$$\widehat{\Gamma} = \widehat{L}^T \widehat{\Delta}^{-1} \widehat{L} : diagonal, \quad \omega_{ii'} = (\Phi \Delta \Phi)_{ii'}.$$

5. Numerical example

We reanalyze the open/closed book data(Mardia, Kent and Bibby[7], Table 1.2.1), which was analyzed in Tanaka and Odaka[13-15]. This set of data consists of $n = 88$ individuals(students) and $p = 5$ variables(tests). Since the estimates and influence functions in LSFA(WLS,SLS) are scale invariant, we may standardize the observations before applying LSFA.

Tables 1a and 1b show the result of analysis with LSFA(WLS) and LSFA(SLS) assuming a two factor model. Values in the parentheses in the column of unique variances indicate the asymptotic standard errors for $\widehat{\Delta}_{jj}$'s.

Table 1a. Results of the LSFA(WLS)
(The open/closed book data, n=88; two-factor model)

Variable	Factor loadings		Unique variances
1	0.2701	0.6790	0.4660(0.2005)
2	0.3603	0.6717	0.4190(0.1634)
3	0.7429	0.5094	0.1886(0.0602)
4	0.7403	0.3166	0.3518(0.0869)
5	0.6981	0.2856	0.4310(0.0907)

(judgement of convergence = 0.00001 ; 59 iterations)

Table 1b. Results of the LSFA(SLS)
(The open/closed book data, n=88; two-factor model)

Variable	Factor loadings		Unique variances
1	0.2718	0.6755	0.4698(0.1977)
2	0.3575	0.6754	0.4160(0.1660)
3	0.7425	0.5102	0.1885(0.0602)
4	0.7415	0.3162	0.3503(0.0874)
5	0.6965	0.2873	0.4324(0.0908)

(judgement of convergence = 0.00001 ; 95 iterations)

To study the influence of each individual we can use the empirical influence curve $\widehat{\Delta}^{(1)}$ and scalar-valued measures $\|\widehat{\Delta}^{(1)}\|$, D_{us} and D_{ms}. Figures 1a and 1b show the index plots of those influence measures. From these index plots it is easy to find, in both of WLS and SLS, that individuals 82 and 75 are influential in the measure $\|\widehat{\Delta}^{(1)}\|$, while individual 56 is very influential in D_{us} and D_{ms}. This discrepancy comes from the influence patterns of those individuals. As shown in Table 2 individuals 82 and 75 exert influence mainly on $\widehat{\Delta}_{11}$ and $\widehat{\Delta}_{22}$, whose standard errors are relatively large. On the other hand the influence of individual 56 is mainly on $\widehat{\Delta}_{44}$ and $\widehat{\Delta}_{55}$, which have smaller standard errors.

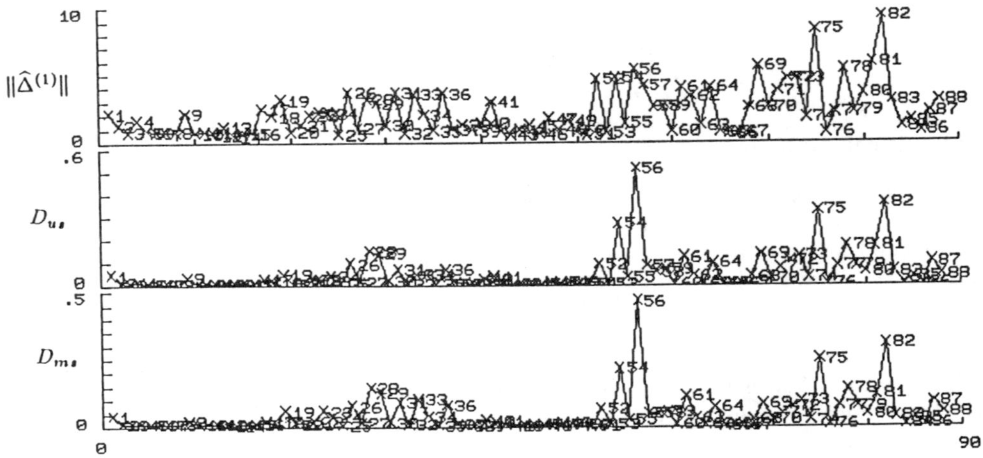

Figure 1a. Index plots of $\|\widehat{\Delta}^{(1)}\|$, D_{us} and D_{ms}. (WLS)

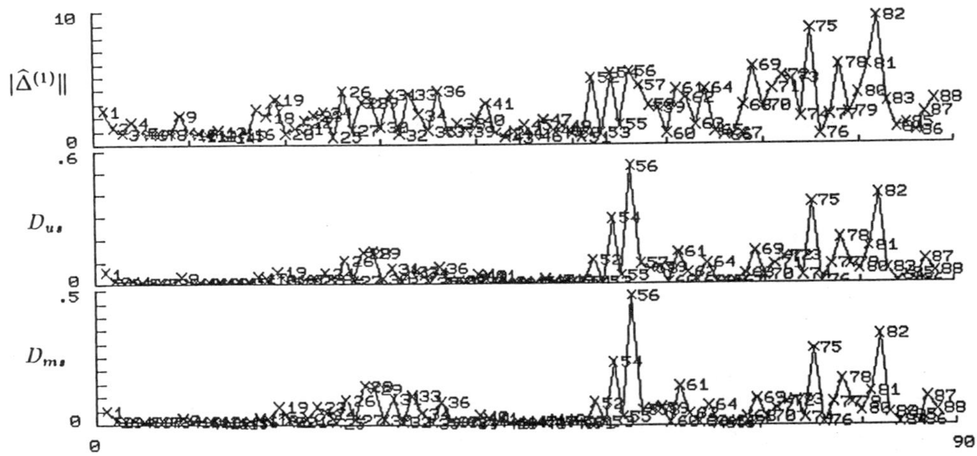

Figure 1b. Index plots of $\|\widehat{\Delta}^{(1)}\|$, D_{us} and D_{ms}. (SLS)

Table 2. Values of $\{\widehat{\Delta}_{jj}^{(1)}\}$ for three influential observations

Individual	$\widehat{\Delta}_{11}^{(1)}$	$\widehat{\Delta}_{22}^{(1)}$	$\widehat{\Delta}_{33}^{(1)}$	$\widehat{\Delta}_{44}^{(1)}$	$\widehat{\Delta}_{55}^{(1)}$	$\|\widehat{\Delta}^{(1)}\|$	D_{us}	D_{ms}
(a)WLS								
82	8.344	-3.773	0.446	1.305	-1.272	9.348	0.362	0.311
75	7.017	-4.106	0.802	1.344	-1.332	8.386	0.329	0.255
56	-1.397	0.458	-1.624	3.905	3.037	5.411	0.519	0.469
(b)SLS								
82	8.275	-4.094	0.502	1.509	-1.582	9.501	0.401	0.335
75	7.024	-4.425	0.846	1.517	-1.580	8.628	0.367	0.283
56	-1.063	0.373	-1.661	3.868	3.201	5.408	0.528	0.480

Table 3 shows the results of WLS and SLS applied to the data set with individuals 75 and 82 omitted and the data set with individual 56 omitted, respectively. As expected from EIC_i's the changes of $\widehat{\Delta}_{11}$ and $\widehat{\Delta}_{22}$ are especially large in the former case and those of $\widehat{\Delta}_{44}$ and $\widehat{\Delta}_{55}$ are large in the latter case.

Table 3. Estimated unique variances and their standard errors based on the data set with individuals 75 and 82 or 56 omitted

Variable	n=86(No.75,82 omitted)		n=87(No.56 omitted)	
	WLS	SLS	WLS	SLS
1	0.1513(0.6011)	0.1516(0.5876)	0.4747(0.1694)	0.4754(0.1693)
2	0.4915(0.1228)	0.4939(0.1224)	0.4095(0.1516)	0.4089(0.1523)
3	0.1723(0.0621)	0.1705(0.0620)	0.2045(0.0571)	0.2045(0.0572)
4	0.3398(0.0885)	0.3335(0.0890)	0.3123(0.0860)	0.3120(0.0862)
5	0.4617(0.0861)	0.4673(0.0874)	0.3957(0.0881)	0.3960(0.0882)

6. Discussion

In this section we discuss two problems. One is the problem related to the validity of our procedure and the other is the relationship with the cases of PFA and MLFA.

First, to confirm the validity of our procedure we investigate the correlation between EIC_i based on our procedure and SIC_i(sample influence curve) defined as

$$SIC_i(\Delta) = -(n-1)(\widehat{\Delta}_{(i)} - \widehat{\Delta}), \quad i = 1, \cdots, n,$$

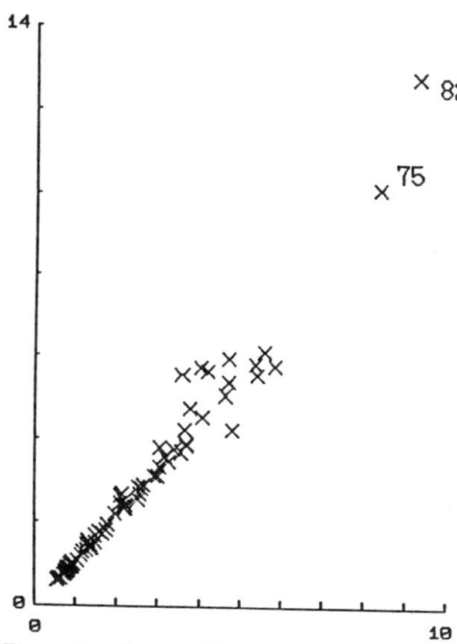

Figure 2a. Scatter diagram of EIC(horizontal) versus SIC(vertical) for the Euclidian norm. (WLS)

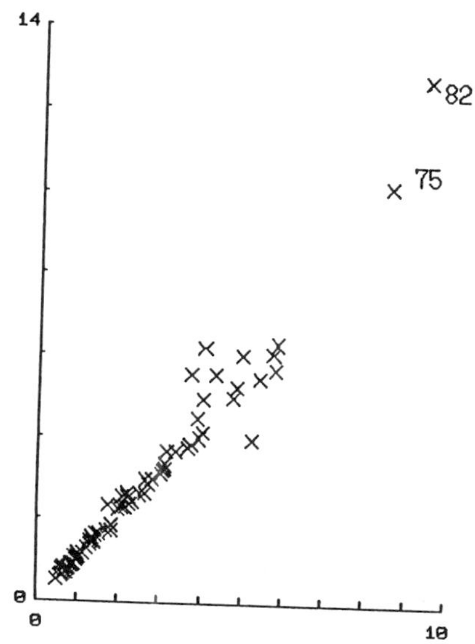

Figure 2b. Scatter diagram of EIC(horizontal) versus SIC(vertical) for the Euclidian norm. (SLS)

where $\widehat{\Delta}$ and $\widehat{\Delta}_{(i)}$ denote the estimates of Δ using the whole data and the data with the ith individual omitted. The measure SIC has an advantage of clear interpretation and a disadvantage of high computing cost. Figures 2a and 2b show scatter diagrams of the Euclidean norms of EIC and SIC for WLS and SLS, respectively. Since the correlations are very high in both cases, EIC can be used in practice instead of SIC.

Now we proceed to the second problem. In our example the estimated $\widehat{L}$, $\widehat{\Delta}$ and also empirical influence curve $\widehat{\Delta}^{(1)}$ are very similar for PFA, MLFA, WLS and SLS(cf. Tanaka and Odaka[13,15]). However, in our experience we sometimes meet with the situations where the results are different among those procedures. We wish to investigate this problem precisely in the future.

REFERENCES

[1] Benasseni, J., Sensitivity of principal component analysis to data perturbation, in: Diday, E.(ed.), Data Analysis and Informatics V(1988).

[2] Browne, M.W., Generalized least squares estimators in the analysis of covariance structures, South Afr. Statist. J. 8(1974) 1-24.

[3] Campbell, N.A., The influence function as an aid to outlier detection in discriminant analysis, Appl. Statist. 27(1978) 251-258.

[4] Critchley, F., Influence in principal component analysis, Biometrika, 62(1985) 531-545.

[5] Kodake, K. and Tanaka, Y., Sensitivity analysis in factor analysis, Proceedings of the Second Japan-China Symposium on Statistics(1985) 137-140.

[6] Lawley, D.N. and Maxwell, A.E., Factor Analysis as a Statistical Method. 2nd ed. (Butterworth,1971)

[7] Mardia, K.V., Kent, J.T. and Bibby, J.M., Multivariate Analysis (Academic Press, 1979)

[8] Okamoto, M., Inshibunseki no kiso [A Mathematical Introduction to Factor Analysis](JUSE Publishing Company,1986) (in Japanese).

[9] Pack, P., Jolliffe, I.T. and Morgan, B.J.T., Influential observations in principal component analysis: a case-study, J. Appl. Statist., 15(1988) 37-50.

[10] Radhakrishnan, R. and Kshirsagar, A.M., Influence functions for certain parameters in multivariate analysis, Comm. Statist., A10(1981) 515-529.

[11] Tanaka, Y., Sensitivity analysis in principal component analysis : Influence on the subspace spanned by principal components, Communications in Statistics, A17(1988) 3157-3175.

[12] Tanaka, Y., Influence functions related to eigenvalue problems and their applications, Contributed Paper, I.S.I. -47th Session, Paris(1989).

[13] Tanaka, Y. and Odaka, Y., Sensitivity analysis in principal factor analysis, Technical Report No.24, Okayama Statisticians Group(To appear Psychometrika)(1988).

[14] Tanaka, Y. and Odaka, Y., Sensitivity analysis in principal factor analysis: A numerical investigation, Technical Report No.29, Okayama Statisticians Group(1988).

[15] Tanaka, Y. and Odaka, Y., Sensitivity analysis in maximum likelihood factor analysis, Technical Report No.34, Okayama Statisticians Group(1988).

[16] Yamane, Y., The influence of the factor analysis, Proceedings of the First Annual Conference of Japanese Society of Computational Statistics 1987 32-35(in Japanese).

DATA BASES

KNOWLEDGE BASED SYSTEMS SUPPORTING EPIDEMIOLOGICAL DATA ANALYSIS

Gianluca Falcitelli

**Istituto di Medicina Sperimentale - Consiglio Nazionale delle Ricerche
Viale Karl Marx, 15 - 00156 Rome, Italy**

Summary

After some notes on the usefulness of a knowledge-based system oriented towards statistical analysis of epidemiological data, the aims, functions and knowledge such a system should have are discussed. Problems of integration between database management, statistical and knowledge management functions are pointed out and some already developed systems are briefly described.
1. Why a knowledge based system for epidemiological data analysis; 2. The basic system data; 3. The system's knowledge; 4. Functionality of the system; 5. Examples of systems used in district health care; 6. Conclusions; 7. Acknowledgments; 8. References.

1. Why a knowledge based system for epidemiological data analysis?

The ongoing trends in a number of countries towards the decentralization of public health management imply the assignment to different types of district structures of a number of health management functions. These functions range from epidemiological monitoring of the state of health of the population, to the ascertainment of risk factors, from the administration and assessment of human and material resources to, what it even more important, namely the programming of health action.

To this end a number of organizations are encouraging the experimental development of information systems to perform epidemiological data analysis which can provide support for practitioners at both the data collection and analysis stage and at that of the production and dissemination of indicators of the state of health, on the use and activity of the services and on the health action carried out in the country.

The aim of the present work is to contribute to defining a sophisticated knowledge based system for epidemiological data management, in particular the management of mortality and morbidity data. An examination will be made of the aspects related to the objectives, functions and type of knowledge that such a system should have. Other aspects, still under investigation, such as the forms in which knowledge can be represented, the interactions between the various knowledge bases, inference procedures and systems architecture, will be treated in subsequent papers.

The difference between a statistical data management system and a more sophisticated knowledge based system lies in the fact that the latter, in addition to its normal DBMS and statistical analysis functions, has the capability, by using suitably formalized knowledge, of providing the user with suggestions as to the steps to follow in the analysis, the interpretation of results and the consequent action to take, whether the latter consists of further analysis or specific action.

The support provided by a knowledge based system can be applied throughout the process of statistical analysis strategy, from the design of the survey to data analysis and result interpretation, following a stepwise approach.

The first stage consists of "intelligent" database management: the imposition of suitable constraints on syntactic and semantic integrity ensures the controlled input of data from individual records, thus avoiding the acquisition of information containing formal errors and/or semantic inconsistencies.

Second, the system should give the epidemiological user indications concerning:

- the choice of survey design from among those available in epidemiology: e.g. longitudinal studies, transversal studies, cohort studies, case-control or experimental studies, which are consistent with the objectives of the analysis as specified beforehand by the users, as well as on the type of data available;
- the choice of the most appropriate and correct statistical functions and indicators in view of the data available and the targets set;
- information of use in interpreting the results;
- any subsequent stages of analysis (data conversion, aggregation, disaggregation, computation of indicators based on converted data, further comparisons, etc.);
- possible specific health action in the field should be taken if certain indicators exceed threshold values.

All these functions are based on a correct formalization of different types of knowledge (medical, statistical, epidemiological-health care knowledge) as well as on the user's capacity to provide the system, in the form of a "short term" knowledge base, with information related to the aims of the study. The production of advices should be the result of the interaction between the different knowledge bases and the data recorded in the system.

2. The basic system data

The basic system data are of two kinds: **microdata** and **macrodata**.

The former consist of individual mortality and/or morbidity records (although they may also take the form of accident, work-related disease or drug-addict records, etc.). These records normally contain the individual's personal data, health data referring to the cause of death or the onset of the disease and other socio-demographic information. These data are obtained from the data sheets normally transmitted by the local health authorities to the country's reference organization (continuous survey), although they may also be in the form of special survey data.

The macrodata consist of aggregated data, mostly in the form of tables, referring to the topic of interest but related to space and/or time contexts with which it could be of interest to make a comparison with individual data. One example of macrodata referring to the causes of death is a table of mortality rates by causes, related to a region or country with which it could be of interest to compare the rates computed at the local level, in order to detect any significant shifts from the mean regional or national values.

Another type of macrodata are represented by basic information, again in aggregated form, which must be taken into account in calculating indicators. One example is the total standard population broken down by age group in calculating the standardized mortality rates, or the number of deaths per age group in the standard population in the case of indirect standardization.

We also consider as macrodata the tables produced by statistical processing carried out by the system on individual records. These tables may be stored in the system and used for further comparisons.

3. The system's knowledge

The system's knowledge can be of at least four different types: knowledge of data (metadata), medical knowledge, statistical knowledge and epidemiological and health care knowledge.

The **knowledge of data** or '**metadata**' consists in the logical characteristics of the microdata and macrodata. In detail, two data dictionaries have to be defined, one for individual records and the other for the reference data stored in tabular form in the system.

For each variable the individual record dictionary must contain information concerning:

- type (dichotomic, nominal qualitative, ordinal qualitative, quantitative);
- length (n. of modalities);
- list of modalities or range of values allowed.

The dictionary of tables must classify the reference tables or, even those resulting from previous processing of individual data, according to:

- content;
- classification criteria or variables;
- table size (n. of rows x n. of columns);
- modality of the variables used as classification criteria (granularity).

Medical knowledge is defined as the sum total of clinical information referring to the phenomenon under investigation which is used to record the case histories.

This knowledge is then transformed into syntactic and semantic integrity constraints which will guarantee the consistency of the information.

The syntactic integrity constraints represent the formally correct set of values that a given variable can have. One example of a syntactic integrity constraint is the one which avoids the sex variable being assigned values other than the allowed M or F.

The semantic integrity constraints are used to define the set of semantically correct values which can simultaneously be taken by two or more variables. In other words, these constraints determine the 'nonsense' areas represented by the copresence of variable values which, although syntactically correct (in the sense that each value lies in the range of values possible for the variable concerned), are meaningless. One example of a semantic integrity constraint is one which prevents the system from accepting reciprocally inconsistent items of information, such as cancer of the prostate being the cause of death in a female subject.

These constraints are active both at the stage of data input, updating and editing, serving to validate the data before they are actually written to the file and preventing the entry of incorrect data, and at the query stage, in order to avoid the pointless implementation of selection and/or display routines. Both types of constraint require a representation and storage system constructed by defining a suitable knowledge base that can also take the form of a data dictionary.

Statistical knowledge consists in rules for applying statistical functions (indicators). Exact rules exist for determining the advisability of using one type of statistical analysis technique or another, together with other rules to indicate the formal possibility of applying it. In both cases these rules are a function of the type of data and of the aims of the study.

A knowledge base should be set up which contains a minimal set of application rules for each of the techniques. By means of interaction between this knowledge base and the dictionary of data to be analysed with the aims of the study and with the study design, the system can suggest which statistical techniques should be correctly applied and/or evaluate the formal and semantic correctness of the application of a given technique (or indicator calculation).

Epidemiological and health care knowledge consists in the set of heuristic knowledge pertaining to the phenomena being investigated, which can be of help in interpreting the observed data. In other words, a survey of the phenomena is envisaged, the interpretation of which may lead to more detailed analysis or the adoption of specific health measures. To achieve this a thematic knowledge base must be organized which contains rules of behaviour for the epidemiologist related to the case in which the above phenomena occur. The interaction between this knowledge base and the results obtained by processing the data can provide the user with useful advices as to what to do next.

It is precisely this epidemiological and health care knowledge which underlies the choice of study design. What must be done in this case is to organize a knowledge base which contains all the types of study likely to be implemented, each of which is classified according to the application rules (aims of the study and type of data). Through the interaction between this base, the user's aims and the type of data available the system will be able to suggest the type of study design best suited for the purpose.

Of course all these knowledge bases can be expanded in time and so the field of application of the system can be extended also to other health and non-health sectors.

4. Functionality of the system

The originality of the system consists in the integration of three different types of function which are normally not present simultaneously in other systems designed to support epidemiological data analysis:

- Database management functions;
- Statistical functions;
- Knowledge management functions.

Taking into account the design stage reached and the applications sector for which the system will be initially intended, i.e. mortality and morbidity data, the following functions have so far been identified for each type.

Database management functions

This type of functions includes: input, updating, deleting, selection and projection of individual records and reference tables.

Statistical functions

The statistical functions envisaged are those referring to the calculation of common epidemiological indicators used to study mortality and morbidity, plus a set of elementary statistical functions which can be used to perform comparatively unsophisticated "ad hoc" analyses.

The epidemiological indicators envisaged are:

Mortality: crude mortality rates; standardized mortality rates broken down by age, sex, cause (direct and indirect method); proportional mortality rates by cause; specific mortality rates by age, sex, cause.

Morbidity: specific morbidity rates; incidence rates; prevalence rates; attack rates; relative risk and possible estimates; assignable risk.

Other statistical functions may consist of: statistical description of a variable (mean, max, min and variance); coefficients of correlation; regression; construction and analysis of two-way and multi-way contingency tables; *chi square* test; Fisher test.

Conventionally included among the statistical functions are also:

- graphic representations of data;
- procedures to compare processed data with reference data.

It is common practice in epidemiology to compare the observed trends in given phenomena with expected trends or trends in other geographic areas in order to identify any unusual pathological phenomena (the socalled "sentinel events") requiring more detailed and specific action. An example of this function is the comparison between mortality rates due to a given cause observed in the area of a local health authority (USL) and the mortality rates for the same cause at the regional or national level.

Knowledge management functions

As regards knowledge management functions, the system should be able to provide the user with advices on:

- the choice of study design (case-control, longitudinal, transversal, cohort, experimental studies);
- the choice of indicators and statistical functions to use;
- the choice of reference data with which to compare the processed data;
- the performance of further data analysis steps;
- special action to be taken with reference to an ascertained significant anomaly (more detailed investigations, health care action).

This type of functions include the set of knowledge management functions used for the syntactic and semantic validation of the data and which come into operation at the stage of individual record input or when accessing the database.

The implementation of these knowledge management functions is directly dependent on three factors:

- the type of knowledge available to the system;
- the forms in which this knowledge is represented;
- the modes of interaction between different knowledge and inference blocks (what is known as a "inferential engine" in an expert system).

In other words, the inference mechanism, and consequently the generation of advices, is implemented by means of interaction between the various knowledge sources or bases, i.e. the aims of the study (as communicated interactively by the user), the logical characteristics of the available data (microdata and macrodata), the type of statistical functions and their formal and semantic applicational requirements, and the epidemiological and health care knowledge concerning the phenomenon.

Only through interaction between these various kinds of knowledge will the system be able to state whether or not, for instance, in a case-control study, the calculation of the relative risk is significant. Or whether, when mortality data aggregated according to certain criteria are available, it is possible to make comparisons with differently aggregated reference data. Or again, whether it is legitimate to perform regression analysis, compute coefficients of correlation, carry out non-parametric tests, etc., when data with certain characteristics are available.

5. Examples of systems used in district health care

The following is a brief outline of several support systems for the treatment of epidemiological data. These systems have already been developed and are being used by several health care structures.

The first two systems are above all database management tools, one of which has also been integrated with statistical functions. Several medical and epidemiological and health care knowledge management functions are included, although they do not contain all the functions outlined above.

These systems in any case represent two significant examples of a tendency towards the combined use of database management, data analysis and knowledge management techniques in the health field.

The first system, known as EPISOFT, is a dedicated system for automated mortality data management. It was jointly developed by Istituto sulla Ricerca e Documentazione Scientifica (ISRDS) of the Consiglio Nazionale delle Ricerche (CNR) and the Osservatorio Epidemiologico della Regione Lazio and is currently in use at several local health authority centres in Lazio.

According to the system for recording the causes of death currently in use in Italy, each Local Health Unit (USL) is required to keep a register of the causes of death which is ordered by name, aggregated at municipal level. Automation of this information system is expected to allow mortality data to be used for health care planning and epidemiological research purposes.

The basic data handled by EPISOFT are the individual records of the cause of death, which contain the individual's personal data plus the medical data referring to the cause of death. In addition to the management and processing of individual records the system also allows the data contained in it to be processed by means of simple graphic and statistical procedures.

EPISOFT has a menu-driven interface which guides the user through the choice of functions. The records are subdivided into two categories: Case history registers and Working registers. The former represent the basic data, and contain the full set of individual records; the latter are an sub-set of interest built up from the original data base by means of selection and projection operators.

EPISOFT comprises a suite of programmes running on an IBM PC or compatible and based on the DBase III file management format.

One line of development of the system, which is still being developed, involves the inclusion of elementary knowledge bases referring to the data.

The second system is the Lombardy Tumour Register (Registro Tumori Lombardia), currently used in the province of Varese. The aim of the register is to record all cases of tumour occurring in the province.

The Tumour Register receives from the hospitals in the province of Varese a personal data-cum-clinical card for each admitted patient suffering from a malignant tumour. The file is divided into individual records each containing a nucleus of essential information for every case of tumour, updated for each new report and in a form that can be rapidly consulted for the generation of statistics, each of which is linked to variable number of detail records, one for each report received from the register or related to the same case.

The system is implemented on a Digital PDP 11/34 computer with 256 Kb of core memory, a mass storage capacity of 200 million characters and an 800 million character tape unit. The software is written in MUMPS, a product developed by the Massachussets General Hospital of Boston and particularly well suited to the management of medical databases running on a mini-computer.

The most interesting aspect of the system is the management of the knowledge related to the topography and morphology of the tumours. In addition to the normal operations of information retrieval from a database, the system also allows a wide range of checks to make on the input data, thereby avoiding the acquisition of incorrect information, as well as providing automatic tumour coding based on a natural language description contained in the input card.

This entailed the construction of a knowledge base using the information referring to the topography and morphology of tumours, which defines the integrity constraints and enables information to be coded automatically.

At the present time the statistical aspects of data treatment have not been fully integrated into the system. It is possible to perform statistical processing using commonly available statistics packages provided that the data have been converted to the required format.

The two systems described above involve the integrated use of database management functions and knowledge management functions. These are sometimes enhanced by statistical functions in order to ensure a more effective utilization of the data for health care and epidemiological research purposes. Nevertheless, in the versions described, these systems seem to be more suitable for preparing epidemiological research and planning health care action than to themselves represent a direct support for this type of action. In other words epidemiological research seems to some extent to lie outside the scope of such systems.

An example of decision support system for epidemiological studies is the MICROIDEA system. The system, still now at the experimental stage, has been developed at the Istituto di Scienze dell'Informazione of the University of Bari, in cooperation with the Epidemiology Laboratory of the Istituto Superiore di Sanità of Rome.

MICROIDEA allows an expert epidemiologist to build consultation models representing the reasoning process followed in specific sectors of epidemiological data analysis. An "unexpert" user may then choose one of the available models, to be guided in the analysis of a specific data set. A first version of MICROIDEA has been used to handle and to analyse data in various types of studies, and has been experimented in some Local Health Unit (USL): it requires very little training for effective use, even for someone who has no experience of computers. When completed and properly tested, the system will make available guidelines on how to perform data analysis in sectors of particular relevance: protocol of inquiry into epidemics of infectious diseases, analysis of mortality data, etc. It will therefore give, to the Regional Health Authority, the possibility of defining a uniform approach to the Annual Health Status Report, at least for the analysis of routine data.

MICROIDEA has been implemented in compiled BASIC under MS DOS, for what concern the database management functions. The knowledge acquisition and advice modules are written in micro-LISP. The system works in MS DOS environment on IBM and compatible PC and requires a minimum of 256 Kbytes of user memory and a hard disk unit.

An interesting example of system integrating statistical functions and database management functions oriented towards epidemiological studies is EPIINFO. This software, non commercial, has been developed by the Centers for Disease Control of Atlanta and it is particularly well suited to the questionnaires data management.

On interesting point of EPIINFO is the easy way to manage the integrity constraints, allowing the user to easily define, check and modify the range values allowed for each variable, as well as the semantic links among variables in the questionnaires. Another valuable feature is the possibility of utilize as input data format the questionnaire itself, withouth predefining a different format.

Powerful statistical analysis are available, also allowing the user to construct "ad hoc" new analysis programs. An interesting development of the system could consist in adding functions to manage statistical databases, i.e. databases containing aggregated data such as statistical reference tables.

The system is working on IBM and compatible personal PC and modules are written in Turbo-Pascal.

6. Conclusions

The study of the interaction between database management functions, statistical functions and knowledge management functions, also by means of artificial intelligence techniques, can lead to significant progress being made in the field of epidemiological data handling. The development of knowledge based systems can be of great help to the epidemiologist in his research as well as to the planner in managing health policies.

Nevertheless the principle on which such systems are based, although highly sophisticated and capable of reproducing part of the human reasoning process, will never allow them to replace the user in the decision making process. Despite the great progress made, they will only be able to guide the user in analysing and interpreting results and in making his decisions. Such systems can indicate a range of possible interpretations and decisions based on several established formal and semantic criteria, although without necessarily exhausting all available possibilities. The final decision will always be made by the user, who will base it on a body of knowledge that is too vast to be represented by a system.

7. Acknowledgments

The present paper is based on an on-going research project by the Istituto sulla Ricerca e Documentazione Scientifica (ISRDS) of the CNR. I wish to express special thanks to Ing. Fabrizio Ricci, director of the above project, and to Dr. Carla Basili for valuable discussions and their precious comments.

8. References

Arcà M. et al., La riorganizzazione del sistema informativo sulla mortalità per causa: la sperimentazione avviata nella Regione Lazio, Medicina e Informatica, N. 3, 1985, Il Pensiero Scientifico Editore, Roma

Armitage P., Il ruolo della statistica nell'epidemiologia, Quaderni di Statistica Sanitaria, II, 2, 1979, Università di Roma

Basili C., Falcitelli G., Ricci F.L. "Il Progetto EPITOOL: un sistema basato su conoscenza per l'analisi di dati epidemiologici", Rapporto Tecnico CNR/ISRDS/RT 1/1989

Colton T., Statistica in Medicina, Piccin Editore, Padova, 1979

De Rosis F., Franich A., Pizzutilo S., Ricci F.L., Handling and analyzing epidemiological data by microcomputer, Lectur notes in Medical Informatics, 24, Springer-Verlag, 1984

De Rosis F., Pizzutilo S., Greco D., MICROIDEA: improving decisions in epidemiological analysis by a microcomputer, in Medical Informatics, Vol. 11, No.3, 1986

Falcitelli G., STIPAM: a system for preliminary management of statistical information oriented towards multivariate data analysis, in Cinquièmes Journées Internationales Analyse des Données et Informatique - Posters, pp. 95-98, INRIA, Versailles, 1987

Falcitelli G., Ricci F.L., Intelligenza artificiale e statistica, Rapporto interno CNR/ISRDS/RI1/88, Roma, 1988

Fortunato E., Rafanelli M., Ricci F.L., Sebastio A., Un modello logico per rappresentare e manipolare tabelle statistiche, Atti di "Giornata di studio su basi di dati statistici", Roma, 1987

Hand D.J., The application of expert systems in statistics, in PHELPS B. (ed.), Interactions in artificial intelligence and statistical methods, Gower Technical Press, Aldershot, 1987

Regione Lombardia - Settore Sanità e Igiene, Incidenza dei tumori e cause di morte in Lombardia, in Notizie Sanità n. 10, 1985

Ricci F.L., Rossi-Mori A., L'impatto futuro dei sistemi basati su conoscenza in sanità, in Medicina e Informatica, Il Pensiero Scientifico Editore, N.2, 1988, Roma

Thisted R.A., Representing statistical knowledge for expert data analysis systems, in Gale W.A. (ed.), Artificial Intelligence and Statistics, Addison-Wesley, Princeton, 1986

A SOFTWARE TOOL FOR PROGRAMMING, TEACHING AND USING DATA ANALYSIS METHODS

A. Hammad*, J.Jockin*, M. Junet**, M. Leonard**, B. Sadeg*, Y. Schektman* and D. Vielle*
*AEGIDE - GRECO CNRS 59 - Université Toulouse le Mirail
5, Allée A. Machado 31058 Toulouse Cedex (France)
** Centre Universitaire d'Informatique (C.U.I)
12, Rue du Lac 1207 Genève (Switzerland)

Key Words : data analysis, matrix calculus, linear algebra calculus, data base management system, software engineering.

1- GENERALITIES

1-1- Who is concerned ?

The SPRADE software described here is used in our University:

(i) by researchers for programming the algorithms of the methods they design. The researchers can use algebraic expressions of matrices which are compiled and valued by the system and many other facilities,

(ii) by students who master linear algebra tools (inner product, euclidean distance, orthogonal projection, ...) and who learn data analysis methods; They can use directly high level operators, such as the orthogonal projection, on small data sets, or get to know more surely these high level notions by programming these operators using simpler functions (matrices multiplication, inversion,...) and finally they can thoroughly master data analysis methods by doing their own programming easily and quickly.

(iii) by users who want a simply and friendly tools for analysing their data on small compatible micro-computers. By using an oriented data analysis system [12] together with a relational data base management system (DBMS) ECRINS [9] in a single system 'SPRADE', the users' data analysis process is made easier.

1-2- What is offered ?

SPRADE provides :

(i) conversational programs for managing data in a relational data base offering basic, associative, complementary relations and subrelations,

(ii) primitives for defining and updating, during a session, a data structure of the data base ,

(iii) extended matrix calculus primitives with the possibility of specifying six kinds of matrices(diagonal, symmetrical,..),

(iv) linear algebraic primitives,

(v) data analysis primitives (mean, correlation, contingency table, ...),

(vi) functions for handling arrays (assignment of names to rows and to columns, concatenation of two arrays, extracting or updating a row or a column or a diagonal, deleting an array,...),

(vii) graphic primitives, ...

1-3- Some miscellaneous informations

Programs and primitives provided by SPRADE are written in FORTRAN 77 and are running on compatible IBM-PC. Users' primitives and system utilitaries are FORTRAN integrated subroutines gathered in nine libraries : they represent about nine hundred subroutines and more than one megabyte of FORTRAN code. SPRADE is a nested system, so it can be used at different levels more or less powerful : we may use it with or without the connection with the DBMS, we may or may not use its compiler of algebraic expressions of matrices, ...

SPRADE belongs to a software family designed for aiding users who have well specifyed problems to solve and who need high level and user-friendly tools. It has similarities with and some differences from softwares like EDA [8], GAUSS [5], CALCUL MATRICIEL [4], S [1], SAS [11], ... Heir to LASEN [13] and LOM [10], we may consider that SPRADE materializes the present state of the researches of our team in this area. So, SPRADE has its own characteristics and particularities comparatively to the softwares previously referred to.

2- PRINCIPAL ELEMENTS OF THE SPRADE SYSTEM

SPRADE is composed of three fundamental elements (fig. 1) :

(i) a programming system for data analysis (SPRAD),

(ii) an evolving relational DBMS (ECRINS-DRT),

(iii) a set of interface primitives for connecting the two previous systems.

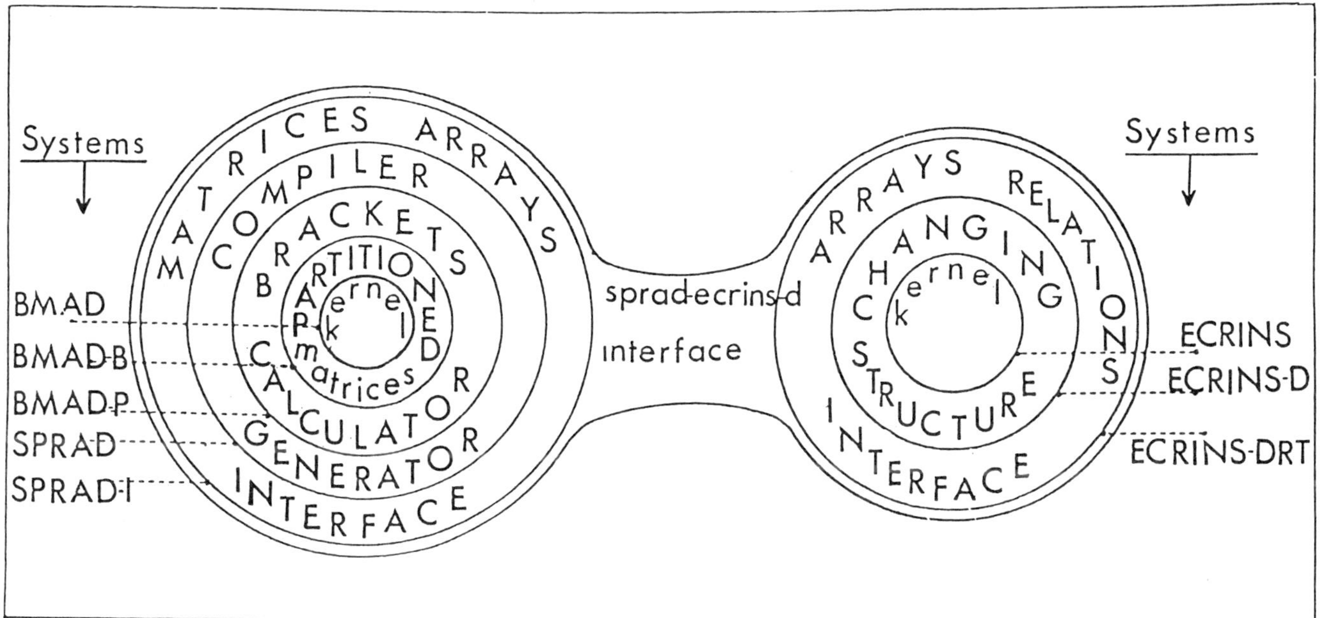

figure 1 : Software Layers of SPRADE

2-1- The SPRAD system

It is composed of a kernel (BMAD [7]), wrapped up by four software layers. As far as the user is concerned, the BMAD kernel provides all mathematics, data analysis, ... primitives which we briefly discussed in the preceding section. It consists of two FORTRAN libraries, providing all the scientific tools for writing programs. In a program, the BMAD subroutines are called in the same way as the pocket calculator keys (without brackets) are pressed : with a pocket calculator, a current scalar is updated by each function key and displayed; the corresponding concept in BMAD is a current matrix .

The first software layer (PARTITIONED MATRICES) gives the possibility of partitioning matrices by blocks. The primitives of this layer are useful for programming methods which analyse simultaneously the associations between several sets of variables.

In a program using the facilities of the second software layer (BRACKETS CALCULATOR), subroutines are called in the same way as the keys of a pocket calculator with brackets are pressed; so this layer provides, in particular, subroutines whose call corresponds to brackets opening or closing, to an expression beginning,....

The third software layer (COMPILER-GENERATOR) provides tool aids for programming : a compiler of the algebraic expressions of matrices and a generator of READ and WRITE FORTRAN instructions, ... At this stage, the SPRAD system (called SPRAD without E) generates a FORTRAN program, calling, in particular, primitives belonging to the system underneath named BMAD-P.

The fourth software layer (MATRICES <--> ARRAYS interface) is concerned with the primitives which transform a data matrix, whatever its kind (diagonal, symmetrical, ...), into a two-dimensional FORTRAN data array. If the matrices rows or columns have names, then these primitives store also these names into one-dimensional FORTRAN arrays.

2-2- Evolving DBMS

The kernel of this part of SPRADE is the relational DBMS ECRINS. It is now only wrapped up by two software layers. ECRINS supports data bases with complex structures. Its data model is built from the first relational model [3], completed by the subrelation notion [14] with its inheritance mechanism, the associative relation notion [2] and the complementary relation defined in the CODASYL network model. ECRINS is an object oriented data base system. For a given application, as it is usually observed, the internal structure, in particular the data base dictionary built by ECRINS does not vary. So, during a session, we cannot create a new relation or a new attribute, delete an existing relation,....

This restriction is extremely awkward in a data analysis process because each data analysis program builds new data arrays i.e. new relations. For example, a basic relation being associated to a given data array :

(i) the elements of the principal components array are stored in a complementary relation to the basic relation,

(ii) the values of the indicator variables defining a clustering are stored in a subrelation to the basic relation,

(iii) the values of a new variable built by coding are stored in a new attribute, etc...

Consequently, the first software layer around ECRINS (CHANGING STRUCTURE) provides primitives for updating the internal structure of a data base during a session. When these primitives are running, the database dictionary is obviously modified and the data are often modified as well. Therefore the database may stay in a critical state during a relatively long time.
To avoid losing the database should any unexpected interruption (power failure,...) happen, the ECRINS-D system is programmed to take into account these difficulties : the security is ensured by automatically saving the data structure to be updated and by restoring the state the database was in when the breakdown happened.

The second software layer (ARRAYS <--> RELATIONS interface) is concerned with the primitives which transform a two-dimensional FORTRAN array into a relation of the required kind (basic, complementary, ...)

2-3- Interface between SPRAD and ECRINS-D

Generally, Data analysis softwares, use a file management facility to store data. It is well-Known that this facility is not user-friendly. It is better to use a DBMS [6,14] because :

(i) when a data array (i.e. a relation) is required, the names of the variables (attributes) and the names of the individuals (entities) come with it automatically,

(ii) during the interpretation stage in a data analysis process, the values of the initial data or the statistics built can be asked in a very user-friendly way.

Consequently, the "SPRAD <--> ECRINS interface" primitives use those of the fourth software layer of the SPRAD system and those of the second software layer of the ECRINS-D system to transform (via FORTRAN arrays) a matrix of any kind into a relation of the required kind and conversely.

3- ILLUSTRATIVE EXAMPLE

A little program written in SPRAD is given in figure 2 : the statements beginning by the letter 'C' are usual comments and the statement beginning by the letter 'P' is the matricial algebraic expression of the multiple linear regression coefficients to be compiled, where :

(i) the elements of CRL are the coefficients,

(ii) the column elements of the matrix X are the predictor values,

(iii) the elements of the column matrix Y are the predictant values,

(iv) Dw is the diagonal matrix of individual weights,

(v) INV, COV1, COV2 are the SPRAD primitives processing the inverse of a matrix and the covariances between one or two sets of variables.

During a session, first the system asks for informations (in a conversational mode) such as the kind, the dimensions, the elements of the matrices, ... (fig. 3) . At the end of this dialogue, SPRAD generates the equivalent FORTRAN program which will be linked with the BMAD-P library and with those which are underneath. In figure 4, we give two pieces of the generated program corresponding, respectively, to a part of the previous dialogue (fig. 3) and to the beginning calculus of CRL : the primitive YDE corresponds to an expression beginning, YOP to a bracket opening, YFP to a bracket closing, YEX to a matrix handling, ...

```
C
C      CALCULUS OF THE MUTIPLE LINEAR REGRESSION COEFFICIENTS
C
P      CRL = INV(COV1(X , Dw)) * COV2(X , Y , Dw)
C
```

fig. 2 : A program in SPRAD system

```
.
.
.

 P      CRL = INV(COV1(X , Dw)) * COV2(X , Y , Dw)

"X"        matrice    ?  (o/n)      o
declaration de X      ?  (o/n)      o
son type              ?  (o/n)      quelconque
nombre de lignes      ?             n
nombre de colonnes    ?             p
lecture de X          ?  (o/n)      o
noms des lignes       ?  (o/n)      o
noms des colonnes     ?  (o/n)      o
affichage de X        ?  (o/n)      o
.
.
.
```

fig. 3 : A piece of dialogue with SPRAD system

```
.
.
write(*,'(a\)') 'lecture de n     : '
read(*,*) n
write(*,'(a\)') 'lecture de p : '
read(*,*) p
call ydq(X, n, p)
write(*,*) 'elements de la matrice X'
call ylirm(X)
write(*,*) 'noms des lignes de la matrice X '
call ynoml(X)
write(*,*) 'noms des colonnes de la matrice X '
call ynomc(X)
write(*,*) 'elements de la matrice X'
call yecrm(X)
.
.
.
call yde
call yop
call yop
call yex(X)
call yfp
call ycovl(Dw)
call yfp
call yinv
.
.
.
```

fig. 4 Two pieces of a program calling BMAD-P primitives

REFERENCES

[1] BECKER & CHAMBERS (1981) "S" A language and system for Data Analysis . BELL Laboratories.

[2] CHEN,P.P.S (1976) The entity Relationship Model : towards a unified view of data. ACM TODS 1,1.

[3] CODD,E.F. (1970) A relational Model for Large Shared Data Banks. CACM, 13, 6.

[4] ESCOZAS,P. & COUDERT,O. (1989) Calcul Matriciel, edition Langages et Informatique.

[5] GAUSS Mathematical and Statistical System. APTECH Systems, inc. 26250 196th S.E. Kent, WASHINGTON

[6] GHOSH,S. (1989) Categoriacl Numerical Relational Operations for Statistical Database Management. In Statistical Software Newsletter. Vol. 15, N 1, 6-8.

[7] HAMMAD,A. JOCKIN,J. SADEG,B. SCHEKTMAN,Y. VIELLE,D. (1987) Bibliothèque Mathématique pour l' Analyse de Données BMAD. In "Data Analysis and Informatics 5th". Paris, INRIA ed. 13

[8] HORBER,E. (1983, 86) EDA : User's Reference Manual, Version 1.9

[9] LEONARD,M. FALQUET,G. JUNET,M. (1986) An extended Entity-Relationship Database Management System and its Semantic Query Language. In Proceedings of the ACM 12th International Conference on Very Large Data Base, Japan.

[10] LEONARD,M. SCHEKTMAN,Y. GALLAND,A. IBRAHIM,A. JOCKIN,J. SNELLA,J.J. TSCHOPP,D. VIELLE,D. (1982) Aided Design and use in Data Analysis Computing. Statistical Software Newsletter. Neuherberg 8,1. 3-14.

[11] SAS Procedure Guide for Personnel Computers, Version 6 ed .Cary.

[12] SCHEKTMAN,Y. HAMMAD,A. JOCKIN,J. SADEG,B. VIELLE,D. (1988) Some General ideas on a software environment for programming and using Data Analysis Methods. In COMPSTAT Denmark p.89-90.

[13] SCHEKTMAN,Y. PASQUIER,P. (1972). Un nouvel outil à la disposition du medecin du travail : le Langage de Depouillement d'Enquète (LASEN). Revue de Medecine du Travail, 1, 2 p.63-69.

[14] SMITH,J.M. SMITH,D.C.P. (1977) Database Abstractions : Aggregation and Generalization . ACM TODS, 2, 2.

[15] SVENSSON,P. (1989) Database Management Systems for Statistical and Scientific applications : are commercially availlable DBMS good enough ? . In Statistical Software Newsletter. Vol. 15, N 1, 15,17

DATA ANALYSIS AND MODELS

DATA ANALYSIS AND MODELS

Kenji Nagasaka
University of the AIR
2-11, Wakaba, Chiba-Shi
Chiba, 260 (Japan)

ABSTRACTS

Emil Borel pointed out the importance of the classification of zero measure sets and proposed the notion of rarefaction. This notion was developed by M. Frêchet and its theoretical basis are well determined. Recently, the author treated again the rarefaction and mentioned its statistical comprehension. Recent developments of the rarefaction will be presented with the results by C. Flamme, W. Just and the author, and their applications on statistical inference will also be discussed.

Second, the author presents two examples of real data. The first example will be a result of sample survey, which was done twice in 1986 and 1988. The used sample designs are almost identical but we found different response rates. We discuss problems for ambiguous and/or difficult questions, applied techniques for getting high response rate, and the role of the incentive explaining different response rates. The second example will be the famous first digits problem, which is sometimes called as the peculiar Benford's law. We analyse this phenomenon and give a simple proof for the logarithmic distribution of leading digits of n-th convergents in simple continued fractions of real quadratic irrationals.

1. STATISTICAL MODELS AND ZERO MEASURE SETS

In the very beginning of statistics, we classify variables into two categories: qualitative or quantitative variables. Among quantitative variables, we distinguish discrete (categorical) variables and continuous variables. Once we set a variable to be continuous, it can be continuous only in mind. Indeed, we are confronted only by finite decimals in statistical data handling, which are at most countable, and then, of course, of measure zero with respect to Lebesgue measure. Actually we consider these data as observed values of continous variables, which is a usual statistical modeling. Thus this statistical modeling seems to be based upon a certain Platonism, which was seriously criticised by L. Wittgenstein [19]. This is the starting point of our research of investigating sets of measure zero.

The class of sets of Lebesgue measure zero is a non-empty σ-ring, and the completeness of this measure asserts that the cardinality of the class of Lebesgue null sets is $2^{\mathfrak{c}}$ where $\mathfrak{c}$ denotes the power of the continum. Therefore the class of Lebesgue null sets is fairly vast and its structure seems to be complex. A. A. Cournot observed that it should be possible to establish a classification of the class of Lebesgue null sets in his publication: *Exposition de la Théorie des chances et des probabilités*, published in 1843.

One century later, E. Borel devoted one chapter in [2] to this subject giving four intuitive definitions of rarefaction. In sixties, his ideas have been examined and deepened by Choquet, M. Fréchet [5], Z. Moszner and L. Urbanek [13].

Recalling first the idea of Borel: Let X , Y be subsets of the unit interval [0, 1], let $\bar{I} = \{I_n\}_{n=1}^{\infty}$ be an open *cover* of X , that is, $\bar{I}$ is a sequence of open intervals such that every $x \in X$ is contained in infinitely many of the intervals I_n , and let $\bar{J} = \{J_n\}_{n=1}^{\infty}$ be an open cover of Y . We say that X is more rarefied than Y if and only if there exits a cover $\{I_n\}_{n=1}^{\infty}$ of X such that for every cover $\{J_n\}_{n=1}^{\infty}$ of Y ,

$$(1) \qquad \lim_{n\to\infty} \frac{\sum_{k=n}^{\infty} |I_k|}{\sum_{k=n}^{\infty} |J_k|} = 0 ,$$

where $|I|$ denotes the length of the interval I .

The above definition is due to Fréchet, a refinement of the original Borel's definition. Fréchet proved in [5] that the Cantor set, and in fact every closed null subset of the unit interval is not less rarefied than a one-point set.

We say that X and Y are of the same order of rarefaction if X is not more rarefied than Y and Y is not more rarefied than X. $\mathfrak{D}$ represents the family of linear null sets in the unit interval with the same order of rarefaction as a countable set. The author [15] shows that $\mathfrak{D}$ contains all nowhere dense sets and also every set of first category with its closure being first category. This signifies that thin sets in the sense of topology are also rarefied sets of the same order of rarefaction as a countable set.

These results leads naturally to the question: Do there exist two nonempty sets X and Y in the unit interval of measure zero such that X is more rarefied than Y?

Claude Laflame [12] extends a definition of rarefaction of null sets due to Borel and shows that a countable set is more rarefied than the Cantor set. Winfried Just gives an affirmative answer to the above question with respect to the definition of rarefaction due to Fréchet.

Original definition of Borel is as follows: X is more rarefied than Y if and only if (1) holds. In order to convert this Borel's intuitive idea into a precise mathematical concept, we need to fix specific covers or to quantify over $\overline{I} = \{I_n\}_{n=1}^{\infty}$ and $\overline{J} = \{J_n\}_{n=1}^{\infty}$. Fréchet's refinement is a choice of quantifiers over $\overline{I}$ and $\overline{J}$. To make precise the expression "not less rarefried" and values of the limit:

$$\limsup_{n\to\infty} \frac{\sum_{k=n}^{\infty} |I_k|}{\sum_{k=n}^{\infty} |J_k|} < \infty , \tag{1}$$

we might also consider all possible choices.

Just and Laflamme [10] propose their variant of the definition of "not less rarefied" and study the structure of sets of measure zero under the resulting partial pre-order. This structure looks arbitrarily complicated if the Continum Hypothesis or Martin's Axiom is assumed. Under different set-theoretical assumptions (related to Blass' NCF principle), the structure becomes simpler and elegant, which indicates the reevaluation of the asumption of the Continum Hypothesis in statistics and probability.

Axiom of Choice is quite essential in measure theory. Indeed, this axiom is indispensable to construct a well-known example of Lebesgue non-measurable sets. Every set is measurable under a certain axiom. It is desirable that we can attribute an appropriate probability for every event in consideration. Martin's Axiom is consistent with ZF+AC. Suppose that we replace Continum Hypothesis or Martin's Axiom by different set-theoretical assumptions related to Blass' NCF principle. If this replacement asserts the measurability for every set, it could be better to adopt these assumptions in statistical inference.

2. SURVEY RESEARCH AND CHOICE OF SURVEY METHOD

In May 1986, the survey research statistics group in the Faculty of Education of Shinshu University and Hachijuni Culture Foundation executed a sampling survey entitled "Survey on Province and Culture" for adults living in Nagano Prefecture in Japan [16]. The purpose of this survey research is to make clear the Nagano Prefecture inhabitants' characteristics from the actual data and to glasp the real situation of cultural activities and the prospect of them. First, we consider a problem of questions.

i) AMBIGUOUS AND/OR DIFFICULT QUESTIONS

In social survey research, the good questionnaires are absolutely indispensable and they must be short and made attractively layout [3]. It is indeed easy to say but difficult to do actually. In our survey research, we succeeded in elaborating the wordings of most questions but it still remains some ambiguous and/or difficult questions. Here we discuss questions of these types.

The first question S to discuss is the follwoing:

S: "What is the main factor to success in social life ?"

The highest answer is "the efforts" 56.7% as usual. The percentages of the answer "academic career" is only 2.8% in our survey research. But this low response rate for "academic career" is very common in many social survey researches in Japan. If you pose the same question S in U.S.A. or in European countries, the results would be different. That is, the response rate for "academic career" is higher than in Japan.

One of our research group analyzed this question S together with several questions on educational problems by MDA method. By MDA method, he grouped the following questions:

1. Academic career is the main factor for success in social life.
2. The talent is tha main factor.
3. The efforts are the main factor.
4. To save money for children's education.

From the intimacy among above questions analyzed by MDA, he concluded that the answer distribution of that question S did not represent the real situation. Thus he still believes that Japan is a society of academic career in spite of the result of the question S.

We agree that the resulting distribution of the question S does not represent the real situation, since the question S is too formal. But "academic career" is not at all considered as the main factor for the success in social life. In order to investigate the real situation, we must break down the question S, specify the situation and make the question clear.

Other questions are the following three questions of which the answer distributions are quite different from other survey researches:

Question M: Will you save money in order to educate your children?

Question O: Do you obey what the old persons said even if you do not completely agree?

Question W: Do you think that the word "stranger: yosomono" still has the real meaning in your living area?

These questions are found in survey research executed by NHK in 1978 by face-to-face interview method. We copied these three questions in our questionnaires in order to compare the results. The response rate of our survey research for "Yes" of the three questions M, O, W are lower than those of NHK.

We want to interpret these differences as follows: The first reason is the difference of survey method between mail survey and face-to-face interview. It might be difficult to say "No" toward interviewers. But we suppose that another reason mainly produced the differences. In questionnaires of NHK, questions for the similar subject are grouped, thus it would be quite possible that the answers of respondents are influenced by the former questions' replies, say after effects. We arranged similar questions far from each other in order to avoid the after effects. We feel that this arrangement of similar questions is better than grouped arrangement, since respondents' replies seemed to reflect well the real situation.

ii) GETTING HIGH RESPONSE RATE IN MAIL SURVEY

It is not so easy to determine the survey method, that is, the choice between mail survey and face-to-face interview. We finally adopted mail survey for our sampling survey. There exist many methodological techniques to improve response rates [3],[4],[6],[7]. We explain what type of techniques are efficient in our social survey research for improving the response rate.

First of all, we wrote the respondent's name by hand on the envelope and also on the cover letter explaining the subject of our social survey. It needs much tedious labour but has the effect of distinguishing our survey questionnaires from direct-mail advertizing letters. We also attached philatelic stamps. These techniques made our survey research letter as a personal one. We do not use more expensive methods of delivering mail, i.e. special express delivery and registered delivery. Since we do not want to force respondents unnecessary pressures.

Secondly we enclosed the post-paid return envelope. The size of envelopes was changed to be bigger than used in the preliminary survey. This change reduced the folding process of questionnaires, when we send them. To respondents, the bigger size envelope makes it easier to fold the questionnaires. Pasted envelopes used in our survey research gave the facility to close envelopes both for senders and respondents.

As for the cover letter, we made a plea for help and put stress on social usefulness of the study and on importance of respondent's reply to obtain the success of the study. We also explain the subject of our social survey and mentioned that the answer is only used for academic purposes and that there exists no danger to reveal privacy. The official sponsorpship by the name of Shinshu University is supposed to give a kind of dignity to our social survey research.

We afford an incentive to all samples that were enclosed in the sending envelopes. The incentive enclosed is a plastic-card case and its colour is wine-red for women and blue for men. We also promise the offer of the summary of the results of the survey research.

For non-returned samples, we made a follow-up by mail which announced the request for reply and we enclosed another questionnaires which are the same as the ones sent previously except the dead-line date. This follow-up was made just after the dead-line of the reply and the second dead-line was moved two weeks later. We certified the follow-up effect both in preliminary and present our social survey in 86 and also in 88.

Other efforts are the elaboration of careful wording of questions which are already mentioned above.

The response rate in the preliminary survey is over 80% and in the present survey in 86, 78.3%. The number of refusals is only one and one bantering reply was collected through the police office as a lost article. We considered these high response rate of our sending mail survey mainly as the results of our efforts expressed above but probably enforced by the honesty of Nagano Prefecture inhabitants' characteristics. This characteristics was not actually certified but we feel it in every day life. The certification of honesty in Nagano Prefecture inhabitants will be the main theme for the next survey research in the near future.

The remaining problem for our social survey was to determine the total samples for analysis, that is to say, how to exclude suspicious samples. In mail survey it is not easy to identify the true respondent, that is, someone may reply instead of the original sample.

In order to detect the true respondent, we set the face sheet at the end of questionnaires. The false respondents are, in general, inclined to forget that they are not the original respondents, then wrong replies for sexes, ages and occupations could be easily found.

We pose another question on the final page of questionnaires outside the column, which asks the true respondent.

Thus we exclude suspicious samples according to the following four principles:

I. The sample who answered that the true respondent is not the original sample.

II. Those who do not answer their sex or their replies about their sex is not identical with those of our sample list.

III. Those who do not answer their ages or their replied ages are different over 10 years from those of our sample list.

IV. Those who have an incredible combination between their sex and their occupation.

Under these principles, we excluded 316 suspicious samples and the effective response rate became 67.8%. We could replace excluded samples by reserved extra samples, but we did not such replacements. Thus the effective response rate is a little lower than 70%, but it still higher than other sending mail survey with almost the same subjects.

iii) ROLE OF INCENTIVE FOR HIGHER RESPONSE RATE

In 1988, we repeat this sample survey again with the same questionnaires by using almost identical sampling design. The only big difference between these two social surveys is the incentive. We are almost sure that the lack of incentive for the survey 1988 makes approximately 10% lower the response rate than the first one.

3. BENFORD'S LAW: AN INTERACTION BETWEEN DATA AND THEORY

There are indeed few examples in which real data structure interacts with its proper theory. Benford's law is actually one of these examples.

Suppose that you look on the front page of the Asahi News Paper (a qualified journal in Japan), and observe all of the numbers which appeared there. Do you think how many of them will begin with one? Some people think about one ninth, since all numbers appear in a haphazard way, so that appearances of one as first digits should be one ninth, which seems to be a natural explication.

It turns out empirically that more numbers begin with one, and in fact it is a very exact proportion of numbers that seem to begin with one, it is .301 . The more you look at it, it is the logarithm of 2 to the base 10. This is an empirical phenomenon and so surprising that many researches have already been done on this fact.

This phenomenon has long been known, that is, the first significant digits k with $k=1,2,\cdots,9$ of data, disregarding the position of their decimal points, do not occur with equal frequency but rather with frequency approximately equal to $\log(k+1/k)$ where we take the base of the logarithm to be 10.

After making many counts from a large body of physical data, Farmer's Almanac, Census Reports, Chemical Rubber Handbook, etc., F. Benford [1] noticed this peculiarity as an anomalous law, hence this logarithmic law for the first significant digits has been called Benford's law.

For the proof of a specific real sequence $\{a_n\}$ to obey Benford's law, we use the uniform distribution mod 1 of the sequence $\{\log a_n\}$ [17]. The fact of which Benford's law holds for so many actual data approximately signifies the randomness of the logarithms of them and not of the actual data themselves. Hence our number system essentially certifies Benford's law for unbounded data. Here the unboundedness means the impossibility of the prediction for the bound.

As a direct application of our results [14] we can prove that the denominators of period 1 of convergents of simple continued fraction expansions of real quadratic irrationals obey Benford's law [11]. For arbitrary real quadratic irrationals, Benford's law holds also by using some excellent techniques [8], [18].

Here we give another simple proof of Benford's law for arbitrary real quadratic irrationals. $\{u_n\}_{n=0}^{\infty}$ denotes a linear recurrence sequence satisfying

(3) $$u_{n+2} = a_{n+1} \cdot u_{n+1} + u_n, \quad n \geqq 0, \quad u_1, \quad u_0 > 0,$$

where $\{a_n\}$ is a given sequence of positive integers with $a_{n+r} = a_n$, $n \geqq 0$. For simplicity, we consider the case $r = 3$ and put:

(4) $$x_n = u_{3n}, \quad y_n = u_{3n+1}, \quad z_n = u_{3n+2}.$$

For (3), we write,

(5) $$\begin{cases} z_n = a_1 y_n + x_n, \\ x_{n+1} = a_2 z_n + y_n, \\ y_{n+1} = a_3 x_{n+1} + z_n. \end{cases}$$

This is a system of linear difference equations with constant coefficients for x_n, y_n and z_n. For solving (5), we put:

$$x_n = A\lambda^n, \quad y_n = B\lambda^n, \quad z_n = C\lambda^n$$

with a real parameter λ. Then we obtain

$$\begin{aligned} -A - a_1 B + C &= 0, \\ \lambda A - B - a_2 C &= 0, \\ -a_3\lambda A + \lambda B - C &= 0. \end{aligned}$$

This has a non-trivial solution for A, B, C if and only if

(6) $$\begin{vmatrix} -1 & -a_1 & 1 \\ \lambda & -1 & -a_2 \\ -a_3\lambda & \lambda & -1 \end{vmatrix} = \mu^2 - e_3\mu - 1 \neq 0,$$

where $\mu = (e_3 + \sqrt{e_3^2 + 4})/2$ with $e_3 = a_1a_2a_3 + a_1 + a_2 + a_3 \geqq 4$. Since

$$e_3^2 < e_3^2 + 4 < (e_3 + 1)^2 ,$$

the root μ_1 , of the equation (6) with equality is irrational and cannot be of the form $10^{p/q}$ with integral $p, q \neq 0$. Since after applying the binomial theorem on the left hand side of

$$\mu_1^q = 10^p ,$$

$\sqrt{e_3^2 + 4}$ has a positive coefficient. Thus $\{x_n\}$, $\{y_n\}$ and $\{z_n\}$ obey Benford's law, consequently $\{u_n\}$ obeys Benford's law. For the case of an arbitrary $r \geqq 3$, we can treat quite similarly and $\{u_n\}$ also obeys Benford's law.

From the Lagrange's theorem of continued fractions, the denominators of convergents of simple continued fractions of real quadratic irrationals satisfy the linear recurrence relation (3). Thus we get an another simple proof.

ACKNOWLEDGEMENTS

The second section of this note is a joint work with Takasuke Muto of the Faculty of Education of Shinshu University, Tatsuzo Suzuki and Kinji Mizuno of the Institute of Statistical Mathematics, Toshiyuki Usui and Hajime Kasahara of Hachijuni Culture Foundation.

The third section of this note is a joint work of Peter Schatte of Sektion Mathematik of Bergakademie Freiberg in G.D.R. and Jau-Shyong Shiue of the Department of Mathematical Sciences of University of Nevada, Las Vegas in U.S.A. This simple proof of Benford's law for arbitrary quadratic irrationals is entirely due to an idea of Peter Schatte. Further generalizations of this result is discussed in "A note on Benford's law for second order linear recurrences with periodical coefficients" by P. Schatte and K. Nagasaka submitted to Math. Nachr.

This study was carried out under the Scientific Research Program of the Ministry of Education, Science and Culture, Japan, Grant-in-Aid for Scientific Research: No. 62530012 with the title "Research on statistical analysis methods for free answers in social surveys", the ISM Cooperative Research Program (86-CRP-65), the ISM Cooperative Research Program (87-CRP-89), the ISM Cooperative Research Program (88-CRP-91), and Special Research Bounties of the University of the Air in 1987, 1988 and 1989.

REFERENCES

[1] Benford, F.: The law of anomalous numbers. *Proc. Amer. Philos. Soc.*, 78 (1938), 551-572.

[2] Borel, E.: *Eléments de la théorie des ensembles.* Editions Albin Michel, Paris, 1949.

[3] Dillman, D. A.: *Mail and Telephone Surveys; The Total Design Methods.* John Wiley & Sons Inc., New York-Chicester-Brisbane-Toronto, 1978.

[4] Erdos, P.: *Professional Mail Survey.* McGraw-Hill, 1970.

[5] Fréchet, M.: Les probabilités nulles et la raréfaction. *Ann. Sc. Ec. Norm. Sup.*, 3 série, 80 (1963), 139-172.

[6] Fujita, Y.: Demand effects on mail survey. (1), (2), (3) and supplementary (in Japanese). *Chuō Chōsahō*, 314 (1983), 2406-2407; 2414-2415; 2422-2423; 2442-2443.

[7] Futatsugi, K.: Professional Mail Survey (in Japanese). *Marketing Research*, 10 (1977), 18-22.

[8] Jager, H. and Liardet, P.: Distributions artihmétiques des dénominateurs de convergents de fractions continues. *Indag. Math.*, 91 (1988), 181-197.

[9] Just, W.: Solution of a problem of Fréchet concerning rare sets of measure zero. Preprint.

[10] Just, W. and Laflame, C.: Classifying sets of measure zero with respect to their open covers. To appear in *Trans. Amer. Math. Soc.*

[11] Kanemitsu, S., Nagasaka, K., Rauzy, G. and Shiue J.-S.: On Benford's law: The first digit problem. Proc. 5-th Japan-USSR Symp. on Prob. Theory 1986. *Lecture Notes in Math.*, 1299 (1988), 158-169.

[12] Laflamme, C.: Sur la raréfaction des ensembles de mesure nulle. Preprint, Univ. of Toronto, 1988.

[13] Moszner, Z. and Urbanek, L.: Comparaison de la raréfaction logarithmique d'un ensemble et d'un sous-ensemble. *Bull. Sc. Math.*, 2, série 91 (1967), 19-24.

[14] Nagasaka, K.: On Benford's law. *Ann. Inst. Statist. Math.*, 36 (1984) 337-352.

[15] Nagasaka, K.: Rarefied sets and their statistical comprehension. *Statistical Theory and Data Analysis*, ed. by K. Matusita, North-Holland (1985), 455-460.

[16] Nagasaka, K. and Muto, T.: *Province and Culture in Nagano Prefecture.* Hachijuni Culture Foundation (1987).

[17] Nagasaka, K., Kanemitsu, S. and Shiue, J.-S.: Benford's law: The logarithmic law of first digits. To appear in *Proc. Colloq. Janos Bolyai.*

[18] Schatte, P.: On Benford's law for continued fractions. Submitted to *Math. Nachr.*

[19] Wittgenstein, L.: *Philosophische Bemerkungen.* Blackwell (1964).

INDECES FOR PROJECTION PURSUIT

I.S. Yenyukov
Central Economics Mathematical Institute of the USSR
Academy of Sciences, Moscow 117418, Krasikova 32 (USSR)

Summary

In this paper we consider several indeces for one-dimensional and two-dimensional projection pursuit.The indeces are based on several properties of one- and two-dimensional standard normal and uniform distributions. In particular the "angular" index exploits the uniformity of the distribution of angle between the random two-dimensional standard normal vector and any fixed direction on the plane.Families of indeces ("radial", "radial weigthed")are suggested.for detecting structurs with some radial unhomogeneity like rings. Computational procedures (indeces value and the gradients estimating) are discussed.

Keywords:projection pursuit,projection indeces,structures in data

1. Introduction

The projection pursuit (PP) technique has been suggested in particular (Friedman,Tukey (1974) to discover any structures in multivariate data with by way of visual analysis of "interesting" low-dimensional linear projections .

More formally,let us have a sample $X^{(n)}=(X_1,...,X_n)$ of p-variate observations and let U be a linear orthogonal projection operator from R^p into R^q (q<<p) and Q $(U,X^{(n)})$ be the sampling value of some projection index (PI). The PI is designed in such a way that its value is connected with the intensity of the explored structure in R^q. We select the q-variate projection $Z=U_1X$ in such a way that

$$U_1 = \underset{\Omega(U)}{\operatorname{argmax}}\, Q(U,X^{(n)}), \tag{1}$$

where $\Omega(U)$ is the set of orthogonal linear projectors from R^p into R^q .

In the paper we discuss some concepts for designing PI (Sections 2,3) and give several new PI for projection pursuit (Sections 3,4). We consider indeces for two-dimensional projection pursuit in Section 5 ("angular" index) and Section 7 ("radial" and "radial weighted" indeces). Section 6 is concerned with the problem of estimating the "angular" index values and its gradient.

2. Structures.

We will regard objects from the data matrix $X^{(n)}$as points in a p-dimensional space .And we are interested in revealing the geometric structures of this set of points. So we want to pick such a linear subspace that accumulates essential features of the data structure in the original space. Then we should define a model of the structure interesting to us in order to design a suitable corresponding index for seecking such subspaces.

There are several kinds of structures for which we can give a satisfactory probability models and use ones for PI designing .

They are, for example

- ellipsoidal cloud structure;
- the same structure with contamination by means of outliers;
- set of ellipsoidal clouds (ellipsoidal cluster structure) with approximately equal ellipsoids;
- concentration of points in the vicinity of nonlinear surfaces or manifolds with some simple constructions.

On more intuitive level we can give some more but without good probability model. For example:

- set of nonellipsoidal clusters;
- chains structure;
- fractal structure like "soap-lather" or "grained" - structure;

3. Homogeneity

But the most general definition we can reach by understanding the structure as something directly opposite to the homogeneity.

We say the cloud of points is homogeneous if all the points are similar in some respect, so we can not prefer one point to any other.

Uniform distribution model of homogeneity

Another model of homogeneity is possible if the cloud of points is apriory situated in some restricted domain of subspace, for example, in the unit cube.

Then we consider the cloud as homogeneous regarding the points similar in the sense that each of them has equal apriory probability to take any position in the domain. The probability model of it is uniform distribution in the domain.

Complete spatial randomness model of homogeneity.

Another model of homogeneity is considered in spatial analysis as the concept of complete spatial randomness (CSR, P.J.Diggle 1976,1981):

(1) the number N(A) of points in any region A of the domain follows a Poisson distribution with mean $\lambda|A|$, where $|A|$ is the area of A.

(2) given N(A)=k, the points in A form an independent random sample from uniform distribution on A . Here λ is the intensity or mean number of events per area unit.

Many tests of CSR or uniformity are available. And any of them is pertinent as a projection index for detecting the subspaces with "structure".

A nearest-neigbour analysis is one possibility. Let $\{d_i|$ i=1=1,...,n$\}$ be distance from each of the projected points to its nearest neighbour. Then we may compare the empirical distribution function F(d) and theoretical F(d) (under CSR and when A is the unit cube $F(d)\approx 1-\exp(-\pi d^2/n)$) for testing CSR (Diggle (1981)).

However, we will use the ratio:

$$Q_{1NN}=\bar{D}/\bar{d} \qquad (2)$$

as some projection index. Here $\bar{D}$ is mean distances among all the pairs of points and $\bar{d}$ is the average nearest neighbour distance

$\bar{d}=\sum d_i/n$.

This PI may have large values also for the projections with fractal structures (like "soap-lather" or "grained" - structure). Such structures may be succesfully interpreted in the spatial analysis which is concerned with natural physical subjects in two - or three dimensional space. But interpretation of the structures may be difficult when they arise as a result of some projecting of multivariate data. We may reduce the probability of discovering such structures if we use the distances not to the nearest neighbours but to the k-th ones; $\{d\ (k),\ i=1,\ldots,n\}$ and then use the ratio

$$Q_{kNN}=\bar{D}/\bar{d}(k) \tag{3}$$

Indeces (2) and (3) are easily calculated for one-dimensional projections. There also are rather fast calculating procedures for seeking the k-th nearest neigbours in the cases of two or three dimensions. However, impossibility to obtain their gradient is some drawback of them because it is necessary use quasioptimisation procedures to solve the problem (1).

Normal distribution model of homogeneity.

If we can not consider that the support set of projected distribution is restricted (i.e.the cloud of points is situated in the restricted domain)but we know that the variance of it is finite, we can generalize the uniform probability model of homogeneity. We know the uniform distribution has the least Shennon entropy among all the distributions on the restricted domain. So we can choose the distribution with minimal Shennon entropy among all the distributions with finite variance as the probability model of homogeneity in this case. Such a distribution is a normal one with independent variables. We can consider further the case with the unit variances only. And any test of normality can be considered as an index for seeking the interesting projections.(The wide list of such indeces is given in Huber (1985),Jones and Sibson (1987)).

The model of the homogeneity is rather important because it is known that linear projections of multivariate data tend to be normally distributed under rather wide assumptions (see details Diaconis,Friedman (1984)).

With statistical point of view the homogeneity plays the role of null-hypothesis and any structure plays the role of an alternative hypothesis.

Certainly, choosing the different indeces for measuring distinction from the homogeneity we can to some extent influence the type of preferably revealed structure. Nevertheless, generally saying, large value of an index for a projection does not necessaraly mean that the found projection contains namely the structure interesting to us. So we may obtain something like "grained" structure. Meanwhile, we are interested in detecting large clusters. Moreover large value may just appear due to sampling local fluctuations. It is more so as the index value is a random one. That is why,when we find a projection with the maximal index value we find in fact the maximal value of some sequences of random values.

4. Indeces for one-dimensional projection pursuit.

Here we consider two indeces for one-dimensional projection

pursuit which are functionals of order statistics. Let $z_{(1)},\dots,z_{(n)}$ be order statistics of the one-dimensional projection with vector U of the original p-dimensional data sample , so $z_i=U'X_i$. Then we can introduce the following indeces:

(i) **Inverse studentized range.** The index is based on the known ratio of the sample range to the sample standard deviation (see, for exampe, David (1969))

$$r_w = W_0/s(U) \qquad (4)$$

where $W_0=z_{(n)}-z_{(n)}$ - the width of the sample

It is known that

$$r_w \leqslant [2(n-1)]^{1/2} \qquad (5)$$

$$r_w \geqslant \begin{cases} 2[(n-1)/n]^{1/2} & \text{if n is even} \\ 2[n/(n+1)]^{1/2} & \text{if n is odd} \end{cases} \qquad (6)$$

The ratio has maximal value for the situation when (n-2) points fall in one position and two points are placed symmetrical to this position. We may regard the situation like some "outliers" pattern. The minimal value of the ratio is obtained when half of the points is in one position and the second half - in another position,distinct from the first one. Such situation is a pattern structure with two clusters.

What about the normal distribution ? We give a short table of expected values EW_0/σ for the normal distribution and of the upper bound EW_0/σ in the set of distributions with the finite variance (we use Table 4.2 ,David (1969))

sample size n	upper bound	normal distribution
50	9.8494	4.4982
100	14.0358	5.0152
1000	44.6878	6.4828

The upper bound is rather close to the bound in (5).The difference between the low bound (6) (the value isclose to 2) and values Er_w/σ increases with the growth of the sample size n. It gives some reason to hope that projections with minimal values of r_w may be connected with a cluster structure and such procedure is not sensitive to outliers. Some drawbacks of the ratio $r_w$as the projection index are impossibility to obtain the gradient and a tendention to seek projections with two symmetrical clusters.

We note that in fact we use the value

$$Q_w=1/r_w=s(U)/w_0 \qquad (7)$$

as the projection index only by reason of similarity of the optimization problem formulation . Despite the above mentioned drawbacks we may use the indexfor the fastseeking good start points for one-dimensional projection pursuit. The index (7) may be modified in several ways. For example, we may try to improve the robustness in relation to outliers if we take some robust estimate of the variance instead of the usual estimate.

(ii) **"Normal scores " index.**

The index has a form of

$$\hat{Q}_{NS}=1-\sum_{i=1}^{n} z_{(i)}E_i/s(u) \qquad (8)$$

where E_i are the scores of standard normal, so $E_i= Ey_{(i)}$ and the y is the random standard normal value.

Some theoretical reason for this index is the Brillinger theorem (1966). The expected value of this index for normal distribution is zero. In practice it is much better to use a robust estimate of variance to preserve the outliers influence. In particular we use the trimmed estimates of the mean and the variance with the trimming level, for example, 0.025.

$$\hat{Q}_{NS}=1-\sum^{n} z_{(i)}E_i/s_{trim}(u) \qquad (9)$$

Then the expected value of $\hat{Q}_{NS}$ will be a negative value.

It is easy to obtain the gradient of the "scores" index by direct differentiation of (8) or (9).

(iii) **One-dimensional "radial" indeces**

Let r_i be the absolute deviations of $z_i(i=1,\dots,n)$ from their mean value $\bar{z}$, so $r_i=abs(z_i-\bar{z})$.

The indeces are

$$\hat{Q}_{1,R}= Er^{\beta}/sd(r^{\beta}),\ (\beta>0)\ , \qquad (11)$$

where Er^{β} is the average of r_i^{β} $(i=1,\dots,n)$,

$sd(r^{\beta})$- the sample standard deviation of them..

The one-dimensional radial indeces reach the maximal (infinite) value in the same situation which maximizes the Q_w index (7), i.e. in the situation with two singular clusters. As distinct from the Q_w and normal scores indeces the radial indeces may be generalized for high-dimensional projections (see, Section 7).

5. Indeces for two-dimensional projection pursuit. "Angular" index.

Sometimes two- or higher- dimensional projection pursuit makes it possible to discover such structures which are impossible to find with the help of the sequentional one-dimentional approach.

At the same time it may be much more expensive in computational relation. For example, the approach given in the work of Friedman (1987) requires the increase of the number of calculations as square of the number of calculations for one-dimensional case. But it is possibie to design the projection index and optimization procedure in such a way that reduces two-dimensional problem to one-dimensional.

Let us seek two linear combinations $Z_1=U'_1X,\ Z_2=U'_2X$.

They must be such that their joint distribution would be the most unhomogeneous. Since we are interested in nonlinear structure we reguire the two linear combinations Z_1 and Z_2

to be uncorrelated, $cor(X_1,X_2)=0$. We consider that our data were preliminary "sphered",i.e. they were transformed with the help of the linear transformation in such a way that the total covariances matrix has become the identity one. For sphered data the uncorrelation is equivalent to orthogonality vectors U_1 and U_2. We also require that the norms $\|U_i\|=1$ (i=1,2).It assures that the variances of any one-dimensional projection will be equal to one.We regard the bivariate standard normal to be the most homogenious joint distribution.

And our indeces are based on the following properties of the two-dimensional normal distribution:

(i) the angle ϕ between some two-dimensional random vector and positive direction of axe z_1 has uniform distribution.

(ii) the norm of the vector $r=\|Z\|$ has χ^2 distribution with two degree of freedom. So it is just exponential distribution.

(iii) the angle ϕ and the norm are independent.

Certainly the first property is true not only for normal distribution but for every spheroidal symmetric distribution. The property can be named "angular" homogeneity and the index that exploits it we shall name the "angular" index. We shall use uniformity of the angle ϕ to design the "angular" index.

For testing the uniformity of angle ϕ we could use a lot of test statistics (see Mardia (1972)).

One PI based on the idea of testing the uniformity was suggested by G.Eslava .Namely , the angle θ between any two points is regarded as the distances between them. And the average nearest-nighbour distance (see definition $\bar{d}$ in expansion (2)) is used as PI (for the minimization problem). Some radial weighting (see Section 7)is also used in this PI: only the points whith $\|Z\|>\alpha R$ ($R=\max(r_i)$ $(i=1,\ldots,n),\alpha>0$) are included in the computation of the PI value.

For our angular PI we use statistic based on the fact that the harmonic moments are equal to zero for uniform distribution. The harmonic coefficients are just Fourier ones:

$$a_i=E\sin(i\phi)=\frac{1}{\pi}\int_{-\pi}^{\pi}p(\phi)\sin(i\phi)d\phi,\quad b_i=E\cos(i\phi)=\frac{1}{\pi}\int_{-\pi}^{\pi}p(\phi)\cos(i\phi)d\phi \quad (12)$$

where $p(\varphi)$ - is density of the angle ϕ on $[-\pi, \pi]$.

And we use as the angular PI

$$Q_{kA}(U)=\sum_{i=1}^{k}(a_i^2+b_i^2) \quad (13)$$

6. Estimating "angular" index values and its gradient

In practice the estimates of a_i and b_i are given by the following expansions

$$\hat{a}_i=\frac{1}{n}\sum_{j=1}^{n}\sin(i\phi_j) \qquad \hat{b}_i=\frac{1}{n}\sum_{j=1}^{n}\cos(i\phi_j)$$

The values $\sin(\phi_j)$ and $\cos(\phi_j)$ are known to us. They are

$$\begin{aligned} \sin(\phi_j) &= z_{j1}/\|Z_j\| \\ \cos(\phi_j) &= z_{j2}/\|Z_j\| \end{aligned} \qquad (14)$$

where z_{ji} (i=1,2) are coordinates of the two-dimensional projected vector for j-th object and $\|Z_j\|$ is the norm of the vector. The values $\sin(i\phi)$ and $\cos(i\phi)$ can be obtained with the help of simple recursion relations.

For efficient optimization it is useful to have derivatives of the indeces.

For this index these ones are easily obtained via recursion

$$\begin{cases} (\sin(i\phi))' = (i-1)[\cos((i-1)\phi)(\sin(\phi))' + \sin((i-1)\phi)(\cos(\phi))'] \\ (\cos(i\phi))' = (i-1)[\cos((i-1)\phi)(\cos(\phi))' - \sin((i-1)\phi)(\sin(\phi))'] \end{cases} \qquad (15)$$

And then the gradient of the index estimate is

$$\frac{\partial Q_{kA}}{\partial U_l} = \frac{2}{n} \sum_{j=1}^{n} c_{jl} X_j$$

$$\text{where } c_{jl} = \sum_{i=1}^{k} (\hat{a}_i (\sin(i\phi_j)'_l + \hat{b}_i (\cos(i\phi_j)'_l) \qquad (l=1,2)$$

$$(\sin(i\varphi_j))'_l = \frac{\partial}{\partial U_l} \sin(i\varphi_j)$$

$$(\cos(i\varphi_j))'_l = \frac{\partial}{\partial U_l} \cos(i\varphi_j)$$

The last derivatives can be easily obtained by means of direct differentiation of the expressions (14) and the use of the reccurent relation (15).

7. "Radial" and "radial weighted" indeces.

The drawback of the angular index is that it loses the information about radial part of the distribution and thus for example the structure like "ring" will not be detected. We can use "radial" and "radial weigthed " indeces for to use the radial unhomogeneity for seeking interesting projections.

Radial indeces.

The indeces have a form of

$Q_R = Eh(r).$,

where $h(r)$ is some positive function. For example,

$$h(r)=r^\beta, \quad h(r)=r^\beta/(Dr^\beta)^{1/2}, \quad h(r)=r^\beta/(c+\alpha r^\beta) \ (\beta>0) \text{ e.t.c.}$$

If $h(r)=r^\beta/(Dr^\beta)^{1/2}$ it is possible to find structures like a ring or a tore (projections with "holes").

Radial weighted angular indeces.

We may try to joint possibilities of angular and radial indeces using, for example, radial weighting for the angular PI (13). We define for it

$$a_i^2(r)=E(h(r)\cos(i\phi)),\qquad b_i^2(r)=E(h(r)\sin(i\phi))$$

And now we define the weigted radial indeces as

$$Q_{RA}=\sum(a_i^2(r)+b_i^2(r)).$$

It easily to find the gradients of the radial and radial weigthed indeces. The estimates of the gradients will be not defined if the $\beta\leqslant1$ and the norm of some projected point is zero.However, it is a rare event and we can avoid it by way of replacing the norm $\|Z\|$ by the value $\|Z\|+\varepsilon$ (ε is some small positive value).

8. Implemenation.

The projection pursuit procedures based on the mentioned above indeces are implemented in SIGAMD (System for Interactive Graphical Analysis of Multivariate Data).Some others indeces (Yenyukov (1986)) and statistical methods (for example ,nonlinear mapping) are included in the system.

The system operates on personal computers of the type IBM XT,AT and compatibale . There are interesting results of using the such projection pursuit to some medical and sociological data.

References.

Brillinger D.R.(1966) An extremal property of the conditional expectation. Biometrika,53,p.594-5

David H.A.(1969) Ordinal statistics. Wiley & S.

Diaconis p.,Friedman D.(1984) Asymptotics of graphical projection pursuit. Ann.Statist.,12,793-815

Diggle P.J., Besag J., Gleaves J.T. (1976). Statistical analysis of spatial point patterns by means of distance methods.Biometrics,32,659-67.

Diggle P.J. (1981) Some Graphical Methods in Analysis of Spatial Point Pattern.In "Interpreting of multivariate data", (Ed. Vic Barnett). Chichester,e.t.c.:j.Wiley & Sons.,p.55-72

Eslava G.(1988)(Privite communication.Oxford University)

Friedman J.H., Tukey J.W.(1974) A projection pursuit algorithm for exploratory data analysis. IEEE Trans. Comput., C-23,p.881-889.

Friedman J.H.(1987) Exploratory projection pursuit. J. Amer. Stat. Ass.,82, p.249-266.

Huber P.J.(1985) Projection pursuit. Ann. Statist., 12, p.433-474.

Jones M.C.,Sibson R.(1987) What is Projection Pursuit. J.Roy.Stat. Soc.,150,ser.A,p.1-38

Mardia K.V.(1972) Statistics of directional data. London & N.Y: Ac.Press

Yenyukov.I.S.(1988) Detecting Structures by Means of Projection Pursuit.In Proceedings of COMPSTAT-1988,p.48-58

APPLICATIONS

NONLINEAR ESTIMATION OF ANIMAL ABUNDANCE IN FISHERIES

Victor Aguirre Torres
IIMAS-UNAM, Apdo Postal 20-726 ADMON NO 20, 01000 Mexico, D.F. Mexico
Enrique Villa Diharce
Escuela de Ciencias Físico Mats., Universidad Autonoma de Sinaloa, Mexico

1. INTRODUCTION

In the management of fisheries that are subject to exploitation, it is important to study the size of the population, as well as the yield obtained when a certain fishing effort is applied. One of the most widely used approach for this problem is Dynamics of Populations where there exists a class of models known as Stock Production Models that consider the size of the population as a function of the natural growth of the population, and the fishing effect. Although this models do not consider other factors like enviroment, relations with other species, or classes of age within the population, every study about a fishery contains a reference to this kind of models. This class contains the models given by Schaefer (1954), Fox (Gulland (1983)), and Pella and Tomlinson (1969).

Traditionally this models are used to obtain the Maximun Sustainable Yield (MSY), which gives the highest yield that can be obtained permanently when a constant effort is applied , year after year, to the population. The MSY is usually computed assuming that the population is exploited under an equilibrium condition, which means that the amount of fish captured in one year is equal to the amount of growth of the population for that year. Among other things this implies that the size of the population remains constant thru the years. Obviously this condition is hardly true, and we will show in this work how the MSY can be estimated without making this assumption. Moreover, using the approach given here one can also obtain a graphical display of the evolution of the size of the population in time.

Section 2 gives the generalized stock production model, a reparametrization and the expression of the MSY in terms of the parameters of the model. Section 3 gives the formulation of the problem in the case of lack of equilibrium. Section 4 gives the numerical aspects required to fit the new model. Section 5 shows the application of the model to the fishery of yellowfin tuna in the eastern tropical pacific ocean. Section 6 gives a summary and discussion of the results.

2. GENERALIZED STOCK PRODUCTION MODELS.

In this kind of models the size of the population is given by its weight which recives the name of BIOMASS, the amount of resources devoted to fishing are called EFFORT. We are going to consider the generalized stock production model (GSPM) given in Pella and Tomlinson (1969) which assumes that the size of the population obeys the following differential equation :

$$dB(t)/dt = aB^n(t) - bB(t) - F(t)\,B(t) \qquad (2.1)$$

where

$B(t)$ = Biomass at time t

$aB^n(t) - bB(t)$ = Population production function. It takes into account natural growth and mortality

$F(t) = q\,f(t)$ = Mortality due to fishing when an effort $f(t)$ is applied

$f(t)$ = Fishing effort

q = Capture coefficient.

Notice that the term

$$dC(t)/dt = F(t)B(t) \qquad (2.2)$$

expresses the instant capture $C(t)$ obtained when an effort $f(t)$ is applied.

The constants a,b,n and q depend on the population, and except for q they bear no physical meaning. Fletcher (1977) gives a reparametrization of model (2.1) such tha the resulting constants have an explicit meaning form the point of view of the management of the fishery. The result is as follows, if we let :

B_o = Biomass at time t = 0

B_∞ = Asymptotic level of the biomass without exploitation

M = Maximum growth rate of the biomass without exploitation

P = Level of the biomass at which M occurs.

then

$$B_\infty = \lim_{t\to\infty} B(t) = (a/b)^{1/(1-n)}$$

$$P = (an/b)^{1/(1-n)}$$

$$M = +-[b(1-n)/n][an/b]^{1/(1-n)}$$

in this last equality the plus sign corresponds to the case $0<n<1$ and the minus sign to the case $n>1$.

After substitution, equation (2.1) becomes

$$dB(t)/dt = GM[B(t)/B_\infty]-GM[B(t)/B_\infty]^n - F(t)\,B(t) \qquad (2.3)$$

where G is a function of n :

$$G = [n^{n/(n-1)}]/(n-1)$$

The solution to (2.3) is

$$B(t) = B_e^{1-n} + A\exp([GM/B_\infty - F(t)][1-n]t)^{1/(1-n)} \qquad (2.4)$$

where

$$B_e = B_\infty[GM/(GM-F(t)B_\infty)]^{1/(1-n)} \qquad (2.5)$$

and

$$A = B_0^{1-n} - B_e^{1-n}$$

The results and expressions for the limiting case when n tends to one in (2.4) are very similar to the ones mentioned here, hence we will not cite them.

So far we have just examined the relationships where the biomass is present, however the data consist of observed efforts and yields. Imposing the condition of equilibrium $dB(t)/dt=0$ into equation (2.3) we get the expression for the biomass in equilibrium (B_e),

$$F\, B_e = GM[B_e/B_\infty] - GM[B_e/B_\infty]^n$$

using (2.2) the total capture per year , under this condition and assuming a constant fishing effort f throughout the year, is

$$C_e = \int_0^1 dC = FB_e\int_0^1 dt = FB_e = qfB_e \qquad (2.6)$$

using expression (2.5) with F(t) constant, and (2.6) we get a formula for the capture in a state of equilibrium as a function of the effort on the resource

$$C_e = q\,f\,B_\infty[1-qfB_\infty/GM]^{1/(1-n)} \qquad (2.7)$$

Maximizing (2.7) with respect to f we get that the MSY is given by

$$MSY = M \qquad (2.8)$$

with a corresponding effort :

$$f_{MSY} = ((n-1)/n)GM/qB_\infty \qquad (2.9)$$

at this point, the usual procedure is to fit (2.7), or a linearization of it to the data in order to estimate the unknown constants of the model and then substitute in (2.8) and (2.9).

It should be clear from here, that the usual procedure is inadequate, since it is hard to believe that for every year the yield corresponded with the amount in which the population was going to grow.

3. LACK OF EQUILIBRIUM.

In order to avoid the deficiency of the usual approach, and taking advantage of widespread of computing power, we formulated the problem in such a way that the equilibrium condition is not required.

To do so, we consider each year separate and in sequence. For i-th year there is a constant fishing effort f_i, for the time interval (t_{i-1},t_i), the biomass developes according to equation (2.4) but now

$$B(t) = [B_{ei}^{1-n} + A_i\exp([GM/B_\infty - F_i][1-n]t)]^{1/(1-n)} \qquad (3.1)$$

where:

$$F_i = qf_i$$

$$B_{ei} = B_\infty [1 - F_i B_\infty / GM]^{1/(n-1)}$$

$$A_i = B_{i-1}^{1-n} - B_{ei}^{1-n}$$

$$B_{i-1} = B(t_{i-1})$$

according to this model the capture for the i-th year should be

$$C_i = \int_{t_{i-1}}^{t_i} dC(t) = \int_{t_{i-1}}^{t_i} F_i B(t) dt \qquad (3.2)$$

where B(t) is given by (3.1). The above integral does not have a closed expression except for n = 2.

From here it is clear that we can write the capture of the i-th year in terms of the implicit equations of the model, the efforts applied to the resource up to the i-th year $f_1, f_2, \ldots, f_i$,and the constants of the model $B_o, B_\infty, M, q,$ and n. That is, if we let

$$\underset{\sim}{\Theta} = (B_o, B_\infty, M, q, n)$$

then

$$C_i = C_i(\underset{\sim}{f}; \underset{\sim}{\Theta}) = C(f_1, f_2, \ldots, f_i; \underset{\sim}{\Theta})$$

where the functional relatioship is given by equations (3.1) and (3.2). It is obvious that the model will not predict the yearly capture exactly, since as mentioned in the introduction there are other factors that influence its behaviour. Therefore if we let $C_1, \ldots, C_k$ to be observed captures the best we can do to estimate the unknown constants of the models is by minimizing the sum of the squared errors

$$E_i = C_i - C_i(\underset{\sim}{f}_i; \underset{\sim}{\Theta}) \qquad (3.3)$$

Formula (3.3) has the typical form of a nonlinear regression, where the parameter to be estimated is $\underset{\sim}{\Theta}$. Once this parameter is estimated we can easily obtain MSY and f_{MSY} from formulas (2.8) and (2.9) without the assumption of equilibrium.

It is important to mention that as a result of this formulation, one can also have an estimation of the biomass of the fishery, by using equation (3.1) for each year. This is important because it gives a better picture of how the resource has been exploited.

4. NUMERICAL ASPECTS.

In order to fit (3.3) we had three main problems, first to obtain an initial estimate of the parameter, second to evaluate the integral in (3.2) and third to compute the derivatives required by the optimization method (modified Gauss-Newton, Hartley (1961)).

To compute (3.2) we used a simple quadrature method which worked well, since the function B(t) is very smooth. To evaluate the derivatives we resorted to difference-type approximations, like those mentioned in Kennedy and Gentle (1980).

The toughest problem was to obtain initial estimates. We used two methods, first a regression method and second a random search method.

The regression method is based on the idea of finding an approximate of the differential equation by a difference. For more details see Villa (1988).

The random search method makes use of previous information that tells us the parameter $\underset{\sim}{\Theta}_o$ may be found in some domain D contained in R^5, the idea is then to randomly (usually uniformly) generate several values of $\underset{\sim}{\Theta}$ and choose the one with the least sum of squares from (3.3). For further discussion on this method see Rubinstein (1981).

5. APPLICATION.

We will show the application of the new model to the yellowfin tuna fishery. The data are reported in Pella and Tomlinson (1969).

Table 5.1. Yellowfin tuna data, 1934-1967.

YEAR	EFFORT	BIOMASS	EST YIELD	OBS YIELD
1	2	3	4	5
1934	5.8790	152.496	61.1214	60.9130
1935	6.295	151.811	67.1359	72.2940
1936	6.771	150.349	71.6137	78.3530
1937	8.233	145.927	84.9686	91.5220
1938	6.830	149.967	71.3626	78.2880
1939	10.488	139.320	104.573	110.418
1940	10.801	138.027	104.872	114.590
1941	9.584	141.502	94.5333	76.8410
1942	5.961	152.440	62.5446	41.9650
1943	5.930	152.919	63.5783	50.0580
1944	6.475	151.280	68.9265	64.8690
1945	9.377	142.591	95.3086	89.1940
1946	13.958	129.255	130.055	129.701
1947	20.383	111.692	166.597	160.151
1948	24.781	100.026	179.710	206.993
1949	23.923	101.390	169.523	200.070
1950	31.856	83.4967	198.288	224.810
1951	18.403	114.038	134.880	186.015
1952	34.834	78.0359	215.282	195.277
1953	36.356	72.5934	189.545	140.042
1954	26.228	93.7559	159.879	140.033
1955	17.198	117.983	133.517	140.865
1956	27.205	94.6965	193.174	177.026
1957	26.768	94.3316	177.388	163.020
1958	31.135	84.5788	191.136	148.450
1959	28.198	90.3207	175.191	140.484
1960	35.841	74.5762	199.434	244.331
1961	41.646	62.4181	193.573	230.836
1962	42.248	60.1393	180.519	174.063
1963	33.303	76.8584	166.306	145.469
1964	42.090	61.8338	196.432	203.882
1965	43.228	58.3828	180.556	180.086
1966	40.393	62.9480	174.033	182.294
1967	33.814	76.1527	170.170	178.944

The observed capture is per year and is given in millions of pounds, the fishing efforts are given in thousands of days/ship. The data are displayed in columns 2 and 5 of table 5.1, they correspond to the period 1934-1967.

From the information given in Pella and Tomlinson (1969) we considered that a feasible domain for a random search could be the following :

$$(100.0,\ 200.0) \quad \text{for } B_o$$

$$(100.0,\ 200.0) \quad \text{for } B_\infty$$

$$(\ 50.0,\ 150.0) \quad \text{for } M$$

$$(0.001,\ 0.100) \quad \text{for } q$$

and taking n to be equal to 2. After 100 random trials uniformly distributed in the domain we obtained the following :

Bo	B∞	M	q	SSE
124.005	113.620	149.968	.089925	49601.1

then we used this as a starting point for a nonlinear fitting, but keeping n = 2, and we got :

	Bo	B∞	M	q	SSE
	150.511	158.457	186.764	.071213	17410.5
STDE	276.128	83.8999	5.77007	.038496	
t test	.545077	1.88864	32.3677	1.84986	

the standard errors are computed to get an idea of the stability of the estimation. The interested reader may check Gallant (1975) for the formulas. They show for for example that the estimation of Bo is not very stable, that is, the initial level of the populaion is quite uncertain, but on the other hand the rest of the parameters are fairly well determined from the data and the model. Specifically, the parameter M which gives the MSY has the smallest relative standard error, and for the purpose of the application this is the most relevant parameter.

From here we may set the optimal levels of capture, effort and biomass

$$MSY = 186.76$$

$$f_{MSY} = 33.10$$

$$B_{MSY} = 79.22$$

After this, we let the n parameter move freely, table 5.2 shows the results of the estimation procedure.

Table 5.2. Estimation of nonlinear GSPM model. Tuna data.

ITERATION	SSE	Bo	B∞	M	q	n
0	17410.5	150.511	158.456	186.764	.071213	2.00
1	17258.8	133.607	180.300	183.203	.066525	1.40253
2	17182.2	135.456	169.951	184.198	.070685	1.51808
3	17179.5	136.763	172.883	184.250	.069906	1.49898
4	17179.4	136.340	172.006	184.287	.070230	1.50368
END	17179.4	136.419	172.182	184.280	.070169	1.50282

Therefore the final results are :

	Bo	B∞	M	q	n
	136.419	172.182	184.280	.070169	1.50282
STDE	213.455	104.355	7.02965	.041182	.803671
t test	.639097	1.64996	26.2146	1.70387	1.86994

CORRELATION MATRIX

	B∞	M	q	n
Bo	.38	-.10	-.42	.02
B∞		-.43	-.97	-.36
M			.35	.47
q				.18

We see from the correlation matrix that Bo is not highly correlated with the rest of the parameters, specially with M. On the other hand B∞ and q are extremely correlated, suggesting that the model is overfitted, however due to the complexity of the model this is hard to check and it is a point for further study.

The optimal levels are :

$$MSY = 184.28$$

$$f_{MSY} = 34.29$$

$$B_{MSY} = 76.58$$

Using the final results we used the model to estimate the evolution of the biomass and the expected value of the capture as a check for the model. The numerical values are given in columns 3 and 4 of table 5.1, and the plot is given in figure I.

FIGURE I. PLOT OF BIOMASS, EFFORT EXPECTED AND OBSERVED YIELD.

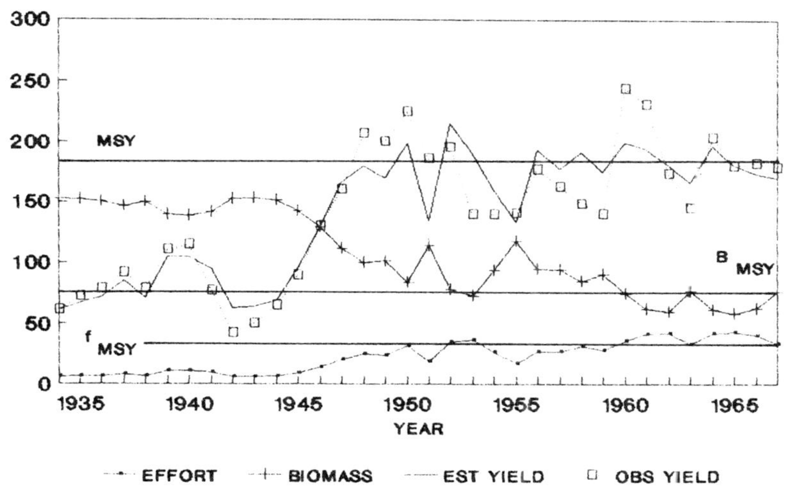

As can be seen from the plot, in the early years the fishery was under-exploited, the level of the biomass was much higher than the optimal, and the levels of capture and effort were much lower than their corresponding optimum. In the late forties the effort starts to raise towards the optimum and it has as a result a significant increase in yield and decrease in biomass. In the sixties the effort has been above the optimal most of the years, this has as a consequence that the capture has been below the optimal too. And whenever the effort is above the optimal the biomass decreases, in 1963 when the effort dropped to its optimal, the biomass recovered and the next year the yield had a very high value, but the effort applied in 1964 was too high and this had the effect of reducing the biomass too much and the next year the yield falled. By the way, this analysis shows that the assumption of equilibrium is completely false. If we compare the plots of capture fitted and observed, we see that there is a good agreement between the two.

6. SUMMARY AND DISCUSSION.

We have given a new approach to the problem of finding the values of the relevant parameters in the management of fisheries. The advantages of this new method are : first, it does not rely on the dubious assumption that the fishery has been exploited in equilibrium, and hence its results are expected to be more reliable; second, it provides a picture of the evolution of the resource, this makes the interpretation of the fluctuations easier to understand; and third, since the procedure is based on nonlinear regression one has an estimate of the stability of the estimation for each parameter.

REFERENCES

FLETCHER, R. I (1978) Time Dependent Solutions and Efficient Parameters for Stock Production Models. Fish. Bull. NOAA/NMFS, 76(2):377-388.

GALLANT, A. R. (1975) Nonlinear Regression. The American Statistician, 29, 73-81.

GULLAND, J. A. (1983) Fish stock assessment, a manual for basic methods. FAO/J. Wiley.

HARTLEY, H. O. (1961) The modified Gauss-Newton method for the fitting of nonlinear regression functions by least squares. Technometrics 3, 269-280.

KENNEDY, W. J. and GENTLE J. E. (1980) Statistical Computing. Marcel Dekker, Inc.

PELLA, J. J. and TOMLINSON, P. K. (1969) A Generalized Stock Production Model. Bull. I-ATTC, 13(3):421-496

RUBINSTEIN, R. Y. (1981) Simulation and the MonteCarlo Method. J. Wiley.

SCHAEFER, M. B. (1954) Some aspects of the dynamics of populations important to the management of marine fisheries. Bull. I-ATTC, No. 1, 25-56.

VILLA, E. (1988) Estimación nolineal de la abundancia de poblaciones de peces explotadas. Master of Science thesis, IIMAS-UNAM, Apdo. Postal 20-726, Mexico 01000, DF, MEXICO.

ORDINAL CODING AND CURVE CLASSIFICATION

A. Faraj*, I.C. Lerman** and N. Valette***
* DER/EDF/GRETS 30 rue de Condé 75006 Paris
** I.R.I.S.A. Campus de Bealieu 35042 Rennes
*** LAA/SLC/EVP C.N.E.T. Lannion 22301 Lannion

1 - INTRODUCTION

The study presented in this paper was carried out on the basis of data measuring the disturbances caused by electrical phenomena. Mathematically, the data takes the form of curves defined over an interval T bounded by R with values in a part of R . They are temporal data. A certain number of methods for analysing this type of data exist, notable due to Deville who, under the name of harmonic analysis, proposed a generalisation of Principal Components Analysis (P.C.A.) to a family of random variables (Xt).

The curves can be considered either as the variables measured at the points t of T or as individuals on which these points t take measures We have here two assumptions that analyse the information contained in a corpus of curves in different ways. The first of them is a morphological analysis and the second is a chronological analysis of the deviations from the trend.

The only thing that our method has in common with harmonic analysis is the descriptive view that is specific to data analysis. It was born from the idea of classifying the curves in two stages.

- The first based on their shape, then within the same class of shape
- The second based on amplitude

The curves that initially considered as quantitative variables are next coded into ordinal qualitative variables (V.Q.O.). We then propose a similarity index between the variables from coding, after which we classify them using an " algorithme de la vraisemblance des liens " (A.V.L.) algorithm. This coding work consists in sampling the set of values reached by the curve at intervals corresponding to the modalities of the coding's qualitative variable.

Transforming a quantitative variable into a qualitative variable is obviously accompanied by a loss of information. This loss could be minimised by associating a modality with each quantity that the quantitative variable attains. The disadvantage would be that we would obtain qualitative variables having a very high number of modalities. The use of A.V.L. is frequent on V.Q.O.s (generally originating from surveys with a small number of modalities (up to 20). The coding of curves into V.Q.O. required, in order to apply it, the development of software making it possible to calculate the reduced centered index between V.Q.O.s having a very large number of modalities (1000 and more)

2 - CODING AND COMPARISON INDEX

2.1 "Hypothèse d'absence de lien (H.A.L.)"

Given two V.Q.O.s f and g defined on the same set T and having, respectively, as modality sets:

$L = \{1,...l,...L\}$
and
$M = \{1,...m,...M\}$

on T, f and g are respectively designed by w_f et w_g respectively and are represented in T x T par $R(w_f)$ et $R(W_g)$

$$R(w_f) = \{(t,t')\in T \times T / f(t) < f(t')\}$$

$$R(w_g) = \{(t,t')\in T \times T / g(t) < g(t')\}$$

The raw index of similarity between f and g is defined by:

s(f,g) = Card (R $(w_f) \cap R(w_g)$)

The "hypothèse d'absence de liens (H.A.L.)" assumption is formulated as follows: the variable g is fixed and a randon variable f' is associated with the variable f in other words w' is associated with w_f in the set:

$\{n(1),...n(l),...n(L)\}$

this set is provided with a uniformly distributed probability with:

$n(l) = Card\{t\in T / f(t) = l\}$

w' is represented in T x T by the random part R(w'), the index associated with s(f,g) is defined by:

S(g) = Card (R (w') $\cap$ R (w_g))

the mean and the variance of which are respectively:

$m_{fg} = \lambda\mu$

$$s^2_{fg} = \lambda\mu + \rho_{11}\delta_{11} + \rho_{22}\delta_{22} + \rho_{12}\delta_{12} + (\theta\xi - \lambda^2\mu^2)$$

The expressions for:

$\mu, \delta_{11}, \delta_{22}, \delta_{12}, xi$

are respectively of the same form as

$\lambda, \rho_{11}, \rho_{22}, rho_{12}, \theta$ the first ones being relative to w_f associated with f and the second ones being relative to w_g associated with g.

If n > 4, we have:

$\lambda = \sum n(l)n(l')/1 \leq l \leq L/\sqrt{n(n-1)(n-2)}$

$\rho_{11} = \sum\{n(l)n1(l)(n1(l)-1)/2 \leq l \leq L\}/\sqrt{n(n-1)(n-2)}$

$\rho_{22} = \sum\{n(l)n_2(l)(n_2(l)-1)1 \leq l \leq L-1\}/\sqrt{n(n-1)(n-2)}$

$\rho_{12} = \sum\{n(l)n_1(l)(n_2(l)/2 \leq l \leq L-1\}/\sqrt{n(n-1)(n-2)}$

$\theta = \sum\{n(l_1)n(l'_1)\{\sum n(l_2)n(l'_2)/1 \leq l_2 \leq l'_2 < L\} + n(l_1) + n(l'_1) - (2n+1)/1 \leq l_1 \leq l_{1'} < L\}/\sqrt{n(n-1)(n-2)(n-3)}$

where $n_1(l) = \sum\{n(l')/l' \leq l\}$

and $n_2(l) = \sum\{n(l')/l' \geq l\}$

The association index between f and g is put in the following reduced centered form:

$$q(f, g) = (s(f,g) - m_{fg}) / s_{fg}$$

and by reference to a scale of probability, using the distribution function of the normal law N (0,1) the final association index between f and g is written as:

$$p(f, g) = q_s(f, g)$$

where $q_s(f,g) = (q(f,g) - moy(q))/\sqrt{var}(q)$

2.2 *Coding a curve in V.Q.O.*

Let f be a curve defined on set T having values in an interval (A,B) bounded by R . We will call a quantified curve, signified by f_q , the curve defined by T having values in:

$\{L = 1,...l,...L\}$

such that, for every element t of T , we have:

$f_q(t) = l$ if and only if

$A + (l-1)(B-A)/L < f(t) < A + l(B-A)/L$

where $(B-A)/l$ is the quantification step. It is a whole number multiple of the measurement error of the equipment providing the values of f. f is said to be coded in V.Q.O. if it induces a total preorder on T .

3 - *EFFECTIVE COMPARISON TECHNIQUE*

This technique is relational and, contrary to the case of comparing two qualitative variables with a small number of modalities, it wasn't pertinent to have a pair the two arguments of which were too far away. Additionally, we initially limited ourselves to a local comparison of two curves, by simply comparing the intervals of strict monotony. The proposed proximity index between two curves takes the form of a sequence of indices standardized by H.A.L relating to the intervals of monotony of the first curve with respect to the second and of the second one with respect to the first. Let f and g be two curves defined on T having values in intervals bounded by R. Let ($T^1, T^2 \ldots T^j$) (resp.($T'^1, T'^2, \ldots, T'^k$)) be a series of intervals of T over which f (resp. g) is strictly monotonic. We shall consider f/T^j and g/T^j (respectively g/T'^k and f/T'^k) as the curves coded in V.Q.O. from restrictions of f and g to T^j for j = 1,...,j (respectively to T'^k for k = 1,...k) having values respectively in:

L = $\{1,\ldots,L\}$

and

$M = \{1,\ldots,M\}$

The similarity index s (f,g) between the two curves f and g coded in V.Q.O. will be equal to the sum of the two following expressions:

$$\sum\{card(T^j) \times q(f/T^j, g/T^j) / j = 1,\ldots J\} / \sum\{card(T^j / j = 1,\ldots,J\}$$

$$\sum\{card(T^k) \times q(f/T^k, f/T^k) / k = 1,\ldots K\} / \sum\{card(T^j / k = 1,\ldots,K\}$$

where q designates the index of association between two reduced centered V.Q.O.s with respect to the "hypothèse d'absence de lien" assumption of absence of a linkage.

4 - *COMPARING THE RESULTS FROM DIFFERENT METHODS*

The coding and the index we are presenting only bring the evolutional aspect of temporal data to light, independently of their amplitude. Can these two aspects be dissociated? To answer this question, we decided to do a comparison from a purely applied point of view of the partitions of a family of curves obtained, firstly, by their classification using V.Q.O coding and, secondly, by their classification using A.V.L. taken as quantitative variables (we used correlation as the proximity index) and their classification, as individuals, using Ward's method (the index employed was the conventional euclidian distance between points in a metric space). The chosen data set consisted of 60 curves with discreet values taken in constant steps at 41 points. The curves represented a speech signal in the frequency domain. Each one of them was the smoothed spectrum associated with a phoneme's window (or instant). There were six phonemes: I AU A Z S AI.

The interest of such data is its ease of classification. An "ideal" method of classification should yield a partition of the family used into six classes each corresponding to a phoneme.

4.1 *Results of classification*

4.1.1 *Using A.V.L. for curves considered as numeric variables*

Here, (see figs. 1 and 2) a curve is considered as a numeric variable. If we use t_i to designate a point on the x-axis, t_i plays the part of the nth individual and $f(t_i)$ is the variable's value on this individual. The reduced centered proximity index between two curves is proportional to the correlation coefficient.
The partition obtained is close to the "natural" partition with this method. This is confirmed by the relative "continuity" of the surface described by the two curves in each class, as can be clearly seen in the three-dimensional representation (figs. 1 and 2).

4.1.2 *Using Ward's method for curves considered as individuals*

Here (see figs 3 and 4) the distance used between two curves is the standard euclidian distance. The more the curves have close (or equal) values over the internal T, the more they will resemble each other. Classifying them in this way brings their resemblance at amplitude level to light.

4.1.3 *Using A.V.L. for curves coded as V.Q.O. variables*

This classification (see figs. 5 and 6) brings the sense of variation of the curves to light, independently of their amplitude. It will be seen that this is true if figures 5 and 6 are examined (principally figure 6), and "discontinuities" on the curves of one class will be noticed.

5 - REFERENCES

DEVILLE J.C.

Méthodes statistiques et numériques de l'analyse harmonique.

Annales de l'INSEE numéro 15 janvier - avril 74

FARAJ A.

Traitement statistique et typologie d'impulsions de foudre recueillies lors de la campagne COPELIA

Thèse de doctorat de l'université de Rennes 1 1988

FARAJ A. VALETTE N.

Analyse factorielle appliquée aux données COPELIA

Note technique NT/LAA/ITP/95 Lannion 1987

LERMAN I.C.

Classification et analyse ordinale des données

Dunod PARIS 1981

LERMAN I.C. -- PETER P.

Elaboration d'un indice de similarité entre objets de types quelconques

Rapport INRIA 1985

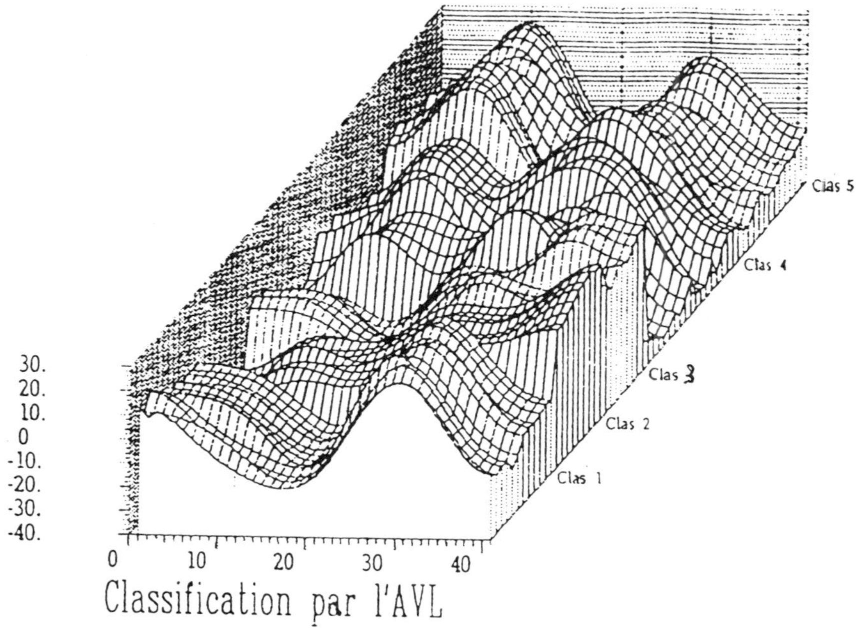
30.
20.
10.
0
-10.
-20.
-30.
-40.
0
10
20
30
40
Classification par l'AVL
Clas 1
Clas 2
Clas 3
Clas 4
Clas 5

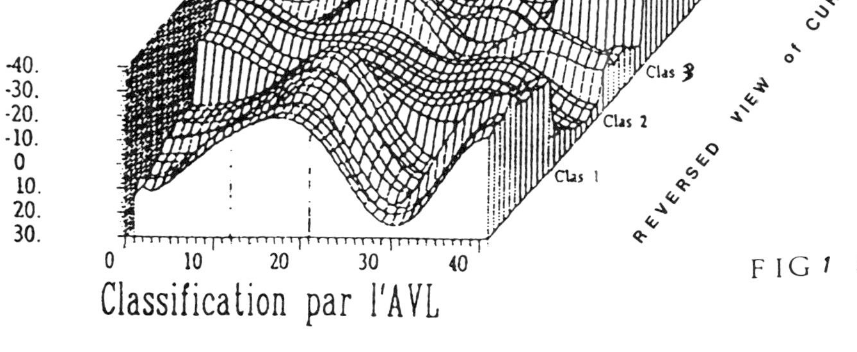
-40.
-30.
-20.
-10.
0
10.
20.
30.
0
10
20
30
40
Classification par l'AVL
Clas 1
Clas 2
Clas 3
Clas 4
Clas 5
REVERSED VIEW of CURVES

FIG 1 2

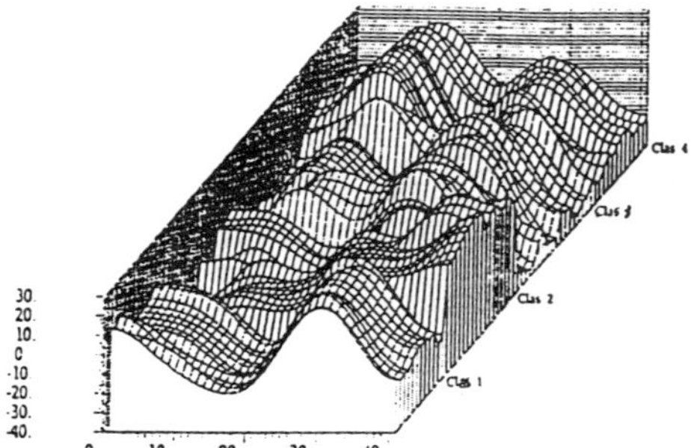

Methode de Ward

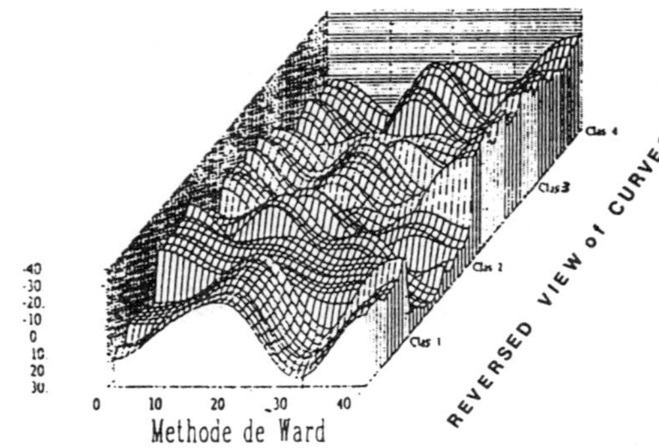

Methode de Ward

FIG. 3-4

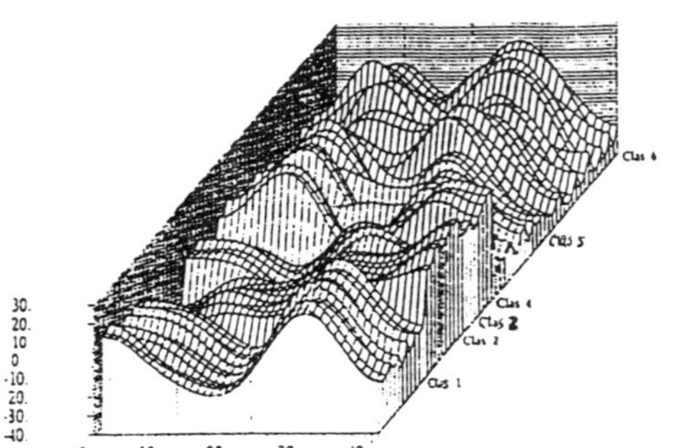

Courbes codees en VQO

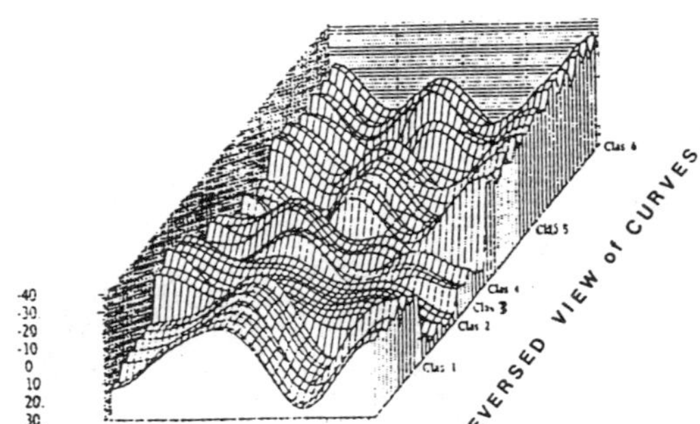

COURBES CODEES en V.Q.O

FIG. 5-6

COMPUTER PROCESSING AND QUANTITATIVE TEXT ANALYSIS: HYPERBASE, AN INTERACTIVE SOFTWARE FOR LARGE CORPORA

Etienne Brunet and Xuan Luong
URl 9:"Etude statistique du Trésor littéraire"
Institut National de la langue française (CNRS)
UER Lettres et Sciences Humaines 98 bd Herriot 06007 Nice (France)

ABSTRACT. The Institut National de la Langue Française is at the head of the largest linguistic data-base available in the world. This paper is devoted to the description of software that makes possible the decentralised exploitation of the data-base without the need to use any data-transmission network. This software, called *hyperbase*, is written in the *hypertalk* object-language. It implements the methods of *hypertext* and emphasizes two main orientations, respectively towards documentation and towards statistics.

Key-words: :
data-base, automatic documentation, computational linguistics, hypertext, CD-ROM.

RÉSUMÉ. L'Institut National de la Langue Française dispose de la plus importante base de données linguistiques qui existe au monde. On s'emploie à décrire ici un logiciel qui en permet l'exploitation décentralisée, sans recours à la télématique. Ce logiciel (HYBERBASE) , réalisé dans un langage-objet (HYPERTALK), met en oeuvre les méthodes de l'hypertexte et privilégie deux orientations dont l'une est documentaire et l'autre statistique.

(Mots-clés: base de données, documentation automatique, linguistique quantitative, hypertexte, CD-ROM

Do all roads lead to the CD-ROM[1] ? At any rate, a long academic tradition leads there. For over twenty years scholars concerned with the application to documentary research of both the resources of the computer and the methods of statistics have multiplied indexing or concordancing programmes and frequency dictionaries. Within the Institut National de la langue française itself, several such achievements have come about, using various programming languages, *PL1, Cobol, Fortran, Pascal, C*. The principle is simple: once the text has been recorded page after page, a direct-access file is created in which each record contains one line of text -or a paragraph or any previously defined unit of discourse-. At the same time, each word is identified and stored with its references (text code, page number, key to line or segment, etc.) in a so-called *reverse* file that has been suitably sorted and sifted. Afterwards, various

[1] Is it necessary to translate the acronym, which refers to the optical reading compact disc (Compact Disk Read Only Memory) as opposed to the Worm (Write Once Read Many times) medium ? Both products share laser technology but only the latter makes writing possible. When the device is associated with magnetic recording (optical-magnetic technique), their advantages are combined , making available both laser's large capacity (a billion characters) and the magnetic medium's own recording and deleting facilities.

possibilities are open, depending on whether they go through the lemmatization[1] stage,whether they take into account the combinatory possibilities of words, of their locations, of their contents, or whether they address questions of syntax or semantics.

These now common products have been very useful. Yet, they have suffered from serious handicaps such as the various forms of the data, the too narrow specialization of the software and the inevitable inadequacy of the former to the latter. They lacked at the same time versatility, power, speed, reliability and standardization.

-I-

Those qualities required by the community of scholars are present in the ***Frantext*** data-base, that the Institut National de la langue française made available two years ago to French and foreign scholars in linguistics or literature. Some specific software, created by Jacques Dendien under the name *Stella,* enables one to know the context of any word or phrase in any text -or set of texts- in the base. Let us dwell here on some advantages that are to be found nowhere else together in applications of this type :

1 - The use is conversational, whereas the packages employed sofar, such as *Cocoa, Jeudemo, Lexicloud*[2], were of the batch type.In the present software the information is distributed in a few seconds.

2 - All that is required is a terminal in a university library, without costly equipment or any technical knowledge.

3 - Yet the scholar has access to very large masses of data, since the Frantext base holds 150 million words, i.e. nearly 3000 complete texts of French literature from 1700 to the present day.

4 - But the scholar needs in no way to feel overwhelmed by such a mass of data for he can choose as he wishes the particular corpus in which he is interested by specifying various criteria such as the date, the genre, the author, the title, or even by imposing context-related constraints. For instance, he can select the works of Voltaire or the whole of XIXth Century novels or the texts containing the word "Revolution".

5 - Just as the scholar chooses his corpus, he also chooses the areas of his research , which can bear on individual forms, lists of words, cooccurrences, phrases. A very large range of combinations that call upon Boolean operators, makes it possible to outline the question with all the desirable precision.

---

[1] One describes as *lemmatization* the bringing together of forms that belong to the same dictionary entry. Thus, the forms *va, allons, irai*, once lemmatized, are brought under the head-word *aller*. Lemmatization is less urgent in moderately inflected languages such as English.

[2] Jeudemo is a University of Montréal product (F. Ouellete creation). Cocoa is also known as OCP (Oxford Concordance Programme), a Susan Hockey creation. as to Lexicloud, it is a package developed at the Saint Cloud Ecole Normale Supérieure by the URL 3 (A. Salem and P. Lafon). The product mentioned last has original properties, as it accounts for the combinations of words and repetition and association phenomena.

6 - Last but not least, the scholar can specify his own preferences as to the nature of the results -contexts, indexes, frequencies- and their final layout. He can at any moment print them, save them or interrupt their display.

- II -

The *Frantext* is the perfect example of what telematics can contribute to literary and linguistic matters. It has no equivalent in the world in this field.

Yet, the evolution of documentary techniques imposes on one to go beyond the present achievement, however remarkable. The compact-disc market, a long time hesitant, is now growing apace in the United States and in Japan, and offers an interesting alternative to telematics. the move is spreading to Europe and at the present time there are fifty documentary projects using this technology which are in progress in France, an example being that recently presented by Emmanuel Le Roy Ladurie for the French National Library[1].

The Institut National de la langue française has long been interested in the possibilities offered by optical memories in terms of storage capacity and ease of diffusion. It is now preparing a portable version of the totality of its data-base, which will allow transfer to new systems and new media, including the optical one.

In order to pave the way for this vast enterprise, a prototype has been experimented and is now running, performing the chief functions of *FRANTEXT* . It is the subject of the present paper.

1 - A search algorithm has been developed and implemented, taking into account CD-ROM dependent constraints, particularly the relatively slow disc access, which for any exploration restricts to two the types of movements of the reading-head. Accesses themselves have been optimized -on the basis of the supposed frequency of requests).

2 - The required speed has not entailed the sacrifices common in documentary products, such as their neglect of function-words. Those words on the contrary are nearly priceless in a linguistic type of research aiming at **exhaustivity**. The reverse file contains all the words and also includes narrative markers, particularly punctuation marks.

3 - To speed and versatility, the HYPERBASE software adds **flexibility** and precision of search. The user is given a choice of both **data** and **types of processing**. He alone determines the sub-set of texts he wishes to consult -by means of a CHOICE OF CORPUS programme which multiplies selection criteria such as genre, author, date, title, contents and applies to them Boolean operators. See figure 1-. He also freely draws up the list of the words he is interested in -CHOICE OF WORDS programme- offering automatic lists based on the selected suffix

[1] *PC Informatique*, september,1 , 1988, n° 46, p. 18.

Figure 1. Choice of corpus

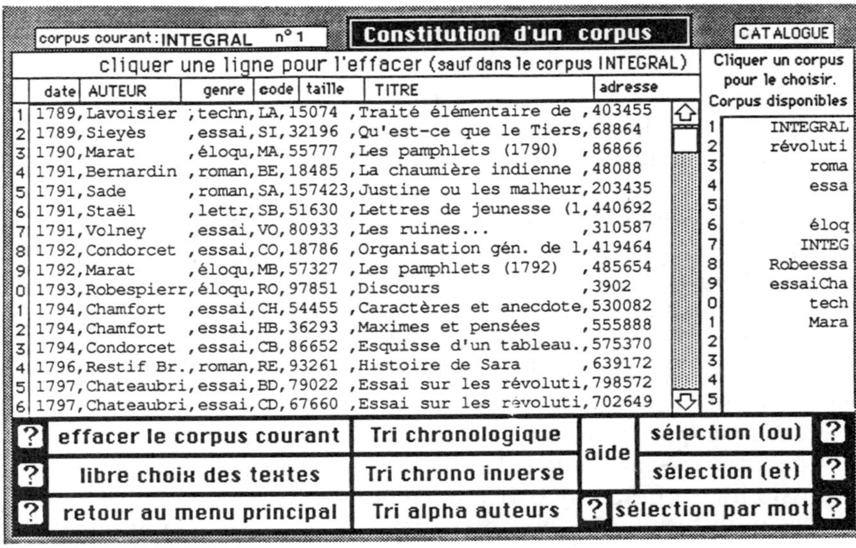

Figure 2. Choice of words

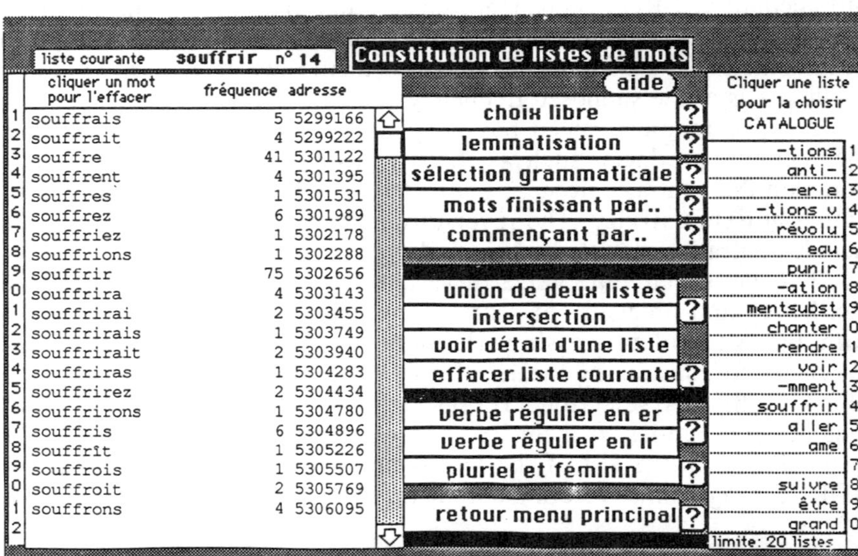

prefix or conjugation and allowing additions, deletions, overlaps. See figure 2-. Finally, he chooses from the menu the tasks to perform, which can lead to an index -INDEX programme-, to a concordance -CONCORDANCE programme-, to a list of phrase-contexts -CONTEXT programme-, or to cooccurrence searches -COOCCURRENCE programme-. He is totally free at any of those stages to supply a single word, a string of words, i.e. a phrase, or a list of words. See figure 3.

Figure 3 : The main menu

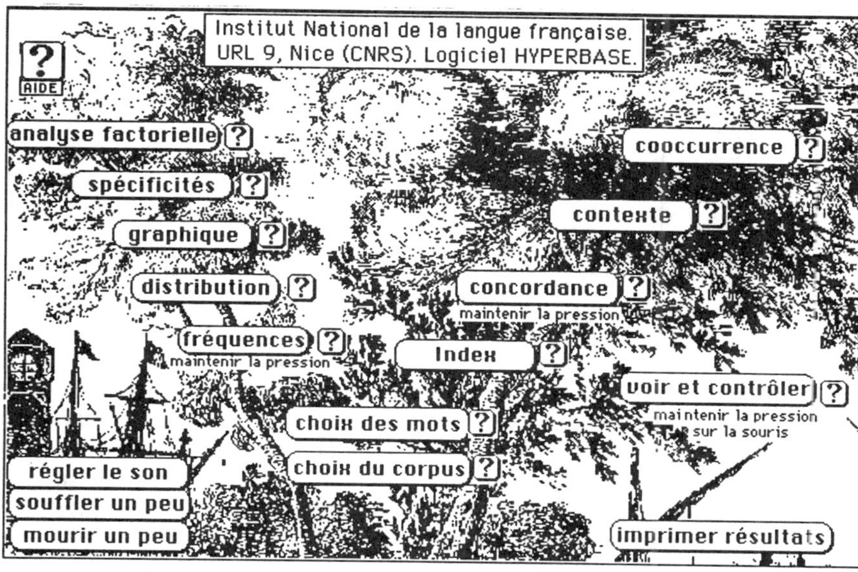

4 - The user can also conduct control operations and examine the texts, page after page -SEE A TEXT programme-, or consult the word file, in alphabetical order or in order of decreasing frequencies -SEE SOME WORDS programme- or pass instantly, by means of a simple "click" from the texts to the words or the words to the texts, according to the methods of hypertext. Should he wish to check the contents of the current list and corpus -which are variable and can be altered at will- a rapid overview will flip through the bibliographical slips of the texts in the corpus or the identity slips of the words in the current list.

The HYPERBASE software also leads towards **quantitative** methods. It can draw up contingency tables -FREQUENCY programme-, and produce graphs comparing the distributions of two words in the various texts, or the profiles of two texts with respect to a set of words -GRAPH programme-. There is also an interface for exploiting such tables by the methods of multidimensional analysis -FACTOR ANALYSIS programme-. Finally, various modules are available that display the distribution of frequency-classes within each text and in the whole corpus -DISTRIBUTION and ZIPF LAW buttons- or that list the specific vocabulary of a text or the specific properties of a word -SPECIFICITY button-.

6 - Lastly, the **results** appear under two forms : on the **screen**, at the very moment when they are obtained, and in a **file** where they are saved before being checked and printed. Both forms are not necessarily identical, the screen alone allowing the display of the complete text. The results then undergo various sorting operations and yield, among other things, concordances that come in various guises : chronological order, reverse chronological order, authors' alphabetical order, sort on the left-hand or right-hand context. An editor -complete with printing module- is supplied which facilitates corrections and layout. Let us note that the results themselves are not the end of the line and that they are amenable to the methods of hypertext : it is for instance enough to point to a line in the concordance to make the corresponding page of the text instantly appear.

7 - All the software has been written bearing in mind **user-friendliness**. Even if it is meant for a specialised public, it has not been given a stern countenance. On the contrary, one has attempted to exploit the **graphic** potentialities of the screen -for instance, the portraits of the writers or the fac-similes of the original editions are displayed-, and full use has been made of the operating facility of scrolling menus or of control buttons and the flexibility of use of the "mouse". Sound itself is welcome and **help** has been provided for each of the functions at the very moment when it is offered : it is enough to activate the question mark that comes with each button.

8 - This concern for user-friendliness and versatility explains the choice of an **object-language** for the whole of this software written in *HYPERTALK* and complemented by external functions and controls. As it is an interpreted[1] language and a modular process, the software is open and admits all kinds of complements.

- III-

1 - Some complements were devised at a later stage. One has endeavoured to supply each word with a grammatical tag -one is even thinking of semantic tagging- in order to make possible the study of grammatical categories, the disambiguation of homographic forms or thematic research. One has taken the option of automatic lemmatization, which opens the way to the treatment of lexemes as opposed to mere forms. One can join together all the forms connected with the same

[1] As a matter of fact, large portions of the processing have been compiled in order to gain both speed and reliability. One has not given up hope of compiling the whole, when the HYPERTALK language, yet too young, has been stabilized and an efficient compiler is available for it.

dictionary entry, either to draw up a list in the CHOICE OF WORDS menu, or to produce the concordance of a given lexeme, or even to find out what lexeme a form is connected to and what other forms are likewise connected to it -LEMMATIZATION button available for each form-. The new version of the programme also supplies for each word -at least for each lexeme- the frequencies observed in the entire TLF corpus and in the Revolution corpus. This therefore makes a very wide backdrop against which the specificities of the 1789 corpus stand out very sharply (z-scores are calculated as soon as permissible).

2 - In its present state, the prototype exploits a base of 24 complete texts of the revolutionary period, from 1789 to 1800. Robespierre rubs shoulders with Sade, Siéyès with Marat, Madame de Staël with Chateaubriand, and Chamfort with Condorcet. If the corpus gives pride of place to essayists -among whom Volney, Marmontel, Bonald-, it does not exclude novelists -for instance Restif de la Bretonne, Madame Cottin or Bernardin de Saint Pierre- or dramatists -Pixéricourt, Lemercier- or scientists -Lavoisier, Monge, Lagrange-. In all, a total of 1 300 000 words are available.

The HYPERBASE software and the associated data have been made accessible to the public of the Centre Georges Pompidou (Beaubourg) as part of the celebrations of the bicentenary of the French Revolution, with the backing of the Apple company who granted the free loan of a MAC IIX. The version on show -which comes in colour- works with a 40 Mo hard disk. The base has also been stored on a WORM compact disc, in order to experiment laser technology. But the CD-ROM version is not planned for the immediate future, because the HYPERTALK language does not seem to be stabilized and because great improvements are expected in the coming months as far as confidentiality and speed[1] are concerned. the HYPERBASE software being under development contract (Apple company), we can hope to be among the first to reap the fruit of future improvements.

3 - As is usual with producers and distributors of data-bases, the interrogation programme is inseparable from the processed data.

a - But copyright and confidentiality problems have led us to envisage offering an empty software, devoid of any data. The user would enter his own texts. It remains to supply him with a chain of data-preparation

[1] We have known four versions of this language within one year. The example of SUPERCARD shows that the handling of multiple windows, of large screens, of colour, is not impossible and that the advantages of compiling can be expected in the near future.

programmes[2]. Such programmes exist since they have made possible the processing of the Revolution texts. But it remains to endow them with the required universality, speed, versatility and reliability[3] .

b - There are also plans for a version of the software that might not display the entire text, when the privileges of copyright oppose any direct publication. The original text would indeed be used to supply the answers, but once the answers given, the text would be, as it were, spirited away. This amounts to the opposite of the present HYPERBASE procedure which waits for the user's questions and looks for the answers in real time. The enormous possibilities of the CD-ROM make it possible to store in advance all foreseeable answers, for instance the concordance to all the words in the corpus and to restore them at the time of questioning, when the original text has been made to disappear. As a matter of fact this product has a fair likeness to what was done in the past using other media such as paper or microfiches. the advantage of the CD-ROM however would be to allow direct access without any manipulation whatsoever. By combining several modes of display of the results -based on preliminary sorts- one might offer various options. Also, by playing on the fields of reference, it would become easy to filter the results and to facilitate their selection.

In short, the two versions envisaged here lie on either side of the data-base watershed. The former is richer, perhaps more exciting because it makes the pleasure of discovery more vivid in the presence of the entire original text. The latter is well-mapped, clearly sign-posted. The routes have been carefully charted and travelling them is quicker. It is the field of structured data bases.

---

[2] Is it useful at this point to go into the detail of the various preparatory phases ? Let us simply say that after the checking operations the texts are segmented into paragraphs and pages and recorded before creating the reverse file and the access-keys -the so-called search algorithm-. Let us specify that these programmes are written in a conventional language, Pascal or C, and that at certain stages of the treatment bridges will be provided towards Hypercard, particularly for knowing the physical address given by Hypercard to each of the created cards.

[3] If a bare software is offered, it is probably not necessary to plan an optical medium, at least not the CD-ROM, whose manufacturing costs cannot be justified for a single copy -which is often the situation of the scholar in the humanities-. If the WORM system can perhaps be suitable, until the optical-magnetic technique is available, the hard disk is the natural medium for such applications. One becomes convinced of this when one looks at the market : the available products all lean towards the hard disk. It is for instance the case of Micro-OCP, created by Susan Hockey, Oxford Computing Service, of TACT (Thematic Analysis of Computer-readable Texts), written by John Bradley and Lidio Presutti, Centre for Computing in the Humanities, University of Toronto and of WordCruncher (Electronic Text Corporation, Provo, Utah), even if this well-known software can now accept the CD-ROM.

SMOOTHING DATA

THE INTERPLAY BETWEEN STATISTICS AND COMPUTING IN DATA ANALYSIS

Dedicated to Friedhelm Eicker

W.Härdle
Wirtschaftstheorie II
Adenauerallee 24-26
Rheinische-Friedrich-Wilhelms Universität
D-5300 Bonn

Abstract

The personal computer is becoming a widely used tool for data analysis. I describe some minimum requirements for a data analysis machine and discuss the theoretical objects (machine data structures) in order to efficiently organize the interplay between statistics and computing. The ideas are shown to be implementable on a desktop system like an IBM AT or PS/2. The particular implementation discussed here is called *XploRe* and is written in TURBO PASCAL 5.0.

1. Statistics and Computing

Detecting structure is a typical task for a statistician interested in model construction and analysis of relations between variables. "Structure" can have many meanings: outliers at specific locations (or indices), clusters of points, linear dependence between variables, etc. Before we can see any kind of structure we have to ask how to detect it and what tools do we need to provide in order to decide whether something is really a structure. Very simple structural elements of a data set may be detected just by looking at the list of values, e.g. a column with small numbers with the exception of one large one will indicate an outlier. On a finer level summary statistics for the variables and the use of graphical techniques like Boxplots help us to see data structures like skewness, for example. Suppose one has detected by these techniques that certain variables have a right skew distribution. A very natural next question is whether the points causing this skewness create other structural features in the other variables as well. Can we efficiently perform this task with one of the "classical packages" ?

Most statistical packages now in use, such as SAS or MINITAB, were designed to operate in a batch-like mode. The user enters commands from the keyboard and the system scrolls subsequent stages of the analysis and displays on explicit request values or plots. Therefore the data analytic task, "investigate the behavior of other variables given the detection of skewness in a particular variable" cannot be done instantaneously. Instead we have to write a separate program. This program has to first identify the indices of the interesting subset and then make several statistics and plots from other variables on the basis of this subset.

The development of hardware in the last years has made it possible to carry out this task and similar ones within a desktop system. It should be emphasized though that the hardware, in principle, gives the "possiblity" to perform such operations efficently at a high "power/price" ratio but in many cases the computing environment (the software) is not capable of doing what a data analyst wants to do. It is the object of this paper to propose some standards for an ideal computing environment and to demonstrate a first approximation to these ideals in an existing computing environment based on a widely used IBM desktop system (IBM AT, IBM PS/2).

The desktop concept allows us to analyze various facets of the data from different viewpoints (windows) at the same time. Stuetzle (1987) has made this evident, stating:

Each window and in particular each plot, corresponds to a sheet of paper on the desktop.

Before computers entered our work we shuffled paper on the desk to compare and evaluate different structural elements of the data. Now the paper shuffling is replaced by window clicking and logical links between the sheets are not only in the statistician's mind but also between defined computer objects inside the machine. These computer objects and their implementation are described below. It is argued in Section 2 that instead of expensive hardware it is a matter of software to bring data analysis into life, especially data analysis in high dimensions. In Section 3 I describe a current approximation to a data analysis computing environment. Section 4 is devoted to an example session with the system *XploRe*.

2. The Corvette versus the Chevette: Do four cylinders suffice ?

In analysing data we are programming, in effect, our view towards the data and the structures we want to and can see. By selecting certain scales we can concentrate our eyes dynamically on local or global features, see McDonald and Pederson (1986). For example, with a single column of numbers, by using LOG–transformations we can sharpen our eye for peaks or other local structures. As a consequence we need a flexible programming and computing environment that allows us to sharpen our eyes if necessary for interesting details on the macro or micro scale. Assisting tools in this task are four "cylinders" (minimum requirements):

⊙ The capability of showing 3D rotation motion.

⊙ The capability of a multi-window system.

⊙ A high resolution display (at least 600×400 pixels).

⊙ A graphical input system like a mouse or a tracker ball.

Even today, high resolution display and the mouse are available at low cost. The important remaining cylinders are the capability of a multi-window system and 3D motion. Having the possibility of a multi-window view on the data enhances our capability of finding structure and the 3D motion allows us to see in an exploratory way features in 3D space. These two cylinders can be driven and supported from different levels of the machine's hard- and software. By adding more cylinders (e.g. hardware), the multi-window technique, for example, can be realized on the operating system level or close to it. More cylinders mean in this case that window operations like opening, closing, moving, reshaping, etc. are available on a (fast) low level. On the other hand we need more than elementary window handling techniques.

Let me give a simple example. Suppose we are looking at similarly scaled 2D data repeatedly by studying a sequence of scatter plots. Our eyes would "detect" artifacts due to different scales unless the scale of the 2D scatterplot pictures are the same for all plots. It is therefore highly desirable to have the possibility of a STATIC2DPICTURE object (possibly of window type) which stays unchanged in subsequent calls. Practically speaking this means that a picture shown in a window system "has to know what scales it had in earlier calls". In a more modern language, we would say that the picture should have the capability of inheriting properties of earlier stages of an analysis.

This capability of inheritance in an Object Oriented Programming System (OOPS) creates extra difficulties in software construction. If we try to realize this concept of a statistical STATIC2DPICTURE object in a eight cylinder vehicle (e.g. the machine with a built in window system) we face extra difficulties since we have to artificially attach statistical meanings to windows constructed for a different purpose. Thus the construction of a "statistical data window" can bring us down to ordinary programming speed. With other words more cylinders can be more expensive and may even slow us down. For this reason some people have called such vehicles "Corvette data analysis machines".

The Corvettes are very powerful, not very frequent among the data analysis vehicles and sometimes have to drive the speed of a Chevette.

A Chevette is a data analysis machine that operates from a lower powered level but has the (dis)advantage of free object programming. Free object programming means that we can define our programming objects by ourselves in one horizontal level without time consuming "vertical diffusions" to other levels. Of course this is an advantage from a pure theoretical point of view but a disadvantage on the programmer's side. The other advantages, such as price and portability, make it clear that the "Chevette data analysis machine" is the preferable choice. Note that portability has two meanings here. First we can export the code more easily since it is written in one level although possibly simulating deeper levels. Second we can implement the code on portable machines (laptops) to show data analysis live to other statisticians ! Summarizing these thoughts leads me to the conclusion that a four cylinder personal computer like an IBM AT or PS/2 is acceptable for interactive data analysis if we tune it correctly e.g. program it correctly. In the following I describe a four cylinder Chevette called *XploRe*.

3. XploRing data

XploRe is an interactive, graphically oriented computing environment designed to analyse various kinds of relations between data and to apply and compare different smoothing methods. *XploRe* is suitable for investigating high dimensional data. It supports the user with the above four cylinders and other sophisticated data management tools such as masking, brushing, labeling and rotation of data. In addition, a wide variety of additive models are available, among them **ACE** (Alternating Conditional Expectations), Breiman and Friedman (1985), **ADE** (Average Derivative Estimation), Härdle and Stoker (1989), **PPR** (Projection Pursuit Regression), Friedman and Stuetzle (1981).

XploRe is designed as an open system. It is basically a framework awaiting more software in the form of user written submodules. The user can write his or her own programs and add them into *XploRe* via an object oriented user interface. The construction of *XploRe* has been influenced by similar systems like S (Becker, Chambers and Wilks, 1988) and ISP (Interactive Scientific Processor). It differs from both systems by the fully graphically oriented user interface. The hard- and software background of *XploRe* consists of: an IBM Personal Computer AT, XT, PS/2 or a closely compatible machine, runnning DOS Version 2.0 or later.

XploRe is an OBJECT oriented system. An object can be of one of four types:

- VECTOR
- WORKUNIT
- PICTURE
- TEXT

A *VECTOR* object is a data vector as a logical unit with which to work. This vector may contain strings or real numbers and can be of variable length. The simplest form of a *WORKUNIT* object is an ordered collection of data vectors. However, a workunit can also include display attributes and a mask vector. Display attributes concern the layout of scatter plots such as data marking symbols, linestyle, line pattern, line thickness etc. A *TEXT* object is necessary to show information as text on the display. This can be data you are analyzing, documentation you want to pin down or system output. A *PICTURE* object contains the viewport characteristics of certain views that you have on data, e.g. the name of the picture and the axes and the scaling of the axes. For 3D rotation the rotation angles, the initial distance from the point cloud, location of the origin and zooming increments are kept in this picture object.

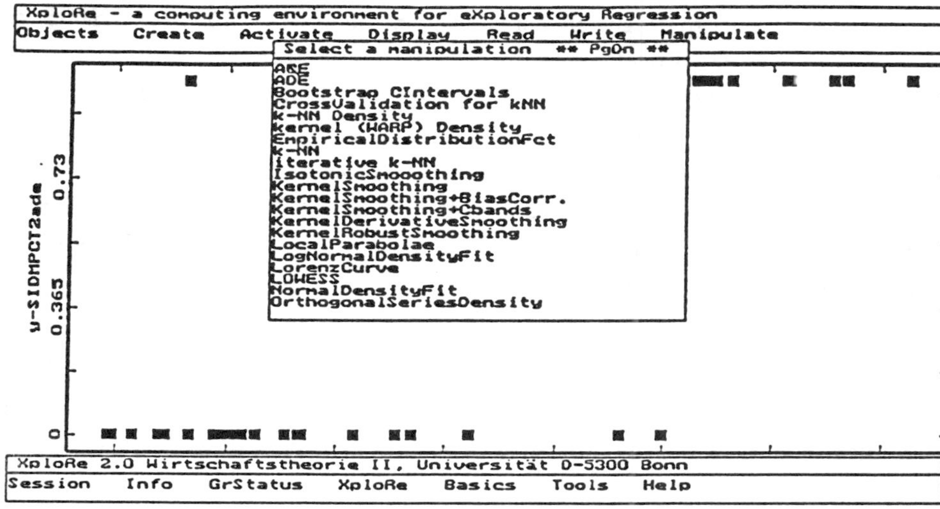

Fig.3.1: The two menu bars of the XploRe main menu after pressing "M". Underlying is a STATIC2DPICTURE scatterplot.

XploRe has a menu structure. Two menu bars will appear on your display (see Fig. 3.1). A third menu bar appears, if you press and hold down the <ALT> key. You can choose an option by typing the capitalized letter of the corresponding menu entry. Whenever a pull down menu appears at the next step, it is in most cases possible to get *QUICK HELP* by pressing **<ALT>+F1**. For example, if you are not familiar with the ACE algorithm, you press this key sequence and a help file explaining the algorithm pops up. The main keys are explained below.

OBJECTS (o,O) After clicking *Objects* you can see an overview of the existing objects with their current names. In addition the type of object (vector, workunit, ...) is indicated.

CREATE (c,C) You are asked which object you would like to create. Clicking *WORKUNIT* will show you a window "select a vector number: 1 or ESC" containing all vectors of the active workunit. In this way you can create other workunits from existing vector objects. You can select as many vectors you want. Clicking *VECTOR* will allow you to make a new vector from an existing vector, e.g. by taking LOGs. After clicking *PICTURE*, you are asked for a desired picture type. By choosing the option *STATIC2DGRAPHICS* you can create a picture object for 2 dimensional static graphics, while the option *DYNAMIC3DGRAPHICS* defines an object for dynamic 3D graphics. The menu entry *DRAFTMANSPLOT* will create up to 25 2-dimensional scatterplots by plotting each of up to five selected vectors (which are kept in the same workunit) against each other of them. After clicking *TEXT* a window "create text" appears on the screen. This means that you have invoked the editor of *XploRe* and are able to write ASCII texts. It is now possible to write datasets or comments on data without leaving *XploRe*.

ACTIVATE (a,A) You are asked which object type you would like to activate. By clicking the object of a certain type (in a window "activation") you activate the object. This means that this object becomes the default data set or picture for following operations.

DISPLAY (d,D) This feature allows you to display any existing object of *XploRe*. Again you will be asked by a window to select an object type to display. Suppose you have created some workunits and you want to display one of them. First you will be asked to choose a corresponding picture object by showing a window. The picture object contains the characteristics of the viewport (axes and origin) of the graphical display. There are three different kinds of display styles available. The option *STATIC2DGRAPHICS* will show you a two-dimensional picture display of two selected vectors of your dataset, whereas *DYNAMIC3DGRAPHICS* shows you a three-dimensional picture display of accordingly three vectors of the workunit. If you would like to display a dataset with more than three dimensions, you have the possibility to display two-dimensional scatterplots of up to five vectors against each other of them. In this case you must choose the option *DRAFTMANSPLOT*. If you want to display a text object (the data vectors of a workunit or the corresponding explanations) you first have to read the workunit or the corresponding text file as a *TEXT* object. After this action you can display the vectors or the explanatory help file.

READ (r,R) You are asked what kind of object you want to read. If you want to display data as text it is neccessary to choose the *TEXT* option. In a next step you have to select the subdirectory of your file . You can choose a standard DOS wildcard mask. After clicking a file *XploRe* creates the reads the desired object.

WORKUNIT INFORMATION (i,I) By clicking the menu option *Info* you first have to select a workunit. After it a window (Fig. 3.2) will be shown.

XploRe - a computing environment for eXploratory Regression

Objects Create Activate Display Read Write Manipulate

--- List of WU and maskarray --- ** PgDn **

	SIDMPCT2_1	SIDMPCT2_2	SIDMPCT2_3	SIDMPCT2_4	mask
1	22.00000	50.00000	98.00000	0.00000	-
2	21.00000	49.00000	160.00000	0.00000	-
3	40.00000	50.00000	134.00000	1.00000	-
4	43.00000	50.00000	142.00000	1.00000	-
5	23.00000	51.00000	118.00000	0.00000	-
6	58.00000	51.00000	143.00000	1.00000	-
7	29.00000	51.00000	77.00000	0.00000	-
8	29.00000	51.00000	184.00000	0.00000	-
9	47.00000	51.00000	100.00000	1.00000	-
10	39.00000	51.00000	188.00000	1.00000	-
11	22.00000	50.00000	162.00000	0.00000	-
12	52.00000	51.00000	151.00000	1.00000	-
13	28.00000	50.00000	181.00000	1.00000	-
14	42.00000	50.00000	158.00000	1.00000	-
15	59.00000	51.00000	168.00000	1.00000	-
16	28.00000	41.00000	128.00000	0.00000	-
17	23.00000	61.00000	268.00000	1.00000	-
18	38.00000	41.00000	76.00000	0.00000	-
19	50.00000	61.00000	185.00000	1.00000	-

XploRe 2.0 Wirtschaftstheorie II, Universität D-5300 Bonn

Session Info GrStatus XploRe Basics Tools Help

Fig.3.2: Workunit Information of a workunit containing the sideimpact data, see section 4.

MANIPULATE (m,M) With this operation you invoke the manipulation part of *XploRe*, see Figure 3.1. At first it is necessary to activate the corresponding object you would like to use by the *Manipulate* option (see *Activate*). Then you can select an operation under the menu entry *Manipulate*. At the next step you have to select an *XploRe* object which should be the "input" of the selected operation. The calculation time depends on the complexity of the procedure being used. The result of this operation is stored in a new workunit which will be the top most entry in the window if you click the menu option *Objects*.

SESSION INFORMATION (s,S) This menu option shows a window containing all important information on the actual *XploRe* session. These are the active objects, time and the remaining memory (number of available bytes).

GRAPHICSTATUS (g,G) This option allows you to alter the graphics driver of your screen display. You have the possibility to select of 8 drivers. These are

- CGA
- MCGA
- EGA64
- EGAMONO
- HERCMONO
- ATT400
- VGA
- PC3270

BASIC STATISTICS (b,B) By this menu option you can select one of the following four basic statistics. The menu option *Boxplot* shows you a parallel boxplot of all vectors which correspond to the workunit you have to select. In the box on the left side of your display you can see a scale (extends from minimum to maximum of all data) and a legend which contains the marker symbols of median, mean, inner- and outer fence. The option *Stem and Leaf Plot* shows you the corresponding display of all vectors after selecting the desired workunit. The option *Data Summary* gives a summary of a selected workunit by showing minimum, maximium, range, mean, median, variance and upper and lower quartile of all vectors. Finally the last option *Correlation Matrix* shows a matrix containing the correlation between all vectors of the workunit.

TOOLS (t,T) You can *Edit workunit display attributes* and *Edit picture display attributes*. If you have created many new objects during an *XploRe* session and you don't want to write all objects separately, it is useful to click the option *Save all* in order to write all existing objects held in memory at this moment. To read all objects of an *XploRe* session back into memory you have to click the option *Load all*.

HELP (h,H) This option informs you about some important keys to get help or to leave *XploRe*.

CLEARSCREEN (<ALT>+c,C) Clears the screen, but leaves the current objects active.

EXIT (<ALT>+x,X) With this option you can leave *XploRe* and return to DOS. Don't forget to save the *XploRe* objects which you want to keep on disk for later use.

OS SHELL (<ALT>+o,O) If you want to use the DOS shell during an *XploRe* session type one of the above keys. To go back to *XploRe* type EXIT and the main menu appears again.

DELETE (<ALT>+d,D) You are asked which object type you would like to delete. By clicking the object type and the object name a deletion of the object is performed.

ENVIRONMENT (<ALT>+e,E) This option shows the content of the pascal source file `typedef.pas` and contains the declarations of the variables types used by the *XploRe* pascal modules.

INVERTSCREEN (<ALT>+i,I) Inverts the actual screen display.

4. An example of an XploRe session

In this section I present an analysis using *XploRe* with the side impact data set given in Table 3, Appendix 2 of Härdle (1989). These data have been gathered by simulating side impacts with Post Mortal Test Objects (PMTO). The response variable is $Y \in \{0, 1\}$ a binary variable denoting fatal injury ($Y = 1$) or non–fatal injury ($Y = 0$). The predictor variables are $X_1 = AGE$, the age of the PMTO, $X_2 = VEL$, the measured speed (in km/h) and $X_3 = T12RM$, the measured accelaration (in g) at the 12th rib. The aim of the analysis is to devise a model for predicting the probability of fatal injury given a certain x.

Figure 4.1a shows a scatterplot of the variables AGE and Y in this side impact data set. One can immediately see that there are a few high AGE variables which seem to fall out of the response pattern for the other observations. How can we investigate this further ? We invoke the cursor and move it to the points we are interested in. In Figure 4.1b this action is shown with the right-most AGE value. By studying the values of the other variables shown in the window on the left top of the display we see that this observations is from an experiment with a velocity of $45km/h$ and received an input of $T12RM = 103g$.

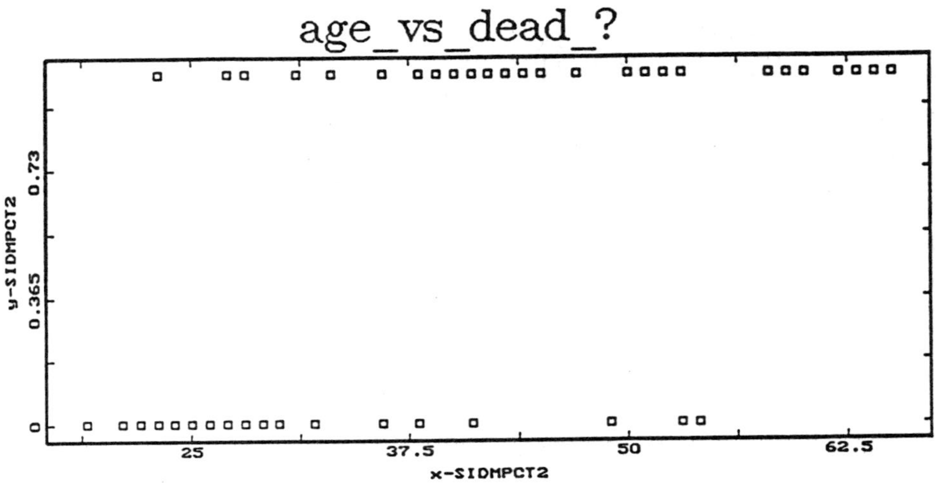

Figure 4.1 a. A scatterplot of the variables AGE and Y.

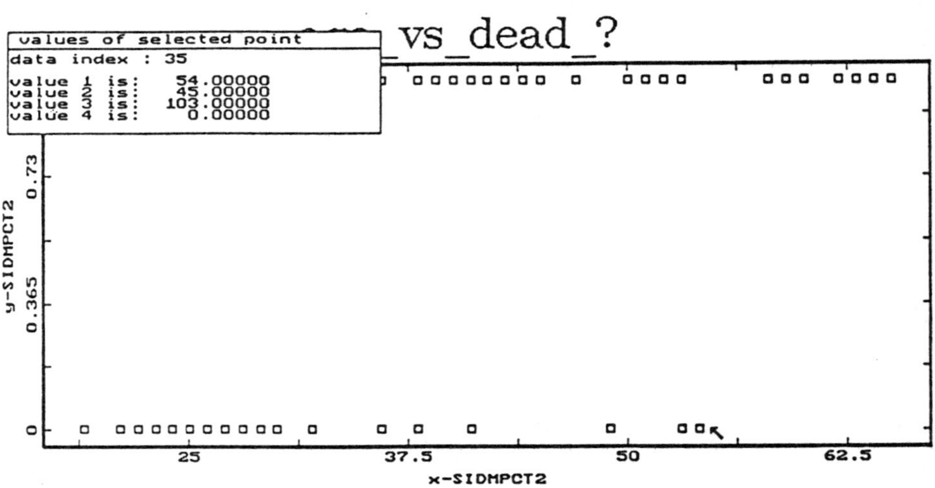

Figure 4.1 b. A scatterplot of the variables AGE and Y with activated cursor for control of values. The cursor is at $(x,y)=(53,0)$.

We have now a hint this specific data value is an outlier in this marginal scatterplot. But how can we see more ? We display the same data set in a DRAFTMAN'S plot, a system of all pairwise scatterplots, see Figure 4.2. By using a brush (in the $Y vs. X_2$ scatter) and by highlighting the values with $VEL = V0 = 45km/h$ we can see that those observations correspond to a nearly uniform AGE distribution in the scatter plot AGE vs.VEL. Brushing is a conditioning technique: by conditioning on the experiments with $VEL = 45km/h$ we see the distribution of the other marginal variables. By moving the brush to the right we can see that the pattern of the highlighted points in the Y vs. AGE plot changes from many "0"s to many "1"s. So we can expect on an intuitive level a positive effect of this influental variable on $m(x) = P(Y = 1|X = x), x \in \mathbb{R}^d$.

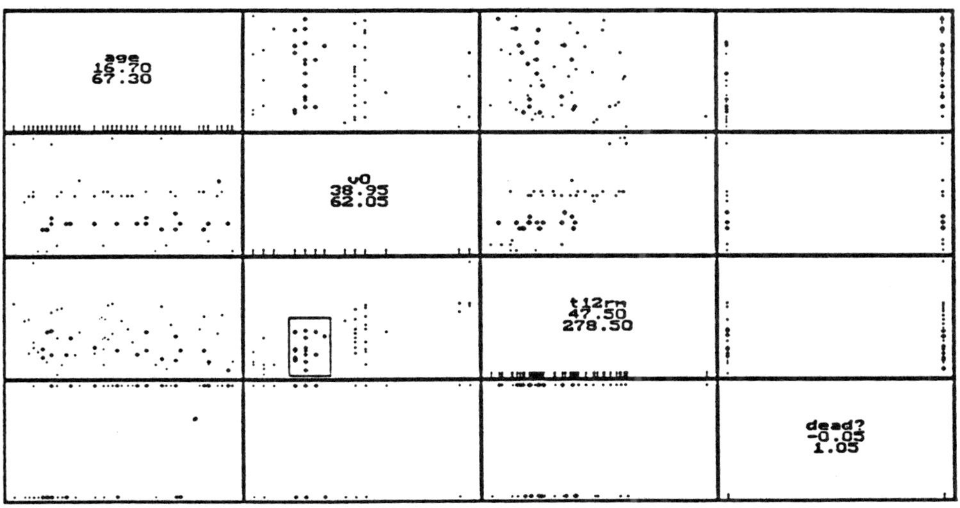

Figure 4.2. A draftman's plot with a brush at the scatterplot $Y vs. X_2$. The brush is at the $45km/h$ group.

To investigate this point more deeply we apply the ADE method. The ADE method is a technique for finding the average derivative

$$\delta = E_X[m'(X)],$$

here $m'(\bullet) \in \mathbb{R}^d$ denotes the gradient vector of $m(\bullet)$. For Generalized Linear

Models with unknown link function, i.e.,

$$m(x) = g(x^T \beta),$$

the average derivate is proportional to β, $\delta = \gamma\beta$ with a scalar γ. Thus, computing γ gives an estimate of β (up to scale) without knowledge of the link function. For this purpose we press "M", for manipulation and have a menu of a variety of smoothing techniques (see Figure 3.1). After pressing the ENTER key when the cursor is on the "ADE" line we are asked to enter the bandwidth h. Using a bandwidth of $h = 1$ we obtain a value of

$$\delta = (0.423, 0.061, 0.137)$$

for the standardized regressors.

The corresponding projection $\hat{\delta}^T X$ vs. Y is shown in Figure 4.3 together with a kernel smooth and confidence bands. One clearly sees the asymmetry of the response function due to a cluster of five points (high AGE, but low $T12RM$). One may now compare this nonparametric link function with more traditional approaches such as logistic regression analysis using the GLIM module of *XploRe* .

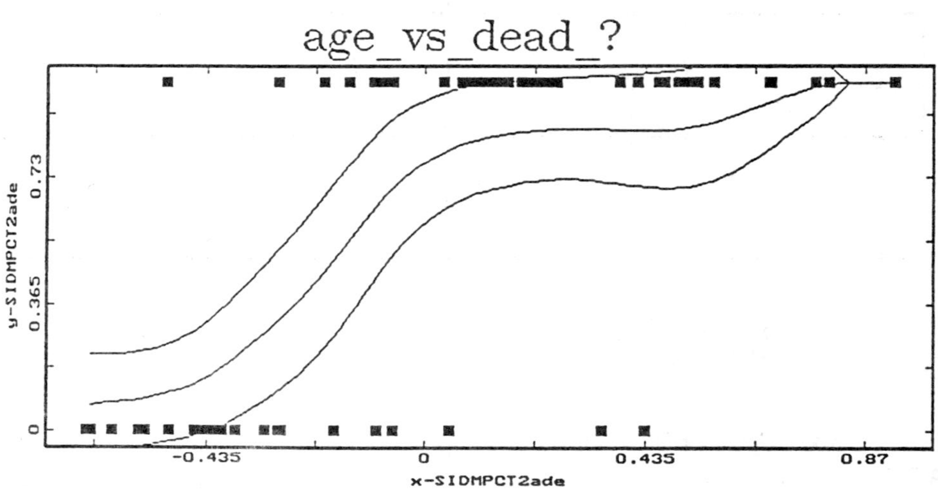

Figure 4.3. The projected data, a kernel smooth plus confidence bands in a STATIC2DGRAPHICS Picture.

Acknowledgements

I would like to thank Ray Carroll and David Scott for numerous fruitful discussions on the subject of statistical computing. They helped sharpening and programming my thoughts on the interplay between computing and statistics.

References

Becker, R.A., Chambers, J.C. and Wilks, A.R. (1988). The New S language: a programming environment for data analysis and graphics. *Wadsworth and Brooks/Cole Advanced Books and Software, Pacific Grove, California.*

Breiman, L. and Friedman, J. (1985). Estimating Optimal Transformations for Multiple Regression and Correlation. *Journal of the American Statistical Association, 80, 580–619.*

Friedman, J. and Stuetzle, W. (1981). Projection Pursuit Regression. *Journal of the American Statistical Association, 76, 817–823.*

Härdle, W. and Stoker, T. (1989). Investigating Smooth Multiple Regression by the Method of Average Derivatives. *Journal of the American Statistical Association, to appear.*

Härdle, W. (1989). Applied Nonparametric Regression. *Econometric Society Monograph Series, Cambridge University Press, in print.*

McDonald, J.A. and Pederson, J. (1986). Computing Environments for Data Analysis: Part 3, Programming Environments. *Laboratory for Computational Statistics, Stanford, Technical Report, 24.*

Stuetzle, W. (1987). Plot Windows. *Journal of the American Statistical Association, 82, 466–475.*

THE DATA ANALYSIS OF VECTOR-VALUED FUNCTIONS

J.O. Ramsay
McGill University
(Canada)

1 Introduction

Multiway designs often involve observation of a vector-valued function x at specific discrete points $t_j, j = 1, \ldots, J$. Multivariate time series analysis concerns itself with such data, but the perspective in this paper is much more general, and assumptions such as mean and covariance stationarity will play no role here. It is assumed only that x is a function mapping a closed real interval T such as [0,1] into a real vector of dimension $K \geq 1$. We can use the notation $x_i(t_j)$ to indicate the ith replication of this multivariate process $i = 1, \ldots, I$.

This paper extends work by Besse and Ramsay (1986) on functional principal components analysis and by Ramsay and Dalzell (1989) on functional data analysis in general. A number of authors including Deville (1978) and Saporta (1981) have considered the data analysis of vectorial temporal processes, and of course there is a vast literature on discrete and continuous stationary time series.

The applied context for this paper is the analysis of monthly means ($J = 12$) for temperature and precipitation ($K = 2$) at 35 Canadian weather stations ($I = 35$) with a view to identifying important components of variation among stations in terms of the joint behavior of these two variables. Figure 1 displays these data for four selected stations. As a part of the analysis, a certain amount of smoothing of the data will be required since monthly means are taken over 30 years and therefore are subject to a noticeable amount of random fluctuation or noise. Moreover, we shall choose to ignore differences between weather stations for either variable due to either yearly level or mean or to simple sinusoidal or harmonic components. Thus, two stations for which temperatures are strictly sinusoidal but with differences in mean, amplitude, or phase will be treated as identitical.

The theory of spline functions (Laurent, 1972; Schumaker, 1981) permits a natural discussion of such data. Both interpolation and smoothing splines presume a partitioning of functional variation of $x(t)$ into two components:

1. a *structural* or model component $u(t)$ which represents predictable or known variation, and

2. a *residual* or unknown component $e(t)$ which is orthogonal to $u(t)$ in a specified sense.

Thus partitioning $x(t) = u(t) + e(t)$ represents a fundamental aspect of the theory of splines which can provide important insights into the behavior of the data. For the weather data,

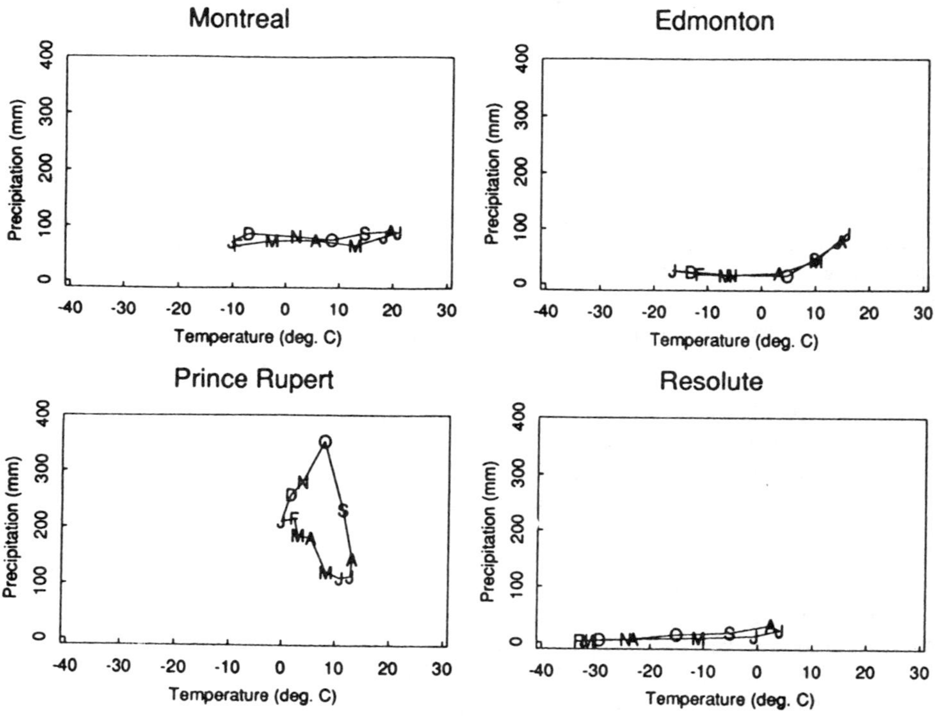

Figure 1: Joint variation in temperature and precipitation monthly averages at four Canadian weather stations.

the structural component will be of the form $u(t) = c_1 + c_2 \sin \pi t/6 + c_3 \cos \pi t/6$, where the coefficient vectors c_1, c_2, and c_3 determining mean, amplitude, and phase vary from station to station. Thus component e represents *anharmonic* variation.

A further component of variation in an observation might arise from noise or observational error at each sampling point, so that an actual datum y_{ij} may arise from the process

$$y_{ij} = x_i(t_j) + \xi_{ij} = u_i(t_j) + e_i(t_j) + \xi_{ij} \tag{1}$$

where the ξ's are presumed to be independent random disturbance terms.

A specific analysis of functional data requires four steps:

1. specification of the function spaces involved
2. definition of the analysis in functional analytic terms
3. representation of the finite data in function space
4. description of the analysis of the representing functions in terms of the original finite set of observations.

Each of these steps will be described in the following sections, although page limits on this exposition prevent detailed discussion or proof of results. A separate manuscript (Ramsay & Dalzell, 1989) provides a more detailed treatment.

2 Notation

The notation used in the paper is summarized as follows:

Variables

- Letters $i, \ldots, n$ will be used to index elements of a set and $I, \ldots, N$ to denote the cardinality of the set. Specifically,

  I The sample size for a sample of observations

  J The number of points at which a function is observed

  K The dimensionality of the range space for vector-valued functions

  M The dimensionality of a subspace spanned by known functions

- ξ denotes a random scalar disturbance or error

Sets and Spaces

- T is a compact subset of the real line $\mathcal{R}$ such as [0,1]. Specific elements in T will be denoted by s and t.
- $\mathcal{L}^2$ is the Hilbert space of square Lebesgue integrable functions defined on T with inner product

$$\ll x_{k1}, x_{k2} \gg = \int \omega(t) x_{k1}(t) x_{k2}(t) dt \ .$$

  where weight function ω is positive and absolutely continuous. $\mathcal{L}^{2K}$ denotes the space of K-vector valued functions with elements in $\mathcal{L}^2$.

- $H^M \subset \mathcal{L}^2$ is the space of K-vector functions x such that $D^M x \in \mathcal{L}^{2K}, m = 0, \ldots, M$, where D^M denotes the taking of the Mth derivative. H_0 is a subspace of H^M with the decomposition $H_0 = H_1 \oplus H_2$.
- G is an arbitrary vector space
- $\mathcal{R}^{IJ\cdots}$ is the space of multiway arrays or tensors of real numbers with dimension $I, J, \ldots$.

Arrays

- $Y \in \mathcal{R}^{IKJ}$ is the three-way array containing I replicated observations of K-vector valued functions at J points in T.
- $W \in \mathcal{R}^{II}$ is a diagonal matrix of weights associated with replications.
- $\Theta \in \mathcal{R}^{KK}$ is a matrix of weights associated with dimensions.

Functions and Mappings

- $x : T \rightarrow \mathcal{R}^K$ is a K-vector valued function defined on T, and u and e will also denote K-vector functions. The kth component of x is x_k.
- $< \cdot, \cdot >$ and $\ll \cdot, \cdot \gg$ are inner products defined on $\mathcal{L}^2$ and $\mathcal{L}^{2K}$, respectively.
- $\boldsymbol{D} : H^M \rightarrow \mathcal{L}^2$ is a linear differential operator of order M, and $H_1 \equiv \ker \boldsymbol{D}$.
- $\boldsymbol{B} : H^M \rightarrow \mathcal{R}^{MK}$ is a matrix-valued functional of codimension M, and $H_2 \equiv \ker \boldsymbol{B}$
- $Z : G \rightarrow H_0$ a known continuous linear mapping which is assumed injective and with a closed image in H_0.
- $\rho_s : H^M \rightarrow \mathcal{R}^K$ such that $\rho_s(x) = x(s)$ is the evaluation functional.
- $A : H_0 \rightarrow \mathcal{R}^{KJ}$ is the continuous linear evaluation mapping which maps x into the finite number of observations $x(t_j), j = 1, \ldots, J$.

3 The Partitioning of Function Space H

The functional analysis of data can be greatly aided by the partitioning of function space into *structural* and *residual* components. That is, each function has the decomposition $x = u + e$. Component u has known structural properties such as linearity, periodicity with specified period, and etc.; and is assumed to be an element of a finite dimensional subspace H_1. Residual component e is an element of an infinite dimensional subspace H_2 defined to be complementary to H_1. The two subspaces and their complementarity are defined by the use of a linear differential operator $\boldsymbol{D} : H^M \rightarrow \mathcal{L}^2$ such that

$$\boldsymbol{D} = \sum_{m=0}^{M} \alpha_m D^m, \alpha_m \in H^1, m = 1, ..., M, \alpha_M \neq 0 \tag{2}$$

and a matrix-valued functional $\boldsymbol{B} : H^M \rightarrow \mathcal{R}^{MK}$. The role of $\boldsymbol{D}$ is to annihilate components in $H_1 = \ker \boldsymbol{D}$ so that $\boldsymbol{D}x = e$ is the residual portion of x. For example, where it is desirable to eliminate or separate out linear effects in x, one would use $\boldsymbol{D} = D^2$, but where sinusoidal components of period γ define H_1, $\boldsymbol{D} = \gamma^2 D^0 + D^2$. For the weather station data, we shall use $\boldsymbol{D} = \gamma^2 D^1 + D^3$ with $\gamma = 2\pi/6$ in order to annihilate mean and harmonic variation with a period of twelve months.

Operator $\boldsymbol{B}$ is defined so that the complementary condition

$$\ker \boldsymbol{D} \cap \ker \boldsymbol{B} = 0 \tag{3}$$

holds. H_2 is defined as $\ker \boldsymbol{B}$. For example, when $\boldsymbol{D} = D^2$, one might use $\boldsymbol{B}x = (x(t_0), (Dx)(t_0))^t$ so that the boundary constraint $\boldsymbol{B}e = 0$ implies that e does not contain linear components in the specific sense of having zero value and slope at point t_0. For a given differential operator $\boldsymbol{D}$ many definitions of $\boldsymbol{B}$ are possible which satisfy this complementarity condition, and which is used in a particular application will depend on substantive considerations and possibly convenience. In the case of the weather station data, the orthogonality conditions

$$\boldsymbol{B}x = (\int x dt, \int x \sin \gamma t dt, \int x \cos \gamma t dt)^t \tag{4}$$

seemed natural, so that H_2 consists of those components of each weather station's temperature-precipitation variation which are orthogonal to the functions $\{1, \sin \gamma t, \sin \gamma t\}$ in this sense.

Finally, H_0 is defined as $H_1 \oplus H_2$ and any one of the three subspaces of H^M will be referred to as $H_\ell, \ell = 0, 1, 2$.

Both H_1 and H_2 are given a Hilbert space structure by using inner products $\ll \cdot, \cdot \gg_1$ and $\ll \cdot, \cdot \gg_2$ defined on $\mathcal{R}^{MK}$ and $\mathcal{L}^{2K}$, respectively. We then define

$$< u_1, u_2 >_1 = \ll B(u_1), B(u_2) \gg_1 \tag{5}$$

and

$$< e_1, e_2 >_2 = \ll L(e_1), L(e_2) \gg_2 \ . \tag{6}$$

By definition of H_0, these inner products then define $< x_1, x_2 >_0 = < u_1, u_2 >_1 + < e_1, e_2 >_2$.

For $\mathcal{R}^{MK}$ we shall use

$$\ll X_1, X_2 \gg_1 = \mathrm{tr} X_1' X_2 \Theta, \quad X_1, X_2 \in \mathcal{R}^{MK} \ , \tag{7}$$

where Θ is an order K symmetric matrix which plays the role of allowing for differences in scale among the K variables as well as for covariation among them. In multivariate statistical work, for example, it is usual to use $\Theta = \Sigma^{-1}$, where Σ is the variance-covariance matrix for the variables. The corresponding functional value for Σ is

$$\sigma_{k_1 k_2} = \sum_i^I < x_{ik_1}, x_{ik_2} >_\ell, \ell = 0, 1, 2. \tag{8}$$

The companion inner product for $\mathcal{L}^{2K}$ is

$$\ll x_1, x_2 \gg_2 = \sum_{k_1}^K \sum_{k_2}^K \theta_{k_1 k_2} \ll x_{k_1 1}, x_{k_2 2} \gg, \quad x_1, x_2 \in \mathcal{L}^{2K} \ . \tag{9}$$

It follows from the regularity of H_0 that the evaluation functional ρ_s is continuous and therefore that there exists a set of functions $r_{sk} \in H^M$ called *representers* for the evaluation functions ρ_s which satisfy the reproducing equation

$$\rho_s(x_k) = x_k(s) = \ll x, r_{sk} \gg_0 \ . \tag{10}$$

Collecting these for all values of s and k yields the *reproducing kernel* $r : T \times T \to \mathcal{R}^{KK}$ such that $r_{sk}(t)$ is the kth column of the order K matrix-valued bivariate function $r(s,t)$. Each of H_1 and H_2 has a reproducing kernel r_1 and r_2, respectively, and the reproducing kernel of H_1 is $r_0 = r_1 + r_2$.

The reproducing kernel plays a central role in representing a finite set of observations in function space through smoothing or interpolation. This is taken up in Section 5. The computation of the reproducing kernel for arbitrary operators $\boldsymbol{D}$ and $\boldsymbol{B}$ can be quite involved, and Ramsay & Dalzell (1989) discuss this problem in some detail.

4 Functional Data Analysis

The familiar techniques of multivariate data analysis such as linear modeling, principal components analysis, and canonical correlation analysis, as well as the many procedures derived from these, can be expressed in functional analytic terms (Dauxois & Pousse, 1976). Here the assumption is that there are replicated functions, $x_i, i = 1, \ldots, I$, which collectively have some structure to be investigated. For example, it may be desirable to approximate each function x_i by an approximation $x_i \approx z_i(\beta)$, where z_i is some known linear mapping from some space G into H_0 and the functional analogue of the regression coefficient vector, β, is an element in that space. One then desires to find some estimate $b \in G$ which minimizes a suitable measure of fit such as $\sum_i \|x_i - z_i(\beta)\|$. Principal components analysis is likewise the decomposition of the functional equivalent of the variance-covariance matrix, where variation may be measured about

- the zero function,
- a mean function $\hat{x}$, or
- about some linear model values $z_i(\beta)$

Boldface $\mathbf{H}_\ell$ will now denote the entire sample space $\otimes_i H_\ell$, and $\mathbf{x} : T \to \mathcal{R}^{KI}$ will indicate the entire sample of functions $(x_1, \ldots, x_I)^t$. That is, $\mathbf{x}$ is a function on T with values in the space of $K \times I$ matrices. Tensor algebra becomes very convenient in discussing the behavior of spatial functions, and in that notation $\mathbf{x}$ is a contravariant tensor of order 2. An inner product for the sample space derived from that for each observation such as $\sum_i w_i < x_{i1}, x_{i2} >_\ell$, where the w_i's are suitable weights, will be denoted by $< \mathbf{x}_1, \mathbf{x}_2 >_\ell, \ell = 0, 1, 2$.

4.1 Principal Components Analysis

Principal components involves the orthogonal decomposition of variation among the replicated functions x_i about some specific components u_i. These components could be 0, a constant function such as $\bar{x}$, or the result of some linear modeling process, $\hat{x}_i$. The partitioning of function space set up in Section 3 permits a simple approach to this problem. The structural components can also be removed or filtered out of the observed functions by the application of an appropriate differential operator D and the PCA can then take place in the residual space H_2 (as well as in H_0 or H_1 if required). Besse and Ramsay (1986) discuss this approach in detail in the context of scalar- valued functions.

The covariance kernel $v : T \times T \rightarrow \mathcal{R}^{KK}$ is $v(s,t) = \mathbf{e}(s)W\mathbf{e}(t)^t$, where $\mathbf{e} \in \mathbf{H}_2$ is the residual component in $\mathbf{x}$. This provides the functional analogue of the covariance matrix in multivariate analysis. We seek a *harmonic* $\zeta \in H_2$ of unit norm which defines a mode of variation among the x_i's about the structural components u_i. The linear mapping $\Xi : H_0 \rightarrow \mathcal{R}$ such that $\Xi(x_i) = < x_i, \zeta >_2$ indicates the amount of the harmonic present in the ith replicate. (Strictly speaking, Ξ is a covariant tensor and cannot appear in an inner product, but in the context of Euclidean space this abuse of notation is harmless.)

Letting v_k denote the kth column of v, we can define the linear operator $V : \mathbf{H}_2 \rightarrow \mathbf{H}_2$ as the mapping for which the kth element in the image is $< v_k, \cdot >_2, k = 1, \ldots, K$. The eigenequation $V\zeta = \lambda\zeta$ defines a series of harmonics having the required properties of specifying, in the first instance, the dominant mode of functional variation in $\mathbf{H}_2$, and in subsequent cases, modes of variation accounting for greatest variation given that they are orthogonal to previous harmonics.

5 Functional Representation of Finite Data

In practical applications the replicated functions x_i are usually unavailable. Rather, some finite amount of information is available about each function, and one important task is to estimate the function itself on the basis of these data. For example, it is common to have observed a function at a finite number of sampling points $t_j \in T, j = 1, \ldots, J$, so that the data for the ith replication consist of the function values $x_i(t_j) = \rho_{t_j}(x_i)$. More generally, what may be observed is $x_i(t_j)$ plus some error component ξ_{ij}, usually assumed to be independently and identically distributed with mean 0. Another variation in the data collection situation is one in which not only function values are collected, but also values of other functionals such as derivative values, values of integrals over subsets of T, and so on.

In general, let operator $A : H_0 \rightarrow \mathcal{R}^{KJ}$ be the continuous linear mapping which maps x into its finite number of observations such as $x(t_j), j = 1, \ldots, J$. Corresponding to A is the operator $\mathbf{A} : \mathbf{H}_0 \rightarrow \mathcal{R}^{IKJ}$ mapping a sample of functions $\mathbf{x}$ into an $I \times K \times J$ array $\mathbf{A}(\mathbf{x})$ of values $x_i(t_j)$. Define array Y to be the corresponding three-way array of data. Note that Y may be only approximately equal to $\mathbf{A}(\mathbf{x})$ due to the presence of the random disturbance terms ξ_{ij}, which are K-vectors. The use of tensor notation is very useful in discussing such multi-way arrays as Y as well as in dealing with the differential geometry of spatial curves. Consequently, the notation Y_j^{ki} indicates that Y can be used as a linear operator to be applied to tensors contravariant in their first index.

The process of representing Y in function space $\mathbf{H}_0$ thus involves finding what amounts to a left inverse $\mathbf{A}^-$ of operator $\mathbf{A}$. Even a sketch of this process is beyond the scope of this paper, and a review of smoothing procedures such as Buja, Hastie, and Tibshirani (1989) can be consulted for details. The reproducing kernels r_ℓ associated with $H_\ell, \ell = 0, 1, 2,$ play a central role in either spline interpolation or spline smoothing.

The reproducing kernel r has an image in the space of order K symmetric matrices. For convenience we now use $r_j^{(k)}$ to indicate the scalar-valued representing function associated with the jth component of $A(x_k)$. However, as noted above, the situation can be simplified by transforming the data. Let $\Theta = L^t L$ and let $z = Lx$. We can always carry out the analysis in terms of the functions z_i and the corresponding data matrix $Z_j^{ki} = L_l^k Y_j^{li}$, where according the summation notation in tensor analysis summation is across the repeated subscript l. When the analysis of the transformed data is completed, one transforms back the original scale by applying the inverse mapping L^{-1}.

Thus, we can assume that these representers will not vary with k, and the superscript k can be dropped. The J-vector valued function $(r(t_1,\cdot),...,r(t_J,\cdot))^t$ will be denoted by $\mathbf{r}$. It can be shown that $\hat{\mathbf{x}} = YS\mathbf{r}$ and $\mathbf{A}(\hat{\mathbf{x}}) = YSR$, where S is an order J matrix defined by the smoothing process (it is assumed that the same smoother is used for all replications) and R is the order J matrix of values $A(\mathbf{r})$. When $A(x_k) = \rho_{t_j}(x_k), j = 1,\dots,J$, as is usually the case, these representers will be $r(t_j,\cdot), j = 1,\dots,J$, where r is the reproducing kernel for H_0. The product YS refers to summation over adjacent indices. In tensor notation the equation $\mathbf{A}(\hat{\mathbf{x}}) = YSR$ becomes $\mathbf{A}(\hat{\mathbf{x}}) = Y_j^{ik} S_m^j R_n^m$, where the indices in this expression are used in a tensoral sense unrelated to their use otherwise in the paper, and summation is across repeated indices.

6 Functional Data Analysis in Terms of Finite Data

Functional data analysis now becomes a matter of analyzing the functional representation $\hat{\mathbf{x}}$ of the sample in $\mathbf{H}_\ell$. Any such analysis can be expressed in terms of an equivalent finite dimensional analysis of the data array Y because of the simple linear relation between Y and $\hat{\mathbf{x}}$ given above. For example, a reader who has followed the development to this point will have no trouble verifying that the evaluation of the location estimate, $A(\bar{x})$, is simply the appropriate average of Y across the first and third indices.

The finite counterpart of the PCA of $\hat{\mathbf{x}}$ requires the analysis of the four-way array $U = RS^tY^tWYSR$ which, with the product Y^tWY involving the third index of Y^t and the first of Y, is an element of $\mathcal{R}^{JKKJ}$. In tensor terms, U can be denoted U_{jm}^{ik}. Using an argument closely following that of Besse and Ramsay (1986), one arrives at the finite dimensional eigenequation

$$(DU)\#(DE) = \lambda E \text{ where } D = (R_0^{-1} R_2 R_0^{-1})^{1/2} .$$

The tensor product $A\#B$ implies summation across the last two indices of A and the first two of B. In tensor notation, this equation becomes $D_i^j U_{jm}^{kl} D_l^n E_{nk} = \lambda E_{im}$. The eigenfunction associated with the eigentensor E_{im} in matrix notation is $\zeta = \mathbf{r}^t R_0^{-1} D^{-1/2} E$. The eigenequation can be put in matrix analysis terms by combining the first pair of indices for DUD into a single index ranging over JK values, doing the same for the final two indices, and representing E as a single vector with JK components. This permits the use of conventional matrix eigenanalysis software to identify E associated with an eigenvalue λ.

7 Analysis of Canadian Weather Data

We apply these techniques to the study of the joint variation of temperature and precipitation monthly means for 35 Canadian weather stations. As indicated in Section 3, we considered the structural component in these data to be sinusoidal variation with differences in annual mean. We defined H_2 by the operator 4. The objective of the analysis was to study variation only in H_2 so that we wished to ignore strictly harmonic variation and annual mean variation.

The first step was to independently smooth the data for the two variables. Using generalized cross-validation to identify the appropriate value for the smoothing parameters resulted in rather more smoothing for precipitation ($\lambda = 1.0$) than for temperature ($\lambda = 0.001$). Figure 2 displays the smoothed joint variation for Montreal along with the actual monthly means.

The next problem was to compute the covariance matrix Σ using 8. This was computed using the smoothed data and the inner product for H_2The inner products for the data analysis were then defined by 5, 6, 7, and 9.

The initial analysis revealed that Prince Rupert was an extremely different station even in H_2 , and as a consequence it exerted a disproportionate influence on harmonics and principal component scores. It was decided to drop this observation, and to base the results on the remaining 34 stations. When this was done, the first three eigenvalues were substantial relative to the remainder (91 % of variance). However, for simplicity results for only the first two harmonics will be displayed here (71 % of variance). Figures 3 show the temperature and precipitation components of the first two harmonics plotted against time, and Figure 4 displays these plotted against each other and thus forming closed loops. A station with a large amount of a harmonic will show a significant departure from purely harmonic variation for times of the year for which the corresponding curve in Figure 4 deviates strongly from the origin. Thus, the first harmonic is associated with strong departures in April and October and departures of an opposite nature in January and June. In the April and October, the precipitation is above that predicted by harmonic variation and the temperature is below. The second harmonic shows a strong lobe in Figure 4 in September when the precipitation is very high but with normal (i.e. harmonic) temperature.

It remains to see how the weather stations themselves are distributed with respect to H_2 variation. Figure 5 plots the inner products of smoothed temperature-precipitation curves with the first two harmonics for each station. We see a curvilinear one-dimensional variation, with the continental and arctic stations at one end and marine stations at the other. Stations along the St. Lawrence river system such as Montreal occupy an intermediate position.

References

[1] Aubin, J.-P. (1979) *Applied Functional Analysis.* New York: Wiley Interscience.

[2] Besse, P. & Ramsay, J. O. (1986) Principal components analysis of sampled functions. *Psychometrika, 51*, 285-311.

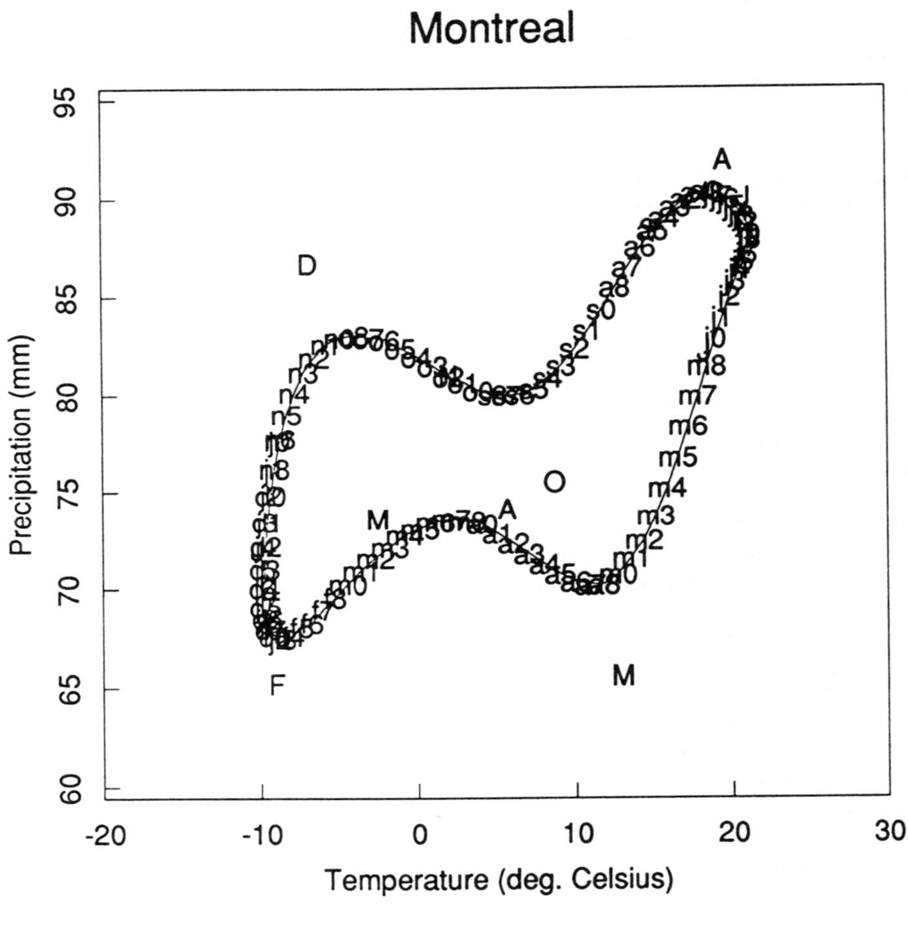

Figure 2: Smoothed temperature-precipitation variation for Montreal

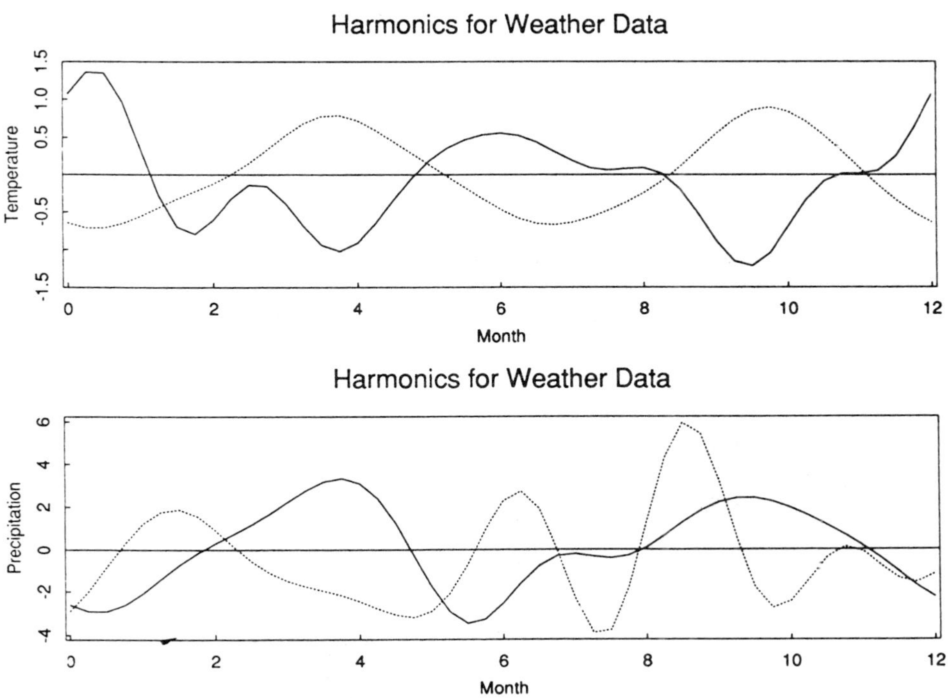

Figure 3: The first two harmonics or principal components of functional variation

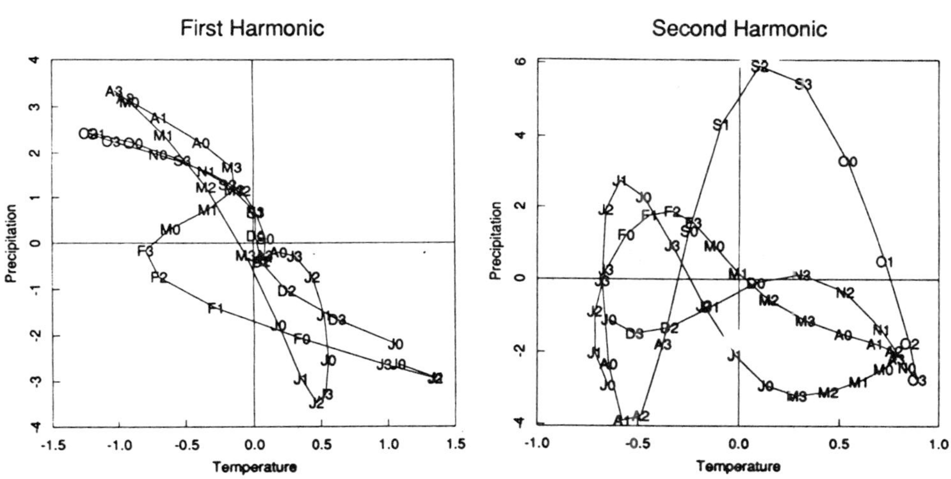

Figure 4: The first two harmonics or principal components of functional variation

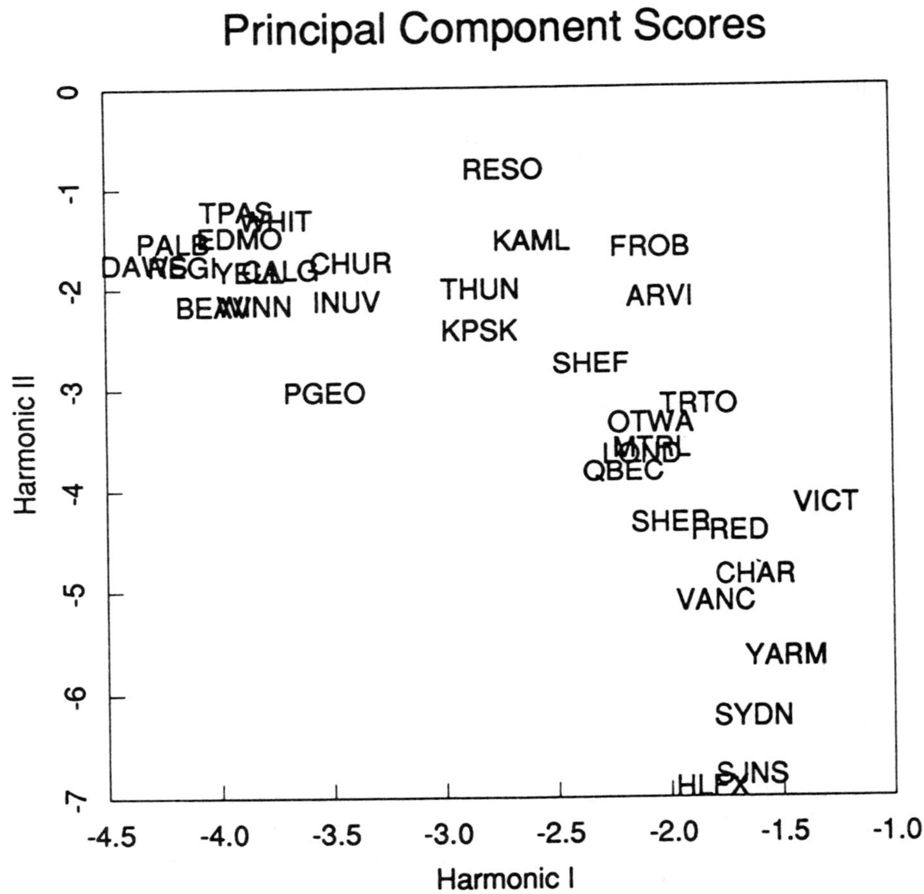

Figure 5: The principal component scores on the first two harmonics.

[3] Buja, A., Hastie, T., & Tibshirani, R. Linear smoothers and additive models. University of Toronto Department of Statistics Technical Report No. 10, 1987.

[4] Dalzell, C. & Ramsay, J. O. (1988) A function space analysis of some non-stationary data. Unpublished manuscript, Memorial University Department of Mathematics.

[5] Dauxois, J. & Pousse, A. Les analyses factorielles en calcul des probabilité et en statistique: Essai d'étude synthètique. Unpublished doctoral dissertation, l'Université Paul-Sabatier de Toulouse, France

[6] Deville, J. C. (1978) Analyse et prevision des series chronologiques multiples non stationnaires. *Statistique et Analyse des Données, 3*, 19-29.

[7] Laurent, P-J. (1972) *Approximation et Optimisation.* Paris: Hermann.

[8] Ramsay, J. O & Dalzell, C. (1989) Some tools for functional data analysis. Unpublished manuscript, McGill University Department of Psychology.

[9] Saporta, G. (1981) Methodes exploratoires d'analyse de données temporelles. *Cahiers du Bureau Universitaire de Recherche Operationelle, 37-38.*

[10] Schumaker, L. (1981) *Spline Functions: Basic Theory.* New York: Wiley.

SMOOTHING NOISY DATA BY TWO EQUIVALENT TECHNIQUES

Christine Thomas-Agnan

Laboratoire de Statistique et Probabilités
U.A. CNRS 745, Université Paul Sabatier,
118, route de Narbonne, 31062 TOULOUSE FRANCE

Keywords : spline functions, interpolation, smoothing, partial splines, reproducing kernel, locally homogeneous random fields, variogram, kriging, ARIMA processes.

Abstract : Two commonly used methods for smoothing noisy data have been shown to give rise to the same solution in certain situations. The first one is smoothing by spline functions and the second uses a stochastic model for the data together with optimal prediction technique. We briefly explore the formal correspondence between partial α-splines and the prediction of locally homogeneous random fields, whose m-spectral measure is absolutely continuous with respect to Lebesgue measure and satisfies some minor assumptions. Then we study concrete examples. In particular, this yields, in the case of rational m-spectral density, a generalization of the prediction theory for discrete and unidimensional "time" ARIMA processes, to continuous and multidimensional "time".

Resumé : On a montré que, dans certains cas, deux méthodes couramment utilisées pour le lissage de données bruitées, conduisaient à la même solution. La première consiste à utiliser des fonctions splines d'ajustement et la seconde à expliquer les données par un modèle stochastique dans le cadre duquel on fait ensuite de la prédiction optimale. Nous exposons brièvement l'équivalence, sur le plan formel, entre α-splines partielles et la prédiction optimale des champs aléatoires localement homogènes, dont la m-mesure spectrale est absolument continue par rapport à la mesure de Lebesgue et satisfait des hypothèses additionnelles faibles.Nous étudions ensuite quelques exemples concrets. Cela conduit, dans le cas de m-densité spectrale rationnelle, à la généralisation de la prédiction des processus ARIMA en "temps" discret et unidimensionnel, au cas du "temps" continu et multidimensionnel.

1.INTRODUCTION

Nowadays, numerical tools from approximation theory such as spline functions are more and more often used in Statistics (cf. Wegman-Wright [17], and Besse and Thomas-Agnan [1]). At the same time, methods inspired from stochastic models are also used in pure approximation problems, like for example kriging in automatic cartography (Dubrule [3]).

There has been some interest in studying the relationships between them (cf. Matheron [10], Kimeldorf and Wahba [5] and [6], Dubrule [3], Salkauskas [12], Watson [16], Thomas-Agnan [15]). A better insight into these connections not only is satisfactory from a theoretical point of view, but could lead, on the computational level, to a cross fertilization of these techniques. In § 2, we give a short description of the techniques and the results exposed in Thomas-Agnan [15]. In § 3, we examine examples of variograms commonly used in the stochastic methods, together with their corresponding splines.

The concrete problem these techniques originated from is to reconstruct a function $f : \mathbb{R}^d \mapsto \mathbb{R}$ from possibly noisy measurements y_i of $f(t_i)$ at a finite number of given irregularly spaced t_i (i=1,...n) in $\mathbb{R}^d$ ($d \in \mathbb{N}^*$).

2. TWO TECHNIQUES AND THEIR EQUIVALENCE

2.1. The numerical technique

We first restrict attention to those multidimensional spline functions which can be obtained as the solution of a quadratic optimization problem in a Hilbert space.

Let H_1 and H_2 be two Hilbert spaces (on $\mathbb{R}$ or $\mathbb{C}$), with scalar products denoted by $(.,.)_1$ and $(.,.)_2$ respectively. Let $h_1,...h_n$ be n linearly independent elements of H_1, $(y_1,...y_n)$ be a vector of $\mathbb{R}^n$, λ a positive real number and T a continuous linear map from H_1 to H_2. Let $v_1,...v_K$ be K ($K\in \mathbb{N}$) given linearly independent functions on $\mathbb{R}^d$ .Then the interpolation problem is :

$$(1) \qquad \begin{cases} \text{Min } \|Th\|_2^2 \\ h \in H_1;\ \theta_k \in \mathbb{R};\ k=1;...K;\ (h + \Sigma\, \theta_k v_k\ ;\ h_i) = y_i;\ i=1;...n \end{cases}$$

The smoothing problem is

$$(2) \qquad \begin{cases} \text{Min} \displaystyle\sum_{i=1}^{n} ((h + \Sigma\theta_k v_k\ ;\ h_i) - y_i)^2 + \lambda \|Th\|_2^2 \\ h\in H_1;\ \theta_k \in \mathbb{R};\ k=1;...K \end{cases}$$

Under additional assumptions on T, $h_1,...h_n,v_1,...v_K$, Laurent [8] proves the existence and uniqueness of the solution of these optimization problems; the corresponding solutions will be called interpolating partial splines for (1) and smoothing partial splines for (2) (inf-convolution splines in France). Similar definitions without the parametric part $\Sigma\ \theta_k\ v_k$ lead to ordinary (not partial) splines. We will refer to $\| Th \|_2^2$ as to the semi-norm associated with the spline.

Since we are dealing here with evaluations of the function at given points, we will work with a reproducing kernel Hilbert space H_1; we recall that it is a Hilbert space of functions on $\mathbb{R}^d$ in which, for all t of $\mathbb{R}^d$, there exists an element $R_t(.)$ in H_1, such that for all f in H_1, $f(t)=(R_t(.),f(.))_1$ and that the kernel of H_1 is the function $R : \mathbb{R}^d \mapsto \mathbb{R}^d$ defined by

$$(3) \qquad R(s,t) = (R_t\ , R_s)_1 = R_t(s) = R_s(t)$$

The concrete examples of splines we will be dealing with in § 3 are the following :

- the D^m-splines, also called Thin Plate splines when d > 1, where the semi-norm is related to the partial derivatives of total order m of the function (see Meinguet [10]),

- the L-splines, where the semi-norm is constructed with a linear differential operator (see Kimeldorf and Wahba [7]),

- the α-splines, where the semi-norm is based on the Fourier transform (hereafter denoted by $\mathfrak{F}$) of the function in suitable spaces (see Thomas-Agnan [14] and [15]). The D^m-splines are a particular case of α-splines, and so are the L-splines when L has constant coefficients and a non-vanishing characteristic polynomial.

2.2. Stochastic filtering

We are refering to the combination of a stochastic model to explain the data with a strategy of prediction called Best Linear Unbiased Estimation (BLUE hereafter).

Let Y(t), $t \in \mathbb{R}^d$, be a stochastic process with second moments (also called random field when d > 1). Let $\varepsilon_1,...\varepsilon_n$ be n random variables independent from the process Y(t), satisfying the usual assumptions for a white noise :

$E\varepsilon_i = 0$ (i=1,...n), $E\varepsilon_i\varepsilon_j = 0$ ($i\neq j$), $E\varepsilon_i^2 = \sigma^2$, where E denotes mathematical expectation.

The observations y_i are explained as one realization of the random variable $Z_i = Y(t_i) + \varepsilon_i$. Moreover, the process Y(t) is often decomposed into the sum of a drift and a fluctuation :

$$(4) \qquad Y(t) = m(t) + X(t)$$

where m(t) is a deterministic component, linear combination of a finite number L of known functions u_i on $\mathbb{R}^d$ with unknown parameters, and where the process X(t) has mean 0 and given covariance structure R(s,t) = E (X(t)X(s)).

The BLUE constructs a linear combination $\hat{Y}(t)$ of the random variables Z_i, which satisfies $E(\hat{Y}(t)) = E(Y(t))$, for any value of the parameters in m(t), and minimizes the predictive mean square error $E(\hat{Y}(t)-Y(t))^2$.

2.3. α-splines and the prediction of locally homogeneous random fields

2.3.1. α-splines

To define the semi-norm for α-splines, one needs an integer $m \in \mathbb{N}$, and a function $\alpha : \mathbb{R}^d \to \mathbb{C}$ verifying the following assumptions :

(A1) $\alpha(\omega) \neq 0$ for all $\omega \in \mathbb{R}^d$,

(A2) $\alpha(\omega) = \alpha(-\omega)^*$,

(A3) $|\alpha|^{-2}$ is locally integrable,

(A4) there exists a ball B of $\mathbb{R}^d$ centered at 0, such that $\omega \to |\alpha(\omega)|^{-2}\|\omega\|^{-2m}$ belongs to $L_1(\mathbb{R}^d \backslash B)$.

For a multiindex $(\beta_1,\ldots\beta_d) \in \mathbb{N}^d$, we denote by $|\beta|$ the sum $\Sigma\beta_i$, by $\binom{m}{\beta}$ the multinomial coefficient $\frac{m!}{\Pi\beta_i!}$ and by D^β the differential operator $\frac{\partial^{|\beta|}}{\partial x_1^{\beta 1}\ldots\partial x_d^{\beta d}}$.

Then let $\mathcal{E}_\alpha{}^m$ be the space of tempered distributions (see Schwartz [13]) h of $\mathcal{S}'(\mathbb{R}^d)$ such that, for all β satisfying $|\beta|=m$, $\mathcal{F}(D^\beta h)$ is locally integrable and $\alpha\,\mathcal{F}(D^\beta h)$ belongs to $L^2(\mathbb{R}^d)$. We proved (see [15]) that $\mathcal{E}_\alpha{}^m$ can be given a structure of reproducing kernel Hilbert space and that its elements are equivalent (equal as a distribution) to continuous functions. Then one can use :

(5) $$\sum_{|\beta| = m} \binom{m}{b} \|\alpha \mathcal{F}(D^{\beta}h)\|^2_{L2}$$

as a semi norm in $\mathcal{E}_{\alpha}{}^{m}$, whose kernel is the set $\mathcal{P}_m$ of polynomials in $t \in \mathbb{R}^d$, of total degree less than or equal to m-1 (its dimension is $M=\binom{d+m-1}{d}$). One finds in [15] explicit formulas for the reproducing kernel of $\mathcal{E}_{\alpha}{}^{m}$ and for the solution, when it exists, of problems (1) and (2) when one uses (5) as the semi-norm and $\mathcal{E}_{\alpha}{}^{m}$ as H_1 (they will be called α-splines of order m). Note that α-splines provide an interpolation method which is translation invariant, and moreover rotation invariant when α is isotropic.

2.3.2. Locally homogeneous random fields

We now specialize the stochastic model described in § 2.2 by assuming that the process X(t) belongs to certain class of stochastic processes convenient for statistical inference.

A theory analogous to the well known theory of stationary random processes has been developed by Guelfand and Vilenkin [4], and Yaglom [19] in the context of generalized random processes. It has been exploited later by Matheron [9] in its theory of intrinsic random functions, restricting to the case where the generalized processes with stationary increments of order m are mere random functions (ordinary processes) with stationary increments of order m. The interest of these processes is that one can develop a spectral analysis of their second order structure by Bochner-type theorems. The role of the function R naturally related to the covariance structure of an ordinary stationary process by R(s,t)=R(s-t) is taken here by the variogram. It is defined, for an ordinary process with stationary increments of order m, as any function V such that

(6) $$E\left(\sum_{i=1}^{n} c_i X(t_i)\right)^2 = \sum_{i;j=1}^{n} c_i c_j V(t_i - t_j)$$

for any $(t_1,\ldots t_n) \in (\mathbb{R}^d)^n$ and any set of coefficients $(c_1,\ldots c_n)$ satisfying $\sum c_i P(t_i) = 0$, for all P in $\mathcal{P}_m$. In the same way as the function R is even and of positive type, the function V is even and conditionally of positive type of order m.

Guelfand and Vilenkin [4] proved that the class of tempered distributions U conditionally of positive type of order m is characterized by the fact that their Fourier transform, in $\mathcal{S}'(\mathbb{R}^d)$, satisfies the following property : for any polynomial P with complex coefficients and total degree less than or equal to m-1, $|P|^2\mathcal{F}U$ is a positive slowly increasing measure. Hence, for an ordinary process with stationary increments of order m, and variogram V, one can define a positive slowly increasing measure by $\|\omega\|^{2m}\mathcal{F}V(\omega)$; since it plays a role parallel to the spectral measure of a stationary process, we will refer to it as to the m-spectral measure.

If one takes for X(t) a second order zero mean ordinary process with stationary increments of order m, and if

(B1) the span of $u_1, \dots u_L$ contains $\mathcal{P}_m$ (hence $L \geq M$),

then the BLUE can be easily computed and depends upon X(t) only through its variogram V.

If we assume moreover that :

(B2) the m-spectral measure $d\mu_x$ is absolutely continuous with respect to Lebesgue measure,

(B3) $(1 + \|t\|^2)^{-m}$ is integrable with respect to the measure $d\mu_x$,

then it is easy to see that the density f_x of $d\mu_x$ can be written $|\alpha|^{-2}$ with a function α satisfying (A1) through (A4). And, we find in [15] :

Theorem : *Under the above assumptions (B1) through (B3), the BLUE of Y(t) in model (3), when one substitutes the data y_i for the random variables $Y(t_i)$, is a partial α -spline of order m, where the functions v_k of (2) span a subspace of the span of $u_1, \dots u_L$ in direct sum with $\mathcal{P}_m$, and where the smoothing parameter λ of (2) is the variance of the noise σ^2.*

3. EXAMPLES

3.1.Stationary case

Let us first note that, if one wants in model (4) a stationary process for X(t) together with a non-zero function for m(t), the previous theorem shows that we need to consider partial splines, rather than ordinary ones.

A particular case of this theorem is to be found in Kimeldorf and Wahba [6] : the dimension is d=1, the process Y(t) is stationary, has mean zero, and spectral density proportional to $|P_L(\omega)|^{-2}$, where P_L is the characteristic polynomial of a linear differential operator L (if $L = \Sigma\, a_j D^j$, then $P_L = \Sigma\, a_j (2\pi i\omega)^j$). By analogy with the discrete time processes vocabulary, we can say that Y(t) is an autoregressive AR(p) process (where p is the degree of P_L). The corresponding spline is then an L-spline.

Yaglom [18] proves that, in dimension 1, a general rational spectral density f_x can be written :

$$(7) \qquad f_x(\omega)=\frac{|Q(2\pi i\omega)|^2}{|P(2\pi i\omega)|^2}$$

where P and Q are real-coefficient polynomials of degree p and q respectively, and where all the zeros of P have positive real part, and the zeros of Q have non-negative real part. One generally makes the additional assumption that the zeros of Q have positive real part. In dimension d, let us define an isotropic ARMA(p,q) process X(t) by the fact that it has a spectral density of the form (7) where one substitutes ω by $\|\omega\|$ in the right hand side. For an ordinary process X(t), f_x must be integrable with respect to Lebesgue measure, hence we assume moreover that $p-q > d/2$. Under these conditions, the function

$$(8) \qquad \alpha(\omega)=\frac{P(2\pi i\|\omega\|)}{Q(2\pi i\|\omega\|)}$$

satisfies assumptions (A1) through (A4) and by the theorem of § 2.3.2, the BLUE of Y(t) corresponds to an α-spline of order 0.

One encounters also non-rational spectral densities in the literature. For instance, a common model of isotropic covariance used in magnetic fields studies (see Chiles and Guillen [2]) is given by :

$$(9) \qquad R_M(s,t)=(1+\frac{\|s-t\|^2}{L^2})^{-3/2}.$$

corresponding to the α function :

$$(10) \qquad \alpha_M(s,t)=L^{-1}(2\pi)^{-1/2}\exp(\pi L\|\omega\|).$$

3.2. Non stationary case

Similarly, let us take now for X(t) an isotropic ARIMA(p,m,q) process, i.e. a second order process with stationary increments of order m, and m-spectral density of the form given by (7), still substituting $\|\omega\|$ for ω on the right hand side, with the condition $p-q > d/2$ replaced by $m+p-q > d/2$. This additional assumption corresponds to (B3). Then, by the theorem of § 2.3.2, the BLUE corresponds to an α-spline of order m, where α is still defined by (8).

A common isotropic variogram used in the kriging methods (see Matheron [9]) is the so called polynomial generalized covariance, given by :

$$V(t) = \sum_{j=p}^{k} (-1)^{j+1} a_j \, B_{dj} \, \frac{\|t\|^{2j+1}}{(2j+1)!} \tag{11}$$

where p and k are integers ($k \geq p$), a_j are real numbers, a_k and a_p are non-zero, $\Sigma\, a_j x^{2(k-j)} > 0$ for all real x, and $B_{dj} = \dfrac{j!\,\Gamma(d/2)}{\sqrt{\pi}\,\Gamma(p+(d+1)/2)}$.

Using classical formulas for Fourier transforms (see Schwartz [13]), one gets :

$$\mathcal{F}V(\omega) = C(d) \sum_{j=1}^{k} a_j \, \frac{1}{\|2\pi\omega\|^{2p+1+d}} \tag{12}$$

where $C(d) = \Gamma(d/2)\, 2^d \pi^{d/2-1}$. Therefore,

$$\mathcal{F}V(\omega) = C(d) \, \frac{\sum_{j=p}^{k} a_j \, \|2\pi\omega\|^{2(k-j)}}{\|2\pi\omega\|^{2k+1+d}} \tag{13}$$

Moreover, Matheron [9] shows that there exists a polynomial Q of degree k-p and real coefficients such that $\Sigma\, a_j \, \|2\pi\omega\|^{2(k-j)} = |Q(2\pi i\omega)|^2$. Then, when the dimension d is odd, setting $m = k + (d+1)/2$, the function $\alpha(\omega) = Q(2\pi i\omega)^{-1}$ satisfies assumptions (A1) through (A4) and $|\alpha|^{-2}$ corresponds to the m-spectral density of the process, which can then be classified as an ARIMA(0,k+(d+1)/2,k-p). In particular, if k=p, this gives $\alpha \equiv 1$, and hence we get the correspondence between the usual Thin Plate splines in odd dimension and the prediction of ARIMA(0,k,0). This specific case was developed with different degrees of generality in Salkauskas [12] and Watson [16].

In order to include also the prediction problem corresponding to Thin Plate splines in even dimension, on has to consider variograms which are linear combinations of functions of the type $\|t\|^{2p} \ln(\|t\|)$ and of the type $\|t\|^{2p}$, with certain restrictions on the coefficients. By similar arguments, one shows that α^{-1} is proportional to a polynomial of degree k-p in $2\pi i\omega$, yielding processes of the type ARIMA(0,k+d/2,k-p). Then for k=p, one gets the correspondance between Thin Plate splines in even dimension and the prediction of ARIMA(0,k+d/2, 0), which is illustrated in [3] for dimension 2.

References :

[1] Besse P. and Thomas-Agnan C. , " Le lissage par fonctions splines en statistique, revue bibliographique ", Statistique et analyse des données (1989), vol 14, nº1, pp. 55-83.

[2] Chiles J.P. and Guillen A. , " Variogrammes et Krigeage pour la gravimétrie et le magnétisme ", in : Royer J.J. (eds) Computers in Earth Sciences for Natural Ressources Characterization (1984), Coll.Int. 9-13, Nancy, France.

[3] Dubrule, O. , " Two methods with different objectives : splines and kriging ", Journal of the International Association for Mathematical Geology (1983), nº 15, pp. 245-257.

[4] Guelfand I.M. and Vilenkin N.Y. , " Les distributions, applications de l'analyse harmonique ", Dunod (eds) (1967), vol 4.

[5] Kimeldorf G. and Wahba G. , " Spline Functions and Stochastic Processes " Sankhya Series A (1970), vol 32, pp.173-180.

[6] Kimeldorf G. and Wahba G. , "A correspondence between Bayesian Estimation on Stochastic Processes and Smoothing by Splines ", Annals of Mathematical Statistics (1970), vol 11, nº 2, pp.495-502.

[7] Kimeldorf G. and Wahba G., "Some results on Tchebicheffian spline functions", Journal of Mathemetical Analysis and Applications (1971), vol 33, pp.82-95.

[8] Laurent P.J., "Inf-convolution splines", Technical report RR 749-M-IMAG (Grenoble) (1988).

[9] Matheron G. , " The Intrinsic Random Functions and their Applications ", Advances in Applied Probability (1973), vol 5, pp. 439-468.

[10] Matheron G. , "Splines and Kriging : their formal equivalence ", in : Merriam D.F. (eds), Syracuse University Geology contribution B (1981) .

[11] Meinguet J. , " Multivariate Interpolation at arbitrary points made simple ", Journal of Applied Mathematics and Physics (1979), vol 30, pp. 292-304.

[12] Salkauskas K. , "Some relationships between Surface splines and Kriging ", in : Birkhäuser, Multivariate Approximation Theory II, International Series of Numerical Mathematics (1982), vol 61, pp.313-325.

[13] Schwartz L. , " Théorie des distributions ", Hermann, (1966), Paris.

[14] Thomas-Agnan C. , " Statistical Curve Fitting by Fourier Techniques ", UCLA thesis (1987).

[15] Thomas-Agnan C. , " Spline Functions and Stochastic Filtering ", (1989) submitted.

[16] Watson G.S. , " Smoothing and Interpolation by Kriging and with Splines ", Technical report 241 serise 2 (1983), Department of Statistics, Princeton University.

[17] Wegman E.J and Wright I.W. , " Splines in Statistics ", Journal of the American Statistical Association (1983), vol 78, nº 382, pp. 351-362.

[18] Yaglom A.M. , " Theory of Stationary Random Functions ", Prentice Hall INS (eds), (1962).

[19] Yaglòm A.M. , " Some Classes of Random Fields in n-dimensional space, related to Stationary Random Processes ", Theory of Probability and its Applications (1957), vol 2, nº 3, pp. 273-320.

A COMPARATIVE STUDY OF SOME SPLINE ESTIMATES OF THE MEAN OF A STOCHASTIC PROCESS

E. BIRITXINAGA[1]

Abstract.: Three spline estimators of the mean function of a stochastic process are considered. They are built from a sample observed at a finite number of knots, either by spline interpolation of the empirical mean or as solution to an appropriate spline smoothing problem or by splines in a convex set defined from confidence intervals. The aim of the paper is to test and compare the behavior of these three estimators in finite sample size and fixed knot-set situation.

Keywords : estimation, spline function, interpolation, smoothing, confidence intervals.

AMS subject classification : Primary : 62 G 05; Secondary : 60 G 12, 65 D 07.

1.Introduction

The main purpose of this paper is to test and compare different methods for non-parametric estimation by spline functions of the mean function m(.) of a stochastic process from independent observatons at a finite number of points. These will be assumed without loss of generality to be in [0,1]. Since any stochastic process can be considered a random variable in an appropriate function space, function space-valued random variables are the natural framework for the problem. Two function spaces in which many stochastic processes take their sample paths are C[0,1] and D[0,1].

D[0,1] denotes the space of real-valued functions on [0,1] wich are right continuous with left limits everywhere on]0,1] . For each u in D[0,1], we consider the norm $\|u\|_\infty=\sup_{0\le t\le 1}|u(t)|$. C[0,1] denotes, as usual, the subspace of D[0,1] whose elements are continuous functions on [0,1]. It is clear that $\|u\|_\infty$ is finite for each u in D[0,1].

Let $(\Omega,\mathcal{A},P)$ be a probability space and let ω be the generic element of Ω. Let $X\equiv(X_t)_{0\le t\le 1}$ be a D[0,1]-valued random variable on $(\Omega,\mathcal{A},P)$ and also let $\{X^i;i=1,...,n\}$ be a set of independent variables distributed as X. We assume that the process X satisfies the central limit theorem and the law of the iterated logarithm in D[0,1] (for details see [7]) but our main hypothesis is that the mean function defind by $m(t) \equiv EX(t)$ $(0 \le t \le 1)$ is "smooth" namely m(.) is in a Sobolev space $H^k[0,1]$ for some k in $\mathbb{N}^*$. Such cases arise frequently: stochastic processes having discontinuous sample paths or having continuous but non-differentiable sample paths and smooth mean function often appear in practical situations (see also Section 3).

$H^k[0,1]$ $(k \in \mathbb{N}^*)$ denotes the Sobolev space of generalized functions (Schwarz distributions) having all their derivatives of order $\le k$ in $L^2[0,1]$. Let (.,.) denote the usual inner product in $L^2[0,1]$. $H^k[0,1]$ endowed with the inner product $((u,v))_k=\sum_{i=0}^{k}(D^iu,D^iv)$ and with the associated norm $\|u\|_k=((u,u))_k^{1/2}$ is then a real separable Hilbert space (cf. for example [1]). In the next Section we shall also use the related semi-norm defined by $|u|_k=(u,u)_k^{1/2}$ on $H^k[0,1]$.

[1]Laboratoire de Statistique et Probabilités UA 1204 CNRS. Faculté de Sciences-IPRA. Av. de l'Université. 64000 Pau . France.

In the framework defined above, if we know a realization in ω of the sample $\{X^i;\ i=1,...,n\}$ at a finite set $T_q=\{t_1,...,t_q\}$ of distinct points in [0,1] (referred to as knots), it seems natural to estimate m by spline functions built from the empirical mean $\overline{X}^n(\omega,.) \equiv (1/n)\sum_{i=1}^{n} X^i(\omega,.)$. Three of such estimates defined either by spline interpolation of the empirical mean or from confidence intervals or by spline smoothing of the empirical mean are presented in Section 2. In Section 3, we test and compare the behavior of these estimates for finite sample sizes and fixed knot-sets. This study is carried out from simulations of two stochastic processes.

2. Definition of the estimates and convergence results

As a first approach to our problem, we can estimate m by the "smoothest" (in the sense of the semi-norm $|\,.\,|_k$) function in $H^k[0,1]$ which interpolates the values $\overline{X}^n(\omega,t_1),..., \overline{X}^n(\omega,t_q)$ of the empirical mean. Hence, we will denote $\hat{m}^{n,q}(\omega,.)$ the solution to the following minimization problem: for given $n\in\mathbb{N}^*$, T_q and $\omega\in\Omega$, find the element $\hat{m}^{n,q}(\omega,.)$ in $H^k[0,1]$ such that

$$\textbf{(P1)}\quad \hat{m}^{n,q}(\omega,.) \in H^{n,q}(\omega) \text{ and } \forall v\in H^{n,q}(\omega),\ |\hat{m}^{n,q}(\omega,.)|_k \leq |v|_k,$$

where $H^{n,q}(\omega)$ is the affine subspace of $H^k[0,1]$ defined by :

$$H^{n,q}(\omega) = \{u \in H^k[0,1]:\ \forall t_i \in T_q,\ u(t_i)=\overline{X}^n(\omega,t_i)\}.$$

The existence and uniqueness of $\hat{m}^{n,q}(\omega,.)$ is guaranteed if $q\geq k$ (cf.[9]) and we will say that $\hat{m}^{n,q}(\omega,.)$ is the interpolation D^k-spline (or simply spline) of the empirical mean.

However, the values $\overline{X}^n(\omega,t_i)$ are only approximations of the corresponding values $m(t_i)$ of the mean function, and moreover, interpolating the empirical mean does not take the random feature of the empirical data into account. This leads us to avoid simple interpolation of the empirical mean . For this purpose we consider first classical confidence intervals $[\alpha_i,\beta_i]$ for the values $m(t_i)$ $(i=1,...q)$, with

$$\alpha_i \equiv \overline{X}^n(\omega,t_i) - c(q)n^{-1/2}\sigma(t_i) \text{ and } \beta_i \equiv \overline{X}^n(\omega,t_i) + c(q)n^{-1/2}\sigma(t_i)\ ,$$

where $\sigma^2(t) \equiv \mathbf{E}(X(t) - m(t))^2$ $(0 \leq t \leq 1)$ denotes the variance function of X and where the real c(q) is choosen in such a way that the global asymptotic confidence level of these q confidence intervals is greater or equal to $1-\alpha$ for a given α $(0<\alpha<1)$. If we note z_α the real such that $P[Z> z_\alpha] = \alpha$, where Z is a standard Gaussian variable, it can be easily verified that we can take $c(q) = z_{\alpha/2q}$ in the above expressions. We can then estimate m by the smoothest spline function that interpolates a point wathever in $\prod_{i=1}^{q}[\alpha_i,\beta_i]$. This leads to the notion of spline function in a convex set introduced by M. Atteia [2]. Note Γ the non-empty convex subset in $H^k[0,1]$ defined by

$$\Gamma= \{u \in H^k[0,1]:\ \alpha_i \leq u(t_i) \leq \beta_i,\ i=1,...,q\}.$$

We consider then the following problem : for given $n\in\mathbb{N}^*$, T_q, $\omega\in\Omega$ and $\alpha\in]0,1[$, find the element $\hat{m}_\Gamma^{n,q}(\omega,.)$ in $H^k[0,1]$ such that

$$\textbf{(P2)}\quad \hat{m}_\Gamma^{n,q}(\omega,.) \in \Gamma \text{ and } \forall v\in\Gamma,\ |\hat{m}_\Gamma^{n,q}(\omega,.)|_k \leq |v|_k\ .$$

Estimation of m by $\hat{m}_\Gamma^{n,q}(\omega,.)$ has been introduced by E.Biritxinaga in [4], where $\hat{m}_\Gamma^{n,q}(\omega,.)$ is called estimation spline.

Existence and uniqueness of spline functions in a convex set has been studied by M. Atteia [2] and P. J. Laurent [9]. By a direct application of their results to our case, it is easy to see that if $\sigma^2(t_i) > 0$ for all t_i in T_q, then the existence and uniqueness of $\hat{m}^{n,q}_{\Gamma}(\omega,.)$ is guaranteed by the condition $\Gamma \cap \mathbb{P}_{k-1}[0,1]=\emptyset$, where $\mathbb{P}_{k-1}[0,1]$ denotes the space of polynomials of degree lower than k, this space being the kernel of the operator D^k. On the other hand, if there are points t_i in T_q for which $\sigma^2(t_i)=0$, we set $T^I_q=\{t_i \in T_q: \sigma^2(t_i)=0\}$. The existence and uniquenss of $\hat{m}^{n,q}_{\Gamma}(\omega,.)$ is guaranteed then by the only condition card $(T^I_q) \geq k$, where card(T^I_q) denotes de number of points in T^I_q.

Now, denote $T\setminus T^I_q$ the complementary of T^I_q in T_q and $H(T^I_q)$ the affine subspace of $H^k[0,1]$ defined by $H(T^I_q)=\{ u \in H^k[0,1]: \forall t_i \in T^I_q,\ u(t_i)=\overline{X}^n(\omega,t_i)\}$.

The last approach we consider in this paper in order to avoid interpolation of the empirical mean consists in constructiong a Penalized Weighted Least Square estimate that we will note $\hat{m}^{n,q}_{\lambda}$, where $\lambda>0$ is a real parameter determining the balance between fit and smoothness.The fitting to data is measured by the weighted sum

$$WLS(v) \equiv \sum_{t_i \in T\setminus T^I_q} \sigma^{-2}(t_i)(v(t_i)-\overline{X}^n(\omega,t_i))^2$$

in such a way that the weight of each data value is inversely proportional to the corresponding value $\sigma^2(t_i) > 0$ of the variance function and smoothness is measured by $|v|^2_k$. In this sense, this approach can be viewed as an intermediate between the interpolation spline of the empirical mean which privileges the fitting to empirical data and estimation spline which privileges smoothness. More precisely, we consider the following problem: for given $n \in \mathbb{N}^*$, T_q, $\omega \in \Omega$ and $\lambda>0$, find the element $\hat{m}^{n,q}_{\lambda}(\omega,.)$ in $H^k[0,1]$ such that

(P3) $\hat{m}^{n,q}_{\lambda}(\omega,.) \in H(T^I_q)$ and $\forall v \in H(T^I_q)$, $WLS(\hat{m}^{n,q}_{\lambda}(\omega,.))+\lambda\,|\hat{m}^{n,q}_{\lambda}(\omega,.)|^2_k \leq WLS(v) + \lambda\,|v|^2_k$

The existence and uniqueness of $\hat{m}^{n,q}_{\lambda}(\omega,.)$ is guaranteed if $q \geq k$ (cf.[3]). We impose that $\hat{m}^{n,q}_{\lambda}(\omega,.) \in H(T^I_q)$ in order that $\hat{m}^{n,q}_{\lambda}(\omega,.)$ interpolates the values $\overline{X}^n(\omega,t_i)$ corresponding to points for which $\sigma^2(t_i)=0$ (this is the case for example when there is a initial condition of type $X(0)\equiv 0$). Notice that if $\sigma^2(t_i)=\sigma^2>0$ for all t_i in T_q, the problem **(P3)** becomes the minimization of

$$\sum_{t_i \in T_q} (v(t_i)-\overline{X}^n(\omega,t_i))^2 + \lambda\,|v|^2_k$$

over $H^k[0,1]$ and leads thus to the classical notion of smoothing spline. For this reason, we will call $\hat{m}^{n,q}_{\lambda}(\omega,.)$ simply smoothing spline of the empirical mean.

Finally, we point out that the use of the three estimates above has been justified by the consistence results established by E. Biritxinaga ([3],[4]). More precisely, define h_{max} and h_{min} by:

$$h_{max}=\sup_{t\in[0,1]}\inf_{t_i \in T_q}|t - t_i| \qquad h_{min}=\min_{t_i \neq t_j}|t_i - t_j|$$

and assume that there is a constant $C>0$ such that

$$\frac{h_{max}}{h_{min}} \leq C \qquad \text{and that} \qquad \lim_{q\to\infty} h_{max}=0.$$

Under these assumptions, it was established in [3] that if X satisfies the law of the iterated logarithm in D[0,1] then $\hat{m}^{n,q}(\omega,.)$ and $\hat{m}^{n,q}_{\lambda}(\omega,.)$ converge almost surely to m in $H^k[0,1]$ when n and q increase in such a way that $q^k(n^{-1}LLn)^{1/2} \rightarrow 0$. A similar result for strong convergence has also been established for $\hat{m}^{n,q}_{\Gamma}$ ([4]) with $q^k n^{-1/2}((Lq)^{1/2} + (LLn)^{1/2}) \rightarrow 0$.

Notice that , in view of the continuous embedding of $H^k[0,1]$ (k de $\mathbb{N}^*$) in C[0,1] (cf. for example [1]), the almost sure convergence of the proposed estimators to m in $H^k[0,1]$ implies the almost sure uniform convergence of the estimators and their derivatives up to the order k-1 to m and to its corresponding derivatives respectively.

3. Numerical results

Although the three estimates proposed in Section 2 have satisfying asymptotic behavior, it is natural to ask how these estimates behave in practical situations, namely for a finite sample size n and a fixed knot-set T_q. The object of this section is to study if these estimation methods lead to correct estimates of the mean function in practical situations and to compare them. So we will test and compare $\hat{m}^{n,q}$, $\hat{m}^{n,q}_{\lambda}$ et $\hat{m}^{n,q}_{\Gamma}$ (that we will note from now on $\hat{m}$, $\hat{m}_{\Gamma}$ and $\hat{m}_{\lambda}$ to simplify the notations) as estimates of the mean function of two simulated processes $(X(t))_{t\geq 0}$.

The first is a non-homogeneous Poisson process with intensity function $\nu(.)$. It is well known (cf. for example [12]) that $X(0)\equiv 0$, that the integer-valued process $(X(t))_{t\geq 0}$ has independent increments and unity jumps and that X(t) is Poisson distributed with mean $m(t) = \int_0^t \nu(s)ds$. For many events, one expects a decreasing intensity of occurence and a form frequently choosen for the intensity function is then $\nu(t) \equiv \alpha e^{-\beta t}$ for some positive constants α and β to be determined empirically. In such a case : $m(t) = \frac{\alpha}{\beta}(1 - e^{-\beta t})$. For simulations we have taken α=1600 and β=4 .

The second simulated process $(X(t))_{t\geq 0}$ is of the form $X(t) \equiv \sin(3\pi t) + B(t)$, where $(B(t))_{t\geq 0}$ denotes the standard Brownian process. Then, $m(t) = \sin(3\pi t)$.

In both cases, the regularity degree of the mean function is clearly greater than that of the sample paths (those of the first process being discontinuous since they are step functions and those of the second being continuous but almos everywhere non-differentiable). More, since the theoretical expression of the mean is known in both cases, not only we can compare but also we can test the three estimates by computing the errors.

We simulate samples (of size n=5, 20, 50) of the process in the points t_i = hi (i=0,...,q) with h = 1/q for differents values of q (q=5,10,15). We built then the spline functions $\hat{m}$, $\hat{m}_{\Gamma}$ et $\hat{m}_{\lambda}$ solution to the problems **(P1)**, **(P2)** and **(P3)** respectively dans $H^2[0,1]$ (that is to say for the operator D^2) from the values of the empirical mean. As is well known, ([9]) $\hat{m}$, $\hat{m}_{\Gamma}$ and $\hat{m}_{\lambda}$ are then piecewise polynomial functions of class C^2, the polynomials being of degree 3.

More, by a direct application of Corollary 4.2 in [7], one deduces that the two processes above satisfy the law of the iterated logarithm in D[0,1]. We know (cf. Section 2) that in this case the three estimates considered here converge almost surely to m in $H^k[0,1]$ when n and q tend adequately to infinity and so, that the estimates (respectively their first derivative) converge to m (respectively its first derivative m') in the sup norm.

It is interesting to notice also that in modeling practical situations by stochastic processes, it often occurs that the derivative m' of the mean function m has a physical interpretation. Unfortunately, we cannot derive the sample paths of the two processes considered here in order to find m' against. However, the estimates being of class C^2, m' can be estimated from the data by differentiating the spline functions $\hat{m}$, $\hat{m}_\Gamma$ or $\hat{m}_\lambda$; this is justified by the convergence results above. So, we will test the proposed estimates comparing them and their first derivatives to m and m' respectively, which are known. For this, we consider relative errors in Euclidean norm. These errors will be computed by the formula

$$e(u,v) = \left[\frac{\sum_{i=1}^{10^3} (u(x_i)-v(x_i))^2}{\sum_{i=1}^{10^3} (v(x_i))^2} \right]^{1/2}$$

where the 10^3 points x_i are equally spaced in [0,1] and where v is m or m' according as we calculate the relative errors of the estimates or those of their first derivatives. The tables containing relative errors for the two examples and some of the related graphical results are given in the end of this section.

The value of the parameter λ has been choosen here by minimizing $e(\hat{m}_\lambda,m)$. This does not solve the problem of the optimal choice of λ in terms of n and q, and this can considerably improve the estimates obtained by spline smoothing of the empirical mean. The problem of the optimal choice of λ in terms of q has drawn the attention of many authors, specially when data are realizations of signal corrupted by white noise (cf. for instance [5], [6], [13], [14]). We think that the Cross Validation method for the choice of λ due to G. Wahba [14] can be applied to the present more general situation but this study is beyond the scope of this paper and we do not deal with here.

Examining relative errors about the first example one can notice that the three proposed methods lead in this case to quite satisfying estimates of the mean function even for very little values of n and q (q=5, n=5). Relative errors concerning $\hat{m}_\lambda$ are slighty smaller than those of $\hat{m}$ for the most part of the values of n and q. In the contrary, those of $\hat{m}_\Gamma$ are slighty greater. However, for an inspection of graphical representations for the first derivative, we can notice that $\hat{m}'_\Gamma$ (as well as $\hat{m}'_\lambda$) lead to nicer graphics than those provided by $\hat{m}'$, since the oscillations of the latter are eliminated (see also the graphics of the first derivatives in the second example).

In the second example, the superiority of the estimates obtained by spline smoothing in comparison with the two other methods appear more clearly. On the other hand, for this example, the value q=5 of knot-number is clearly insufficient to take the oscillations of the mean function m(t) = sin(3πt) in [0,1] into account. This fact explains the impossibility of improving in this case the estimate $\hat{m}$ by the use of $\hat{m}_\lambda$. For all of the other cases $\hat{m}_\lambda$ improve clearly the estimate $\hat{m}$, which is better than the estimate $\hat{m}_\Gamma$. The latter is particularly improper for the values q=5, n=5 and q=10, n=5 although it is substantially improved as n and q increase.

Finally, we point out that, in both cases, for fixed q, the estimates get better as n increases but we notice also that increasing q while n remains fixed do not lead necessarily to better estimates. This fact shows that we need to link both n and q in order to improve the estimates (see the improvement of the relative errors along the diagonal in each table) and illustrates the convergence results mentionned in Section 2.

RELATIVE ERRORS ; EXAMPLE 1

Means (order 10^{-2})

INTERPOLATION

	N=5	N=20	N=50
Q=5	4.44	1.50	1.19
Q=10	4.62	1.09	0.69
Q=15	4.35	1.11	0.73

SMOOTHING

	N=5	N=20	N=50
Q=5	3.75	1.13	1.19
Q=10	4.24	1.09	0.65
Q=15	4.13	1.12	0.68

ESTIMATION

	N=5	N=20	N=50
Q=5	3.93	3.45	1.94
Q=10	6.10	1.29	1.05
Q=15	4.70	1.35	0.86

Derivatives (order 10^{-1})

INTERPOLATION

	N=5	N=20	N=50
Q=5	1.41	1.17	1.12
Q=10	1.34	0.65	0.64
Q=15	1.60	0.60	0.68

SMOOTHING

	N=5	N=20	N=50
Q=5	1.48	1.69	1.12
Q=10	1.36	0.64	0.58
Q=15	1.20	0.57	0.49

ESTIMATION

	N=5	N=20	N=50
Q=5	1.84	1.45	1.18
Q=10	1.91	0.83	0.57
Q=15	1.28	0.85	0.45

RELATIVE ERRORS; EXAMPLE 2

Means (order 10^{-1})

INTERPOLATION

	N=5	N=20	N=50
Q=5	2.44	1.67	1.04
Q=10	4.27	2.07	0.69
Q=15	4.04	1.43	0.96

SMOOTHING

	N=5	N=20	N=50
Q=5	2.44	1.67	1.04
Q=10	3.43	1.43	0.66
Q=15	2.80	1.24	0.54

ESTIMATION

	N=5	N=20	N=50
Q=5	9.94	4.62	4.08
Q=10	8.48	4.25	3.15
Q=15	2.50	1.40	1.57

Derivatives (order 10^{-1})

INTERPOLATION

	N=5	N=20	N=50
Q=5	2.90	1.79	1.16
Q=10	7.42	4.04	1.76
Q=15	1.03	3.96	2.87

SMOOTHING

	N=5	N=20	N=50
Q=5	2.90	1.79	1.16
Q=10	5.15	2.27	1.62
Q=15	3.82	1.40	0.53

ESTIMATION

	N=5	N=20	N=50
Q=5	9.94	5.96	5.04
Q=10	8.44	5.40	4.27
Q=15	3.38	1.22	1.87

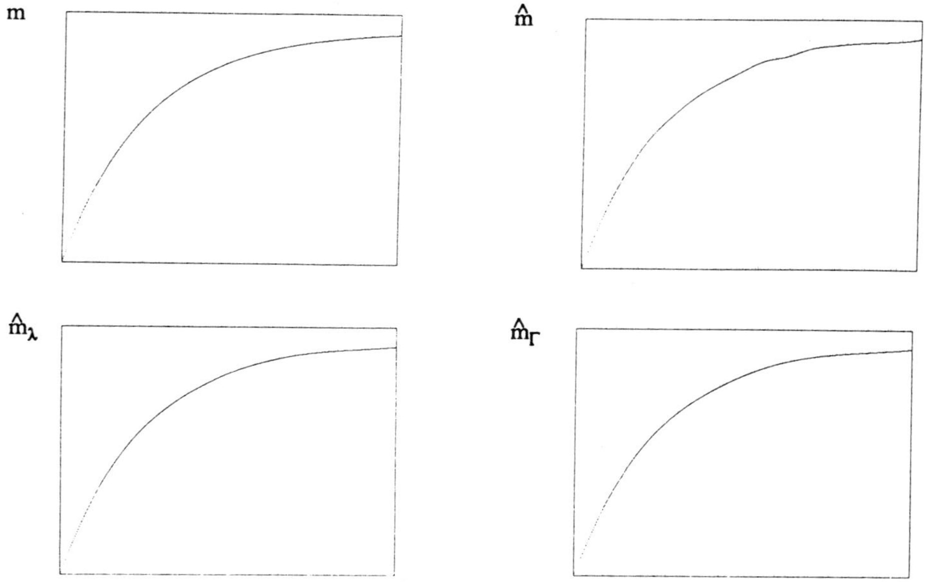

Means

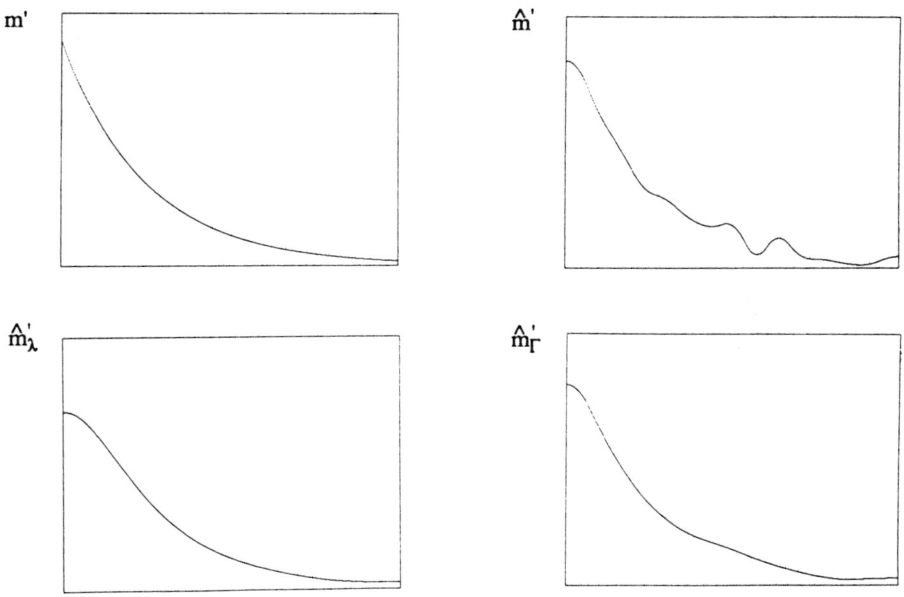

Derivatives

EXAMPLE 1 n=50,q=15

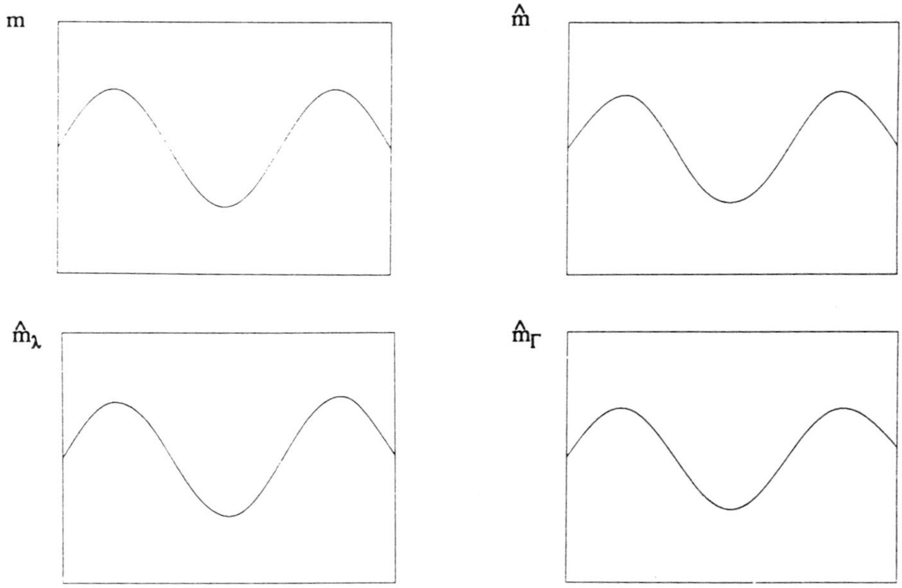

Means

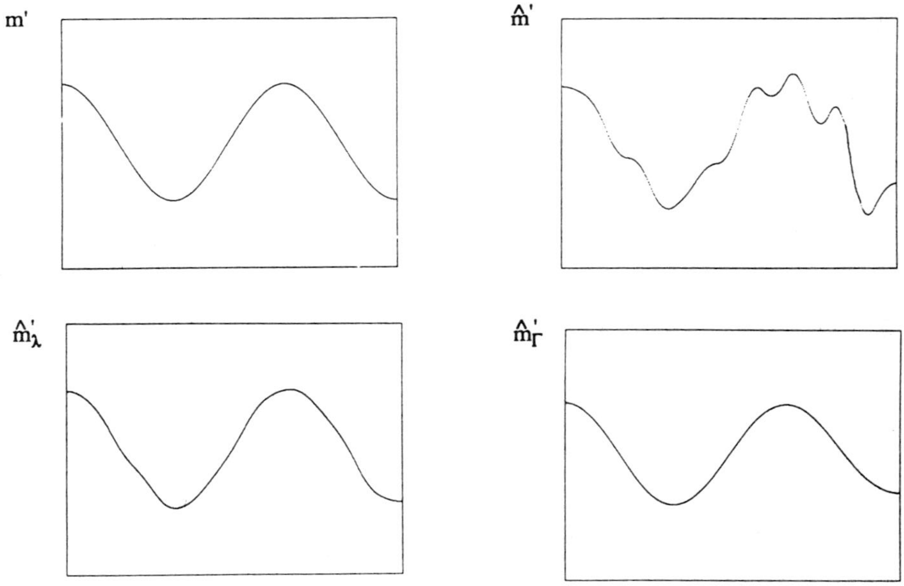

Derivatives

EXAMPLE 2 n=50,q=15

To summarize, we can say that if the parameter λ is adequatelly choosen, smoothing spline of the empirical mean will lead generally to better estimates of the mean function than interpolation spline of the empirical mean or estimation spline. The latter is unadequate for little values of n and q if from a priori information we know that the mean function has frequent oscillations in the interval under study. Otherwise $\hat{m}_{\Gamma}$ leads to smooth graphical representations but its relatively large errors and its implementation expensive in comparison with the other estimates lead to decide in favor of $\hat{m}_{\lambda}$.

REFERENCES

[1] ADAMS R.A.(1975) "Sobolev Spaces", Academic Press, New York.

[2] ATTEIA M.(1968) "Fonctions spline définies sur un ensemble convexe", Numer. Math., 12, 192-220.

[3] BIRITXINAGA E.(1989) "Sur la convergence de deux estimateurs spline de la moyenne d'un processus stochastique", Pub. n° 89/3 UA 1204 CNRS.

[4] BIRITXINAGA E.(1989) "Intervalles de confiance et estimation spline de l'espérance mathématique d'une variable aléatoire à valeurs dans D[0,1]", Publ. n° 89/4 UA 1204 CNRS.

[5] CRAVEN G., WAHBA G.(1979) "Smoothing noisy data with spline functions", Numer. Math. 31, 377-403.

[6] GOLUB J.H., HEATH M., WAHBA G.(1979) "Generalized Cross-Validation as a method for choosing the right ridge parameter", Technometrics 21, 2, 215-223.

[7] KUELBS J.(1976) "A strong convergence theorem for Banach valued random variables", Ann. Prob. 4, 5, 744-771.

[8] KUELBS J.(1976) "The Law of the Iterated Logarithm in C[0,1]", Z. Wahrscheinlichkeitstheorie verw. Gebiete 33, 221-235.

[9] LAURENT P.J.(1972) "Approximation et Optimisation", Hermann, Paris.

[10] LEWIS P.A.V., SHEDLER G.S.(1976) "Simulation of nonhomogeneous Poisson processes", Biometrika, 63, 3, 501-505.

[11] MORIN M.(1969) "Méthodes de calcul des fonctions spline dans un convexe", Thèse de 3° cycle, Grenoble.

[12] PARZEN E.(1962) " Stochastic Processes", Holden-Day Inc.,

[13] UTRERAS F.(1980) "Sur le choix du paramètre d'ajustement dans le lissage par fonctions spline", Numer. Math., 34, 15-28.

[14] WAHBA G.(1977) "Practical approximate solutions to linear operator equations when the data are noisy", SIAM J. Num. Anal., 14, 651-667.

TIME SERIES

A BILINEAR REPRESENTATION OF SEASONAL TIME SERIES

Iheanyi S. Iwueze
Department of Mathematics and Computer Science
Federal University of Technology
Owerri, Nigeria

ABSTRACT

A bilinear representation of time series whose only non zero autocorrelation is at the seasonal lag is considered in this paper. The conditions for stationarity and invertibility of the model are derived and the second-order moments are obtained. Simulated examples are presented to show the properties of the bilinear model. An analysis of Umudike mean monthly minimum temperature is included.

Key Words: Bilinear time series, seasonality, Stationarity, invertibility, identification, estimation, Umudike mean monthly minimum temperature.

1. INTRODUCTION

It is not yet known what types of bilinear time series models can best describe periodic behaviour in time series. A series exhibits periodic behaviour with period s, when similarities in the series occur after s basic time intervals. An example is when the basic time interval is one month and the period s=12 months. Other common examples are s=4 for quarterly data and s=13 for four weekly data showing seasonal effects within years. A series exhibiting periodic behaviour with period s is called seasonal time series.

Let e_t be a sequence of independent, identically distributed random variables with $E(e_t) = 0$ and $E(e_t^2) = \sigma^2$ The general bilinear autoregressive moving average model X_t, with respect to e_t, satisfying

$$X_t = \sum_{j=1}^{r} a_j X_{t-j} + \sum_{j=1}^{h} b_j e_{t-j}$$

$$+ \sum_{i=1}^{m} \sum_{j=1}^{v} \theta_{ij} X_{t-i} e_{t-j} + e_t \qquad (1.1)$$

with suitable values for the coefficients $a_1, a_2, \ldots, a_r, b_1, b_2, \ldots, b_h, \theta_{ij},\ 1 \leq i \leq m,\ 1 \leq j \leq v$ is entirely adequate to describe many non linear seasonal time series. The problem is to choose a suitable system of parsimonious parametirization for such models in conformity with the periodic behaviour of the time series.

When we have a series exhibiting nonstationary and seasonal behaviour with period s, we always proceed with the simpler analysis of

$$X_t = (1-B)^d (1-B^s)^D Y_t \qquad (1.2)$$

where $B^j Y_t = Y_{t-j}$ and Y_t is the original series. The variable X_t are formed from the original series Y_t by differencing to remove trend and seasonality. The simplest class of seasonal pattern in time series are those the only non zero autocorrelations of X_t represented by (1.2) are at lag s. A bilinear model that seem representationally useful for this type of seasonal series is the model

$$X_t = (b + \sum_{j=0}^{m} \theta_j X_{t-s-j}) e_{t-s} + e_t, \quad s \geq 1 \qquad (1.3)$$

In this paper, we consider the basic bilinear representation (1.3) for seasonal time series X_t whose only non zero autocorrelation is at lag s. We also consider the problems of stationarity, invertibility, identification and estimation which arise in relating such models to actual data.

2. CONDITIONS FOR STRICT STATIONARITY AND INVERTIBILITY

The model (1.3) is a special case of the model

$$X_t = \sum_{j=1}^{r} a_j X_{t-j} + (b + \sum_{j=0}^{m} \theta_j X_{t-s-j}) e_{t-s} + e_t \qquad (2.1)$$

Iwueze (1988) after obtaining the vectorial representation of (2.1) and using the result of Akamanam, Bhaskara Rao and Subramanyam (1986), showed that all bilinear models having that vector form are strictly stationary, ergodic and unique under suitable conditions on a matrix built on $a_1, a_2, .., a_r$, b, $\theta_o, \theta_1, \ldots, \theta_m$ and σ^2. A sufficient condition for the series X_t satisfying (1.3) to be second-order stationary is given in section 3.

Iwueze (1988) also showed that the strictly stationary and ergodic process X_t satisfying (2.1) is invertible if and only if

$$E \log \left| b + \sum_{j=0}^{m} \theta_j X_{t-j} \right| < 0 \tag{2.2}$$

From (2.2), a sufficient condition for invertibility of (2.1) is

$$b^2 + 2b\mu \sum_{j=0}^{m} \theta_j + \sum_{i=0}^{m} \sum_{j=0}^{m} \theta_i \theta_j E(X_{t-i} X_{t-j}) < 1 \tag{2.3}$$

where

$$\mu = E(X_t) = \sigma^2 \theta_o / (1 - \sum_{j=1}^{r} a_j) \tag{2.4}$$

It follows from (2.3) that a sufficient condition for invertibility of (1.3) is

$$b^2 + 2\sigma^2 b \theta_o \sum_{j=0}^{m} \theta_j + \sum_{i=0}^{m} \sum_{j=0}^{m} \theta_i \theta_j E(X_{t-i} X_{t-j}) < 1 \tag{2.5}$$

3. SUFFICIENT CONDITION FOR STATIONARITY AND EXPRESSIONS FOR THE COVARIANCES

In obtaining the expressions for the covariances in this section, we have made the assumption that the random variables e_t are Gaussian with $E(e_t) = 0$, $E(e_t^2) = \sigma^2 < \infty$, so that $E(e_t^3) = 0$ and $E(e_t^4) = 3\sigma^4$. We also used the fact that by expression (1.3), e_t is assumed to be independent of X_s, $s < t$. Based on these assumptions, it can easily be checked that

$$E(X_t e_t) = \sigma^2 \tag{3.1}$$

$$E(X_t) = \mu = \sigma^2 \theta_o \tag{3.2}$$

$$E(X_t e_t^2) = \sigma^4 \theta_o \tag{3.3}$$

$$E(X_t^2 e_t) = 2\sigma^4 \theta_o \tag{3.4}$$

$$E(X_{t-i}e_{t-i}X_{t-j}e_{t-j}) = \sigma^4 \quad , \; i \neq j \qquad (3.5)$$

$$\left.\begin{aligned} E(X_t e_t^2 X_{t-j}) &= \sigma^2 E(X_t^2) + 2\,\sigma^4, \; j = 0 \\ &= \sigma^4(b + \sigma^2\theta_o(2\theta_o + \theta_1 + \theta_2 + .. + \theta_m)), j = s \\ &= \sigma^6 \theta_o^2, \; j \neq 0, \; s \end{aligned}\right\} (3.6)$$

and

$$E(X_t^2) = \sigma^2 \sum_{j=0}^{m} \theta_j^2 E(X_{t-s-j}^2) + 2\theta_o \sum_{j=1}^{m} \theta_j E(X_{t-s}e_{t-s}^2 X_{t-s-j})$$

$$+ \sigma^2(1 + b^2 + 2\,\sigma^2\theta_o^2 + 2\,\sigma^2 b\theta_o(\theta_o + \theta_1 + \theta_2 + .. + \theta_m))$$

$$+ 2\,\sigma^2(\theta_1\theta_2 + \theta_2\theta_3 + ... + \theta_{m-1}\theta_m)E(X_t X_{t-1})$$

$$+ 2\,\sigma^2(\theta_1\theta_3 + \theta_2\theta_4 + ... + \theta_{m-2}\theta_m)E(X_t X_{t-2})$$

$$+ ....$$

$$+ 2\,\sigma^2(\theta_1\theta_{m-1} + \theta_2\theta_m)E(X_t X_{t-m+2})$$

$$+ 2\,\sigma^2\theta_1\theta_m E(X_t X_{t-m+1}) \qquad (3.7)$$

The expressions (3.1) through (3.5) are needed for the evaluation of $E(X_t X_{t-k})$, $k = 0, \pm 1, \pm 2, ...$. We state the results

$$E(X_t X_{t-k}) = \begin{cases} \sigma^4\theta_o^2; \; k = \pm 1, \pm 2, \; ..., \; \pm(s-1) \\ \sigma^2(b + \sigma^2\theta_o(\theta_o + \theta_1 + ... + \theta_m)) + \sigma^4\theta_o^2, \; k = \pm s \\ \sigma^4\theta_o^2, \; k = \pm(s+1), \; \pm(s+2), ... \end{cases} \qquad (3.8)$$

The autocovariances are given by

$$R(k) = \begin{cases} E(X_t^2) - \mu^2, \; k = 0 \\ \sigma^2(b + \sigma^2\theta_o(\theta_o + \theta_1 + ... + \theta_m)), \; k = \pm s \\ 0 \text{ otherwise} \end{cases} \qquad (3.9)$$

We see that the bilinear model (1.3) has its only non zero autocovariance at lag s.

Using (3.6) and (3.8) we can write (3.7) in the form

$$E(X_t^2) = \sigma^2 \sum_{j=0}^{m} \theta_j^2 E(X_{t-s-j}^2) + c \qquad (3.10)$$

where c is a constant defined in terms of b, $\theta_o, \theta_1, \ldots, \theta_m$ and σ^2. We can write (3.10) as a first-order vector difference equation as follows. Let

$$p = s+m$$

$$\underset{p\times 1}{\underline{W}_t} = (E(X_t^2), E(X_{t-1}^2), \ldots, E(X_{t-p+1}^2))^T$$

$$\underset{p\times 1}{H} = (1,0,0,\ldots, 0)^T$$

$$\underset{p\times p}{D} = \begin{bmatrix} \overbrace{0\ 0\ 0\ \ldots 0}^{(s-1)} & \phi_o^2 & \phi_1^2 & \ldots \phi_{m-1}^2 & \phi_m^2 \\ 1\ 0\ 0\ \ldots 0 & 0 & 0 & \ldots 0 & 0 \\ 0\ 1\ 0\ \ldots 0 & 0 & 0 & \ldots 0 & 0 \\ \cdot\ \cdot\ \cdot\ \quad \cdot & \cdot & \cdot & \cdot & \cdot \\ \cdot\ \cdot\ \cdot\ \cdots\cdot & \cdot & \cdot & \cdots & \cdot \\ \cdot\ \cdot\ \cdot\ \quad \cdot & \cdot & \cdot & \cdot & \cdot\cdot \\ 0\ 0\ 0\ \ldots 0 & 0 & 0 & \ldots 1 & 0 \end{bmatrix} \qquad (3.11)$$

where $\phi_i = \sigma^2\theta_i$, i=0,1,...,m and T stands for operation transpose of a matrix. Then

$$\underline{W}_t = D\underline{W}_{t-1} + cH \qquad (3.12)$$

For the solution of $\underline{W}_t$ to converge, it is sufficient that all the eigenvalues of the matrix D have moduli less than unity. This is the condition for asymptotic stationarity of (1.3) formally stated in Theorem 3.1

THEOREM 3.1. Let e_t be a sequence of independent real random variables and each e_t is distributed $N(0, \sigma^2)$. Then the process X_t satisfying

$$X_t = (b + \sum_{j=0}^{m} \theta_j X_{t-s-j}) e_{t-s} + e_t$$

is asymptotically second-order stationary if all the eigen values of the matrix D defined by (3.11) have moduli less than unity.

COROLLARY 3.2. For the process X_t satisfying (1.3) the asymptotic second-order stationarity condition implies that the roots (in modulus) of the equation

$$Z^p - \phi_o^2 Z^m - \phi_1^2 Z^{m-1} - \ldots - \phi_{m-1}^2 Z - \phi_m^2 = 0$$

lie inside the unit circle.

Proof: The characteristic roots of the pxp matrix D are the solutions to the determinant equation

$$\left| D - ZI_p \right| = 0 \tag{3.13}$$

By partitioning (3.13), it is easy to show that the characteristic polynomial is

$$\left| D - ZI_p \right| = Z^p - \phi_o^2 Z^m - \phi_1^2 Z^{m-1} - \ldots - \phi_{m-1}^2 Z - \phi_m^2 = 0$$

4. IDENTIFICATION

The identification of (1.3) depends upon the fact that the autocorrelations for all lags except at lag s are zero. In fact, the autocorrelations can be represented as

$$\rho_k = \begin{cases} 1 & k = \pm 1 \\ \rho_s & k = \pm s \\ o & \text{otherwise} \end{cases} \tag{4.1}$$

The partial autocorrelation function of time series models (linear and non-linear) with autocorrelation function (4.1) is another tool for identification. The partial autocorrelation function is given by

$$\phi_{kk} = \begin{cases} \rho_1, & k = 1 \\ |A_k^*| \,/\, |A_k|, & k \geqslant 1 \end{cases} \tag{4.2}$$

where

$$A_k = \begin{bmatrix} 1 & \rho_1 & \rho_2 & \cdots\cdots \rho_{k-1} \\ \rho_1 & 1 & \rho_1 & \cdots\cdots \rho_{k-2} \\ \rho_2 & \rho_1 & 1 & \cdots\cdots \rho_{k-3} \\ \rho_{k-1} & \rho_{k-2} & \rho_{k-3} & \cdots\cdots 1 \end{bmatrix} \tag{4.3}$$

and A_k^* is the matrix composed of the first k-1 columns of A_k with the kth column replaced by $(\rho_1, \rho_2, \ldots, \rho_k)^T$.

The partial autocorrelation of time series models with autocorrelation function (4.1) is zero everywhere except at the multiples of the seasonal lag. Using (4.2), it can easily be shown that

$$\phi_{kk} = \begin{cases} \rho_s, & k=s \\ -\rho_s^2/(1-\rho_s^2), & k=2s \\ \rho_s^3/(1-2\rho_s^2), & k=3s \\ -\rho_s^4/(1-3\rho_s^2+\rho_s^4), & k=4s \end{cases} \tag{4.4}$$

For correct identification of models with autocorrelation function (4.1), we look at the first few values of ϕ_{kk}, particularly the first seasonal lag with special characteristic $\phi_{ss} = \rho_s$, and see if any are significantly different from zero.

REMARK

The optimal linear model having the autocorrelation function (4.1) and non-zero mean value is given by

$$X_t = \mu + \beta a_{t-s} + a_t \tag{4.5}$$

where $E(X_t) = \mu$ and a_t is a sequence of independent, identically distributed random variables with $E(a_t) = 0$ and $E(a_t^2) = \sigma_a^2 < \infty$.

5. ESTIMATION

To obtain estimates of the parameters of (1.3), we proceed as in Subba Rao (1981), Gabr and Subba Rao (1981) and apply the method of least squares to minimize

$$S(\underline{\theta}) = \sum_t e_t^2 \tag{5.1}$$

with respect to the parameters $\underline{\theta} = (b, \theta_0, \theta_1, \ldots \theta_m)^T$.

The choice of the value of m is made on the basis of the information criterion of Akaike (1977), which is giving by

$$AIC = N\log \hat{\sigma}^2 + 2 \text{ (independent number of parameters)} \tag{5.2}$$

where

$$\hat{\sigma}^2 = S(\hat{\underline{\theta}})/N \tag{5.3}$$

and N is the number of observations used for calculating $S(\hat{\theta})$.

When minimizing $S(\underline{\theta})$ with respect to $\underline{\theta}$, the normal equations

$$\frac{\partial S(\underline{\theta})}{\partial \underline{\theta}} = 0$$

are nonlinear in $\underline{\theta}$. The solution of these equations requires the use of a nonlinear algorithm such as Newton-Raphson. The variance of $\underline{\theta}$ is estimated from

$$V(\hat{\underline{\theta}}) \simeq 2\, \sigma^2 H^{-1}(\hat{\underline{\theta}}) \tag{5.4}$$

where

$$H(\underline{\theta}) = \frac{\partial^2 S(\underline{\theta})}{\partial \theta_i \partial \theta_j} \tag{5.5}$$

is the qxq matrix of second-order partial derivatives and $q = m + 2$

6. SIMULATION RESULTS

This section presents some simulation results to illustrate the properties of (1.3) given in sections 2, 3 and 4. In what follows, the random variables e_t are independent and each e_t is distributed $N(0, \sigma^2)$. The three strictly stationary, ergodic and invertible processes considered are

$$X_t = \theta_o X_{t-13} e_{t-13} + e_t \quad (s=13) \tag{6.1}$$

$$X_t = (b+\theta_o X_{t-12})\, e_{t-12} + e_t \quad (s=12) \tag{6.2}$$

$$X_t = (b+\theta_o X_{t-4} + \theta_1 X_{t-5}) e_{t-4} + e_t \quad (s=4) \tag{6.3}$$

For each of the three models, we have generated 100 observations with $\theta_o = 0.8$, $\sigma^2 = 0.49$ for model (6.1); $b = 0.5$, $\theta_o = 0.4$, $\sigma^2 = 1.0$ for model (6.2) and $b = 0.5$, $\theta_o = 0.3$, $\theta_1 = 0.2$, $\sigma^2 = 1.0$ for model (6.3). The mean, variance, autocorrelation and partial autocorrelation were estimated for each simulated series and Table 1 gives a comparison of the theoretical (TH) and estimated (ET) results. Our simulation results **confirm the fact** that the only nonzero autocorrelation is at the seasonal lag and the partial autocorrelation at lag s is significantly different from zero.

Table 1: Theoretical and Estimated mean, variance, autocorrelation and partial autocorrelation of Simulated series.

Series	Model (6.1) TH	Model (6.1) ET	Model (6.2) TH	Model (6.2) ET	Model (6.3) TH	Model (6.3) ET
Mean	0.39	0.29	0.40	0.34	0.30	0.27
Variance	1.01	0.61	1.90	1.51	1.74	1.68
ρ_s	0.15	0.21	0.35	0.27	0.37	0.33
ϕ_{ss}	0.15	0.19	0.35	0.27	0.37	0.31
$\phi_{2s,2s}$	-0.02	-0.05	-0.14	-0.08	-0.16	-0.33

On fitting model (4.5) to our simulated series, the estimates given in Table 2 were obtained.

Table 2: Summary of linear models fitted to simulated series

Series	Fitted Model	$\hat{\sigma}_a^2$	% Increase
Model (6.1)	$X_t = 0.2742 + 0.2440a_{t-13} + a_t$ $(\pm 0.0923)\ (\pm 0.0976)$	0.5734	17*
Model (6.2)	$X_t = 0.2797 + 0.4627a_{t-12} + a_t$ $(\pm 0.1606)\ (\pm 0.1053)$	1.3144	31*
Model (6.3)	$X_t = 0.2037 + 0.6600a_{t-4} + a_t$ $(\pm 0.1827)\ (\pm 0.0789)$	1.2622	26*
	*Percentage increase in error variance		

The summary of bilinear models fitted to our simulated series is given in Table 3

Table 3: Summary of bilinear models fitted to simulated series

Series	Fitted Models	$\hat{\sigma}^2$
Model (6.1)	$X_t = 0.7634\ X_{t-12}e_{t-12} + e_t$ (± 0.0637)	0.4880
Model (6.2)	$X_t = e_t + (0.4922 + 0.4608X_{t-12})e_{t-12}$ $(\pm 0.0624)\ (\pm 0.0519)$	0.9873
Model (6.3)	$X_t = e_t + (0.6403 + 0.2790X_{t-4}$ $(\pm 0.0575)\ (\pm 0.0499)$ $+ 0.1175X_{t-5})e_{t-4}$ (± 0.0443)	0.9519

7. AN EXAMPLE OF REAL TIME SERIES DATA

The data presented in Table 4 show the minimum air temperature at Umudike, in Nigeria, from 1974 to 1987, where the individual observations are averaged over periods of one month. The data are plotted in Figure 1 and it is evident that there is a regular seasonal effect with lowest temperatures in December and January (Harmattan Season).

We use the first 144 observations (1974-1985) for model fitting and the next 24 observations (1986-1987) for forecasting. Table 5 shows the estimated autocorrelations and partial autocorrelations of the 144 observations differenced with respect to months. Table 5 suggests that model (1.3) with s=12 is worthy of further investigation for the differenced series $X_t = (1-B^{12})Y_t$, where Y_t denotes the observed minimum temperature in month t.

First we obtain the estimates of the parameters of the linear model (4.5) that fits the data. The following linear model is identified and estimated:

$$X_t = \underset{(\pm 0.0576)}{-0.6704}\,a_{t-12} + a_t \qquad (7.1)$$

$$\sigma_a^2 = 0.7275$$

The resulting residuals suggest model inadequacy. The correlogram of the estimated residuals took very 'small' values, except for lag 12 with a value of -0.21. The correlogram of the squares of the residuals took very small values, except for the first with a value of 0.35. The resulting residuals from (7.1) suggest model inadequacy and non-linearity in the series.

TABLE 4: Mean Monthly Temperature (^{o}C) at Umudike (1974-1987)

YEAR	JAN	FEB	MAR	APR	MAY	JUN	JUL	AUG	SEPT	OCT	NOV	DEC	AV
1974	19	22	23	22	22	22	22	22	22	22	22	18	22
1975	18	22	23	22	22	22	22	22	22	22	22	18	21
1976	19	22	22	23	23	22	22	22	22	22	22	21	22
1977	22	22	23	23	22	22	22	22	22	22	22	19	22
1978	19	23	23	23	22	22	22	22	22	22	21	21	22
1979	21	23	23	23	23	23	22	23	22	23	23	19	22
1980	22	23	23	23	23	23	21	22	22	22	23	21	22
1981	19	22	23	23	23	22	22	22	22	22	21	21	22
1982	22	22	22	23	22	23	22	22	22	22	22	22	22
1983	19	23	24	24	23	23	22	22	22	23	23	22	23
1984	21	23	23	23	23	23	22	23	22	22	22	20	22
1985	23	22	23	23	22	22	22	22	22	22	23	20	22
1986	21	23	22	23	23	23	22	22	22	22	23	19	22
1987	21	23	23	24	23	23	22	22	23	23	23	22	23
AV	20	23	23	23	23	23	22	22	22	22	22	20	

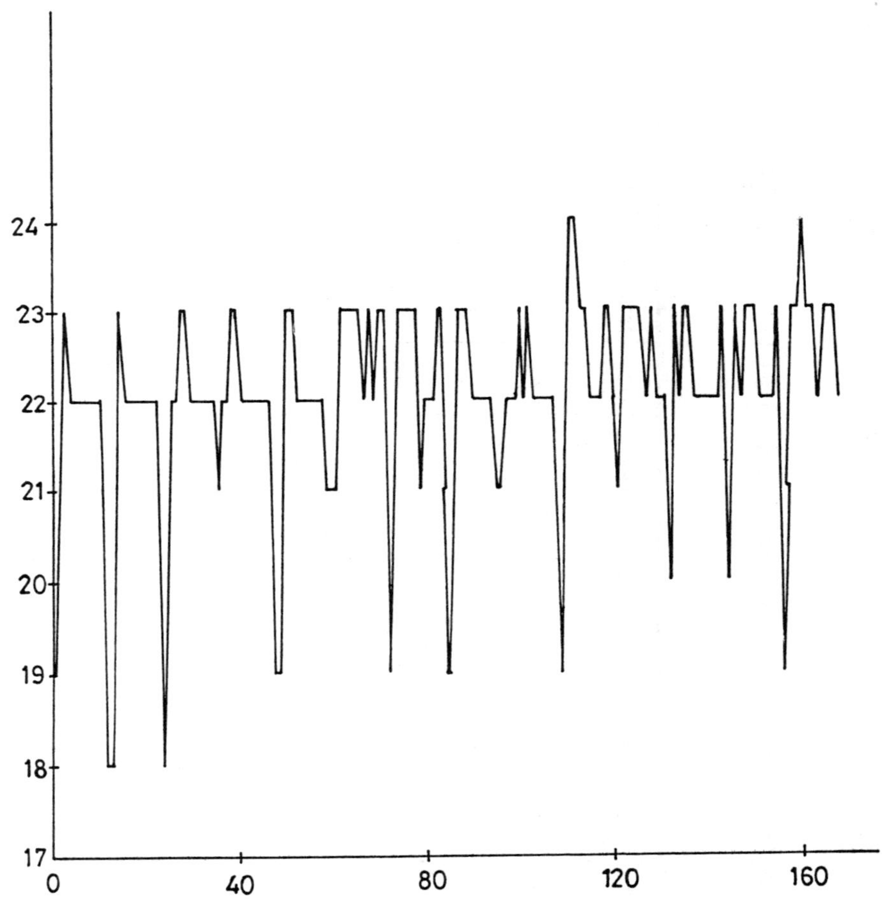

Figure 1 : Mean monthly minimum temperature (°C) at Umudike, Jan. 1974 - Dec. 1987.

Table 5. Estimated autocorrelations and partial autocorrelations of $X_t = (1-B^{12})Y_t$ of the Umudike Temperature data

Autocorrelations

Lags									
1- 9	0.05	-0.12	0.07	-0.06	0.11	0.06	0.05	0.18	-0.06
10-18	0.13	-0.09	-0.52	0.02	0.14	-0.01	-0.02	-0.15	-0.06
19-27	-0.16	-0.14	-0.01	-0.09	0.09	0.05	-0.17	-0.05	-0.02
28-33	-0.01	0.06	-0.03	0.17	0.05	-0.05			

Partial Autocorrelations

Lags									
1- 9	0.05	-0.12	0.09	-0.09	0.14	0.02	0.09	0.17	-0.05
10-18	0.19	-0.19	-0.50	-0.06	0.03	0.01	-0.06	-0.04	-0.01
19-27	-0.11	0.01	-0.07	0.05	0.04	-0.27	-0.11	0.11	0.10
28-33	-0.07	-0.03	-0.07	0.08	0.02	-0.06			

Note: Y_t denotes the observed temperature in month t.

The following bilinear model was identified and estimated for the data:

$$X_t = (-0.8506 + 0.1037X_{t-12}$$
$$(\pm 0.0594)(\pm 0.0209)$$
$$-0.0401X_{t-13})e_{t-12} + e_t \qquad (7.2)$$
$$(\pm 0.0360)$$

with $\sigma^2 = 0.6334$, reducing residual variance by about 13%. The resulting residuals from (7.2) did not suggest any model inadequacy.

For the next 24 months out of the sample, both models (7.1) and (7.2) were used to forecast. The mean square of various step forecast errors are giving in Table 6.

Table 6. m-step forecast errors of models (7.1) and (7.2)

m	Model (7.1) (1)	Model (7.2) (2)	% Reduction $\frac{(1)-(2)}{(1)} \times \frac{100}{1}$
1	0.49	0.50	-2
2	0.64	0.61	5
3	1.09	0.98	10
4	1.20	1.03	14
5	0.10	0.08	20
6	1.64	1.31	20

It is thus seen that the bilinear model seems to be better when it is used for forecasting several steps ahead.

8. CONCLUDING REMARKS

We have considered the properties of a class of bilinear models which are of value in representing stationary and invertible seasonal series. The problems of identification and estimation were also considered and applied to simulated series and the Umudike temperature data.

Model (1.1) is entirely adequate for the description of many non-linear seasonal time series. For optimal description, we need a suitable system of parsimonious parametirization of (1.1) in conformity with the periodic behaviour (of the autocorrelation function) of the time series. These are the theoretical problems still to be solved in modeling non-linear seasonal time series which the author hopes will attract the attention of researchers in due course.

ACKNOWLEDGEMENT

The author is grateful to the Director, National Root Crops Research Institute, Umudike and Mr. A.O. Ubbor for permission to use the Umudike temperature data.

REFERENCES

1. Akaike, H.(1977). On entropy maximisation principle, in. P. R. Krishnaiah (Ed). Applications of Statistics, pp. 27-41, North-Holland, Amsterdam.

2. Akamanam, S.I, Bhaskara Rao, M. and Subramanyam, K.(1986). On the ergodicity of bilinear time series models. Journal of Time Series Analysis, 3, pp. 157-163.

3. Gabr, M.M. and Subba Rao, T.(1981). The estimation and prediction of subset bilinear time series models with applications. Journal of Time Series Analysis, 3, pp. 155-171.

4. Iwueze, I.(1988). On the invertibility of bilinear time series models. To appear in STATISTICA.

5. Subba Rao, T.(1981). On the theory of bilinear time series models. Journal of the Royal Statistical Society Series B, 43, pp. 244-255.

GRAPHS AND MODELS

GRAPHICAL MODELLING: OUTLINE THEORY AND APPLICATIONS

Joe Whittaker, University of Lancaster (UK)

Abstract: Graphical models represent multivariate interaction by means of a finite graph, in which nodes represent variables and edges represent association; more precisely, the absence of an edge represents a conditional independence relationship. Such graphs were introduced by Darroch, Lauritzen and Speed (1980), where the variables are discrete; their continuous counterpart was examined by Speed and Kiiveri (1986); and more recently, the mixed discrete and continuous case has been discussed by Lauritzen and Wermuth (1989). The conditional independence graph has a characteristic Markov property which gives an immediate visual accessible interpretation to the multivariate pattern of interaction, and has important consequences in application, for instance to the development of probabilistic expert systems, Lauritzen and Speigelhalter (1988). There are extensions to directed graphs and graphs with a mix of directed and undirected edges.

Independence graphs form a natural tool of applied statistical modelling, a very simple form of which is choosing a subset of predictors to include in a linear regression equation, and more complicated models include the path analysis models suggested by Sewell Wright some 70 years ago. The positive benefits that graphical modelling techniques bring to data analysis are: a probability framework to assess interaction, phrased in terms of the original untransformed variables; visual display; a framework which naturally includes both discrete and continuous variables; and a statistical model to assess goodness of fit.

This paper outlines the basic theory of independence graphs and graphical modelling and gives some illustrations.

Keywords : conditional Gaussian model; covariance selection; Gaussian model; graphical models; loglinear model.

1 Introduction

The essential ingredients of a theory of graphical model are the notion of conditional independence, standard graph theory, the conditional Gaussian distribution, and maximum likelihood estimation. Our intention in this paper is to outline the basic theory, without derivation, and to report some practical applications of the techniques that are to appear in Whittaker (1990).

There is now a fairly large published literature on the theory of graphical models and their applications. The material described here is heavily dependent on the lecture notes of Speed (1978) to the Institute of Mathematical Statistics in Copenhagen, some of which later appears in Speed and Kiiveri (1986); the seminal Annals paper of Darroch, Lauritzen and Speed (1980); and the elegant exposition by Lauritzen (1982) of the ideas for contingency tables; more recently, the family of conditional Gaussian distributions for mixed variables has been introduced by Lauritzen and Wermuth (1984, 1989) and generalised by Edwards (1989). A theory for directed independence graphs, based on the ideas of Wermuth and Lauritzen (1983), and Kiiveri, Speed and Carlin (1984), is developed by limiting the conditioning set to the 'past'. This material is very much related to the original path analysis ideas of Sewell Wright (1921, 1923). Finally the extension is made to variables that only satisfy a weaker partial ordering in the form of chain independence graphs.

2 Conditional Independence Graphs

Let X denote a k-dimensional random vector, $X = (X_1, X_2, \ldots, X_k)$ where X_i is the i-th coordinate of X. Put $K = \{1, 2, \ldots, k\}$, the index set consisting of the full set of suffices, and let $a = \{i_1, i_2, \ldots, i_p\}$ denote an arbitrary subset of K. Define the random vector X_a as the ordered tuple

$$X_a = (X_{i_1}, X_{i_2}, \ldots, X_{i_p}) = (X_i;\ i \in a).$$

Then $X_{a\cup b}$ and $X_{a\cap b}$ are well-defined, and so are other set operations, in particular, $X_{K\setminus\{i\}}$ denotes the the sub-vector of X obtained by excluding X_i. The full vector is $X_K = X$, and X_ϕ is empty.

The density function f_K of X is defined as the joint probability density function of $X_1, X_2, \ldots, X_k$, and is more compactly denoted by $f_{12\ldots k}$. We shall assume that the density is positive on its support, and sufficiently smooth, and we make no claim to supply very general statement of results. In particular, we point out that the theory has to be extensively reworked if the assumption of positivity is not met. We use the same notation for the probability density function for continuous random variables and the probability mass function for discrete random variables. The marginal density function f_a is similarly defined as the density function of X_a, for $a \subseteq K$; and, if a and b disjoint, the conditional density of X_b given X_a is $f_{b|a} = f_{a\cup b}/f_a$, defined as f_a is positive.

As we assume the existence of an everywhere positive density function for the random vector, we define the relationship of conditional independence between sub-vectors of X in terms of the density. Using the independence notation due to Dawid (1979), for disjoint sets a, b and c of K, $X_b \perp\!\!\!\perp X_c | X_a$ if and only if the conditional density function $f_{bc|a} = f_{b|a} f_{c|a}$ for all values of x.

2.1 Independence Graphs

Let $X = (X_1, X_2, \ldots, X_k)$ denote a vector of random variables, and consider a graph, in the sense of Harary (1969) or Berge (1973), with k vertices representing each variable.

Definition. The *conditional independence graph* of X is the undirected graph $G = (K, E)$ where $K = \{1, 2, .., k\}$ and (i, j) is *not* in the edge set E if and only if $X_i \perp\!\!\!\perp X_j | X_{K\setminus\{i,j\}}$. □

The graph is a conditional independence graph, or more loosely, an independence graph, if there is no edge between two vertices whenever the pair of variables is independent given all the remaining variables. The vector of the remaining variables is sometimes referred to as the *rest.* This definition is the *pairwise Markov property*; and because of its Markov properties, a better name might be a *Markov* graph, but unfortunately this term is already used in the theory of random graphs. We use the shorthand, for example $1 \perp\!\!\!\perp 2 | \{3, 4\}$ for $X_1 \perp\!\!\!\perp X_2 | (X_3, X_4)$, so that the independence of X_i and X_j given the rest can be written as $i \perp\!\!\!\perp j | K\setminus\{i, j\}$. The resulting undirected graph gives a picture of the pattern of dependence or association between the variables.

It is often easy to construct the independence graph if we are given the joint density function of X by repeated application of the factorisation criterion, Dawid (1979), for conditional independence.

EXAMPLE 2.1 Take $k = 4$ and consider the density function

$$f_K(x) = \exp(u + x_1 + x_1x_2 + x_2x_3x_4),$$

of $X = (X_1, X_2, X_3, X_4)$ on the 4 dimensional cube, $\{x; x = (x_1, x_2, x_3, x_4),\ 0 < x_i < 1,\ i = 1, 2, 3, 4\}$, where the constant u ensures the density integrates to 1. The function factorises into $f_K(x) = g(x_1, x_2)h(x_2, x_3, x_4)$, and it is easy to verify that

$$X_1 \perp X_4 | (X_2, X_3) \quad \text{and} \quad X_1 \perp X_3 | (X_2, X_4).$$

Consequently the conditional independence graph is

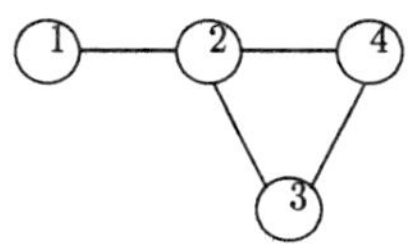

The graph uses the fact that vertex 1 is not adjacent to either 3 or 4 for construction but highlights the fact that the *cliques* of the graph are $\{1, 2\}$ and $\{2, 3, 4\}$. Note that this independence graph is identical to the interaction graph constructed by drawing an edge between variables that occur together in the interactions terms x_1x_2 and $x_2x_3x_4$, of the log-linear expansion of the density function. □

We remark on the importance of conditioning; there is no suitable theory for graphs constructed from pairwise marginal independences.

2.2 Markov Properties

Inspection of the independence graph in the example above suggests that conditionally on X_2, X_1 and (X_3, X_4) are independent. It is a remarkable feature of independence graphs that is true in general: more specifically that with respect to a given graph on a finite number of vertices, the following properties are equivalent:

- the pairwise independence property: that non-adjacent pairs of variables are independent conditional on the remaining variables;
- the local Markov property: that conditional only on the adjacent variables, any variable is independent of all the remaining variables; and
- the global Markov property: that any two subsets of variables separated by a third is independent conditionally only on variables in the third subset.

The strongest part of this assertion is the separation theorem, that the pairwise Markov property implies the global Markov property.

Since these properties are equivalent, any one could be taken as the definition of the conditional independence graph. But from the applied point of view of statistical modelling, it is most convenient to use the pairwise independence property. There are two essential reasons for this choice: firstly, the list of requirements to be verified in constructing the independence graph to model a given data set, is the least stringent, while the interpretations that follow are the strongest. Secondly, the pairwise definition of the independence relationship naturally corresponds to the pairwise graph-theoretic definition of the edge set in a graph.

2.3 Chain Independence Graphs

It often happens that the set of variables under study naturally partitions into groups, where some the variables can be considered response variables and some as explanatory variables or *covariates*. As the adjectives suggest, the responses and the covariates do not enter into the model on the same footing, and this assymmetry has to be incorporated into the model; 'treatment-response', 'exogoneous-endogonous' and 'independent-dependent' are three other synonyms for this partition.

We shall be slightly more general and suppose that the full set of vertices, K, can be partitioned into disjoint sets, or blocks, $b_1, b_2, \ldots, b_m$, which are themselves completely ordered as a chain. The elements in b_1 are potential causes of the elements in b_2, the elements in $b_1 \cup b_2$ are potential causes of the elements in b_3, and so on. This structure induces a partial ordering, $\preceq$, on the vertices. The parents of a vertex $i \in b_r$, the r-th block are drawn from the 'past', $b_1 \cup b_2 \cup \ldots \cup b_{r-1}$, and are joined to i by directed edges or arrows. The conditioning sets for each pairwise independence statement consist of the 'past' and all the remaining variables in the 'present'. The resultant graph is known as a chain independence graph and provides the conditional independence framework necessary for discussion of multivariate regression and simultaneous equation models.

EXAMPLE 2.2 Consider an 8 variable system $K = \{1, 2, \ldots, 8\}$, partitioned into subsets $b_1 = \{1, 2, 3\}$, $b_2 = \{4\}$, $b_3 = \{5, 6\}$, and $b_4 = \{7, 8\}$, with an edge set defined by the edges in the diagram:

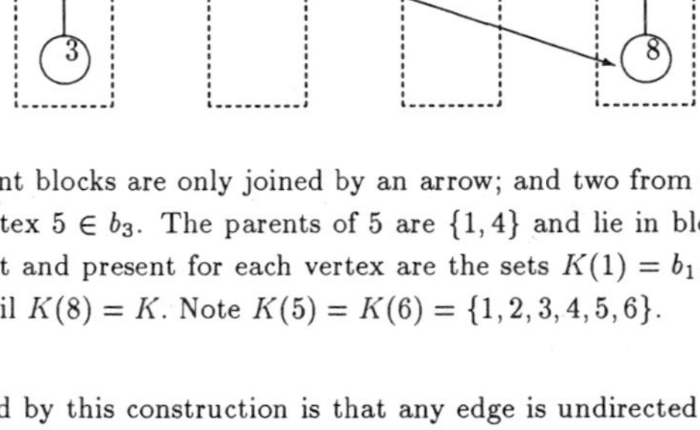

Any two elements from different blocks are only joined by an arrow; and two from the same block are only joined by a line. Consider vertex $5 \in b_3$. The parents of 5 are $\{1, 4\}$ and lie in blocks preceding b_3, while 6 is a neighbour of 5. The past and present for each vertex are the sets $K(1) = b_1$, $K(2) = b_1$, $K(3) = b_1$, $K(4) = b_1 \cup b_2$, and so on, until $K(8) = K$. Note $K(5) = K(6) = \{1, 2, 3, 4, 5, 6\}$. □

The essential property satisfied by this construction is that any edge is undirected for intra-block vertices, and directed for inter-block vertices with direction determined by the ordering on the blocks. This block formulation excludes graphs with directed cycles.

We may now define the chain independence graph from pairwise conditional independences taking the conditioning set for each statement as $K(i)$,the set of all 'past' and 'present' variables defined with respect to vertex i, that is, $K(i) = b_1 \cup b_2 \cup \ldots \cup b_{r(i)}$ where $r(i)$ is the index of the block containing i.

Definition. The *chain independence graph* of X is the graph $G^{\preceq} = (K, E^{\preceq})$, where $K = \{1, 2, \ldots, k\}$, $K(i) = \cup_{l \leq r(i)} b_l$ and the edge (i, j) with $i \preceq j$, is *not* in the edge set $E^{\preceq}$ if and only if $j \perp\!\!\!\perp i | K(j) \backslash \{i, j\}$. If this condition fails, and $i \prec j$, then the edge is directed and only $(i,j) \in E^{\preceq}$; otherwise it is undirected and both $(i,j) \in E^{\preceq}$ and $(j,i) \in E^{\preceq}$. □

Such graphs are also called *block recursive* by Lauritzen and Wermuth (1984, 1989).

Instead of specifying a block structure, an apparently more general definition is to allow a mixture of directed and undirected edges subject to the constraint that no cycle containing one or more directed edges is permissible. Under this hypothesis it is fairly straightforward to demonstrate the existence of a block representation which separates vertices into blocks such that inter-block edges are arrows, intra-block edges are lines, and such that the blocks are completely ordered. However, the matter is a little more complicated, because the block representation need not be unique. For example, with vertex set $K = \{1,2,3\}$ and an empty edge set, $E = \phi$, three alternative block structures are: (i) $b_1 = \{1,2\}$, $b_2 = \{3\}$ or (ii) $b_1 = \{1\}$, $b_2 = \{2\}$, $b_3 = \{3\}$ or (iii) $b_1 = \{1,2,3\}$. Clearly the initial list of pairwise independences is different, for instance: as neither $(1,2)$ nor $(2,1)$ is in the edge set, then under (i), $1 \perp\!\!\!\perp 2$; while under (iii), $1 \perp\!\!\!\perp 2|3$. It is shown by Frydenberg (1986) that all consistent block representations are probabilistically equivalent, who goes on, Frydenberg (1989), to give a fairly definitive answer to the question of when two chain graphs are equivalent.

2.4 Markov properties of chain graphs

The Markov properties of chain independence graphs can be deduced from the following considerations.

Definition. A chain graph satisfies the *Wermuth condition* if no vertex, has parent vertices that are not directly connected in the graph: or equivalently, no subgraph has the configuration

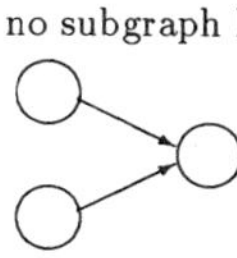

or, more generally, if no pair of connected vertices which are in the same block, have parent vertices that are not directly connected in the graph. □

The *moral graph*, G^m, associated with the chain graph $G^{\preceq}$ is the undirected graph with the same vertex set as $G^{\preceq}$, and edge set constructed by augmenting the edge set of $G^{\preceq}$ so that any forbidden configuration is eliminated. Thus 'moralising' a chain graph means dropping directions and 'marrying' parents.

EXAMPLE 2.3 The vertex set $K = \{1,2,3,4,5\}$, is partitioned into two blocks, $b_1 = \{1,2\}$, and $b_2 = \{3,4,5\}$ and the defining independences are $1 \perp\!\!\!\perp 2$, $3 \perp\!\!\!\perp 2|\{1,4,5\}$, $4 \perp\!\!\!\perp 1|\{2,3,5\}$, $5 \perp\!\!\!\perp 1|\{2,3,4\}$, and $5 \perp\!\!\!\perp 2|\{1,3,4\}$. The diagram of the graph is given on the left. The graph fails the Wermuth condition because there are two vertices, 3 and 4, in the same block which are connected, but with parents that are not joined.

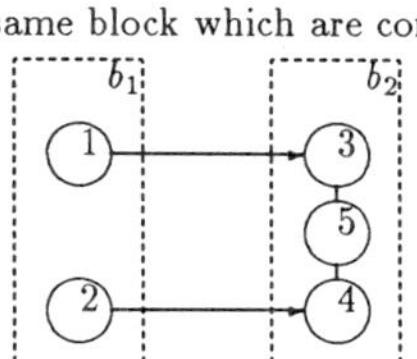

chain independence graph

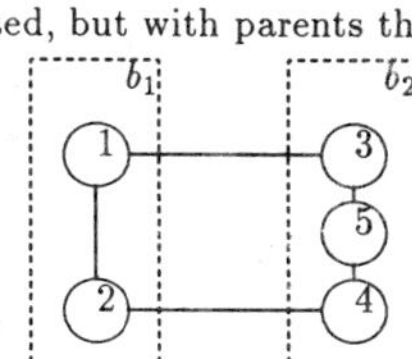

its moral graph

The moral graph on the right joins these parents. Note that the diagonal vertices are not joined: even though vertex 5 is required in the conditioning set for the independence of 1 and 4, this does not make 5 a parent of 4. The blocks, of course, are redundant for reading the moral graph. □

The Markov properties of the chain independence graph are neatly summarised by the following assertion:

Theorem 2.1 *The directed independence graph $G^{\preceq}$ posseses the same independence interpretations as its associated moral graph, G^m.*

The proof is relatively straightforward; fundamental is the recursive factorisation identity applied to the joint density function, expressed in the terms of the blocks:

$$f_K = \prod_{r=2}^{m} f_{b_r|b_1\cup b_2\cup\ldots\cup b_{r-1}}.$$

This allows the elicitation of the pairwise conditional independence statements that characterise the undirected conditional independence graph.

EXAMPLE 2.4 Consider the chained block graph of the first example in this section. Its defining independence statements are

$1 \perp\!\!\!\perp 3 \mid 2$	$5 \perp\!\!\!\perp 2 \mid \{1,3,4,6\}$	$7 \perp\!\!\!\perp 1 \mid \{2,3,4,5,6,8\}$	$8 \perp\!\!\!\perp 1 \mid \{2,3,4,5,6,7\}$
$1 \perp\!\!\!\perp 2 \mid 3$	$5 \perp\!\!\!\perp 3 \mid \{1,2,4,6\}$	$7 \perp\!\!\!\perp 2 \mid \{1,3,4,5,6,8\}$	$8 \perp\!\!\!\perp 2 \mid \{1,3,4,5,6,7\}$
$4 \perp\!\!\!\perp 1 \mid \{2,3\}$	$6 \perp\!\!\!\perp 1 \mid \{2,3,4,5\}$	$7 \perp\!\!\!\perp 3 \mid \{1,2,4,5,6,8\}$	$8 \perp\!\!\!\perp 3 \mid \{1,2,4,5,6,7\}$
$4 \perp\!\!\!\perp 3 \mid \{1,2\}$	$6 \perp\!\!\!\perp 2 \mid \{1,3,4,5\}$	$7 \perp\!\!\!\perp 4 \mid \{1,2,3,5,6,8\}$	$8 \perp\!\!\!\perp 5 \mid \{1,2,3,4,6,7\}$
	$6 \perp\!\!\!\perp 3 \mid \{1,2,4,5\}$	$7 \perp\!\!\!\perp 5 \mid \{1,2,3,4,6,8\}$	$8 \perp\!\!\!\perp 6 \mid \{1,2,3,4,5,7\}$
	$6 \perp\!\!\!\perp 4 \mid \{1,2,3,5\}$		

The recursive factorisation identity expressed in the terms of the blocks is

$$f_K = f_{b_4|b_1\cup b_2\cup b_3} f_{b_3|b_1\cup b_2} f_{b_2|b_1} f_{b_1},$$

and simplifies to

$$\begin{aligned} f_{12345678} &= f_{87|46} f_{56|14} f_{4|2} f_1 f_{23} \\ &= f_{87|46} f_{6|5} f_{5|14} f_{4|2} f_1 f_{23}. \end{aligned}$$

Regarding the right hand side as a factorisation into coordinate projection functions, we have

$$f_K = g_{4678} g_{456} g_{145} g_{24} g_{23}.$$

There are two independences, that between 8 and 6 and that between 7 and 4, that have not been applied to this factorisation. Taking these into account generates the independence graph:

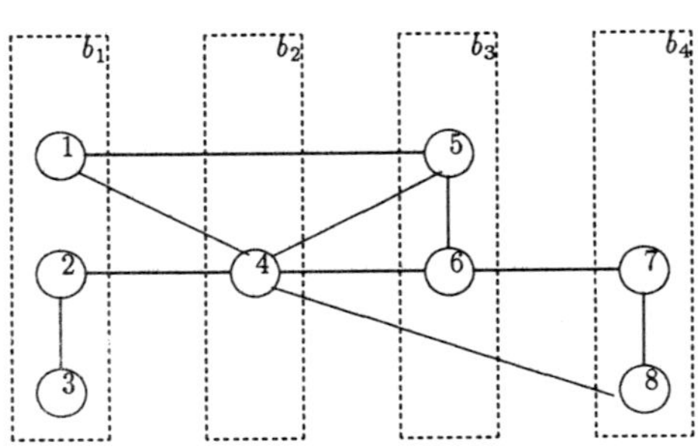

which is the moral graph G^m. □

3 Distributions

The conditional independence graph or chain graph reflects the interaction structure of a set of variables. In order to use these techniques in applications with data we need to build statistical models in which the conditional independence constraints can be incorporated. More specifically we need a parameteric family of distributions this is rich enough to model discrete and continuous variables, easy to fit and satisfies certain forms of closure under marginalisation. The prime candidates are the family of multivariate normal distributions for continuous variables, the family of cross-classified multinomial distribution for categorical variables, and the family of conditional Gaussian distributions for the mixed variables, that contains the normal and multinomial families as special cases.

3.1 The normal distribution

The density function of the normal distribution is

$$f_X(x) = (2\pi)^{-k/2}\det(D)^{1/2}\exp\{-\frac{1}{2}(x-\mu)^T D(x-\mu)\}, \quad \text{for } x \text{ in } R^k$$

where μ is a fixed k-dimensional vector and the $k \times k$ matrix D is symmetric and positive definite. In fact the parameter D is the *inverse variance* $D = V^{-1}$, and V is the variance of the k-dimensional vector X. A way to extract its properties is by conformably partitioning $X = (X_a, X_b)$, D and V and showing that

$$D_{bb} = V_{bb|a}^{-1},$$

where $V_{bb|a} = V_{bb} - V_{ba}V_{aa}^{-1}V_{ab}$ is the partial variance, sometimes termed the Schur complement of the matrix V. This relationship leads to the following result.

Lemma 3.1 If a, b and c partition K then the normal random vectors X_b and X_c are independent conditional on X_a, $X_b \perp\!\!\!\perp X_c | X_a$, if and only if the block of the inverse variance $D_{bc} = 0$. In particular, if X_b and X_c are both 1-dimensional then

$$X_r \perp\!\!\!\perp X_s | X_{K\setminus\{r,s\}} \quad \text{if and only if} \quad d_{rs} = 0.$$

So the normal distribution gives a simple parametric criterion for pairwise conditional independence on the remaining variables. The distributional model based on sets of these conditional independence constraints is called a *graphical Gaussian model*, or a covariance selection model, following Dempster (1972) who first suggested fitting models with zeros in the inverse variance. (Actually though, it is the partial covariances that are selected to be zero.) There are $2^{\binom{k}{2}}$ graphical Gaussian models corresponding to the $\binom{k}{2}$ off-diagonal elements in the inverse variance matrix.

3.2 The cross-classified multinomial distribution

Suppose that the k-dimensional random vector I represents observations taken on k categorical variables, and the density function, f_K, of I, is determined by the k-way table of probabilities $P(I = i) = p_K(i)$. Following Goodman (1970), Bishop et al. (1975), the standard way write the density function of the k-dimensional cross-classified multinomial distribution (of size 1) is through the log-linear expansion

$$\log f_K(i) = u_\phi(i) + \sum u_i(i) + \sum u_{ij}(i) + \sum u_{ijl}(i) + \cdots + u_{12\ldots k}(i)$$

where i takes values in a finite set given by a Cartesian product of the permissible values of each coordinate. There are 2^k u-terms. Each u-term is a coordinate projection function with the coordinates indicated by the suffix; for example, $u_{24}(i) = u_{24}(i_2, i_4)$. To ensure that the expression is identifiable a condition such as the sum of the u-terms is zero, or that $u_a(i) = 0$ whenever $i_j = 0$ and $j \in a$, has to be imposed.

There is a neater way to write this expansion in terms of subsets of $K = \{1, 2, \ldots, k\}$. The *log-linear expansion* of the cross-classified multinomial density function f_K is

$$\log f_K(i) = \sum_{a \subseteq K} u_a(i)$$

where the sum is taken over all possible subsets a of $K = \{1, 2, \ldots, k\}$. The expansion is linear in the u-terms. The relationship between the two parameters p_K and $\{u\}$ in the multinomial distribution parallels the relationship between the two parameter V and D in the normal distribution. It is particularly easy to list the conditions for independence as conditions on the u-terms.

Lemma 3.2 If (I_a, I_b, I_c) is a partitioned multinomial random vector then $I_b \perp\!\!\!\perp I_c | I_a$ if and only if all u-terms in the log-linear expansion with one or more coordinate in b *and* with one or more in c, are zero.

When a is empty, this is a proposition about independence. For pairwise independence of variables I_r and I_s conditioned on the remaining variables, the dimension of I_a is $k - 2$ and $I_b = I_r$ and $I_c = I_s$ are both 1-dimensional. The proposition requires $u_{\{r,s\} \cup t} = 0$ whenever $t \subseteq a$. There are 2^{k-2} such terms. Contrast this with the corresponding condition on the Normal density function where only a single coefficient in the inverse variance is required to be 0.

3.3 The CG Distribution

We now turn to the general case of mixed variables. The full vector of k random variables is $X = (I, Y)$ where I is the p-dimensional vector of discrete variables and Y is the q-dimensional vector of continuous variables. The full set of k vertices K is correspondingly partitioned into $K = \Delta \cup \Gamma$. A typical value taken by (I, Y) is denoted (i, y) and a marginal vector of (I, Y) as (I_a, Y_b) where $a \subseteq \Delta$ and $b \subseteq \Gamma$.

The conditional Gaussian distribution is defined by the following construction: suppose that the vector of p categorical variables, I, has the cross-classified multinomial distribution with parameters $p(i)$, the table of probabilities, and that the conditional distribution of the q continuous variables Y given I is multivariate Normal with expectation $E(Y|I = i) = \mu(i)$ and variance $\text{var}(Y|I = i) = V(i)$. The joint distribution is obtained by multiplying these functions together according to $f_{IY}(i, y) = f_I(i) f_{Y|I}(y; i)$ and is called the CG distribution because of this conditional Gaussian construction. Explicitly, the density function of the $p+q$-dimensional random vector $X = (I, Y)$ with the CG distribution is

$$f_{IY}(i, y) = p(i)(2\pi)^{-q/2} \det(V(i))^{-1/2} \exp\{-\frac{1}{2}(y - \mu(i))^T V(i)^{-1}(y - \mu(i))\}.$$

for $y \in R^q$ and i taking values in the p-fold Cartesian product of the index sets for each categorical variable. The CG distribution is indexed by the *moment parameters* $p(i)$, $\mu(i)$ and $V(i)$. The family of all CG density functions with a given dimension is $CG(p, q)$ and we write $f \in CG(p, q)$ to denote a member.

There is a variety of ways to parameterise the CG distribution: most intuitive is the moments parameterisation incorporated in the definition, but equally useful is the reparameterisation in exponential family form. Write f_{IY} above as

$$f_{IY}(i,y) = \exp\{\alpha(i) + [\beta(i), y] - \frac{1}{2}[y, D(i)y]\}.$$

The *canonical parameters* $\alpha(i)$, $\beta(i)$ and $D(i)$, are in one to one correspondence with the moment parameters: and, for instance, it is easy to verify that $D(i) = V(i)^{-1}$ and $\beta(i) = V(i)^{-1}\mu(i)$. The $D(i)$ are inverse variance matrices and must be symmetric and positive definite; they determine the interaction between two continuous variables, conditionally on the discrete variables. Note that $D(i)$ may also be an ingredient of discrete-continuous and of discrete-discrete interaction.

The other two sets of canonical parameters are more difficult to interpret. The β parameters which relate the continuous variables to the discrete ones do not have the intuitive simplicity of means or regression coefficients. Furthermore, it is not particularly easy to list the conditional independence criterion in terms of this parameterisation, though checking for the pairwise independence of two continuous variables is easiest. In parallel to the purely discrete case, we need a hierarchical interaction expansion, Lauritzen and Wermuth (1989), Edwards (1989):

$$\alpha(i) = \sum_{a\subseteq\Delta} \lambda_a(i), \quad \beta(i) = \sum_{a\subseteq\Delta} \eta_a(i) \quad \text{and} \quad D(i) = \sum_{a\subseteq\Delta} \psi_a(i),$$

where the sum is taken over all subsets of the set of discrete indices Δ. Because $\beta(i)$ is a vector of dimension q then so is $\eta_a(i)$ for each a, and similarly $\psi_a(i)$ is a matrix of order $q\times q$.

We may now re-express f_{IY} above as

$$f_{IY}(i,y) = \exp\{\sum_{a\subseteq\Delta} \lambda_a(i) + \sum_{a\subseteq\Delta} \eta_a(i)^T y - \frac{1}{2}\sum_{a\subseteq\Delta} y^T\psi_a(i)y\}.$$

We use the same constraints adopted for the u-term expansion in log-linear expansion of the multinomial density function, so that whenever one element of i is zero, say the j-th so that $i_j = 0$, then $\lambda_a = 0$ for all $j \in a$. This convention extends to the parameters η and ψ.

EXAMPLE 3.1 CG(1,1). Consider specialising to the case of one categorical and one continuous variable. With $p = 1$ and $q = 1$, the CG distribution of (I, Y) is the joint distribution

$$f_{IY}(i,y) = \exp\{u_\phi + u(i)\}\frac{1}{\sqrt{2\pi}\sigma(i)}\exp\{-\frac{1}{2}(\frac{y-\mu(i)}{\sigma(i)})^2\},$$

where the support of y is the whole real line, and the support of i is a finite set of integers. In general, both the mean and variance of Y depend on i. Taking logs gives

$$\log f_{IY}(i,y) = \text{const} + u(i) - \log\sigma(i) - \frac{1}{2}\frac{\mu(i)^2}{\sigma(i)^2}\frac{\mu(i)y}{\sigma(i)^2} - \frac{1}{2}\frac{y^2}{\sigma(i)^2},$$

which, by appropriately identifying coefficients we may write as

$$\log f_{IY}(i,y) = \lambda_\phi + \lambda(i) + (\eta_\phi + \eta(i))y - \frac{1}{2}(\psi_\phi + \psi(i))y^2$$

There are three distinct components to this log density function: the discrete part indexed by the λ terms; the linear part, indexed by the η terms; and the quadratic part indexed by the ψ terms. To make this expression identifiable requires $\lambda(0) = 0$, $\eta(0) = 0$ and $\psi(0) = 0$; and to ensure it integrates to unity, requires $\psi_\phi + \psi(i) > 0$ for all i, and a suitable choice of the norming constant λ_ϕ.

The one-way analysis of variance model in which the number of units allocated to each treatment is determined at random is a special case. The standard ANOVA model supposes that $E(Y|I = i) = \mu(i)$ and

$\text{var}(Y|I) = \sigma^2$ for all i. To ensure variance homogeneity the canonical parameter $\psi(i) = 0$; the parameter $\psi_\phi = \sigma^{-2}$; and, by identifying the coefficients of the linear interaction, $\eta_\phi + \eta(i) = \mu(i)/\sigma^2$. The hypothesis of no treatment effect is equivalent to the condition that $\eta(i) = 0$ for all i. □

The following lemma gives conditions for pairwise conditional independence bewteen coordinates of the CG distribution.

Lemma 3.3 If the random vector (I, Y) has the CG distribution above then

1. $I_j \perp\!\!\!\perp I_k | (I_{\Delta\setminus\{j,k\}}, Y_\Gamma)$ if and only if $\lambda_a = 0,\ \eta_a = 0$ and $\psi_a = 0$ whenever $j, k \in a \subseteq \Delta$.
2. $I_j \perp\!\!\!\perp Y_r | (I_{\Delta\setminus\{j\}}, Y_{\Gamma\setminus\{r\}})$ if and only if $\eta_a^r = 0$ and $\psi_a^{rs} = 0$ whenever $j \in a \subseteq \Delta$ and for all $s \in \Gamma$.
3. $Y_r \perp\!\!\!\perp Y_s | (I_\Delta, Y_{\Gamma\setminus\{r,s\}})$ if and only if $\psi_a^{rs} = 0$ for all $a \subseteq \Delta$.

We use the convention that ψ^{rs} is the (r, s)-th element of the matrix ψ, so that superscripts refer to continuous variables and subscripts to discrete variables.

EXAMPLE 3.2 $CG(2,2)$. When $p = 2$ and $q = 2$ the density is

$$\begin{aligned}\log f(i_1, i_2, y_1, y_2) &= \lambda_\phi + \lambda_1(i_1) + \lambda_2(i_2) + \lambda_{12}(i_1, i_2)\\ &+ [\eta_\phi + \eta_1(i_1) + \eta_2(i_2) + \eta_{12}(i_1, i_2)]^T y - \tfrac{1}{2} y^T [\psi_\phi + \psi_1(i_1) + \psi_2(i_2) + \psi_{12}(i_1, i_2)] y\end{aligned}$$

where $y^T = (y_1, y_2)$. While this expansion appears more complicated than in the α, β, D set-up, this parameterisation permits a flexible specification of hypotheses; for instance, if the variance of Y is not depend on I_2 then the matrices ψ_2 and ψ_{12} must both be zero.

Suppose I_1 and I_2 are binary. When $p = 2$ and $q = 2$ the possible pairwise conditional independences are summarised by

1. $I_1 \perp\!\!\!\perp I_2 | Y_{\{1,2\}}$ if and only if $\lambda_{12} = 0,\ \eta_{12} = 0$ and $\psi_{12} = 0$. There are $1 + 2 + 3 = 6$ parameters set to zero.
2. $I_1 \perp\!\!\!\perp Y_1 | (I_2, Y_2)$ if and only if $\eta_1^1 = 0,\ \eta_{12}^1 = 0$, and $\psi_{12}^{11} = 0,\ \psi_{12}^{12} = 0,\ \psi_1^{11} = 0,\ \psi_1^{12} = 0$. There are 6 parameters set to zero.
3. $Y_1 \perp\!\!\!\perp Y_2 | I_{\{1,2\}}$ if and only if $\psi_\phi^{12} = 0,\ \psi_1^{12} = 0,\ \psi_2^{12} = 0,\ \psi_{12}^{12} = 0$. There are 4 parameters set to zero.

□

The analysis of the mixed case is more difficult than the separate treatments of continuous and of discrete variables for one reason: closure. The marginal distributions of the CG distribution are not necessarily members of the CG family of distributions, and so the family is not necessarily closed under marginalisation; and consequently it may not be closed under conditioning. In particular, the marginal distribution of I comes by integrating out Y from the joint density function, $f_{IY} \in CG(p,q)$, and, trivially from the definition of the CG density, establishes that $f_I \in CG(p,0)$. On the other hand, the marginal distribution of Y, $f_Y(y) = \sum_i f_{IY}(i,y)$, is a mixture of Normal distributions with mixing weights $P(I = i)$ and in general is not itself Normal, so that $f_Y \notin CG(0,q)$. It is well-known that mixture distributions pose awkward estimation problems.

3.4 Likelihood inference

The observations in the sample are directly related to the probability model under consideration by the likelihood function. The standard results and techniques of maximum likelihood estimation and likelihood ratio tests (Cox and Hinkley, 1974) are then available to the graphical modeller.

We give a brief outline of such an approach here, but not the details, which can be found in Whittaker (1990). First specify the family of distributions for X: the family of k-dimensional CG distributions, and its parameter set. Determine the graphical model of interest: each pairwise independence constraint specified by the independence graph generates a constraint on the parameters. Construct the likelihood function. We assume simple random sampling, so that the sample observations are mutually independent and each has the same density function. The likelihood is maximised over the unknown parameters, subject to the constraints that determine the particular graphical model. Test the goodness of fit of the fitted graphical model. Twice the generalised log-likelihood ratio test that compares the fitted graphical model with the saturated model corresponding to the complete graph, and called the *deviance,* is calculated. The comparison of the observed deviance to the percentage point of the relevant chi-squared distribution gives an assessment of whether a particular graphical model provides an adequate representation of the data. An important aid to graphical model selection is the preliminary computation of the edge exclusion deviances for each edge in the model.

4 Applications

In this section we discuss four examples taken from Whittaker (1987, 1990) of fitting graphical models to data sets with just continuous variables, just discrete, mixed and with a block recursive structure.

4.1 Mathematics Marks

This data set comes from Mardia, Kent and Bibby (1979) and consists of the examination marks of 88 students in the five subjects mechanics, vectors, algebra, analysis and statistics. Mechanics and vectors were closed book examinations and the remainder were open book.

If we assume that the distribution of the vector of mathematics marks is multivariate Normal then there is no loss of information by condensing these data into the sample mean vector and the sample variance-covariance matrix. We report the correlation matrix here, only the lower triangle is printed as the matrix is symmetric.

Table 1: The sample correlation matrix of the mathematics marks

mech	1.0				
vect	.55	1.0			
alg	.55	.61	1.0		
anal	.41	.49	.71	1.0	
stat	.39	.44	.66	.61	1.0
	mech	vect	alg	anal	stat

That all the entries are positive can be interpreted as a reflection that good students tend to do well on all subjects and bad ones badly. The range of values varies from a high of 0.71 between algebra and analysis to a low of 0.39 between statistics and mechanics. Apart from this it is difficult to discern much pattern in these correlation coefficients as they stand.

If an element in the inverse variance matrix is zero then the corresponding variables are conditionally independent given the remaining variables. This immediately motivates the calculation of the inverse of the correlation matrix:

mech	1.60				
vect	-0.56	1.80			
alg	-0.51	-0.66	3.04		
anal	0.00	-0.15	-1.11	2.18	
stat	-0.04	-0.04	-0.86	-0.52	1.92
	mech	vect	alg	anal	stat

Each diagonal element of the inverse correlation matrix is related to the proportion of variation in the corresponding variable explained by regressing on the remaining variables. More explicitly each diagonal element equals $1/(1-R^2)$ where R is the multiple correlation coefficient between that variable and the rest. For example, the proportion of explained variation for mechanics is $R^2(\text{mech; rest}) = (1.60-1)/1.60 = 37.5\%$ while it is $R^2(\text{alg; rest}) = (3.04-1)/3.04 = 66.7\%$ for algebra. In fact algebra is the most predictable mathematics mark and mechanics is the least.

The inverse correlation matrix is now scaled to have unit entries on the diagonal, in the same way as a variance matrix is scaled to give a correlation matrix:

mech	1.0				
vect	-0.33	1.0			
alg	-0.23	-0.28	1.0		
anal	0.00	-0.08	-0.43	1.0	
stat	-0.02	-0.02	-0.36	-0.25	1.0
	mech	vect	alg	anal	stat

The off diagonal elements of this scaled inverse correlation matrix are the negatives of the partial correlation coefficients between the corresponding pair of variables given the remaining variables. The correlation between algebra and analysis persists after adjusting for the other variables but the correlation between mechanics and analysis does not.

Unlike the original correlation matrix of the mathematics marks, there is a discernible structure to this matrix: the elements in the lower lefthand block are all near zero, which suggests that the matrix can be approximated by one with the following structure:

mech	*				
vect	*	*			
alg	*	*	*		
anal	0	0	*	*	
stat	0	0	*	*	*
	mech	vect	alg	anal	stat

where * denotes a non-zero entry.

We have constructed the independence graph for the mathematics marks in a probability model based on the normal distribution:

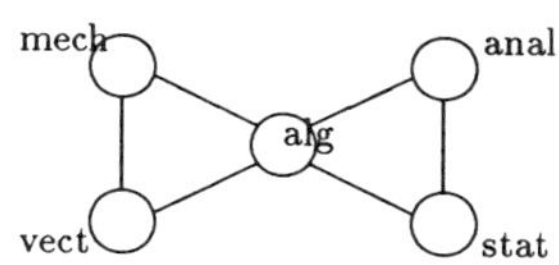

Figure 1: Independence graph of the mathematics marks

Of the ten possible edges four are missing: mechanics and analysis are independent conditional on the other remaining variables, as are mechanics and statistics, vectors and statistics and finally vectors and analysis. The most important tool for interpreting independence graphs is the global Markov property: variables are independent conditioned on the separating set alone. For example, as the partial correlation between mechanics and analysis is zero then these variables are conditionally independent given the variables, vectors, algebra and statistics. But since algebra alone separates mechanics and analysis it follows that these two variables are independent conditionally on algebra alone.

The graph resembles a butterfly dividing into two segments with its wings separated by a single vertex at the body. Application of the global Markov property gives the result that (mech,vect) is independent of (anal,stat) conditionally on algebra alone. This is an important conclusion for several reasons:

- it reduces the complex five dimensional object into two simpler three dimensional objects;
- it groups the variables into two sets (mechanics, vectors, algebra) and (algebra, analysis, statistics);
- it highlights algebra as the one crucial examination in analysing the interrelationship between different subjects in exam performance;
- it asserts that algebra and analysis alone, will be sufficient to predict a statistics mark and that algebra and vectors will be sufficient to predict mechanics; but that all four marks are needed to predict algebra.

More generally mathematical educationalists should note that algebra has the highest average score, the smallest variance, that it occurs in the best predictor of every other variable and finally that it turns out to be the most predictable of all of these examination marks.

We fit the model determined by the independence statement

$$(\text{mech},\text{vect}) \perp\!\!\!\perp (\text{anal},\text{stat}) \mid \text{alg}$$

which has the graph given above. The observed and fitted partial correlations are

mech	1.000				
vect	.329	1.000			
alg	.230	.281	1.000		
anal	- .002	.078	.432	1.000	
stat	.025	.020	.357	.253	1.000
	mech	vect	alg	anal	stat

and

mech	1.000				
vect	.332	1.000			
alg	.235	.327	1.000		
anal	- .000	.000	.451	1.000	
stat	- .000	.000	.364	.256	1.000
	mech	vect	alg	anal	stat

and compare well. The overall goodness of fit test for this model is the deviance which takes the value 0.895 to be compared against the chi-squared distribution with 4 df. The model provides an excellent fit to the data.

4.2 Unemployment in Rochdale

In this section we illustrate the use of fitting a two-way interaction log-linear model in an exploratory analysis of a survey of households in Rochdale, conducted to elicit information about the factors effecting the pattern of economic life and how these have changed in recent years. The survey was performed under an initiative sponsored by the ESRC entitled 'Social Change and Economic Life', see Davies (1987) and Penn (1988), for more details; the data are available for further analysis on request.

A great deal of information was collected but a particular subset of eight binary variables relating to women's economic activity and husband's unemployment has especial interest for the investigation. The data consist of a 2^8 cross-classification of the N=665 households, classified by

A, wife economically active; B, age of wife > 38; C, husband unemployed; D, child ≤ 4; E, wife's education, Olevel or more; F, husband's education, Olevel or more; G, Asian origin; H, other member of household working.

This is a large number of dimensions and the resulting 8-way cross-classification has 256 cells, of which 165 are empty, and 217 have 3 or less observations; while on the other hand, there are a few cells that contain some 30, 40 and even 50 observations.

We fit the all two-way interactions model, defined by setting $u_a = 0$ whenever the cardinality of a is greater than 2 in log-linear expansion above. The sufficient statistics of this model are the two-way marginal tables and the entries in these are quite respectable. This model has just 37 free parameters, 28 of which are interactions, but for which there is a sample of 665 independent observations available. In regression analysis such a ratio of parameters to observations in commonplace.

The maximum likelihood estimates of the two-way interaction terms are given in the next table.

Table 2: Parameter estimates from the two-way interaction model.

inter.	est.	s.e.	(est./se.)^2	inter.	est.	s.e.	(est./se.)^2
$A.B$	-0.38	0.24	2.50	$C.E$	-1.25	0.36	11.92
$A.C$	-1.33	0.30	19.87	$C.F$	-0.71	0.29	6.18
$A.D$	-1.49	0.25	34.12	$C.G$	1.09	0.37	8.78
$A.E$	0.63	0.21	9.23	$C.H$	0.28	0.34	0.67
$A.F$	0.09	0.20	0.20	$D.E$	0.24	0.25	0.93
$A.G$	-2.17	0.47	21.30	$D.F$	0.38	0.25	2.37
$A.H$	0.22	0.25	0.82	$D.G$	1.32	0.43	9.32
$B.C$	-0.17	0.34	0.24	$D.H$	-1.08	0.45	5.80
$B.D$	-2.97	0.34	77.23	$E.F$	1.03	0.18	32.96
$B.E$	-0.78	0.20	14.53	$E.G$	-0.82	0.43	3.66
$B.F$	0.01	0.20	0.00	$E.H$	-0.18	0.22	0.61
$B.G$	-0.56	0.46	1.50	$F.G$	-1.39	0.38	13.26
$B.H$	1.74	0.27	41.04	$F.H$	-0.24	0.21	1.21
$C.D$	0.01	0.37	0.00	$G.H$	-0.06	0.50	0.01

The standard errors of these estimates are reasonably small, in line with the reasonable overall sample size.

Here we shall adopt the quick model selection method of selecting interactions for which the square of the standardised parameter estimate exceeds 3.84. There are just 14 such terms and the reduced model that contains these interactions is

$$A.(C+D+E+G)+B.(D+E+H)+C.(E+F+G)+D.(G+H)+F.(E+G).$$

The difference in deviance between the all two-way and this reduced two-way interaction model, is a generalised likelihood ratio test between two efficiently estimated models; and hence has a chi-squared distribution. An increase in deviance of 16.17 with 14 df is not significant which suggests that the reduced model that excludes 14 of the 28 possible two-way interaction terms provides a good representation of the data.

The conditional independence graph associated with the reduced model is

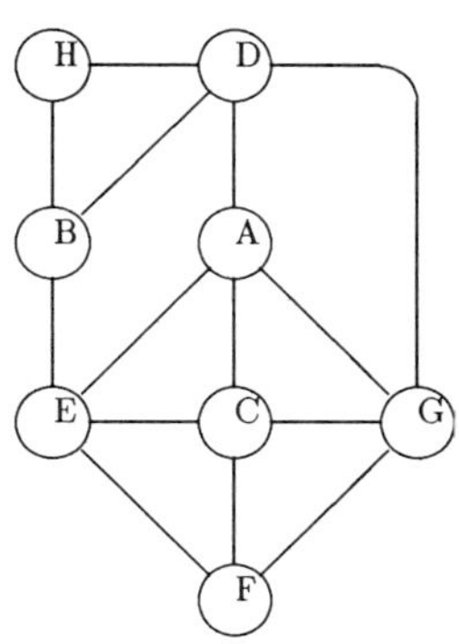

The graph is planar and systematises the information about the non-zero interactions rather succintly. This graph gives a good starting point to any further model selection procedure. All of the standardised interaction estimates of the included terms in this reduced model are significant, and so no further term can be dropped. Incorporating each of the deleted interactions now only involves 14 model fits, but does not affect the selected model.

On the whole, the graph makes sociological sense, connecting variables such as wife's and huband's education, E and F, which one might expect to be strongly related. The strongest interaction is the edge (B, D), wife's age and a child under 4; the next is (B, H), wife's age and the presence of another working working member of the family. There is also a strong interaction on the edge (A, G) indicating a different level of wife's economic activity in the Asian and non-Asian communties.

4.3 College plans

Sewell and Shah (1968) studied the relationship between a student's intention to attend college, parental encouragement and three other factors, social class, sex and a measure of intelligence. The data derives from 10,318 Wisconsin high school seniors, and has been analysed by Fienberg (1977) and Cox and Snell (1981).

The names and levels taken by the variables are X_1, sex: male/female; X_2, social class: low /lower middle/upper middle/high; X_3, intelligence: low /lower middle/upper middle/high; X_4, parental encouragement: low/high; X_5, college plans: yes/no. In this standard order, with variable 5 varying most quickly, the counts in the 128 cells are The principal questions of interest are how the three background variables effect parental encouragement, and how all four variables effect college plans, so we analyse this data as a recursive system consisting of three blocks, $b_1 = \{1, 2, 3\}$, $b_2 = \{4\}$, and $b_3 = \{5\}$.

An analysis of the 3 variables in the first block, based on the edge exclusion deviances in the three-way marginal table quickly shows that the model 1 + 2.3, in which sex is independent of both social class and intelligence is the only model possible. The deviance of the model is 20.25 on 15 df, while the deviance to explain is 864.85 on 24 df.

The total deviance to explain in the system is the deviance against the complete independence of the five variables, 8203.45 on 118 df; its large magnitude is not so much a reflection of the magnitude of dependence in the table but of the large sample used. While the subsequent analysis is thus able to detect some very small interactions, we wish to give a broader picture, and so we sometimes neglect relatively small interactions even if significant.

An omnibus test of linearity in the logistic regression is the deviance of the fitted linear logistic regression containing all explanatory variables. In the second block with 3 explanatory variables and one response, X_4, the model $1.2.3 + (1 + 2 + 3).4$ has a deviance of 55.82 on 24 df; and, in the third block with 4 explanatory variables, and one response, X_5, the model $1.2.3.4 + (1 + 2 + 3 + 4).5$ has a deviance of 73.82 on 55 df. Thus we feel fairly confident about the ability of linear regression to represent the data faithfully. In relation to the other values, the 1.5 interaction deviance difference is small and may be neglected in explaining X_5, but all 3 terms need to be included in the model for X_4. The corresponding chain graph is

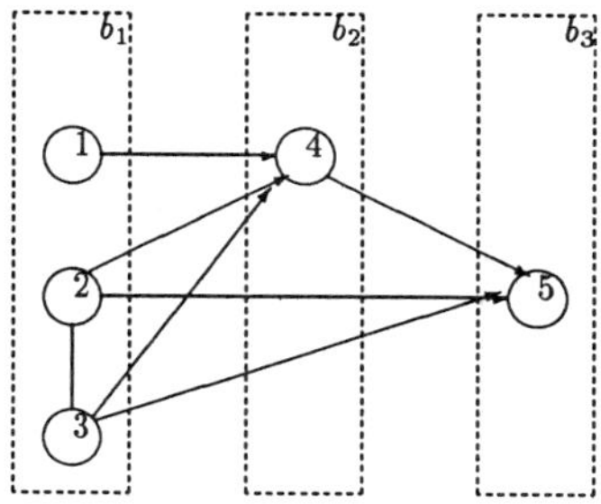

The overall fit of the model may be summarised in the analysis of deviance table

	residual		total	
model	dev	df	dev	df
$() + 1 + 2.3$	20.25	15	864.85	24
$(1.2.3) + (1 + 2 + 3).4$	55.82	24	2841.09	31
$(1.2.3.4) + (2 + 3 + 4).5$	89.50	56	4497.51	63
total	166.57	95	8203.45	118

The vast majority of the variation is accounted for.

4.4 A Drug Trial

For our final example, we take some data analysed by Edwards (1987, 1989) and Whittaker (1989), and previously reported by Morrison (1976). It is an example with mixed variables with a chain graph structure. Again we show how deviance decompositions generalise the classical analysis of variance table and give a concise summary of the source of variation in a data set.

The randomised trial measured weight loss of male and female rats under three drug treatments. The continuous variables X and Y represent weight loss during the first and second week respectively, A is sex and B treatment. One might conjecture that, X is independent of A given B, and because of the temporal relationship, that Y should be independent of both A and B given X. The corresponding chain independence graph is

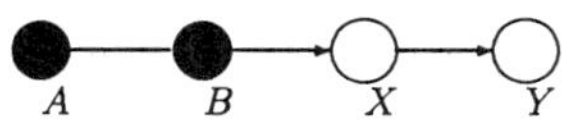

An informative breakdown of this hypothesis is

attributable deviance	dev	df	P-value
$X \perp\!\!\!\perp A \mid B$	4.37	6	0.629
$Y \perp\!\!\!\perp A \mid B, X$	19.86	9	0.019
$Y \perp\!\!\!\perp B \mid X$	12.88	6	0.045
equal residual variances of Y on X for B	1.11	2	0.580
parallel regression lines for each B	1.45	2	0.512
common intercepts for each B	10.32	2	0.006
Total $Y \perp\!\!\!\perp (A,B) \mid X$, $X \perp\!\!\!\perp A \mid B$	37.11	21	

The low P-value of the hypothesis $Y \perp\!\!\!\perp A \mid B, X$ in this table suggests that there may be sex effects not accounted for in the model with the graph above. The second part of this table breaks down the interaction between Y and B measured by the deviance against $Y \perp\!\!\!\perp B \mid X$ into subhypotheses: equal residual variances of Y on X for B, parallel regression lines for each B and common intercepts for each B. It is only the last that is significant.

REFERENCES

Andersen, A. H. (1974). Multidimensional contingency tables. *Scand. J. Statist.*, **1**, 115-127.

Andersen, E. B. (1980). *Discrete statistical models with social science applications.* North Holland: Amsterdam.

Asmussen, S. and Edwards, D. (1983). Collapsibility and response variables in contingency tables. *Biometrika*, **70**, 567-578.

Barndorff-Nielsen, O.E. (1978). *Information and Exponential Families in Statistical Theory.* Wiley: New York.

Bartlett, M. S. (1935). Contingency table interactions. *J. Roy. Statist. Soc. Suppl.* 2, 248-252.

Bartlett, M. S. (1954). A note on the multiplying factors for various chi-squared approximations. *J. Roy. Statist. Soc. B*, **16**, 296-298.

Benedetti, J. K. and Brown, M. (1978). Strategies for the selection of log-linear models. *Biometrics*, **34**, 680-686.

Berge, C. (1973). *Graphs and Hypergraphs.* Transl. from French by E. Miniecka. North Holland: Amsterdam.

Birch, M. W. (1963). Maximum likelihood in three-way contingency tables. *J. Roy. Statist. Soc.*, **25**, 220-223.

Bishop Y. M., Fienberg, S. and Holland, P. (1975). *Discrete Multivariate Analysis.* M.I.T. Press: Cmabridge, Mass.

Cox, D. R. (1970). *The Analysis of Binary Data.* Methuen: London.

Cox, D. R. (1984). Interaction. *Int. Statist. Rev.*, **52**, 1-31.

Cox, D. R. and Hinkley, D. V. (1974). *Theoretical Statistics.* Chapman Hall: London.

Darroch, J. N., Lauritzen, S. L. and Speed, T. P. (1980). Markov fields and log linear interaction models for contingency tables. *Ann. Stat.*, **8**, 522-539.

Davies R. B. (1987). The relationship between husband's unemployment and wife's economic activity: some new evidence. Lancaster University, Social Change and Economic Life Intitiative, Working Paper 25. (to appear in Urban Studies.)

Dawid, A. P. (1979). Conditional independence in statistical theory (with discussion). *J. Roy. Statist. Soc. B*, **41**, 1, 1-31.

Dawid, A. P. (1980). Conditional independence for statistical operations. *Ann. Statist.*, **8**, 598-617.

Dempster, A. P. (1972). Covariance Selection. *Biometrics*, **28**, 157-175.

Edwards, D. E. (1987). A guide to MIM. Research Report 87/1. Statistical Research Unit, University of Copenhagen.

Edwards, D. E. (1989). Hierarchical interaction models, (with discussion). *J. R. Statist. Soc B* **51**, 3.

Edwards, D. E. and Havranek, T. (1985). A fast procedure for model search in multidimensional contingency tables. *Biometrika*, **72**, 2, 339-51.

Edwards, D. E. and Kreiner, S. (1983). The analysis of contingency tables by graphical models. *Biometrika*. **70**, 3, 553-565.

Frank, O. and Strauss, D. (1986). Markov graphs. *J. Amer. Statist Assoc.*, **81**, 832-842.

Frydenberg, M. (1989). The chain graph Markov property. *Research Report, 186*. Department of Theoretical Statistics, University of Aarhus.

Golumbic, M. C. (1980). *Algorithmic Graph Theory and Perfect Graphs*. Academic Press: New York.

Goodman, L. A. (1970). The multivariate analysis of qualitative data: interaction among multiple classifications. *J. Amer. Statist. Assoc.* **65**, 226-256.

Goodman, L. A. (1973). The analysis of multidimensional contingency tables when some variables are posterior to others: a modified path analysis approach. *Biometrika*, **60**, 179-192.

Gower, J. (1966). Some distance properties of latent root and vector methods in multivariate analysis. *Biometrika*, **53**, 315-328.

Haberman, S. J. (1974). *The Analysis of Frequency Data*. Univ. Chicago Press: Chicago.

Harary, F. (1969). *Graph Theory*. Addison-Wesley: Reading, Mass.

Jeffers, J. N. R. (1967). Two case studies in the application of principal component analysis. *Appl. Statist.*, **16**, 225-36.

Kiiveri H. and Speed, T. P. (1982). Structural analysis of multivariate data: a review. *In* Leinhardt, S. (Ed). *Sociological Methodology*. Jossey Bass: San Francisco.

Kiiveri, H., Speed, T.P. and Carlin, J.B. (1984) Recursive causal models. *J. Aust. Math. Soc.*, **36** 30-52.

Kreiner, S. (1987). Analysis of multidimensional contingency tables by exact conditional tests: techniques and strategies. *Scand. J. Stat.*, **14**.

Lauritzen, S. L. (1982). *Lectures on Contingency Tables*. (2nd Ed.) University of Aalborg Press: Aalborg.

Lauritzen, S. L., Dawid, A. P., Larsen, B. N. and Leimer, H. G. (1988). Independence properties of directed Markov fields. Research Report. R-88-32. Institute of Electronic Systems. Aalborg University.

Lauritzen, S. L. and Spiegelhalter, D. J. (1988). Local computations with probabilities on graphical structures and their application to expert sustems (with discussion). *J. Roy. Statist. Soc. B*, **50**, 2, 157-224.

Lauritzen, S. L. and Wermuth, N. (1984). Mixed interaction models. Research Report. R-84-8. Institute of Electronic Systems. Aalborg University.

Lauritzen, S. L. and Wermuth, N. (1989). Graphical models for associations between variables, some of which are qualitative and some quantitative. *Ann. Statist.*, **17**, 31-57.

Morrison, D.F. (1976). *Multivariate Statistical Methods*. McGraw-Hill: New York.

Nelder, J. A. and McCullagh, P. (1983). *Generalised linear models.* Chapman Hall: London.

Nelder, J. A. and Wedderburn, R. W. M. (1972). Generalised linear models. *J. Royal Statist. Soc. A*, **135**, 370-384.

Pearl, J. (1988). *Probabilistic Reasoning in Intelligent Systems.* Morgan and Kaufman: San Mateo.

Penn, R. (1988). Ten hypotheses about male unemployment and female activity. Lancaster University, Social Change and Economic Life Intitiative, Working Paper 31.

Porteous, B. T. (1985). Improved likelihood ratio statistics for covariance selection models. *Biometrika*, **72**, 97-101.

Simpson, C. H. (1951). The interpretation of interaction in contingency tables. *J. Roy. Statist. Soc. B*, **13**, 238-241.

Speed, T. P. (1978). *Graph-theoretic methods in the analysis of interaction.* Lecture Notes, Institute of Mathematical Statistics, University of Copenhagen.

Speed, T. P. and Kiiveri, H. (1986). Gaussian Markov distributions over finite graphs. *Ann. Statist*, **14**, 1, 138-150.

Spiegelhalter, D. J. (1986). Probabilistic reasoning in predictive expert systems. *In* Kanal, L. N. and Lemner, J. *Uncertainty in Artificial Intelligence.* 47-68. North-Holland.

Sundberg, R. (1975). Some results about decomposable (or Markov-type) models for multidimensional contingency tables: distribution of marginals and partitioning of tests. *Scand. J. Statist.*, **2**, 71-79.

Wermuth, N. (1976). Analogies between multiplicative models in contingency tables and covariance selection. *Biometrics*, **32**, 95-108.

Wermuth, N. (1976). Model search among multiplicative models. *Biometrics*, **32**, 253-264.

Wermuth, N. (1980). Linear recursive equations, covariance selection and path analysis. *J. Amer. Stat. Ass.*, **75**, 963-972.

Wermuth, N. and Lauritzen, S. L. (1983). Graphical and recursive models for contingency tables. *Biometrika*, **70**, 3, 537-552.

Wermuth, N. and Lauritzen, S. L. (1989). On substantive research hypotheses, conditional independence graphs and graphical chain models. *J. Roy. Statist. Soc. B*, **51**, 3.

Whittaker, J. (1982). Glim syntax and simultaneous tests for graphical log-linear models. *In* Gilchrist R. (Ed). *GLIM 82. Lecture notes in Statistics*, **14**, 98-108. Springer Verlag.

Whittaker, J. (1984). Fitting all possible decomposable and graphical models to multiway contingency tables. *In* Havranek, T. et al. (Eds). *Compstat*, Physica-Verlag: Vienna. 401-406.

Whittaker, J. (1986). A mixed derivative measure of interaction. Technical Report, Department of Mathematics, University of Lancaster.

Whittaker, J. (1987). Graphical models in survey analysis: Case Studies. Report to the ESRC. University of Lancaster Technical Report.

Whittaker, J. (1989). Discussion contribution to van der Heijden, de Falguerolles and de Leeuw. *Appl.Statist.* **38**, 2, 278-279.

Whittaker, J. (1989). Invited contribution to Edwards, and Wermuth and Lauritzen. *J. Roy. Statist. Soc. B*, 1989, **51**, 3.

Whittaker, J. (1990). *Independence Graphs in Applied Multivariate Statistics.* To be published by Wiley.

Whittaker, J., Iliakopoulis., T. and Smith, P. (1988). Graphical modelling with large numbers of variables: a comparison with principal components. *In* Edwards, D.G. and Raun, N. E. (Eds). *Compstat 88.* Physica Verlag: Heidelberg. 73-80.

Whittemore, A. S. (1978). Collapsibility of Multidimensional Contingency Tables. *J. Roy. Statist. Soc. B*, **40**, 328-340.

Wright, S. (1921). Correlation and causation. *J. Agric. Res.*, **20**,557-585.

Wright, S. (1923). The theory of path coefficients: a reply to Niles' criticism. *Genetics,* **8**, 239-255.

Wright, S. (1934). The method of path coefficients. *Ann. Math. Statist.*, **5**, 161-215.

COMPUTER SYSTEMS

COMPUTERIZED SUPPORT FOR NONLINEAR MULTIVARIATE ANALYSIS

Ron A. Visser & Gerda M. van den Berg
Leiden University

summary
The feasibility is studied of consulting systems for various forms of nonlinear multivariate analysis (simple and multiple correspondence analysis, nonlinear principal component analysis etc.). Problems and solutions in three stages of the analysis process are discussed separately (choice of methods, specification of control parameters, interpretation of results). Results from several pilot studies are brought together, and an inventory is made of what is presently possible and where future research is desired. Knowledge elicitation appears to be a major problem in all stages. In addition we argue for more behavioral studies of the consulting process.

1. introduction

We will discuss the possibility of computerized support facilities for the exploratory analysis of multivariate data. The background of this discussion is the fast technical and commercial development of information technology which has caused a general proliferation of software for complex forms of data analysis, like multiple correspondence analysis and nonlinear principal component analysis. The population of users and the number of applications of such programs has grown considerably. On the other hand, the number of statistical consultants who could support these new applications, is not increasing in the same pace. It is generally believed that this process forms a threat to the quality of data analysis. Of all thinkable solutions to this problem, the construction of supporting software to be connected to the analysis programs, seems to be the most logical because the problem rises from the spread of software.

In the last few years a number of studies have been published on the subject of computerized support in statistics, usually indicated as "statistical expert systems" (Gale, 1988b; Nelder, 1988; and references there). Despite this great interest in the subject, there is no general agreement on strategy and methodology, or even on the fundamental possibility of statistical expert systems as such (e.g. Streitberg, 1988). Generally speaking the field is still in an early stage of reconnaissance. Several prototypes are available, but few or none mature consulting systems are functioning in a production environment. (Note: the term "expert systems" is open to ambiguity and inflation; we will use "consulting systems" as the overall indication for supporting systems in data analysis, cf. Chambers, Gale & Pregibon, 1988.)

Consulting systems are not evenly spread over all segments of data analysis. Relatively much is known, e.g., on regression analysis (Pregibon & Gale, 1984; Gale, 1986b) and time series analysis (Streitberg & Naeve, 1985; Hietala, 1986, 1988; Gale, 1988a). One of the ways to progress from the present state of affairs is, in our opinion, to investigate the possibilities of consulting systems for other specific segments of data analysis. The aim of such studies should be to make an inventory of the demand for support in the specific field, and an inventory of methods and software which are (1) already available, or (2) could be built given the present knowledge, and (3) what research has still to be done.

In this paper we try to make such an inventory for the field of nonlinear multivariate analysis (NLMVA). The term NLMVA is used to indicate a family of methods generating a geometrical representation of various forms of dependence in non-numerical multivariate data. This is accomplished by assigning (in some sense) optimal scores to categories of variables. Examples are (multiple) correspondence analysis, nonlinear principal components analysis and nonlinear canonical correlation analysis (cf. Gifi, 1981; Young, 1981; Greenacre, 1984). NLMVA methods are mainly used for exploratory analyses; they may be considered as complex forms of descriptive statistics.

We will present, firstly, a general outline of the problems related to consulting systems for NLMVA. This is followed by a more detailed discussion of three stages of the analysis process (choice of methods, specification of parameters and interpretation of results).

2. the aim of support

In this section the aims and limits are described of the system we have in mind. We try to provide support for a majority of applications as they occur in practice. A typical user of NLMVA is the (applied) researcher with general knowledge of methodology and statistics, and with knowledge of NLMVA on the level of an introductory course. This choice seems reasonable because many researchers seeking advice, fall in this category. While, on the other hand, it may be argued that for persons with substantially more or substantially less knowledge of the methods, it might be very hard to provide automated guidance. So, we consider a consultancy system for "the non-expert who is no fool". A consultancy system for this type of user may have several functions: (1) advise about the effectiveness of a user-proposed action, pointing out dangerous side-effects, and presenting alternatives to reach the same or a nearby goal, (2) present tools to facilitate the interpretation of the output, and (3) suggest possible follow-up actions.

In a typical application of NLMVA, data are ordered in a matrix of (say) objects by variables or categories of variables. The user chooses a method, and consequently specifies input parameters. The input parameters describe properties of the data (like level of measurement and number of categories), and properties of the required representation (like the number of dimensions, restrictions on the solution, etc.). The program then produces graphic displays and tables of parameter values. From these, the user infers (mainly qualitative) properties of the dependence in the dataset. Usually, this process is repeated (partially) in a cyclic way before a satisfactory description of the data is found. In each phase of the process the analyst has to make specific decisions. In order to support decision making different kinds of information are needed. In this respect we find it useful to distinguish four phases in the application of NLMVA methods:

(1) the general choice of a method for the analysis;
(2) specification of the precise mode of the method (i.e. the values of the input parameters);
(3) calculation of the solution and presentation of the results on video display or paper;
(4) interpretation of the output; i.e. translating it in an answer to the research question, or in a decision to perform an other analysis.

Comparable devisions are found in literature (Hand, 1986; Gale, 1985). Traditionally computers play a major role in phase (3) only; in the other phases support takes the form of (human) expert advice. In the following we discuss computerized support for phases (1), (2) and (4). For each phase we will discuss what forms of computerized support could be given, what conditions should be satisfied to do so, and what could be realized given the present state of the art.

There are some further restrictions. We will emphasize context-sentive support and will give no attention to the usual help systems mainly directed towards explanation of term and concepts. The consulting system we have in mind supports the application of a series of related NLMVA-methods, and is connected to existing software. It may be seen as a highly user-controlled interactive "knowledge-based front-end system" (cf. Nelder, 1988).

3. phase 1: choice of methods

The choice of the most appropriate statistical procedure for a given problem is not a traditional topic in courses or textbooks on statistics. Yet, it is one of the main issues in statistical consultation, and therefore it should be an important part of a consulting system for data analysis. Such a system should be build around an algorithm that generates the name of the (or better: an) appropriate method. Generally speaking the input for such an algorithm should be information about the data, about the research design and about the kind of question to be answered. However, an explicit theory about this choice of method is not available, and so we will have to study and model the behavior of expert consultants, and disclose their implicit theories (cf. Chambers, Gale & Pregibon, 1988).

Based on the few available studies on statistical consultation (cf. Hand, 1984) the choice of method can be described as a diagnostic process. Supposing that we can make explicit the diagnostic rules involved in the decision process during consultation, an advice giving system could fundamentally be based on the fitting of production rules of the form "if...then...". Research so far indicates that this could be possible.

However, some points need further clarification. The first of these concerns the differences between consultants. A consulting system should preferably accumulate the knowledge of several experts since not all human consultants are experts in the same segment of data analysis, and we would like to unite their knowledge. A system that exceeds the knowledge of one particular expert, supposes common concepts and rules for a community of experts, and we do not know whether such a common view exists. We need more information about the communalities and differences between consultants. One approach to the study of this problem may be based on the methodology used in research on "person perception". Instead of describing differences between persons, here methods of data analysis can be compared (cf. Van den Berg & Heiser, 1989). The degree in which consultants share ideas about the structure of statistical methods, influences the range of applications of a consulting system for the choice of an analysis method.

A second major point in the construction of a consulting system concerns the question of the dialogue between the system and the user. A system as indicated above does not contain knowledge about the domain where the data come from. This is a serious handicap in consultation (cf. Molenaar, 1988) since it leads to a very abstract dialogue. The success of the dialogue will depend heavily on the user's understanding of the system's classification scheme for statistical methods. Consultation is easier when the client and the consultant share a common view on the task. A consulting system differs from a human consultant, however, in the flexibility to change the level and topic of discussion. Behavioral experiments must show if, and with what prerequisites at the side of the user, such an abstract dialogue will produce useful results in practice.

A last important condition for a system that gives advice about the choice of a statistical method to apply, is the system's potentiality to suggest a method that is outside the range of methods it was planned for. E.g., when we have a system supporting decisions about the choice of NLMVA methods, and the problem of the user essentially concerns a test of hypothesis about the difference in mean of numerical variables, then the system should at least suggest to go elsewhere for advice. The necessity of this condition for practice seems obvious, its implications for the complexity of the knowledge in the system are not so obvious, however.

A tentative conclusion may be that support for the choice of method in data analysis requires, above all, insight in the cognitive processes of consultants and in the dialogue between client and consultant. This is not a subject that is much studied, and experience in other fields of expert knowledge indicates that it is a hard problem in general. We are, nevertheless, optimistic about the possibility to get at least a rough idea about the reasoning of some consultants about some segments of data analysis.

4. Phase 2: specifications of parameters

Once a NLMVA method for the analysis of data has been chosen, a computer program will be started by specifying a number of control parameters, like the measurement level of the variables, the number of dimensions in the representation and the precision of the solution. The input may accumulate to so many options that users find it hard to specify the appropriate parameters for their problem. The specification of input parameters might be considered as "fine tuning" of the choice of method. The difference with the choice of an appropriate method lies in the much more prominent role of the details of the data and the properties of the computer program. In the same way, practical considerations, like the size of the data set, are more important.

One part of the knowledge that is necessary for the choice of input parameters can be deduced directly from the data-analytical theory and from properties of the software involved. For this part a front-end system to the data-analytical program could check properties of the data and ask the user for information. In addition a "help system" may be implemented to explain concepts which appear in the dialogue. Such a system functions preferably in a number of layers with increasing detail, under control of the user.

Another part of the knowledge needed to specify input parameters does not follow directly from statistical theory or software constraints, but reflects the experience and craftsmanship of the consultant. An example is the advice not to use a certain form of graphical output when many variables are involved, because the multitude of variables would make the display uninterpretable. It appears that at least part of this kind of knowledge can be represented in the form of production rules.

Both kinds of knowledge may be united in a rule-based front-end system. An example is PRINCE (Duijsens, Duijkers & Van den Berg, 1988) that serves as a consultant for the input of the computer program PRINCALS for nonlinear principal component analysis (Gifi, 1981). PRINCE derives facts by backward chaining. When a fact can not be generated from the rules, the user is asked to provide it. In this way all input parameters are specified in a logical order. If the user chooses a "dangerous" option, the system indicates the consequences and presents an alternative. The system contains a help system and a "why-function" which explains the reasoning that led to a question or an advice. Both may be activated by the user at any moment in the interaction process.

The rules on which this process is built, were derived from repeated interviews with a few experts on PRINCALS. Even though the domain is very restricted, it was not possible to derive a consistent and strict set of rules which was accepted unanimously by the experts. Experts considered many choices as a matter of taste. This needs not be surprising. In the field of NLMVA there are many schools who in a general sense agree on methods, but disagree on approach, applications and interpretation. The more we get into the details of an analysis, the more these differences will come to surface.

PRINCE shows that it is certainly feasible to construct consulting systems supporting the input for NLMVA programs. Such systems may be partly straightforward help systems, and partly rule-based front-end systems. The hard point, again, is the knowledge acquisition that is necessary to get the input information for this type of system.

5. Phase 4: interpretation of results

All methods of nonlinear multivariate analysis produce a simplified approximate representation of relationships in the data. This representation takes the form of a number of tables and two dimensional graphical displays where objects and variables (or categories of variables) are represented as points. In practice, the simplified representation often appears not to be so simple: it may be difficult to find a correct and useful interpretation. And critics of NLMVA methods point at the subjective conclusions these methods evoke. The problem of interpreting NLMVA output may be split in two parts. One is the recognition of patterns in the graphical display, and translating these back to coherence in the data. A second question is how seriously an interpretion should be taken. I.e., how large should differences in the graphical patterns be to report them as "interesting"? We will discuss both problems separately.

How should we look at the graphic displays of NLMVA? Literature gives some indications (e.g. Greenacre, 1984), but not very much is known about the craftmanship in practice. In a pilot study on the interpretation of the output of correspondence analysis, Slooff (1989) asked four experts on this method to interprete output from a program for correspondence analysis while "thinking aloud". In a first session she presented the output without any information on the content of the data; in a second session the meaning of the variables was revealed. For the interpretation of graphical patterns, the consultants used several paper-and-pencil aids, like lines in the display, colours on the labels, and so on. In an interactive system such aids could be represented by graphical tools (line segments connecting points, drawing lines from the origin to certain points, highlighting labels, or changing their size). A few of such tools are already available in some NLMVA programs, but usually not in an interactive form (cf. Becker et al., 1987).

Much attention was given by the interviewed consultants to the fit of the overall solution and of individual points. They were able to formulate rules in the form like: "Be careful with a point, if the value of that parameter is relatively high". Sometimes it is necessary to check combinations of figures from several tables for indication of dubious points. Again, several computerized tools are thinkable based on windowing systems where related information is highlighted simultaneously.

Concerning the overall handling of the computer output, there where large differences between the consultants in the order in which they looked at the tables and displays of the output (although each had his own explicit preference). So, there is not one optimal order, and a supporting user-interface should make it easy to move around in the output. In many cases the consultants combined several tables and/or displays. In an interactive system such simultaneous inspection may be obtained by windowing techniques.

The second question in output interpretation is, how serious an interpretation should be taken as a description of the empirical world. This may be interpreted as the problem of the stability of the representation of the data. We should be cautious about a representation, if a small variation in the data produces a large difference in the representation. To make this (in)stability visible we may apply the-bootstrap method to generate nonparametric confidence regions for the points in the solution (Efron, 1982; Diciccio & Romano, 1988). The generation of such confidence regions could be implemented as an option in a user interface. As a pilot project, a prototype was constructed of a front-end system for the computer program HOMALS (Van de Geer, 1985) for multiple correspondence analysis, see Borsboom & Visser (1988) and Borsboom (1988). The user of that system may choose different forms of confidence regions for the category points. In many cases clear insight in the stability is obtained. From the displayed confidence regions it may be recognized, for example, that a certain order of points in one dimension reflects an interesting structure of the data, while in an other dimension the order may be pure chance. First results with the prototype are promising. Present research is directed towards evaluation in practice, and the study of the cases in which application of the bootstrap is warranted.

In summary, we may say that for the support of the interpretation process the availability of a rich user interface can be as important as the addition of expert knowledge in the form of a rule based system. Such an interface could be based on general modern interface techniques, where specific numerical and graphical tools are added.

6. discussion

We have shown a number of opportunities and problems related to the construction of consulting systems for the application of nonlinear multivariate analysis. We have also indicated the smooth transitions between (a) AI like supporting systems, (b) the addition of numerical and graphical procedures as tools in analysis, and (c) user-interfaces, which are effective in the sense of cognitive ergonomy. Clearly, several practical additions to data analysis packages could be designed and constructed on the basis of present knowledge. Examples are the rule-based system PRINCE and the sketched graphical tools for the interpretation of output of correspondence analysis. For other aspects of support in the application of NLMVA more research is necessary. Examples are the application of the bootstrap to indicate stability of solutions, and the cognitive structures involved in choices of methods by consultants.

In the discussion of all three stages of data analysis, we observed the need for behavioral studies. What is known about consultation and the cognitive process of data analysis, is relevant, but far too imprecise and unvalidated for the present stage of design of consultancy systems. What we need are carefully designed and specific empirical studies, as was put forward by Molenaar (1986, 1988). These studies may concern different aspects of consultancy. Some could be classified as mental model research in cognitive science (Van den Berg & Heiser, 1989, is a modest example). Another important topic is the structure of the consulting dialogue. What do parties demand and expect from each other? How do they negociate? How is social status involved in this process? We know from other fields, like computer supported education, that such questions are related to the utilizations of computerized support systems. Likewise, the evaluation of prototypes should be a form of behavioral research. Here too, we know from other fields (e.g. computerized psychological tests), that the behavior of the user can not be predicted from the system design as such.

We do not claim that empirical studies will be easy, or produce highly generalizable results immediately. But we need the empirical knowledge because there is no formal theory of statistical advice available, and because consultation is essentially a social process that can not be computerized effectively on the basis of a prescriptive theory. One may wonder why so relatively few empirical work is done on statistical consulting systems, compared to, for example, the study of intelligent tutoring systems. Perhaps it is the traditional mathematical background of data analysts that has restrained empirical approaches.

Concerning the actual construction of consulting systems, we believe that in the future two types of systems could be profitable in NLMVA. One system should give advice on the choice of methods. It has a loose connection with the data and with the data analytic software. It is an expert system in the sense that it contains a model of expert reasoning and gives expert advice. The other system is a knowledge-based front-end system to NLMVA programs, supporting both parameter specification and interpretation of output. It should function as an environment for nonlinear multivariate analysis, and it contains not only support facilities but is integrated with a rich user-interface with windows, output management, and interactive graphical tools.

7. references

Borsboom, G. (1988). *Confidence regions for multiple correspondence analysis using bootstrap methods.* Research report 88-03, Vakgroep Sociaal-Wetenschappelijke Informatica, Leiden University.

Borsboom, G. & R.A. Visser (1988). Interpreting analyses of qualitative multivariate data using bootstrap methods. In: Faulbaum & Uehlinger (1988), p. 319-327.

Becker, R.A., W.S. Cleveland & A.R. Wilks (1987). Dynamic graphics for data analysis. *Statistical Science*, **4**, 355-395.

Chambers, J.M., W.A. Gale & D. Pregibon (1988). On the existence of expert systems; comment on Streitberg's remarks on artificial intelligence in statistics. *Statistical Software Newsletter,* **14**, 63-66.

De Antoni, F., N. Lauro & A. Rizza (eds.) (1986). *COMPSTAT 1986; Proceedings in computational statistics.* Heidelberg: Physika Verlag.

Diciccio, T.J. & J.P. Romano (1988). A Review of Bootstrap Confidence Intervals. *J. R. Statist. Soc.* **B**, **50**, 338-354.

Duijsens I.J., T.J. Duijkers & G. M. van den Berg (1988). *PRINCE; an expert system for non-linear principal components analysis.* In Edwards & Raun (1988), p. 149-153.

Edwards, D. & N.E. Raun (eds.) (1988). *COMPSTAT 1988; Proceedings in computational statistics.* Heidelberg: Physica-Verlag.

Efron, B. (1982). *The jackknife, the bootstrap and other resampling plans.* Philadelphia: SIAM.

Faulbaum, F. & H.M. Uehlinger (eds.)(1988). *Fortschritte der Statistik- Software*

1, Stuttgart: Fischer.
Gale, W.A. (1986a). Knowledge representation in data analysis. In: E. Diday (ed.). *Data analysis and informatics IV*. Amsterdam: North-Holland. pp. 703-719.
Gale, W.A. (ed.)(1986b). *Artificial intelligence and statistics*. Reading: Addision-Wesley.
Gale, W.A. (1988a). *Microstrategies: formal models of simple statistical decisions*. Paper presented on the 1988 Illinois Interdisciplinary Workshop on Decision Making.
Gale, W.A. (1988b). Statistical applications of artificial intelligence and knowledge engineering. *Knowledge Engineering Review*, 4.
Gifi, A. (1981). *Nonlinear multivariate analysis*. Leiden: Dept. of Data Theory, Leiden University.
Greenacre, M.J. (1984). *Theory and applications of correspondence analysis*. London: Academic Press.
Hand, D.J. (1984). Statistical expert systems: design. *The statistician*, **33**, 351-369.
Hand, D.J. (1986). Patterns in statistical strategy. In: Gale (1986), pp. 335-388.
Haux, R. (ed.)(1986). *Expert systems in statistics*. Stuttgart: Fisher.
Hietala, P. (1986). How to assist an inexperienced user in the preliminary analysis of time series; first version of the ESTES expert system. In: *De Antoni et al.* (1986).
Hietala, P. (1988). Inside a statistical expert system: statistical methods employed in the ESTES system. In: Edwards & Raun (1988), pp. 163-168.
Molenaar, I.W. (1986). Computer graphics and data presentation: a first step towards a cognitive and ergonomic analysis. In: De Antoni *et al.* (1986).
Molenaar, I.W. (1988). Statistical consultants and statistical expert systems. In: Edwards & Raun (1988), pp. 187-192.
Nelder, J.A. (1988). The role of expert systems in statistics. In: Faulbaum & Uehlinger (1988), pp. 175-182.
Pregibon, D. & W.A. Gale (1984). REX: an expert system for regression analysis. In: *COMPSTAT 1984*. Vienna: Physika-Verlag.
Slooff, N. (1989). *A pilot study for the development of an expert system to support correspondence analysis*. (in Dutch). Research Report 89-02, Vakgroep SWI, Leiden University.
Streitberg, B. (1988). On the non-existence of expert systems; critical remarks on artificial intelligence in statistics. *Statistical Software Newsletter*, 14, 55-62.
Van de Geer, J.P. (1985). *HOMALS*. (users guide). Department of Data Theory, Leiden University.
Van den Berg, G.M. & W.J. Heiser (1989). *Exploring agreement and disagreement in methods catalogues of statistical experts*. Second Workshop on Artificial Intelligence and Statistics, Ford Lauderdale, 1989.
Young, F.W. (1981). Quantitative analysis of qualitative data. *Psychometrika*, **46**, 357-387.

PEELING A LARGE SET OF INDIVIDUALS IN THE FACTORIAL SPACE COMING FROM CORRESPONDENCE ANALYSIS

Micheal Petitjean (*), and Gilbert Saporta
Conservatoire National des Arts et Metiers,
Departement de Mathematiques, 292 rue Saint-Martin, 75003 (France)

KEY WORDS : Correspondence Analysis, Convex Hulls, Peeling Algorithm.

INTRODUCTION

A set of individuals is described with p categorical variables, where m1, m2, ..., mp are the respective numbers of categories. The correspondence analysis is performed, and we consider the factorial space of the individuals. Let N be the number of individuals in a group defined with a supplementary variable, or, if none, the total number of individuals. Then, we want to compute the successive convex hulls of the group, either to get a partial ordering (1,2), or to visualize the shape of the group and find possible outliers (3,4). The problem is : how peel quickly the set, when N becomes large.

THE USUAL PROCEDURE

When the coordinates of the individuals are computed in the factorial space (or their projections in a factorial plane), a convex hull algorithm is selected, giving the extremal individuals of the external hull. These individuals are then removed, and the next convex hull is computed. This procedure is repeated until there is no remaining individual (some faster algorithms are available for mono and bi-dimensional sets (6), but not for higher dimensions).

Note : if few extremal individuals take identical coordinates during an iteration (it means that the representative point is weighed), only one is removed at a time; this choice is explained so : the simultaneous removal would loss the information carried on the weights of the points (of course, the coordinates are always depending on these weights), and a major interest of the hulls is precisely to get a visualization of the density of the individuals.

When N becomes large, the following problems arised :

- We need N coordinates (this is usually an O(N) computation)
- The complete peeling is worst-case lower-bounded with an $O(N^2)$ (5)

USING LARGE SETS OF INDIVIDUALS

When few individuals are described with the same p-uple of categories, it is important to see that their coordinates are identical; then, it follows than the maximum number of individuals theoretically described with the p variables is M0 = m1 * m2 * ... * mp ; the value of M0 is often large, but becomes quickly smaller than N when large databases are involved (see further); so, it is sufficient to compute one time the M0 positions of the theoretical individuals, and to read in the contingency table the weights of the corresponding points; it is also to note than only the M coordinates of individuals having a non-zero weight (M < M0) are needed by the algorithm.

Then, it may be seen that the removal of extremal individuals in a given convex hull is performed when the weight of each corresponding theoretical individual is decreased by 1 unit, and then, if no decreased individual reached a null weight, the next hull will be exactly identical to the actual one : this will be true as long as no weight becomes null.

Algorithm description :

- Compute the coordinates of the M theoretical individuals (and not those whose corresponding weight is zero : M < M0)
- While there is remaining individuals, repeat :
    - Compute the convex hull with any algorithm; (see (4,6) to get a description of some of them)
    - Remove the extremal individual having the minimal weight; let Emin be this minimal weight; (if few, perform a simultaneous removal)
    - Decrease by Emin units the weights of the remaining extremal individuals;

  End while;

End of algorithm description.

Advantages of this algorithm are :

- The computation of each hull is performed with at most M input points, and not N input points, with M << N.
- The number of different (non-superimposed) hulls is at most M (it does not depend of N) : there is at most M individuals, and at least 1 is removed during each iteration, then we get at most M iterations.

- The algorithm has a performance depending on M, and not of N >> M.

In fact, the unique high-consuming step is the initial read of the database, giving the weights and the Burt array; the computation of eigen values and vectors is not affected by the high value of N, and all subsequent operation need only the theoretical individuals with their associated weights.

APPLICATION

The Chemical Abstract Service file is considered (7); it contained 4 552 475 compounds and 200 537 176 atoms during 1979, february 18-th. Each atom is represented by one of the 103 atomic symbols defined in the periodic classification of the elements : H, He, ..., No, Lr ; the weights of each of these 103 elements has been published (7).

The 200 537 176 individuals (atoms) are then described with two categorical variables :

- The chemical family : 32 categories
- The period : 7 categories

Crossing these two variables give the Mendeleiev periodic table, containing 32*7 - 103 = 121 empty boxes.

The contingency table is then submitted to correspondence analysis, then the peeling of the 200 537 176 individuals is performed in the first principal plane, giving 94 403 140 hulls with only 103 iterations (the usual procedure would have required 94 403 140 iterations !).

Eigenvalues and associated percent inertia :

0.997486	38.173 %
0.665908	63.657 %
0.533667	84.080 %
0.233831	93.028 %
0.136700	98.260 %
0.045475	100.000 %

Extract from output in the first principal plane (symbols are trigonometrically ordered in each hull; see also figure 1) :

Number and list of extremal elements	Minimal weighed symbol to be removed	Cumulated number of hulls
3 : H Np C	Np	988
4 : H Fr U C	Fr	1068
3 : H U C	U	13669
. . .	. . .	. . .
3 : H Li C	Li	25234989
2 : H C	C	76131591
1 : H	H	94403140

CONCLUSION

The previous example shows clearly the help given by the new algorithm to compute the successive convex hulls in the factoriel space, when the number of real individuals becomes very large, compared to the theoretical number of individuals defined with the set of the categories of the variables. Unfortunately, this approach is limited to correspondence analysis, and is not possible for PCA, because the number of theoretical individuals is infinite.

REFERENCES

(1) BARNETT V.,
"The Ordering of Multivariate Data",
Journal of the Royal Statistical Society, Series A,
1976,139(3),318-355.

(2) BARNETT V.,
"Interpreting Multivariate Data",
Chap. 1 : "Peeling Bivariate Data",
John Wiley & Sons, 1981.

(3) HOLMES-JUNCA S.,
"Outils Informatiques pour l'Evaluation de la Pertinence d'un Resultat en Analyse des Donnees",
3-rd cycle Thesis, Montpellier II University, 1985, november 16-th.

(4) PETITJEAN M.,
"Utilisation des ellipses de confiance et des enveloppes convexes en analyse factorielle",
Engineer Memoir, 1988, november 15-th,
Conservatoire National des Arts et Metiers, Paris.

(5) PETITJEAN M., Unpublished results (1988).

(6) PREPARATA F.P., SHAMOS M.I.,
"Computational Geometry : an Introduction",
Texts and Monographs in Computer Science,
Springer-Verlag 1985.

(7) STOBAUGH R.E.
"The Chemical Abstracts Service Chemical Registry System. VI. Substance-Related Statistics",
Journal of Chemical Information and Computer Science,
1980,20(2),76-82.

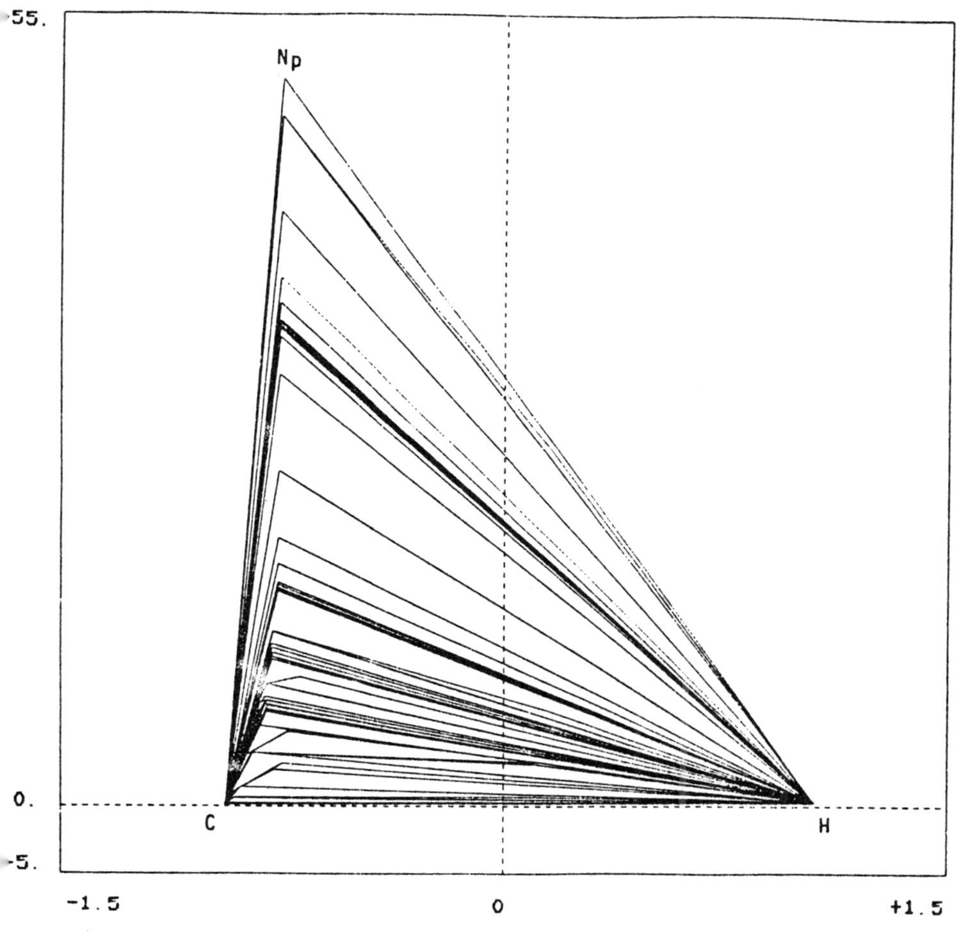

FIGURE 1

Peeling of the 200 537 176 individuals in the first principal plane (axis 1 horizontal; for clarity, only external symbols are plotted)

THE "TAKCOH" METASYSTEM

Zagoruiko N.G., Chirkin E.A. Novosibirsk (USSR)

The taxonomy algorithms as well as any other class of data analysis algorithms are connected with different problem situations. A student - studies the existing algorithms, an analyst - investigates the features of the algorithms known and compares them with each other, a deviser - develops new algorithms and their versions, a user - chooses the necessary algorithm and applies it to his data.

Each of the above situations requires its own software facilitating the work of a man. We call such software program-tool complexes.

The program -tool complex or the "TAKCOH" metasystem worked out in Novosibirsk University and designed for the taxonomy algorithms investigation is described below [I].

By the taxonomy we mean a group of methods and corresponding algorithms and programs designed for partitioning a set of objects into a finite number of classes by certain criteria of "similarity" and "difference" of the objects. The object for the analysis is usually presented as a vector of its features. In the space of features the object is mapped into a point, and a set of similar objects forms a compact bunch of points in this space. The taxonomy objective is to find and single out these bunches (taxons, clusters).

The metasystem is based on the following taxonomy programs: FOREL, SKAT, SKIF, COLAPS, PULSAR, KRAB, KRAB-2.

Let us give their brief descriptions.

The FOREL algorithm. The hypersphere placed in the space of features moves to the gravity centre of points got into it till a complete stop. Points got inside the hypersphere are included into the first class. The procedure is repeated on the remaining points untill all the points of a working access get into the hypersphere. The algorithm can give more than one solution, the number of taxons depends on the hypersphere radius.

The SKAT algorithm. When the FOREL algorithm has made the partition, the taxons discovered are tested for stability on the initial set of points. The FOREL program starts from the points - centres of the taxons selected. Some spheres remain in their places showing the taxons "stability", others - gravitate towards "stable" taxons showing their "instability". Usually, the algorithm gives fewer taxons in comparison with the FOREL one. The future of "splashes" of isolated taxons is determined by means of the acceptable recognition algorithm.

The SKIF algorithm. is a combination of SKAT and FOREL algorithms. Hyperspheres of a chosen radius start from all the points of a working access by turns selecting stable taxons. The algorithm gives only one solution, as the taxonomy results for a fixed radius of a hypersphere do not depend on the access examination order.

The COLAPS algorithm. Unlike the FOREL algorithm the Hypersphere radius increases by a quantity of after stop. As a result of capturing new points the hypersphere starts moving and stops in the centre of another local extremum of the point density. The procedure is repeated until a sharp overfall of the point density connected with the taxon boundary exit into emptiness is discovered. If the increase of the radius does not set the hypersphere in motion, the radius is reduced with the step of ΔR until a sharp loss of points discovers a taxon boundary.

The PULSAR algorithm. Unlike the FOREL algorithm the hypersphere radius changes in each intermediate positron preserving a maximum quantity of the taxon "brightness" we mean the ratio of the number of points contained in the sphere to the radius of this sphere.

The KRAB algorithm singles out taxons of arbitrary form. The algorithm is based on the procedure of constructing the shortest non-closed pass on the points of the working access.

The KRAB-2 algorithm. For a large number of cbjects the construction of the shortest non-closed pass is quite difficult. Thus, at the first stage the FOREL algorithm is used which singles out spherical taxons. Then, the centres of these taxons serve as objects for taxonomy by means of the KRAB algorithm.

Apart of the above algorithms the system can contain other taxonomy algorithms.

The "TAKCOH" complex is realized on a terminal system (ТЕБУС). ТЕБУС includes a personal computer containing a micro-computer, a keyboard, an alpho-digital colour graphic monitor. FORTRAN-4 is the programming language.

It is supposed that those who work with the complex know the basic definitions and concepts of the pattern recognition theory.

The complex structure is a set of independent functional programs. Each of the programs is called at the desire of a student (see fig.I). After fulfilling its function the program passes the management to the menu. The authors consider this structure convenient and most acceptable for creating adaptable program complexes.

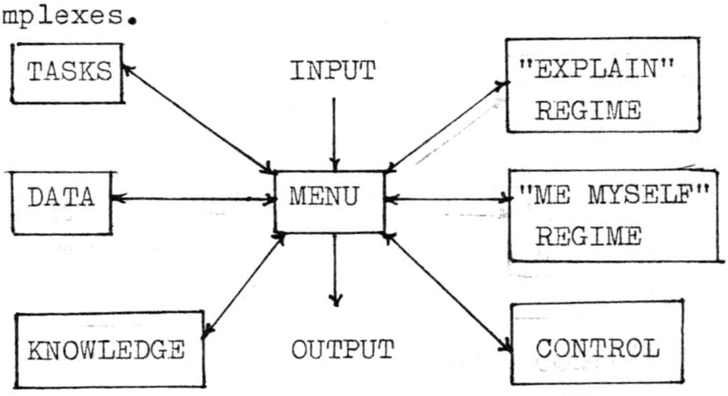

Fig.I. Block-scheme of the complex structure

Let us briefly describe the functions of the blocks.

<u>TASK.</u> The text of a task is called to the screen of the alpho-digital monitor at the command → T. It contains: purpose and contents of the task, posing of the problem, methods of solving the latter and field of application. It is a usual text file. It is created by means of a system text editor. The only restriction is the following: one frame of a screen(24 lines, 80 symbols in each line) should contain the whole text. This file is not transmitted, it does not take part in composing a program, however it is obligatory as well as a problem loading modulus.

<u>DATA.</u> Data model is determined by the input parameters. At the command → G the computer asks after the taxon parameters and generates a set of normally distributed points according to the input values. The taxon is mapped onto the screen of a graphic monitor, the coordinates of the points are written to the computer's memory.

At the command → D the last taxon is removed from the screen and from the computer's memory as well.

At the command → R a random intermixing of data takes place (randomization). A data matrix obtained is a population model.

Usually, only a part of the whole population - the working access is required for work. At the command → V the working access volume is asked after. The input of an integer results in the removal of the population points and the appearence of the working access ones on the screen of a grafic monitor.

The command → E brings us back to the menu.

KNOWLEDGE. This part of the program is a set of reference data about the algorithm under study: key words and basic concepts, the description of the algorithm and its peculiarities (the command → A). At the command → L a list of teaching literature appears on the screen of the alpho-digital monitor. The list can be printed if it is desired.

The set of reference data can be enriched. The only restriction is the following: the text file should not exceed the screen size.

The command → E brings us back to the menu.

"EXPLAIN" REGIME. This part of the program is a kind of a training cartoon.

The program demonstrates the algorithm work making comments on every step. These comments can have a form of a "running line" on the screen of a graphic monitor, or they can be recorded in sound if a speech synthesizer block is available.

Thus, the complex allows to watch the algorithm work in dynamics.

"ME MYSELF" REGIME. A student is invited to make input data taxonomy by hand with the help of program tools.

At the command → O classification "window" parameters are asked after, and a hypersphere of a given radius appears on the screen of a graphic monitor.

At the command → P the hypersphere centre moves to the gravity centre of the points got into it till a complete stop.

At the command → D points got inside a taxon are

removed from the screen and excluded from further consideration.

At the command → F a picture of the population appears, a quality functional value of the taxonomy made is calculated and printed. The command → E brings us back to the menu. Students' mistakes are commented on in the process of work.

<u>CONTROL.</u> Great importance is attached to knowledge control. A student is invited to answer some control questions. The computer gives students marks in correspondence with the number of correct answers and the degree of difficulty of the questions asked.

At the beginning control is considered as an access to laboratory work fulfilment, at the end - as a mark for knowledge of the algorithm under study.

At the command → E(exit) exit from the program takes place and the information

END OF THE TASK

appears on the screen.

To make the description of the complex complete we give the examples of the tasks:

I. Study the work of the FOREL algorithm. Find the dependence between the taxonomy results and a hypersphere radius as well as the access examination order.

2. Study the work of the SKAT algorithm. Check the taxons found by the FOREL one for "stability".

3. Study the work of the SKIF algorithm. See for yourself that the algorithm gives the only solution for a fixed radius of a hypersphere.

4. Study the work of the COLAPS algorithm. What are the advantages of this algorithm over the FOREL one?

5. Construct the shortest non-closed pass for the points of the working access. Find the edges along which the KRAB algorithm will draw boundaries between the taxons.

Astudent should fulfil the tasks with the help of the program tools in the process of the independent work with a terminal.

The "TAKCOH" metasystem includes tool means some of which are useful for support and other problem situations mentioned above.

At present work is carried out of supplementing the complex by software tools for INVESTIGATION, CREATION, COMPARISON, CHOICE and APPLICATION of the taxonomy algorithms.

Let us give a brief description of such software.

2."INVESTIGATION".

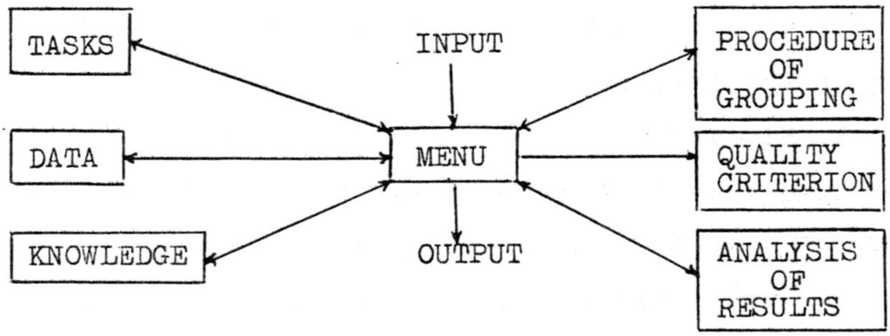

An analyst has the opportunity of investigating the properties of any taxonomy algorithm testing it under various conditions, changing: a working access volume, a hypersphere radius, an access examination order, a quality criterion, etc. Thus, checked is the dependence between a decision and parameters, accuracy and other characteristics.

3. "CREATION"

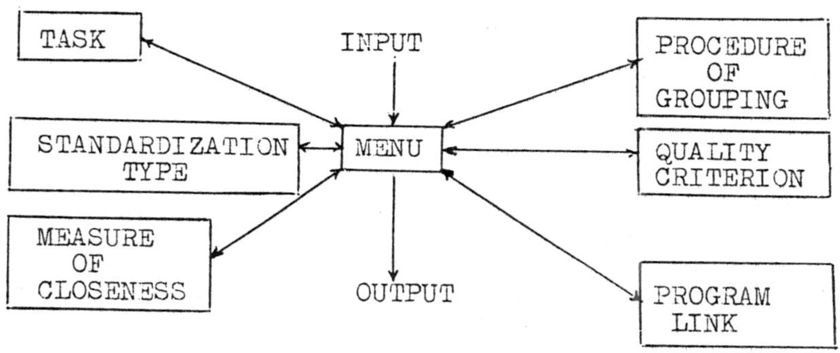

A deviser can combine the elements available in the system: standardization type, measure of closeness, procedure of grouping and quality criterion using them as "bricks" for constructing an algorithm.

The system gives convenient means for adding any element of a new type to it.

4. "COMPARISON"

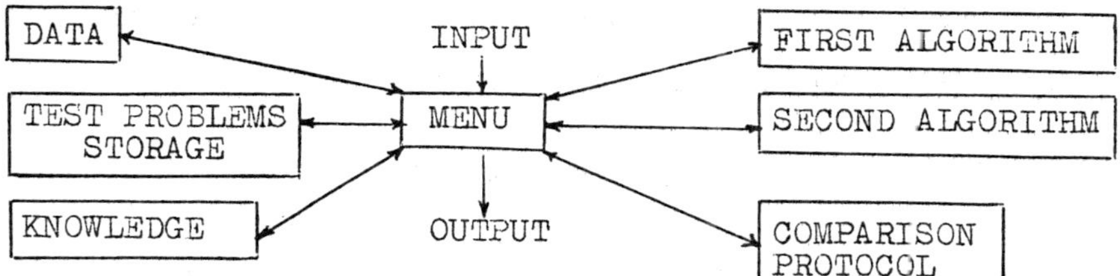

The system includes the taxonomy algorithm set,

test problems starage and the results of the application of these algorithms for the solution of test problems (knowledge). One can generate a new problem and pass it through all the algorithms. One can introduce a new algorithm into the system and pass all test problems through it.

The system gives out the protocol of comparative test results including problem decision time, the volume of the memory used and the quality of the decision obtained.

5. "APPLICATION"

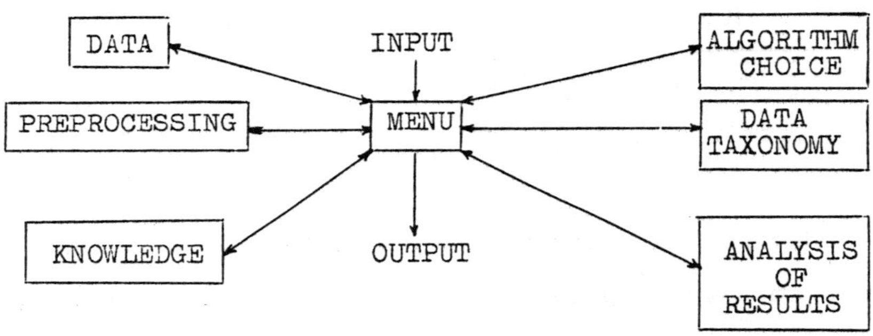

In dialogue with a user the system clears up the a priori information: the quantity of objects, the number and type of features, the desired number and form of taxons, etc. On the basis of this information the system chooses the appropriate taxonomy algorithm and solves the problem.

REFERENCES

I. Zagoruiko N.G., Chirkin E.A. Program-tool Complex for the Taxonomy Algorithm Investigation. In: "Computer in Educational Process in a High School".

Nov. State Univ., Novosibirsk, 1987, pp.71-78

2. Zagoruiko N.G., Yolkina V.N., Emelianov S.V., Lbov G.S. "ОТЭКС " Package of Applied Programs, "Finance and Statistics", Moscow, 1986.

3. The Algorithms of Empirical Laws Discovery. "NAUKA", Novosibirsk, 1985, pp.28-30

SOME SOFTWARE TOOLS FOR DATA ANALYSIS

Marc Csernel (Inria, Université Paris IX Dauphine)
INRIA, Project CLOREC
Domaine de Voluceau Bp 105
F-78153 le Chesnay Cedex (France)

This paper presents a set of tools which have been created to help the development and maintenance of FORTRAN programs. All these tools have been written in FORTRAN, and this provides them with a great degree of portability . Most of them have been written using a parser, FGRAMM also written in FORTRAN. All the tools described in this paper are working and can be provided at INRIA.

1. INTRODUCTION

May be because of its old age FORTRAN [ANSI 78] has given very little enthusiasm to software engineers, and then, the tools able to help a programmer to write and maintain FORTRAN programs are rare except such tools as those provided by the TOOLPACK [OST 83] project, but this kind of tools are nearer to a complete environment than to a toolbox in which the user can choose to use one single tool or another.

We have developed a set of tools which can be used separately one from the other. They have been defined especially to suit the needs of the system of clustering analysis SICLA [RAL 86, BOC 88] and those of the MODULAD [MOD 87] club, but due to their general design, they can be used by any scientific program written in FORTRAN. One of their specificity is to have been written in FORTRAN and thus to be easily transportable on any computer supplied with a FORTRAN compiler.

The purpose of a project such as MODULAD is to collect and integrate a set of programs all written in FORTRAN, but developed on different places, and different kinds of computers. Nevertheless all of them must be able to satisfy some standard of programming. To allow all the developers to co-ordinate their efforts, it could be very useful to be able to provide on each place the same set of tools giving the possibilities :

- of a certain kind of verification of the programming standard
- of a uniform presentation of the program .

Therefore the portability of the tools that we have developed is an important advantage to reach such a goal.

Most of these tools are using a descending parser FGRAMM [CSE 88] also written in FORTRAN and very near in its design to LL(1) parser [AHO 77] . All the tools are very easy to use, and each of them can be used without the others, and none of them needs a special environment.

The tools, we are providing, are performing the following functions:

- automatic indentation
- automatic building of calling trees

- documentation help
- search and impression of characters strings

II. AUTOMATIC INDENTATION

Automatic indentation increases the program's readability, showing the program's structure using the source file presentation.

We have created INDAUTO, automatic indentator of FORTRAN programs using a parser built by our LL(1) parser FGRAMM. The need of a parser to build such a tool is due to the necessity to have a correct cutting of program's line , and we shall see more precisely why, later in this document.

INDAUTO : - indents the program
- re-numbers the labels

INDAUTO does a right shift of the source text after every DO and IF THEN instruction it meets; it does a left shift after each label corresponding with a DO and after each END IF. When an ELSE IF is met, a temporary left offset is proceeded just for the ELSE IF instruction.

The proceeding of such a treatment looks very simple at first sight, but it becomes rather complicated when an instruction cannot be written on a single line. In this case you have to choose the place where an instruction has to be cut without destroying the readability of the text. A very simple way to do this could be to cut where the nearest comma or arithmetic operator before the optimal cutting point stands. Unfortunately such a method is insufficient as the following example shows.

Example: given the FORTRAN instruction

```
      WRITE(IPRI,1000) 'VALEUR ELEMENT 1',
&     ITAB(IMAX , IDEF + I), 'VALEUR ELEMENT 2 ', TAB( I + J )
```

If, due to the different shifts made by INDAUTO, the instruction must be cut a bit farther than the identifier IMAX, a cut done immediately after the comma following IMAX, will entail an instruction which is certainly correct, but not readable at all because the cut point has been placed in the middle of the description of an array element:

```
      WRITE(IMP,1000) 'VALEUR ELEMENT 1', ITAB(IMAX ,
&     IDEF + I), 'VALEUR ELEMENT 2 ', TAB( I + J )
```

You can hardly admit such a miss from a tool supposed to improve the readability of a program.

Because of this, to be able to deal correctly with the problem, one has to execute a syntax analysis of the source text by a parser. Such an analysis can provide the structure of the instruction and allows a well fitted cut.

In the previous example, the parser will recognize the text ITAB(IMAX,IDEF + I) as an array element and the routine in charge of the cutting will avoid establishing it in the middle of this element. With the preceding example the right cutting such as the one done by INDAUTO will give:

```
      WRITE(IMP,1000) 'VALEUR ELEMENT 1', ITAB(IMAX , IDEF + I),
&     'VALEUR ELEMENT 2', TAB( I + J )
```

In a more general way, INDAUTO builds a syntaxic tree to do the cutting. The tree is projected upon the text of the instruction. When a cut is needed, it is done following the structure of the tree.

At the moment INDAUTO is using the '&' as continuation character, but it could, if needed, be easily modified to use numbered continuation lines.

In short the way the cutting is done depends on:

- the structure of the grammar used by INDAUTO
- the explicit parameters of indentation

The actual grammar seems to fit well to our needs, but it can be modified by an experienced user, if he whishes.

The three explicit parameters that can be modified are:

- the shift value used after each IF, END IF ...
- the offset value used after each ELSE IF
- the maximum number of blank characters allowed at the end of the line when a cut is needed. This parameter indicates how deep in the syntactic tree the program must go, to cut within a syntactic element.

These parameters are initialized by semantic insertions within the grammar. They can be modified without changing the grammar structures. But the grammar must be re-compiled by FGRAMM if the user wants the changes to become effective.

To re-number the labels, INDAUTO needs to perform two passes upon the source file. The labels are re-numbered according to two different sequences.

- the first one concerns the ordinary labels
- the second one concerns the input-output labels

In each sequence, the new labels generated are growing by tens, in the first one the sequence begins at 10, in the second one at 1000. All the labels are right justified within the column 1 to 5. All references to the re-numbered labels within the source file are changed according to the new one.

We must notice than INDAUTO is not working in the same way as some other tools as MENTOR [MEL 85] or PARADIS [DES 85], such tools are completely re-organizing the source text, providing a standard number of blank characters between each element of a FORTRAN instruction. On the contrary INDAUTO is not modifying the source file other than by the cuttings provided.

The complete re-organization of the program's presentation can in certain circumstances embarrass a user. But it can also be very useful, especially by helping for character string search within a text editor. If you wish to search with a text editor a string such as

SUBROUTINE   CHARAC

your success depends on the number of blank characters placed between SUBROUTINE and CHARAC. The usual solution consists in searching the string CHARAC, but the numerous CHARACTER declarations will be found before the text we are looking for. Hence we consider the possibility to add such a tool to the present set.

III. BUILDING OF CALLING TREES.

The building of calling trees is performed in two steps:

- The first one consists in finding in the source files all the references of a subprogram or function calling. This step is performed by a program called GENAPP, which is built using a parser generated by FGRAMM.
- The second one is to build the tree, using the information collected during the first step.

At first sight, the first step seems easy to perform, because each subroutine call can be found very easily by looking at all the "call" instructions within the source file. It is true for subprogram, but functions calls need a really more sophisticated approach and then, the use of a parser such as FGRAMM becomes crucial.

As a matter of fact, if a "call" can be easily recognized by a simple lexical analyzer, a function call is more difficult to detect, and it can be easily confused with an array reference. To be able to deal with such a problem, the parser must recognize all the array declarations and store them in its table. After this the parser is able to make the difference between a function call and an array declaration.

Then GENAPP is building a file containing:

- a header for each file processed
- a header for each program, subprogram or function processed
- a reference for each calling met within each program, subprogram or function

By the way GENAPP is also filling the file with some other information which will be used for program documentation and this will be dealt with more precisely in the next section.

GENAPP is able to process several source files within one call, but all the informations corresponding to these different input files are put in the same output file.

The output file of GENAPP must be used by another program ARBR to produce the calling tree. ARBR is able to deal with one or several input files. The source file name is associated with each subprogram name mentioned in the tree. When the source of a subprogram has not been process by GENAPP or when the corresponding file was not been taken into account by ARBR, the source file name appears as **unknown** and no sub-tree is produced.

The user of ARBR can get a tree more or less deep by giving the program more or less files to process.

ARBR displays, after the tree, a list of all the subroutines used by the the program, and a second list of all the subroutines unused. Those lists, especially the second one, can be very useful to maintain a library, or to discover unused subroutines which are overcrowding source files.

IV. HELP TO DOCUMENTATION.

Our tools are able to help program documentation in two ways:

- The first one consists in building automatically a set of information for each subprogram. For each of the subprogram's parameters the set will contain :
    - the parameter's name
    - the parameter's type

- the length if the parameter's type is character
- 'scalar' or the dimension if the parameter is an array
- an indication upon the use of the parameter: input, output, input-output, not used

The output file built by GENAPP contains information able to help to build the subprogram headers that we have mentioned in describing the first point. The program ENTET using this file is able to produce the informations mentioned above. The informations appear as FORTRAN comments and then can easily be included within the subprogram related to them. Generally the header is completed by hand with semantic information.

The information provided by ENTET can obviously be created by hand, but when a great number of programs must be processed, it provides a systematic description and then helps the programmer not to forget any significant information.

- The second one consists in building a listing made up of all program and subprogram headers. The listing is completed by a table of contents where the name of all subprograms mentioned within the listing appear: sorted by alphabetic order or unsorted.

This second step is performed by the program COMMENT. COMMENT can collect in a single output file the headers of one or more source file. Then if the subprogram headers are correctly filled, by hand or by ENTET, COMMENT is able to provide a short but convenient description of all the elements of a set of programs.

V. DISPLAYING OF CHARACTERS STRING.

There is a twofold advantage to a systematic search and displaying of character strings in a program:

- it enables to check the consistency of all the messages together
- it permits to get easily a complete list of all messages if a translation is needed.

We have built a very simple program called CHLANG which is able to provide a complete list of all the instructions of a FORTAN program containing a character string. Each instruction is displayed with its origin file and its line number within this file. All the constant character strings within the instruction are displayed one by one on the next lines.

CHLANG could be easily modified to allow an on-line modification of the string displayed, but this feature does not seem useful because the modification of a single string can induce a lot of subsequent modifications:

- the change of a string's length can cause an unwanted change of the display format
- of the declarations of the variable supposed to contain the string.

VI. THE PARSER FGRAMM.

A large number of parser generators already exists, allowing a user to generate parsers in a rather easy way. We must specially mention the YACC [JOHN 75], LEX [LESK 75] pair, closely associated with UNIX, and SYNTAX [BOU 84], which have been developed at INRIA, and have very good error recovery performance. The inconvenience of those very good software tools is to be very closely linked with the UNIX system (for the pair YACC-LEX) or to have been written in a language very different from FORTRAN. That is why we have developed FGRAMM.

FGRAMM is a parser generator completely written in full ANSI 77 FORTRAN, that we have developed to build our tools. As we mentioned it before, this allows FGRAMM a great deal of portability, but in turn, it involves the use of very unsophisticated data structures, and the necessity to avoid recursivity in programming.

FGRAMM is self-defined, by this we mean that FGRAMM uses, to process its source file, a parser built by FGRAMM itself. This implies that some FGRAMM features, such as the error procedure, will be used exactly in the same way by FGRAMM, and by the tools as INDAUTO or GENAPP built with FGRAMM.

As for all the parser's generators, to get a tool using a parser built by FGRAMM we need to provide FGRAMM with a grammar. This grammar can contain semantic insertions i.e some parts of program which are executed when associated elements of the grammar are recognized by the generated parser.

When it is used FGRAMM reads the grammar source file. This file is in a format close to the B.N.F. Because FGRAMM is self-defined the format of the input file can be changed by the user by changing the grammar of FGRAMM. FGRAMM tests if the grammar presented is syntactically correct, then FGRAMM tests if the grammar is semantically correct and produces a program and a file containing the tables needed by the program. The resulting program must be compiled and linked.

FGRAMM is generating descending parsers, this implies a certain number of constraints upon the grammar that FGRAMM is able to deal with. In particular the different alternatives of a grammar rules must begin with different 'first' as the example below can show:

```
<if-instruction>        ::= if (<logical-expression>) <instruction > |
                            if ( <logical-expression > ) then
```

can not satisfy to the condition we mentioned before, because both of the alternatives begin with the key-word 'if'. To be satisfactory the rule should be put under the form:

```
< if-instruction >      ::= if (<logical-expression>) <following–if>
<following-if>          ::= <instruction> | then
```

If the grammar is not satisfying this condition, an attention message is produced, indicating an ambiguity within the grammar. A parser is generated anyway, but it will never consider the second alternative of the rule.

This kind of constraints could be embarrassing if you try to build a real compiler with FGRAMM, but to build the kind of tools under consideration, they are acceptable.

Compared with other tools of the same kind FGRAMM is not very sophisticated but it is easy to use:

- FGRAMM allows semantical insertions before and after any element in the right part of a rule. This could be very useful to have some initialization within the semantic insertions.
- FGRAMM enables to handle error automatically. Due to the structure of the grammar it uses, FGRAMM provides the parser it has generated with an automatic error routine. This routine is able to deal with the most common syntactic error. For each error it provides a message indicating:
    - The instruction where the error was found, the name of the corresponding file, the line number of the instruction within the file. The instruction containing the error is underlined till the precise error point.
    - The character string which has caused the error.
    - A list of elements suitable in place of the wrong one.

- Due to its self-definition, FGRAMM is able to provide generally clear error messages to its own users. Those who have used YACC or LEX can understand how interesting such a feature is.
- FGRAMM has an integrated lexical analyzer, and all key-words used within the grammar are automatically recognized by the lexical analyzer generated.
- FGRAMM is dealing with the notion of instruction.
- FGRAMM provides a special descriptor allowing to avoid a complete description, of an instruction. That can be useful when one does not want to describe a grammar completely.

VII. THE LEXICAL ANALYZER.

As we said it earlier FGRAMM generates also a lexical analyzer. Most of the descriptions generally needed to build it are provided by the key-words described within the syntactical part of the grammar. FGRAMM uses regular expressions to describe the lexical elements of the language which are not key-words. FGRAMM uses the KLEENE's operator to proceed to this description and is able to deal with a lookahead operator to precise the context in which certain elements appear. FGRAMM provides also a limited number of generic descriptors as letter, digit, any, to do the lexical description more easily.

The example below shows the description of a FORTRAN integer, the lookahead operator represented by the character '\' is there to show that an integer followed by '.EQ.', '.NE.' ... is still an integer, and not a real.

Example : &INTEGER = (DIGIT) + \ ('.EQ.' | '.NE.' | ......)

A certain number of routines are provided by FGRAMM to help to maintain an identifiers dictionary, including a hash-coding function. Those routines must be used as semantic insertions within the lexical description. The user can deal with these routines or with his own one's.

We must remark that the lexical analyzer generated by FGRAMM is not able to deal with a pure FORTRAN text where the blank characters are unsignificant. We have built our tools to provide a better presentation of FORTRAN programs. A program where blank does not appear as separator seems to have very little chance to became well presented. Thus in regard to the work needed to provide such a feature, and the complication that it can introduce within the complete tools, we did not put in into FGRAMM.

VIII. CONCLUSION.

The use of a parser written in FORTRAN provides an easy way to realize portable software tools dealing with that language. Other tools can be considered like a macro pre-processor such as RATFOR. All these tools provide to statistical software programers a way to realize better programs with a substantial time saving.

The tools we have described have been developed under MULTICS and are at the present time working under UNIX on APOLLO and SUN. They can be provided at INRIA as TAR cartridge or as PC floppy disk.

REFERENCES

AHO 77] Alfred V. AHO, Jeffrey D. ULLMANN
Principles of Compiler Design
Addisson-Wesley 1977

ANSI 77] American National Standard Institute
Programming Language FORTRAN
ANSI X3.9 1978

BOC 89] S.BOCHI, G. CELEUX , G. GOVAERT, Y. LECHEVALIER, H. RALAMBONDRAINY, Y. OK
SICLA: Manuel utilisateur
CESIA 1989

BOU 84] Phillipe BOULIER
Contribution à la Construction Automatique d'Analyseurs Lexicographiques et Syntaxiques
Thèse d'état Universitée d'Orlèans 1984

CSE 88] Marc CSERNEL
Des outils logiciels pour l'analyse de données
Revue MODULAD N° 2 1988

DES 85] Phillippe DESCHAMP, Pierre BOULLIER
PARADIS un système de paragraphage dirigé par la syntaxe
INRIA Rapport de Recherche N° 455 1985

JOH 75] Stephen C. JOHNSON
YACC: Yet another compiler compiler
Bells Labs, Comp. Sc.Tech.Rpt 32 1975

LESK 75] M.E. LESK
LEX a lexical analyzer generator
Bells Labs, Comp. Sc.Tech.Rpt 39 1975

MEL 85] B. MELESE, V. MIGOT,D. VEROVE
The Mentor-V5 documentation
INRIA Rapport Technique N° 43 1985

MOD 87] MODULAD Bochure Utilisateur V 2
INRIA 1987

OST 83] Leon J. OSTERWEIL
Tolpack an Exeperimental Software Developement Environnment Research Project
IEE transaction on Software Engineering Vol SE-9,.N° 6 1983

OST 83] Leon OSTERWEIL, Geoffrey CLEMM
An Extensible Toolset and Environment for the Production of Mathematical Sofware
Conférence Internationale: Outils,Méthodes et Langages adaptés au Calcul scientifique.
Paris 1983

RAL 86] H. RALAMBONDRAINY
Le Système SICLA
Thèse d'état, Université Paris IX Dauphine 1986

RYD 75] B.G. RYDER, A.D. HALL
The PFORT Verifier
Bells Labs, Comp. Sc.Tech.Rpt 12 1975

PART 2

LEARNING SYMBOLIC AND NUMERIC KNOWLEDGE

SYMBOLIC REQUIREMENTS FOR THE DEFINITION OF NUMERIC MEASURES OF EXPLANATORY PROOFS, ABDUCTIVE RECOVERY, AND ANALOGY.

Yves Kodratoff

LRI, bât. 490, Equipe Inférence et Apprentissage, UA 410 du CNRS, Université Paris-Sud, F-91405 Orsay
George Mason University, AI Center, Fairfax, Virginia 22030, USA

SUMMARY

Artificial intelligence (AI) has developed a number of inference techniques, such as explanatory proofs, abductive recovery, and analogy. This paper will show that, even in the "simple" case of deductive inference, the problems met in practice forbid the use of a unique law of combination of uncertainties. It follows that when performing more complex inferences, like the inductive inferences of abduction and analogy, it would be very surprising that one finds only one combination law of uncertainties. This paper does not intend to propose explicit solutions, as for instance in Collins and Michalski (1986, 1989), but will rather describe the structure of these reasonings. Any numeric technique will have to take into account this structure.

1 - INTRODUCTION

We have been already advocating for a number of years in favor of a clarification of the links between numeric and symbolic approaches, as in (Kodratoff, Perdrix and Franova, 1985; Ganascia and Kodratoff, 1985; Duval and Kodratoff, 1986, 1989; Kodratoff, Manago and Blythe, 1987; Manago and Kodratoff, 1987; Kodratoff, 1988). We would like to recall here briefly the incredible amount of resistance we have been meeting in this task, as well from the numeric as from the AI community.

For instance, it seems quite obvious that performing modus ponens, i.e., [from A and A $\Rightarrow$ B, infer B], which, in clear, means: look for some A, look for some implication A $\Rightarrow$ B, and then you can safely add B into your knowledge base, is not some kind of special case of a statistical rule when the probabilities are equal to 1. The very "business" of looking for true assertions, of comparing different expressions for A and recognizing their equivalence etc. is not at all numerical even if it deals with numbers. For instance, the application of a theorem relative to rational numbers to an arithmetic expression requires the recognition that the value of the expression is indeed a rational number. Since there will always be rational that are not computable on a specific machine, you are forced to have some kind of symbolic recognition method to recognize its nature. More generally, modus ponens is used in AI under its resolution form which reads: [from A' and A" $\Rightarrow$ B, and if there exists a substitution σ such that $\sigma \circ A' = \sigma \circ A''$, then add safely $\sigma \circ B$ to the knowledge base]. Again, the business of finding such substitutions is purely symbolic. It amounts to the recognition that A' and A" share similar features. Unless all the possible features (which are usually in an infinite number, even in very simple cases) have been indexed, simple numeric methods are not applicable. We apologize for insisting on such simple ideas but, along those years where we have been advocating for the clarification of the links between numeric and symbolic approaches, we have been meeting again and again scientists from the numeric side that seem

to be unaware of these simple facts and that keep repeating absurdities such as "everything symbolic can be handled numerically", or that "logics is statistics with probabilities 0 or 1".
Now, we want to stress that the situation is not so good on the symbolic side. For instance, we have been advocating also for years that handling uncertain knowledge, from the point of view of deductive inference, leads to handling contradictory data bases, as the following section 2 will illustrate. Knowing the difficulty for performing correct deductive inferences in the case of tricky but perfectly coherent knowledge bases, we have been proposing to spend some work on the problem of the incoherent ones. As a whole, the community has been answering that the interesting problem is not this last one, but the one of combining belief coefficients. In practice, however, the experts of the field hardly associate numbers to their beliefs. As noted by Lebbe, Vignes and Darmoni (1988), for instance, typicality has three different meanings in medicine, the first one is "more frequent", the second "belonging exclusively to the case, but not specially frequent", and the third one "following the descriptions in the books, that is, maybe never met as such" (see section 2 below). Each meaning entails a special logics for its combination with others, and a special computation. Work as the one done by Lebbe et al., which seems the very first step before attempting any numeric computation, is seldom performed, or not reported by the authors. We claim that all the difficulties met by expert systems, once they step out of the research and development departments of the companies, are at least partly explained by the behavior of the AI community, showing a strange lack of interest for the symbolic representations and the logical consequences of uncertainty (or belief, noise, etc.).
Next section will illustrate how uncertainty on deductive inference demands handling apparently incoherent data. We shall then show how inductive inference asks for a special numeric treatment, especially when the abductions concern instantiated facts. Finally, section 4 describes how analogies are performed, and how their similarity and causality components must combine the beliefs in very different ways.

2 - UNCERTAIN DEDUCTIVE INFERENCE AND EXPLICABLENESS

2.1 - Explanations provided by expert systems

Let us quickly recall that expert systems use belief coefficients, as in MYCIN for instance (Shortliffe, 1976). While reasoning proceeds, these coefficients are combined in order to obtain a final coefficient expressing the belief in which the conclusion is held. Actually, these coefficients express belief when they are given positive values and disbelief when they are given negative values. When a reasoning session uses both belief and disbelief, the final result is a mixture of them, from which it is almost impossible to find the reasons why a final value of belief has been computed. As shown by Clancey (1983), when expert systems are used as tutorial media, i.e., when the reason why a conclusion has been reached is very important to know, then the choice of an appropriate knowledge representation, giving unmixed reasonings, becomes mandatory in order to provide understandable explanations to the students.
As we shall now see, finding such an appropriate knowledge representation might be just impossible. In this case, one should keep track of the reasoning leading to belief independently of the one leading to disbelief.

2.2 - Uncertain rules and explanations

As shown in Duval and Kodratoff (1986), uncertainty forces to handle syntactically contradictory knowledge, even though it is semantically coherent. Before coming to this point, let us first illustrate that indeed uncertainty does not always causes apparent incoherence. This will allow us also to show that explanations can then be found in the trace of proofs done in a coherent knowledge base.

2.2.1 - In a coherent knowledge base, explanations are inference traces

Let us analyze the following toy medical knowledge base written in an Edinburgh PROLOG style, with the added feature that each clause starts with an expression in brackets, expressing the belief in the clause. We shall call **modulation** this belief.
Suppose that our base states that vomitings imply sometimes gall bladder problems, that vomitings did occur. It reads
C1: [*sometimes*] gall-bladder-problems :- vomitings
C2 : [*certainly*] vomitings :-
Suppose know that we want to ask if the patient has gall-bladder-problems, we shall ask the question
C3: :- gall-bladder-problems
By resolution of C3 and C3 into
C4: [*sometimes*] :- vomitings
and C4 with C2, we shall therefore prove that the patient has gall-bladder-problems, with a belief that we shall write here as [*sometimes*, *certainly*]. Notice that we do not give our full belief to our conclusion. Conversely, given this knowledge base, we know with full certainty that the only explanation why our patient should have gall-bladder-problems is that he has vomitings. This means firstly that we are using the trace of our proof to build up an explanation, and secondly that the belief in the result and in its explanation are different: each different proof may provide a new explanation, even in the case where each proof would be 100% certain. In this last case, one has shown (Mitchell and al., 1986; DeJong and Mooney, 1986) that explanations are actually generalization of proof traces.

2.2.2 - An incoherent knowledge base is not fit to explaining

Consider now a knowledge base stating that vomitings imply sometimes gall-bladder problems, that vomitings often imply digestive problems, that fever and digestive problems generally imply appendicitis. One will introduce *potential* contradictions each time some state is recognized as impossible. For instance, when adding to the basis that, typically, appendicitis does not occur with gall bladder problems.
All this provides the (still coherent) uncertain data basis :

C1: [*sometimes*] gall-bladder-problems :- vomitings
C2: [*often*] digestive-problems :- vomitings
C3: [*generally*] appendicitis :- fever, digestive-problems
C4: [*typically*] :- gall-bladder-problems, appendicitis

If one is concerned by a patient who actually has fever together with vomitings, one introduces a contradiction in the data base since C1, C2, C3, C4 together with

C5: [*certain*] fever :-
C6: [*certain*] vomitings :-

leads to the empty clause as shown below by the sequence of resolution steps :
(C1/C4 -> C7) :- vomitings, appendicitis
(C6/C7 -> C8) :- appendicitis
(C3/C8 -> C9) :- fever, digestive-problems
(C5/C9 -> C10) :- digestive-problems

(C2/C10 -> C11) :- vomitings

and C11 contradicts C6.

The problem with this data basis is that one can prove A and ¬A from it[1]. For instance, if one asks if the patient has appendicitis, one will get an empty clause by adding

:- appendicitis

to the basis (that proves that there is appendicitis with a modulation *[generally, [certain,[often, certain]]]*), but also by adding

appendicitis :-

and that proves that there is no appendicitis with a modulation *[typically,[sometimes, certain]]*.

When these modulations are expressed by numbers, the relevant problem becomes to express by combinations of number the following properties.

1 - associate a reasonable number with each of the qualifier (this is easily performed),

2 - find combination laws for sequences, like *[sometimes, certain]*,

3 - find combination laws for sequences of sequences, like *[generally, [certain,[...]]]*,

4 - find combination laws for positive evidence and negative evidence.

Once all those combination laws have been chosen, unless all the combination laws are reversible operations which is unlikely to happen, it will be impossible to backtrack the way a particular result was obtained. Besides, it will be often the case that the semantics of the domain will be difficult to fit with the semantics of number combinations. For instance, suppose that the domain is such that linear reasoning (corresponding to simple sequences of modulations) are preferred to branching reasoning (corresponding to sequences of the form [...[...[...]]]). The system will therefore need computing a kind of "branching index", and use it in its computations. The more complex the numerical representation, the best it will allow explanations.

On the contrary, more complex representations, using strings of symbols may better convey the semantics of the domain. For instance, in the above appendicitis example, the two strings of symbols we obtained may be interpreted as follows. Choose to believe first that there is appendicitis (because all evidence in favor of it is strong, while one branch of the reasoning against it is a bit weak) but beware that at the first show of weakness in new data confirming appendicitis, then both hypotheses become equivalent, and the whole reasoning should be done again with the new evidence included.

Including such a procedural interpretation of strings of symbols is heavily domain dependent, and should be part of the information requested from human experts.

2.3 - Conclusions

From the above examples one can draw three conclusions.

1 - Knowledge bases can be syntactically incoherent because of modulations without containing any real contradiction from the human expert's point of view. Notice that one cannot hide the contradiction by writing clauses like

---

[1] Notice that the PROLOG "NOT" and logical negation are not equivalent. The difference is the following. In PROLOG, NOT A means that one is unable to prove that A is True. In logics, ¬A means that A is False. This difference can be used to hide contradictions as the above. In order to avoid it, it is forbidden in PROLOG to introduce "NOTs" in the left hand-side of a clause. Nevertheless, a programmer can always go around this protection by writing clauses like

```
NO(bidon)      :-      conditions, !
```

and then define the predicate "NO" in the same way as "NOT" is defined, i.e.,

```
NO(P)      :-      P, !, FAIL
NO(P)      :-
```

In that case, one will not find a contradiction by adding the clause

```
bidon      :-
```

It is clear that this is a programmer's trick which does not actually remove the contradiction in the base.

C4': [*typically*] NOT gall-bladder-problems :- appendicitis
because PROLOG forbids NOTs in the left-hand-sides of the clauses (for more details, see the footnote above), and using a logical negation like in

C4": [*typically*] ¬ gall-bladder-problems :- appendicitis

This leads us back to full resolution which will unify C4" with C1 exactly as above, leading to the same resolution. It follows that the contradiction cannot be hidden, and is a consequence of clause modulation.

A first proposal on how handling such knowledge bases is found in Duval and Kodratoff (1986).

2 - One cannot find explanations when the modulations are combined by a classical numeric technique because the clauses that conclude to A and those concluding to ¬A are mixed up. In order to illustrate this last statement, let us suppose that in place of the above modulations, one would see numeric coefficients. Just for the sake of precision, let us suppose that one associates 1 to [*certainly*], 0.8 to [*generally*], O.7 to [*typically*], 0.6 to [*often*], and 0.3 to [*sometimes*]. Suppose also that the combination law is simply multiplication.

Our mode of reasoning would then show that one can prove *appendicitis* with a belief of 0.48, and ¬ *appendicitis* with a belief of 0.21. The expert would then have to interpret these figures.

On the contrary, in a classical expert system, in which the theorem prover is unable to make the difference between proofs of belief and proofs of disbelief, the combined belief is around 0.11. The user does not know if this low value comes from the combination of contradictory information (as is the case here), or if it comes from a combination of low positive belief values.

3 - Actually, explanations depend very much on the user himself, and on his understanding of the modulations. Including this meaning into the way computations are performed is mandatory if one wants ever to see expert systems go out of research and development departments of the companies. An interesting account on this topic comes from Lebbe et al. (1988), who have been studying the meaning of *typical* in the context of medical knowledge. These authors show that it actually takes three very different meanings. The first meaning is the one of relative frequency, as it is usually understood by AI specialists. The second one is not linked at all to any frequency. It means that none other case shows this kind of "typical" property. For instance, measles shows a trait called *typical* because no other sickness than measles shows it, but it is seldom shown by measles itself, and it is even less often observed because it does not last in time. The third meaning of *typical* refers to scholarly descriptions, that is to a concatenation of all the possible features, and they are seldom, if ever, observed all together. In this last case, typical maybe associated with null observation frequency.

It is obvious that these three kinds of understanding for *typical* are widely different and should not be blindly combined. Recognizing their existence is the work the AI specialist, and telling the way they combine should be part of the standard knowledge acquisition procedures.

3 - INDUCTIVE INFERENCE

This section defines induction and abduction. In case of uncertainty, the inductive inference rules given here for induction cannot be combined with the deductive ones without a careful examination. We refer to Collins and Michalski (1986, 1989) for a more complete description.

3.1 - Induction/deduction

We give characterizations and different rules for induction, and describe their relations with the preservation of falsity which opposes neatly induction to deduction.

3.1.1 - Notations and definitions

Let us denote a (inductive or deductive) inference rule by **II–**.
In general, the word *theory* may convey two meanings. The first one, views a theory as a minimal set of axioms and inference rules, the second one views a theory as an exhaustive enumeration of all the theorems it contains. This paper will use a third definition, which becomes necessary when dealing with inductive inferences. We call **theory** a set of axioms and inference rules, and all the consequences that have been drawn "up to now" from the axioms. Let Th_1 be such a theory. Suppose that we now apply an inference process to Th_1, thereby inferencing the new statement B. We will consider $Th_1 \cup B$ as a new theory, called Th_2, and note that by Th_1 **II–** Th_2. In the case where inference is deductive, it is clear that the difference between Th_1 and Th_2 is artificial. In the inductive case, however, Th_2 may be quite different from Th_1, and this point of view becomes very useful.

General definition of **Truth-preserving**. For any inference A **II–** B, **II–** is said to be **truth-preserving** if whenever A is true, then B is also true.
General definition of **Falsity-preserving.** For any inference B **II–** A, **II–** is said to be **falsity-preserving** if whenever B is false, then A is also false.

Induction and deduction. We shall say that **II–** is a **deductive inference**, noted **I–**, if and only if **II–** is truth-preserving. We shall say that **II–** is an **inductive inference**, noted **I–** $_{IND}$, if **II–** is falsity-preserving. In other words, truth-preserving is a necessary and sufficient condition for defining a deduction, while falsity-preserving is simply necessary to define induction. We are well aware that induction is more often defined by: B **I–** $_{IND}$ A if and only if A **I–** B. This new definition is slightly more general than the classical one, and is "humbler" in that sense that it will demand more details about the way inductive steps are actually performed.

Let Th_1 and Th_2 be two theories, we will say that Th_2 can be deduced from Th_1 when, using some deductive inference rules, we will be able to infer Th_2 from Th_1. This is written Th_1 **I–** Th_2. We will say that Th_1 can be induced from Th_2 when, using some inductive inference rules, we will be able to infer Th_1 from Th_2. This is written Th_2 **I–** $_{IND}$ Th_1.

3.1.2. - First definition: Induction of theories from facts, Inventing theories

Given Th_2, **inducing** Th_1 from Th_2, noted Th_2 **I–** $_{IND}$ Th_1, is finding (or "inventing") Th_1 such that Th_1 **I–** Th_2.
This definition is the classical one. It is clearly falsity preserving since, if Th_2 is false, Th_1 being True would contradict Th_1 **I–** Th_2. Since it might well happen that some **I–** $_{IND}$ is falsity preserving but that one cannot deduce Th_2 from Th_1, our new definition is indeed more general than the classical one. Nevertheless, let us again emphasize that our purpose is not generalizing definitions, but showing their consequences in practice.
This definition shows clearly that induction is a kind of reverse for deduction. There is no reason why this inversion should be unique for each deductive inference rule. This explains

why from this definition, we shall be able to provide a characterization only of induction, instead of rules, as the next abstract example shows.

Abstract example: Modus ponens

Let the truth-preserving inference rule allowing to do Th_1 **I**– Th_2 be modus ponens, i.e., [from A and A ⇒ B, infer B]. Then, applying the above definition leads to

C_I : [from B_1, .., B_n, infer A such that A ⇒ B_1&..& B_n].

This condition prescribes what task has to be performed to perform a valid induction of Th_1 from Th_2. This not a rule that can be applied in the same way as modus ponens, since it simply tells that one can choose any A, as long as it respects the implication A ⇒ B_1&..& B_n, but does not give any "rule" to choose A in the first place.

All rules respecting condition C_I are falsity-preserving since, if B_1&..& B_n is false, then (A and A ⇒ B_1&..& B_n) is false. Let us prove this point as follows.

A is either false or true.

If A is false then (A and A ⇒ B_1&..& B_n) is false, and we have proven our point.

If A is true then A ⇒ B_1&..& B_n cannot be true since by hypothesis B_1&..& B_n is false, therefore A ⇒ B_1&..& B_n is false, and we have proven our point.

3.1.3 - Second definition: Abduction and repairing theories

The abduction process is certainly basic to induction, which can be seen (also) as a sequence of abductions, followed by justifications of the abductions. Nevertheless, the process of abduction, is essentially used to "repair" a theory, that is, to modify a theory in which a proof is currently failing so as to insure the future success of this proof. In other words, and that starts justifying our title, abduction-generated induction is more humble than C_I, in that sense that it does not invent theories from facts, it just repairs failing existing theories.

Reminder.

Let us also recall that the classical rule for abduction, as given by Peirce (1934) (see also Addis, 1987), is the "inductive version" of modus ponens, i.e.,

I_0: [from B and (A ⇒ B), infer A].

In other words, for inducing A, we look for some implication of which A is the premise, and look if the conclusion of the implication is true. The difference between C_I and C_A is that in C_A, (A ⇒ B) is supposed to exist before the inference is done. In C_I this implication has to be proven after the induction is done.

Classical abduction and truth/falsity-preserving

If B is true, (A ⇒ B) will hold when A is false. Therefore, one can induce from [B & (A ⇒ B)] = True, something, namely A, which is False. It follows that abduction is not truth-preserving, as one could guess.

Suppose now that B is False, and if (A ⇒ B) is True then A is False. Suppose now that B and (A ⇒ B) are false, then A is True, and classical abduction is not falsity-preserving.

Falsity-preserving abduction (fp-abduction)

As we did in section 3.1.2, instead of giving a simple abduction rule, we shall give a necessary condition for rules to be abduction, called C_A (Condition for Abduction) in the following. The above remark leads us to

$$\mathbf{C_A: [from\ B\ and\ If\ (A \Rightarrow B) = True,\ infer\ A]}$$[1].

It is now trivial to prove that all rules respecting C_A are falsity-preserving.

Abstract example of fp-abduction

In practice, one often uses fp-abduction when, inside a theory Th_2, the proof of a theorem fails (Cox and Pietrzykowski, 1987). This proof was attempting to build in Th_2 a chain of facts, $(F_1, .., F_n, True)$ such that each link F_i is deduced from the preceding link, F_{i-1}. Let us suppose that our proof fails, i.e., we are not able to built the above chain. Let us then suppose that by any means we are able to built a chain $(F_1, .., F_q, F'_q, .., F_n, True)$ but that F'_q is not part of Th_2. We shall then fp-abduct Th_1, such that $Th_1 = Th_2 \& (F_q \Rightarrow F'_q)$. In Th_1 the proof will succeed. In other words, this first use of fp-abduction looks for a "missing" implication in a theory, and adds it.

In this use of fp-abduction, "A" is the implication "$F_q \Rightarrow F'_q$", and "B" is "$\vdash F_1$", which means "one can prove F_1". In other words, this use of fp-abduction contains a meta-statement on the provability of F_1.

3.2 - Refinement of incomplete theories

A great deal of work has been done on the topic of error recovery by explanations, which is quite similar to explanation improvement. The interested reader should consult Carbonell and Gill's, DeJong and Mooney's, and Hirsh' chapters in Kodratoff and Michalski (1989), and also (Hill, 1987; Rajamoney and DeJong, 1987; Mostow and Bhatnagar, 1987).

For the management of incomplete theories, the central idea is to complete the failure proofs by an abduction mechanism guided by analogical reasoning about explanations and enables to discover new rules that refine the domain theory.

We shall illustrate this by the following example drawn from DeJong & Mooney (1986). In this example, our aim is learning a definition of the concept of suicide Kill(x,x). The domain theory TH_0 contains the following rules

```
KILL(a,b)        :-   HATE(a,b), POSSESS(a,c), WEAPON(c)
HATE(w,w)        :-   DEPRESSED(w)
POSSESS(u,v)     :-   BUY(u,v)
WEAPON(z)        :-   SHOT-GUN(z)
```

where a, b, c, w, u, v, z are variables.

The training instance TR1 is a suicide, described by the following facts.

TR1

```
DEPRESSED(JOHN)  :-
```

[1]It is well-known that Peirce made the difference between abduction and induction as: induction infers theories, abduction infers instantiated facts. We clearly do not follow this definition since we make the difference as: induction infers facts or theories and afterwards proves the implication, while abduction checks the implication before inferring.

BUY(JOHN,OBJ1) :-
SHOT-GUN(OBJ1) :-
:- KILL(JOHN, JOHN)

DeJong and Mooney have shown that applying EBL to the training instance leads to build the following generalized proof of the fact that John has been committing suicide, that we shall from now on call S (Source) in which x and c are variables.

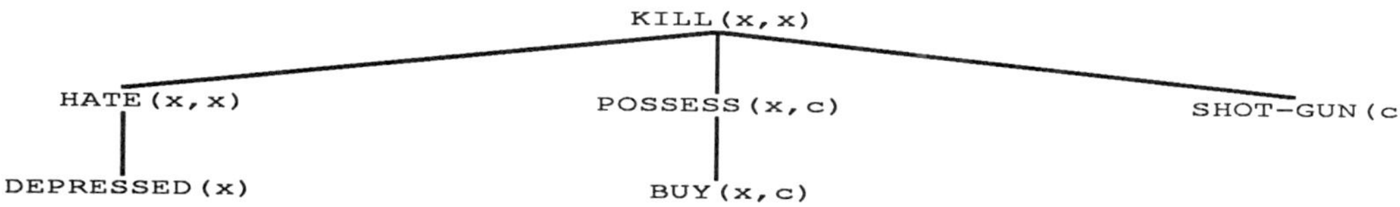

Figure 1. Generalized proof that John has been committing suicide

This produces the rule
KILL(x,x) :- DEPRESSED(x), BUY(x,c), SHOT-GUN(c)
which is "added" to TH_0 in order to define the concept C of suicide. Nothing is actually added since this new rule is a consequence of TH_0. It happens that this rule seems to be more operational to define suicide that applying KILL(a, b) with a=b.
Suppose now that the system is provided with an example TR2 of concept C which is not recognized by the theory. In other words, some oracle will have been telling the system that TR2 does belong to the suicide case, but the PROLOG solver failed to prove it. Nevertheless, it may happen (this is the case we are actually considering here) that partial proofs might be obtained. They will provide partial proof trees that may match parts of S. Let us call Ti (Target incomplete) incomplete proof trees that can be viewed as partial explanations as explained with more details in (Duval & Kodratoff 1989). Our aim will then be to complete each of these Ti into Tc (Target complete) in such a way that S and Tc are as "close" as possible.

As an illustration of such a Ti, consider an other suicide instance TR2 described by the following facts.
TR2
DEPRESSED(MARY) :-
BUY(MARY,OBJ1) :-
SLEEPING-PILLS(OBJ1) :-
PRICE(SLEEPING-PILLS, 6) :-
BUY(MARY, OBJ2) :-
BOOK(OBJ2) :-
PRICE(BOOK, 5) :-
. . .
where OBJ1 and OBJ2 are constants.
We will be unable to prove KILL(MARY, MARY) because she has no shot-gun. Nevertheless we obtain one, and only one in this case, partial proof:

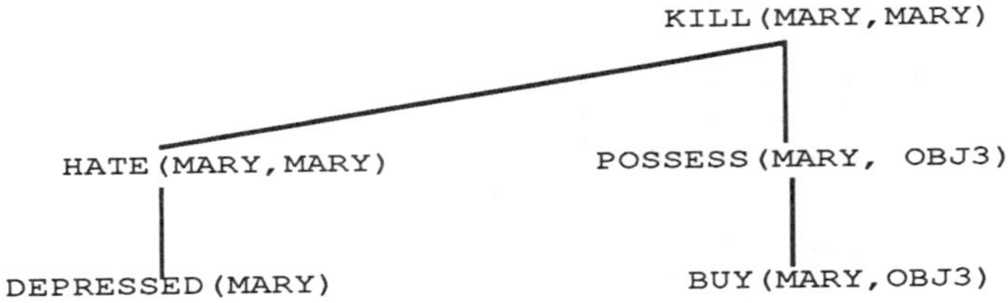

Figure 2. Ti: Partial explanation of Mary's suicide

Ti obviously matches a sub-tree of S.

In order to achieve the completion of Ti into Tc, and to achieve it so that Tc is "close" to S, we propose to use two different abductions. In general, the mechanism underlying this abduction process is analogy. We shall not speak of analogy here, see (Duval and Kodratoff, 1989) for more details.

First abduction

We attempt to complete Ti by viewing it as an instance of S (this is one of the possible definitions of "closeness"). Therefore, Ti will be completed by taking the missing pieces from S, appropriately instantiated.
In our example, such a forced matching leads to Tc1

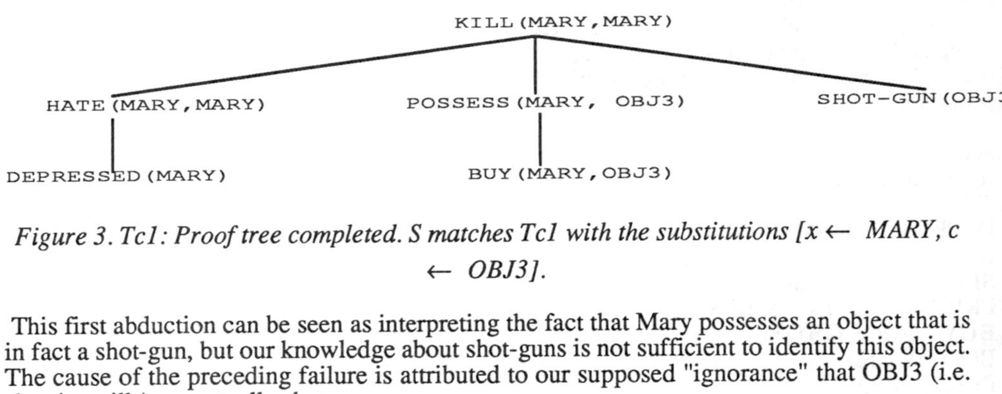

Figure 3. Tc1: Proof tree completed. S matches Tc1 with the substitutions [x ← MARY, c ← OBJ3].

This first abduction can be seen as interpreting the fact that Mary possesses an object that is in fact a shot-gun, but our knowledge about shot-guns is not sufficient to identify this object. The cause of the preceding failure is attributed to our supposed "ignorance" that OBJ3 (i.e. sleeping pills) are actually shot-guns.
Notice that this first mechanism is the only one considered in other works about abduction such as Cox and Pietrzykowski (1987), and note how easily it can lead to absurd results, as here.

Second abduction

In this case, we try to add a new rule that will allow to complete the proof. One easily understand why this mechanism has not been taken into account so far; in principle one can add so many ridiculous rules that this approach seems to be hopeless.
For instance, in our example adding to TH_0 the rule
KILL(x,x) :- PRICE(SLEEPING-PILLS, 6)

vill indeed allow to prove Mary's suicide. But it means that everyone will suicide when the price of sleeping pills reaches the value 6, which is totally irrelevant to the preceding suicide case, notwithstanding its strange meaning.

n order to avoid adding such ridiculous rules as above, we define a new notion of distance between S and Tc. Given two possible completed proof trees, Tc and Tc', consider an attempt to match S and Tc, and S and Tc'. In this second abduction, we do not consider the case where they match (this would be the first kind of abduction). We shall therefore collect he mismatches between S and Tc on the one hand, and between S and Tc' on the other hand. We shall say that that Tc is closer to S than Tc if the number of mismatches between Tc and S s less than the number of mismatches between Tc' and S. When they are equal, we shall say hat Tc is closer to S than Tc' when the conceptual distance (supposedly defined) between the nismatches is less for Tc than for Tc'.

n our example, it is clear that the number of mismatches between S and the proof tree obtained by using

KILL(x,x) :- PRICE(SLEEPING-PILLS, 6)

o prove Mary's suicide is very high.

On the contrary, completing Ti by SLEEPING-PILL(OBJ3), obtaining thus the proof tree Tc2

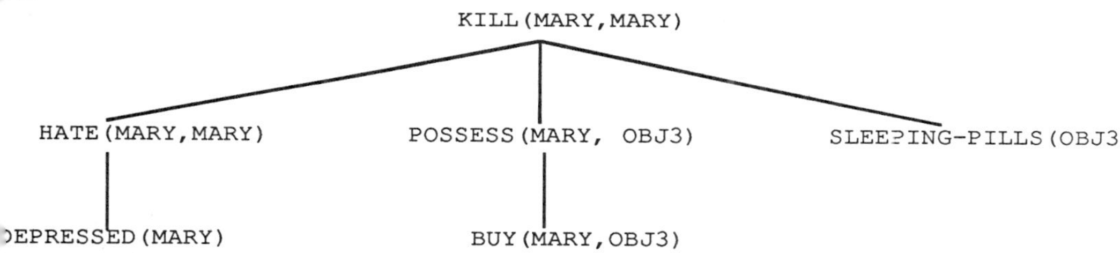

Figure 4. Sleeping-pills are viewed as the cause of Mary's death.

eads to one mismatch only, namely [SHOT-GUN(OBJ3) ← SLEEPING-PILLS(OBJ3)]. Notice that we speak here of mismatch since SHOT-GUN(OBJ3) is not a variable. Applying EBG to this proof leads to add the new rule

KILL(x,x) :- DEPRESSED(x), BUY(x,c), SLEEPING-PILLS(c)

o TH_0.

One can complete it as well by BOOK(OBJ4), obtaining the proof tree Tc3

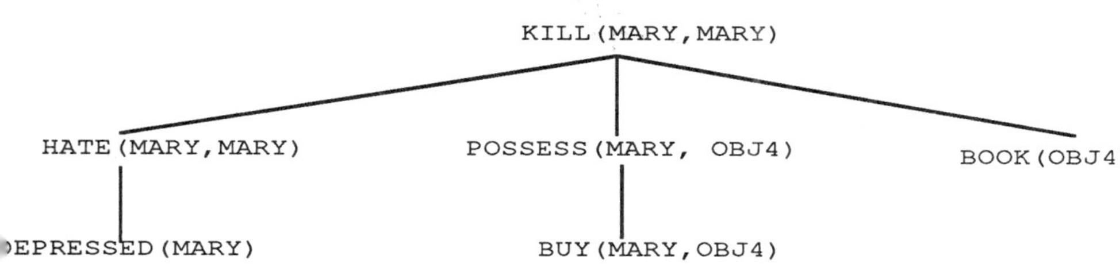

Figure 5. Books are viewed as the cause of Mary's death.

Applying EBG to this proof leads to add the new rule

KILL(x,x) :- DEPRESSED(x), BUY(x,c), BOOK(c)

o TH_0.

From our semantic knowledge, it is clear[1] that Tc2 is the good answer, but this result is not so easy to obtain. It needs to have some ways to prove that sleeping pills are more likely to kill someone than books. Proving this would ask a large amount of extra-knowledge not contained in TH(). Anyhow, we simply suppose here that one has been able to set up some way of evaluating the semantic distance between SHOT-GUN and SLEEPING-PILLS, and between SHOT-GUN and BOOK, and that the first is less than the second. Then, we would choose Tc3 as other possible abduction.

3.2 - Application of numeric techniques

We will not speak presently of the case of uncertain abductions. One should again follow very carefully what have been the logical steps for finding a conclusion.
On the other hand, two numeric techniques will be necessary for an effective implementation of the above ideas. The first one is the way to measure the semantic distance between concepts, which can well make use of a large number of numeric features. The second one is to choose between the results of the above two kinds of abductions. A a simple hint, let us point out that the new rule should be checked against other other knowledge of different granularity, for instance finding if shot-guns and sleeping-pills are very "distant" or not.
Further work is needed to explore these possibilities. In this paper, we have been concentrating on showing how and why these possibilities exist, rather than making them explicit, which can be done in a variety of ways.

4 - COMBINING SIMILARITY AND CAUSALITY IN ANALOGY

In Gentner's Structure-mapping theory (Gentner, 1983), "the central intuition is that an analogy is the mapping of knowledge from one domain (the base) into another (the target) which conveys that a system of relations known to hold in the base also holds in the target" (Falkenhainer, Forbus, and Gentner, 1986). This describes analogy as made of two different components, namely **similarity** and internal dependency, the last one often amounts to **causality**. We stress that analogy is made of two phases. The first phase is analytic: similarity and causality are carefully separated in this first phase of the analogy. The second phase is synthetic, similarity and causality are then combined again to draw the analogy. We shall be very brief on the numeric implications of our scheme since they were largely illustrated already in section 2 and 3. Similarly to section 2, our scheme for analogy shows that one should carefully avoid mixing up beliefs for causality and beliefs for similarity. As in section 3, our scheme for analogy shows that one should use numeric information in order to measure similarity.

We shall mainly use three different examples to illustrate the concepts introduced. They are inspired from Winston's Macbeth scheme (Winston, 1982), from Carbonell's transformational analogy (Carbonell, 1983, 1986), and from Hofstadter & Mitchell's COPYCAT (Hofstadter and Mitchell, 1988; Mitchell and Hofstadter, 1989).
The analogy scheme we shall use is described as follows. Let us suppose that we dispose of a piece of information, the **base**, that can be put into the form of a doublet (A, B) in which it is known that B depends on A. This dependency will often be causal, and it does not need to be very formal nor strict. In the following, we shall call this relation β, and refer at it as the

[1]Consider however that a French writer published recently a book entitled "A cheap guide to happy suicide". He is regularly sued by families of suicidees arguing that his book is the cause of the suicidee's death, and the families win always their case.

causality of the analogy. Suppose now that we find an other piece of information, the **target**, (A', B') that can be put into the same form, and such that there exists some resemblance (similarity) between A and A'. In the following, we shall call this relation α, and refer at it as the **similarity** of the analogy. Let us call β' the causal dependency between A' and B', and α' the similarity between B and B', as shown in the figure below.

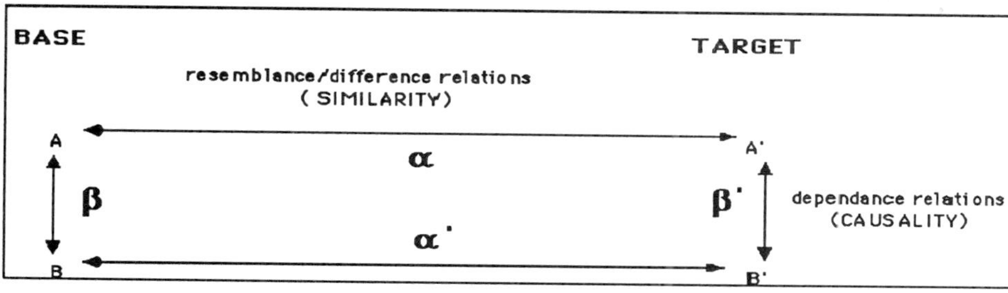

Figure 6. The general scheme of analogy.

The analogy problem amounts to find a missing part of this diagram, knowing some others of its parts. For instance, we shall see that, in Winston's view of analogy, A, A', and B' are known, α is not given in advance but it is a partial matching of A and A', B is the result of this partial matching, β are the causalities inside B, and one assumes that β' = β. In other words, we present Winston's view of analogy as a progressive completion of the diagram above.

The most frequently met opinion about what analogy is about, is the one presented in COPYCAT. In this kind of analogy, one tries to "invent" a B' that will fit with A, B and A'. More precisely, COPYCAT is a system that can give answers to question of the like "If *abc* changes to *abd*, then what does *pqrs* change to? Let us call the couple *(abc, abd)* the base, then the function that transforms *abc* into *abd* is the "causality" β. The problem is then to find a β' that one can apply to the partially known target, here *pqrs*, in order to obtain the looked for answer, B'. Their view of analogy can be summarized by the following scheme.

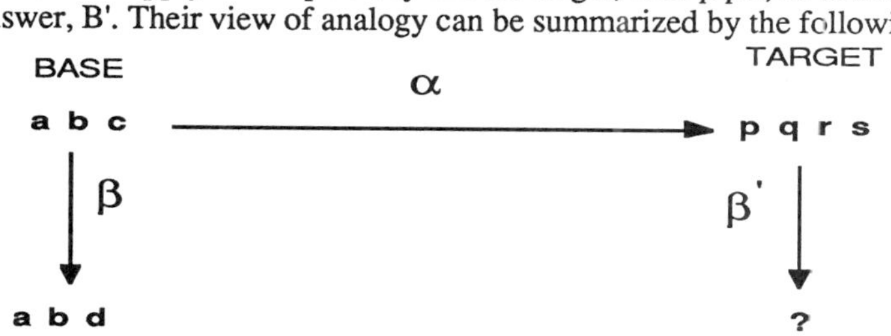

Figure 7. Hofstadter and Mitchell's scheme of analogy.

Winston's view of analogy is more oriented towards the discovery of possible causalities, i.e., the invention or retrieval of a relevant β'. In (Winston, 1982), one is given a text A (e.g., *Macbeth' story*), an exercise A' (e.g., *a short story between two other individuals, say, Tom and Jane*), and a question about A', B' (e.g., *Is Tom likely to be evil?*). The analogy problem is to answer B' (and to justify that answer), by considering the causal relationships in A. The similarity between A and A' is defined as their partial matching, i.e.,

α is the partial matching of A and A'. For instance, let us suppose that *Tom and Jane are married, Tom is weak, and Jane is greedy*. This matches partially Macbeth' story, since the same links hold between Macbeth and Lady-Macbeth. The part of A that matches A' is B (e.g., *Macbeth and Lady-Macbeth are married, Macbeth is weak, and Lady-Macbeth is greedy*), and the causality links inside B form the β of our scheme. Suppose that in Macbeth' story one assumes that *his weakness and his wife's greediness are cause of his evil*. One then postulates that the same causality links hold between A' and B', i.e., $\beta = \beta'$. If these causality links are relative to the question B' (e.g., here, the causality links are relative to an explanation of Macbeth' evil), then one answers 'yes' to B', using these causality links (e.g., "*Yes, Tom may be evil, because of his weakness, and his wife's greediness*").
In conclusion, we can summarize Winston's approach by the following scheme.

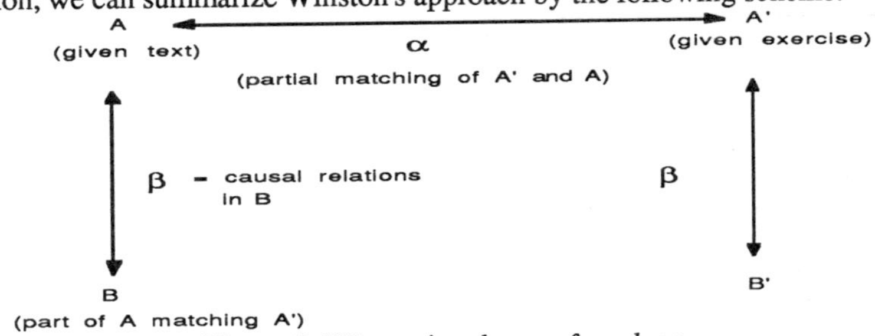

Figure 8. Winston's scheme of analogy.

In (Carbonell, 1983, 1986), the goal is also the invention of β', but the means are very different. One is given a completely solved problem that can be decomposed into its means A, and its goals B. In the following we shall elaborate an example coming from (DeJong and Mooney, 1986). The means will be *abducting a rich person's child*, and the goals be *kidnapper becomes rich*. We are given a plan (the one of DeJong and Mooney, 1986) that solves the problem of achieving goal B by means A. This plan gives the cause why it is possible to achieve goal B by means A, it will therefore seen here as the causality β linking A and B. One is also given an other problem, i.e., goals B' and means A' (e.g., A' may be: *abducting a famous politician*, and B' may be: *terrorists advertise their political cause*). The analogical problem is then to use the plan β in order to invent a new plan β' that will achieve B' by A' (e.g., *how terrorists can get advertisement by abducting a famous politician?*). In short, Carbonell gives A, B, A', B', β, and asks analogy to provide reasonable β'. He describes sets of possible transforms that can be applied to b in order to generate candidate β'.
In conclusion, one can summarize Carbonell's approach by the following scheme.

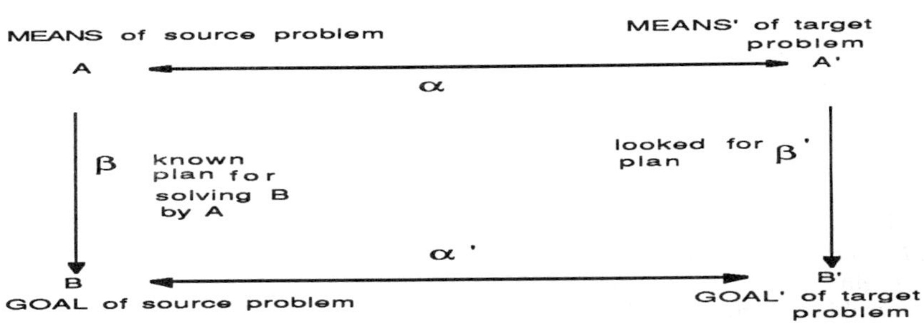

Figure 9. Carbonell's scheme of analogy.

In this case, one knows A, B, A', B', and β. The analogy problem consists in finding α and β' that fit together. The similarity α does not need, as in Winston's, to be partial matching.

The analysis of these three examples of non trivial analogies shows very clearly that in all cases, one is looking for β', the causality in the target that will allow to "understand" and draw the analogy. In COPYCAT, this is necessary for finding B', in Winston's approach one retrieves a $\beta = \beta'$, and in Carbonell's the looked for plan is β'. In short, these approaches illustrate the importance of causality in drawing meaningful analogies.

4 - GENERAL CONCLUSION

This paper shows that AI systems usually perform three kinds of inferences, namely deductive, inductive or analogical inferences. A careful analysis of each of them shows that numeric computation are necessary for practical applications, either for combining uncertainties, or for choosing among several possible inferences. It shows also that the structure of the inferences should be conserved by all numeric techniques.

REFERENCES

Addis T. R., "A Framework for Knowledge Elicitation", Proc. 1st European Knowledge Acquisition Workshop (EKAW-87), Sept. 87, Addis, Boose & Gaines eds, Reading Univ. 1987.

Carbonell J.G., "Learning by Analogy: Formulating and Generalizing Plans from Past Experience" in R.S. Michalski, J. G. Carbonell, T. M. Mitchell (Eds.), *Machine Learning: An Artificial Intelligence Approach,* Morgan Kaufmann 1983, pp. 137-159.

Carbonell J.G., "Derivational Analogy: A Theory of Reconstructive Problem Solving and Expertise Acquisition", in R.S. Michalski, J. G. Carbonell, T. M. Mitchell (Eds.), *Machine Learning: An Artificial Intelligence Approach, Volume II*, Morgan Kaufmann 1986, pp. 371-392.

Clancey W. "The Epistemology of a Rule-Based Expert System: A Framework for Explanation", *Artificial Intelligence 20*, 1983, 215-251.

Collins A., Michalski, R. S., "The Logic of Plausible Reasoning: An Advanced Report," *ISG Reports*, No. 86-3, UIUCDCS-F-86-951, Department of Computer Science, University of Illinois, Urbana, February 1986.

Collins A., Michalski R. S. "The Logic of Plausible Reasoning: A Core Theory", to appear in *Cognitive Science*, 1989.

Cox, P.T., Pietrzykowski, T., "General Diagnosis by Abductive Inference," Proc. 1987 Symposium on Logic Programming, San Francisco, CA, IEEE, 1987, pp. 183-189.

DeJong G., Mooney R., Explanation-Based Learning: An alternative View. *Machine Learning,1*, p.145-176. Kluwer Academic Publishers.

Duval B., Kodratoff Y. "A Tool for the Management of Incomplete Theories: Reasoning about explanations" to appear in *Machine Learning, Meta-Reasoning, Logic,* P. Brazdil eds, Kluwer, Boston, 1989.

Duval B., Kodratoff Y., "Automated Deduction in an Uncertain and Inconsistent Data Basis", Proc. ECAI-86, Brighton 1986, pp. 101-108.

Falkenhainer, B., Forbus, K. D., Gentner, D., "The Structure-Mapping Engine", Report N° UIUCDCS-R-86-1275, DCS, Univ. of Illinois at Urbana-Champaign, May 1986. See also Proc. AAAI-86.

Ganascia, J. G., Kodratoff, Y., "Symbolic Uncertain Inference: A Study of possible Modalitites", Expert Systems 85, Martin Merry (Ed), Cambridge University Press 1985, pp. 169-179.

Gentner, D., "Structure-Mapping: A theoretical Framework for Analogy", Cognitive Science 7, pp. 155-170, 1983.

Hill, W. L., "Machine Learning for Software Reuse", Proc. IJCAI-87, pp. 338 - 344, John McDermott editor, Morgan Kaufmann, Los Altos, Ca., 1987.

Hofstadter, D. R., Mitchell M., "Concepts, Analogy, and Creativity", Proc. Canadian Soc. for Computational Intelligence, Edmonton, Univ. of Alberta, 1988.

Kodratoff Y. *Introduction to Machine Learning*, Pitman Publisher 1988.

Kodratoff Y., Manago M., Blythe J. : "Generalization and Noise" *Int. J. Man-Machine Studies* (1987) 27, 181-204.

Kodratoff, Y., Michalski R. S., (eds.), Machine Learning: An Artificial Intelligence Approach, Volume III, Morgan Kaufmann Publishers, 1989 (to appear).

Kodratoff, Y., Perdrix, H., Franova, M., "Traitement symbolique du raisonnement incertain", Actes Congrès AFCET *Matériels et Logiciels pour la 5ème génération*, Paris, March 1985, pp. 33-45.

Lebbe J., Vignes R., Darmoni S. "Objets symboliques qualifiés, application médicale", Actes 2èmes Journées Symbolique/Numérique, Orsay, Dec. 88. An English version is found in the Proceedings of this conference.

Manago M., Kodratoff Y. : "Noise and Knowledge Acquisition", Proc. IJCAI-87, Milan Aug. 87, pp. 348-354.

Mitchell M., Hofstadter, D. R., "Conceptual Slippage and Analogy-Making: A Report on the Copycat Project", Proc. 10th Annual Conf. of the Cognitive Science Soc. Hillsdale, NJ: Lawrence Erlbaum Associates.

Mitchell T., Keller R., Kedar-Cabelli S. Explanation-Based Generalization: A Unifying View. *Machine Learning Journal,1,* p. 47-80., Kluwer Academic Publishers.

Mostow J., Bhatnagar N., "Failsafe - A Floor Planner that Uses EBG to Learn from its Failures", Proc. IJCAI-87, pp. 249 - 255, John McDermott (ed), Morgan Kaufmann, Los Altos, Ca., 1987.

Peirce, C. S., *Collected Papers of Charles Saunders Peirce*, A.W. Burks (Ed), 1934

Rajamoney S., DeJong G., "The Classification, Detection and Handling of Imperfect Theory Problems", Proc. IJCAI-87, pp. 205 - 207, John McDermott (ed), Morgan Kaufmann, Los Altos, Ca, 1987.

Shortliffe E. H. Computer-based medical consultations: MYCIN, Elsevier, New York, 1976.

Winston P.H. "Learning New Principles From Precedents and Exercises", *Artificial Intelligence 19,* (1982), 321-350.

KNOWLEDGE ACQUISITION BY METHODS OF FORMAL CONCEPT ANALYSIS

Rudolf Wille

1. Formal Representation of Conceptual Knowledge

In this paper we restrict our considerations to conceptual knowledge and its representation by methods of formal concept analysis. In particular, we use the set-theoretical model for concepts and conceptual hierarchies as it is introduced in [Wi82] (see also [Wi84] and [GW90]). Let us recall that our model mathematizes the philosophical understanding of a concept as a unit of thoughts consisting of two parts: the extension and the intension (comprehension); the extension covers all objects (or entities) belonging to the concept while the intension comprises all attributes (or properties) valid for all those objects (cf. [Wa73]). The relationship between objects and attributes which is constitutive for a concept, is modelled by the basic notion of formal concept analysis, the *(formal) context*: it is defined as a triple (G, M, I) where G and M are sets and I is a relation between G and M; the elements of G and M are called *objects* and *attributes*, respectively, and gIm (i.e. $(g,m) \in I$) is read "the object g has the attribute m". In a context (G, M, I), the relation I between the object set G and the attribute set M gives rise to a *Galois connection* (cf. [BM70]) formed by the following two operators:

$$\begin{array}{lll} A \mapsto A' := \{m \in M | gIm \text{ for all } g \in A\} & \text{for } A \subseteq G, \\ B \mapsto B' := \{g \in G | gIm \text{ for all } m \in B\} & \text{for } B \subseteq M. \end{array}$$

Now, a *(formal) concept* of the context (G, M, I) can be defined as a pair (A, B) with $A \subseteq G$, $B \subseteq M$, $A' = B$, and $B' = A$; A and B are called the *extent* and the *intent* of the concept (A, B), respectively. The hierarchical relation *"subconcept-superconcept"* has to be defined for a context by $(A_1, B_1) \leq (A_2, B_2) :\Longleftrightarrow A_1 \subseteq A_2$ ($\Longleftrightarrow B_1 \supseteq B_2$). The set of all concepts of (G, M, I) with this order relation is a complete lattice called the *concept lattice* of (G, M, I) and denoted by $\mathfrak{B}(G, M, I)$. In this lattice, a family (A_t, B_t) $(t \in T)$ of concepts has as its *infimum* the concept $(\bigcap_{t \in T} A_t, (\bigcup_{t \in T} B_t)'')$ and as its *supremum* the concept $((\bigcup_{t \in T} A_t)'', \bigcap_{t \in T} B_t)$.

A formal context may be considered as an elementary model for knowledge representation. For example, fig.1 shows a context which contains the basic knowledge underlying an educational film on "living beings and water" (see [BGMT79], [Ta84]); learning of this knowledge was tested in such a way that children were asked to reproduce the context after seeing the film. The context and its concept lattice, which is represented by a line diagram in fig.2, were also used for planning and designing the film. In general, formal contexts with their concept lattices are useful tools for representing knowledge. They allow a great variety of evaluations of the represented knowledge; in particular, the structural insight via the concept lattices may clarify

	a	b	c	d	e	f	g	h	i
leech	×	×					×		
bream	×	×					×	×	
frog	×	×	×				×	×	
dog	×		×				×	×	×
spike-weed	×	×		×		×			
reed	×	×	×	×		×			
been	×		×	×	×				
maize	×		×	×		×			

a: needs water to live, b: lives in water, c: lives on land, d: needs chlorophyll to prepare food, e: two little leaves grow on germinating, f: one little leave grows on germinating, g: can move about, h: has limbs, i: suckels its offsprings.

Figure 1: Context of an educational film "Living Beings and Water"

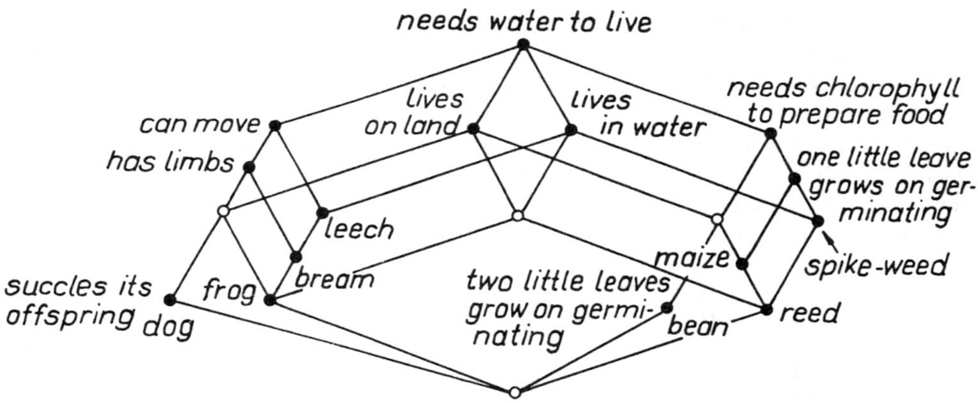

Figure 2: The concept lattice of the context in fig.1

connections and interpretations (cf. [Wi87]). Representing concept lattices by diagrams is an effective method for communicating knowledge and establishing a high level of transparency (cf. [Wi84], [Wi89]); various connections and interpretations of the knowledge may be visualized by different drawings of a concept lattice. For formal representation of conceptual knowledge one can also use many-valued contexts and their concept lattices derived from conceptual scaling (cf. [GSW86], [GW89]); but for our discussion of knowledge acquisition we consider in this paper only formal contexts as they are defined above.

The central idea of knowledge acquisition in the frame of formal concept analysis lies in the assumption that in the field of exploration the conceptual knowledge can be thougt to be represented by a formal context and its concept lattice; such a context is called a *universe* and denoted by $\mathsf{U} := (G_{\mathsf{U}}, M_{\mathsf{U}}, I_{\mathsf{U}})$. The exploration of knowledge starts with some partial information about the universe U and acquires more information by phrasing questions which are answered by experts. For this procedure it is a main concern not to ask questions which can already be answered by the acquired knowledge. In the following two sections we discuss two types of explorations: one starts with a set of attributes, the other with a set of concepts of the universe.

2. Attribute Exploration

It seems the best to explain the *attribute exploration* by an example. In the summer term 1988, the students of my course on formal concept analysis at the TH Darmstadt had to refresh their knowlegde in lattice theory. For this we used the attribute exploration as it is programmed in [Bu87]. For the universe we chose as objects all finite lattices (up to isomorphism) and as attributes the lattice properties from the index in [Bi67] which are listed in fig.3.

To start the exploration program, one has to input the names of the attributes; furthermore, the program allows to give as information specific objects g together with the attribute sets $\{g\}'$ (formed in the universe). As such an information, the students of my course produced the first four rows of the table in fig.3. Then we started the interactive procedure of questions and answers. With the given information the program phrases the following first question concerning finite lattices:

> Does the property "uniquely complemented" imply all the other properties?

The answer "yes" (see [Bi67], p.122) must be given back to the program. The second question is then:

> Does the property "Stone lattice" imply the properties "Brouwerian", "distributive", "dually semimodular", "graded", "metric", "modular" and "semimodular"?

Again the answer is "yes" (see [Bi67], p.130f). The third question of the program is:

> Does the property "relatively complemented" imply the properties "atomistic", "complemented", "graded" and "sectionally complemented"?

After getting the answer "no", the program asks for a finite lattice (an object of the universe) which fulfills the premise of the implication in the third question, but does not satisfy at least one property of its conclusion. For this my students proposed the fifth lattice in fig.3 which is not graded. After the input of the fifth row in the table of fig.3, the program rephrases the third question as follows:

> Does the property "relatively complemented" imply the properties "atomistic", "complemented" and "sectionally complemented"?

	a	b	c	d	e	f	g	h	i	j	k	l	m	n	o
1				×											
2	×			×				×			×		×		
3*			×		×	×		×	×	×		×		×	
4*	×			×		×	×	×	×	×	×	×	×		
5*	×			×							×		×		
6								×				×			
7	×			×				×			×				
8*	×			×			×	×			×	×	×		
9*	×					×		×							
10*			×		×	×		×	×	×		×			
11*	×			×				×							
12				×				×							
13*				×				×				×			
14*	×			×		×		×			×				
15*	×			×		×		×			×		×		

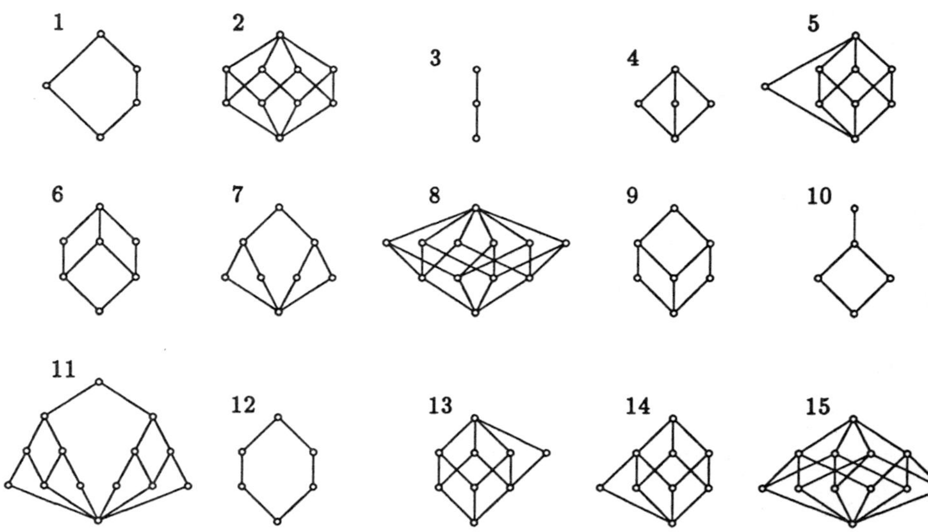

a: atomistic, b: Boolean, c: Brouwerian, d: complemented, e: distributive, f: dually semimodular, g: geometric, h: graded, i: metric, j: modular, k: sectionally complemented, l: semimodular, m: relatively complemented, n: Stone lattice, o: uniquely complemented.

Figure 3: The resulting context of an attribute exploration of finite lattices

```
 1 :   unco   ==>   atom   Bool   Brou   comp   dist   duse
              geom  grad   metr   modu   seco   semo   reco
              Ston
 2 :   Ston   ==>   Brou   dist   duse   grad   metr   modu
              semo
 3 :   reco   ==>   atom   comp   seco
 4 :   semo   ==>   grad
 5 :   seco   ==>   atom   comp
 6 :   modu   ==>   duse   grad   metr   semo
 7 :   metr   ==>   duse   grad   modu   semo
 8 :   geom   ==>   atom   comp   grad   seco   semo   reco
 9 :   duse   ==>   grad
10 :   duse   grad  semo   ==>    metr   modu
11 :   dist   ==>   Brou   duse   grad   metr   modu   semo
12 :   comp   duse  grad   ==>    atom   seco
13 :   Brou   ==>   dist   duse   grad   metr   modu   semo
14 :   Bool   ==>   atom   Brou   comp   dist   duse   geom
              grad  metr   modu   seco   semo   reco   Ston
              unco
15 :   atom   grad  semo   ==>    comp   geom   seco   reco
16 :   atom   Brou  comp   dist   duse   geom   grad   metr
              modu  seco   semo   reco   ==>    Bool   Ston
              unco
```

Figure 4: Basis of the implications in the context of fig.3

Now, the answer is "yes" (see [Bi67], p.17). The attribute exploration continues in this way in accepting implications and producing counterexamples until all implications between the attributes can be determined by the information given to the program. In my course, the students ended their work after several hours with the context in fig.3 and the list of implications in fig.4 which is a basis for all valid implications in the chosen universe. Fig.5 shows the concept lattice of the context in fig.3 which is isomorphic to the concept lattice of the universe. The lattice has only ten join-irreducible elements so that ten counterexamples would have been enough; the ten objects g, for which $\gamma g := (\{g\}'', \{g\}')$ is join-irreducible, are marked with * in fig.3, and in fig.5 the line diagram of the lattice g is attached to the little circle which represents the concept γg. The name of an attribute m occurs in fig.5 at the little circle of the concept $\mu m := (\{m\}', \{m\}'')$. This convention allows to read all valid implications from the line diagram: for instance, a complemented modular finite lattice can be seen to be always geometric because the infimum of $\mu(\textit{complemented})$ and $\mu(\textit{modular})$ is a subconcept of $\mu(\textit{geometric})$.

Let us shortly sketch the mathematical background of the attribute exploration. In a context (G, M, I) an implication $P \rightarrow C$ with $P, C \subseteq M$ is valid if $P' \subseteq C'$, i.e., each object having all attributes of P has also all attributes of C. V.Duquenne and J.-L.Guigues have shown that a basis for all implications valid in (G, M, I) is given by $\{P \rightarrow (P'' \setminus P) | P \textit{ pseudo-intent}\}$ where a pseudo-intent is recursively defined as follows: a set P of attributes is a *pseudo-intent* in (G, M, I) if $P \neq P''$ and $Q'' \subseteq P$ for all pseudo-intents Q with $Q \subset P$ (see [DG86], [Ga87], [Ma83]). B.Ganter has de-

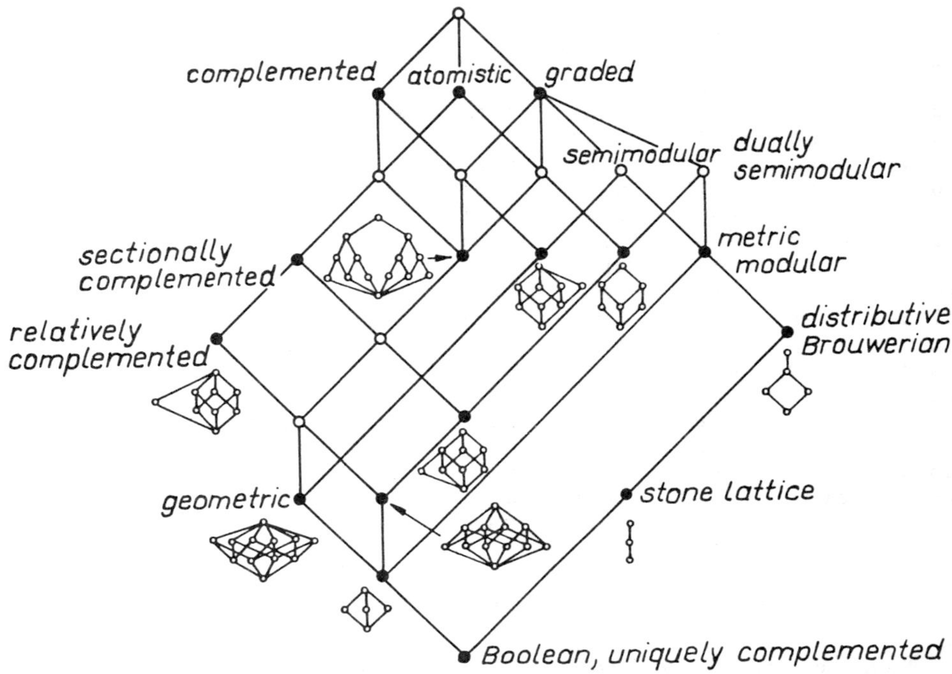

Figure 5: The concept lattice of the context in fig.3

scribed an algorithm to determine lexicographically all pseudo-intents; the algorithm is based on the fact that the intents together with the pseudo-intents form a closure system (see [Ga87]). If the context (G, M, I) is the start information for the attribute exploration (G might even be the empty set), the first question is phrased with the lexicographically first pseudo-intent P_1 of (G, M, I):

Does P_1 imply $P_1'' \setminus P_1$ in the universe?

In the sequel the program lexicographically constructs further pseudo-intents always with respect to the actual context which might have been grown by counterexamples. If the question formulated with such a pseudo-intent P has a positive answer, P is also a pseudo-intent of the universe. The procedure stops if the accepted implications form the basis of Duquenne and Guigues for all implications valid in the universe. Recently, P.Burmeister has extended the attribute exploration to universes representing only incomplete knowledge (see [Bu89]).

The attribute exploration is a powerful tool for examining a restricted number of attributes with respect to a fixed set of objects. Its procedure is optimal in such a way that there is no question which can already be answered with the knowledge acquired by the other questions. The acquisition of the knowledge, which is represented in the resulting context, often requires intense efforts from the answering experts so that the exploration may uncover deeper insights into the accessable knowledge. The attribute exploration with its interactive procedure has turned out to be quite stimulating; in particular, it has been proven successful for learning and repeating knowledge (and even for establishing selfconfidence in the ability of understanding which was a main effect in my course on formal concept analysis). Of course, the attribute exploration may be applied in almost every area of knowlege. For instance, we have used it in content analysis to explore the conceptual structure on the categories of a fixed dimension; here we understood the questions of the program in such a way that they ask dually how the given categories can be separated by specific attributes.

A concrete example shall show the result of the attribute exploration in a non-mathematical field. In the summer term 1987, we discussed in the music seminar of the philosophical institute at the TH Darmstadt how one can verbally describe musical pieces. Besides other literature we studied the report [Ba76] on an experiment which has examined how musicians assign to presented musical compositions characteristic adjectives chosen from a given list. Since the experiment did not convince us because of the selection of musical pieces, we carried out the attribute exploration with the eleven most discriminating adjectives of the report. After four hours of intense work, answering always by consensus, the result was the context in fig.6 (marks indicating the 20 irreducible objects) and the list of accepted implications in fig.7; the concept lattice is shown in fig.8 by a graded line diagram (cf. [Wi84]). Although the choice of musical pieces was completely free, one might criticize that the set of adjectives was fixed for the exploration. To overcome this more frequently mentioned argument against a fixed set of attributes, an exploration has been developed in which not only the set of objects but also the set of attributes may increase; this so-called "concept exploration" is discussed in the next section.

	abgerundet	ausgewogen	dramatisch	durchschaubar	durchstrukturiert	kräftig	lebhaft	munter	rhythmisierend	schnell	spielerisch
1. Beethoven: Romanze für Violine und Orchester F-dur	×	×		×	×						×
2. Bach: Contrapunctus I	×	×		×	×	×					
3. Tschaikowsky: Klavierkonzert b-moll, 1. Satz			×			×					
•4. Mahler: 2. Sinfonie, 2. Satz					×		×	×	×	×	×
•5. Bartok: Konzert für Orchester						×	×	×	×		×
6. Beethoven: 9. Sinfonie, 4. Satz, Presto			×			×			×	×	
•7. Bach: WTK 1, Präludium c-moll			×		×	×	×			×	
8. Bach: 3. Brandenburgisches Konzert, 3. Satz	×	×		×	×		×	×		×	×
9. Ligeti: Continuum				×			×			×	×
•10. Mahler: 9. Sinfonie, 2. Satz (Ländler)					×	×	×		×		
•11. Beethoven: Mondscheinsonate, 3. Satz			×	×		×	×		×	×	
•12. Hindemith: Kammermusik Nr.1, Finale				×		×	×	×	×	×	×
•13. Bizet: Arlesienne Suite	×	×		×	×	×			×		×
•14. Mozart: Ouvertüre zu Figaro	×	×		×	×	×	×	×		×	×
•15. Schubert: Wandererfantasie			×			×	×		×	×	
16. Beethoven: Frühlingssonate, 1. Satz	×	×		×	×		×	×			×
17. Bach: WTK 1, Fuge c-moll	×	×		×	×		×	×	×		×
•18. Schostakowitsch: 15. Sinfonie, 1. Satz					×	×	×	×			×
•19. Wagner: Vorspiel zu den Meistersingern	×	×		×	×	×	×	×	×		×
•20. Beethoven: Streichquartett op. 131, Schlußsatz			×	×	×	×			×	×	
•21. Joh. Strauß: Frühlingsstimmenwalzer	×	×		×	×		×	×	×	×	×
•22. Mozart: "Oh, wie will ich triumphieren" (Entführung)			×	×	×	×	×			×	×
23. Bach: Matthäus Passion, Nr. 5 (Chor)			×	×	×	×	×				
•24. Brahms: Intermezzo op. 117, Nr. 2	×	×		×	×	×	×		×	×	
•25. Wagner: Walkürenritt			×	×	×	×	×		×		
26. Mozart: "Der Hölle Rache tobt ..." (Zauberflöte)			×	×	×	×	×			×	
•27. Mendelsohn: 4. Sinfonie (Ital.), 4. Satz (Saltarello)	×	×		×	×	×	×	×	×	×	×
•28. Brahms: 4. Sinfonie, 4. Satz	×	×	×	×	×	×					
•29. Beethoven: Große Fuge op. 133	×	×	×	×	×	×	×		×		
30. Goretzky: Klageliedersinfonie	×	×	×	×	×	×	×				
•31. Verdi: Requiem, Dies Irae	×	×	×	×	×	×	×			×	

Figure 6: The resulting context of an attribute exploration of musical compositions

```
 1  schnel spiele   ==>  lebhaf
 2  munter   ==>  lebhaf spiele
 3  lebhaf rhythm spiele   ==>  munter
 4  kraeft lebhaf schnel spiele   ==>  dschau
 5  dstruk kraeft rhythm schnel   ==>  dschau
 6  dramat   ==>  kraeft
 7  dramat dstruk kraeft rhythm   ==>  dschau
 8  dramat kraeft spiele   ==>  dschau dstruk lebhaf schnel
 9  ausgew   ==>  abgeru dschau dstruk
10  abgeru   ==>  ausgew dschau dstruk
11  dschau dstruk rhythm spiele   ==>  abgeru ausgew
12  abgeru ausgew dschau dstruk schnel   ==>  lebhaf
13  dschau dstruk lebhaf rhythm schnel   ==>  abgeru ausgew
14  dschau dstruk lebhaf munter spiele   ==>  abgeru ausgew
15  abgeru ausgew dschau dstruk lebhaf spiele   ==>  munter
16  dstruk kraeft rhythm spiele   ==>  abgeru ausgew dschau
17  abgeru ausgew dramat dschau dstruk kraeft rhythm   ==>  lebhaf
18  abgeru ausgew dramat dschau dstruk kraeft lebhaf rhythm schnel
       ==>  munter spiele
19  abgeru ausgew dramat dschau dstruk kraeft lebhaf munter schnel
     spiele   ==>  rhythm
```

Figure 7: Basis of the implications in the context of fig.6

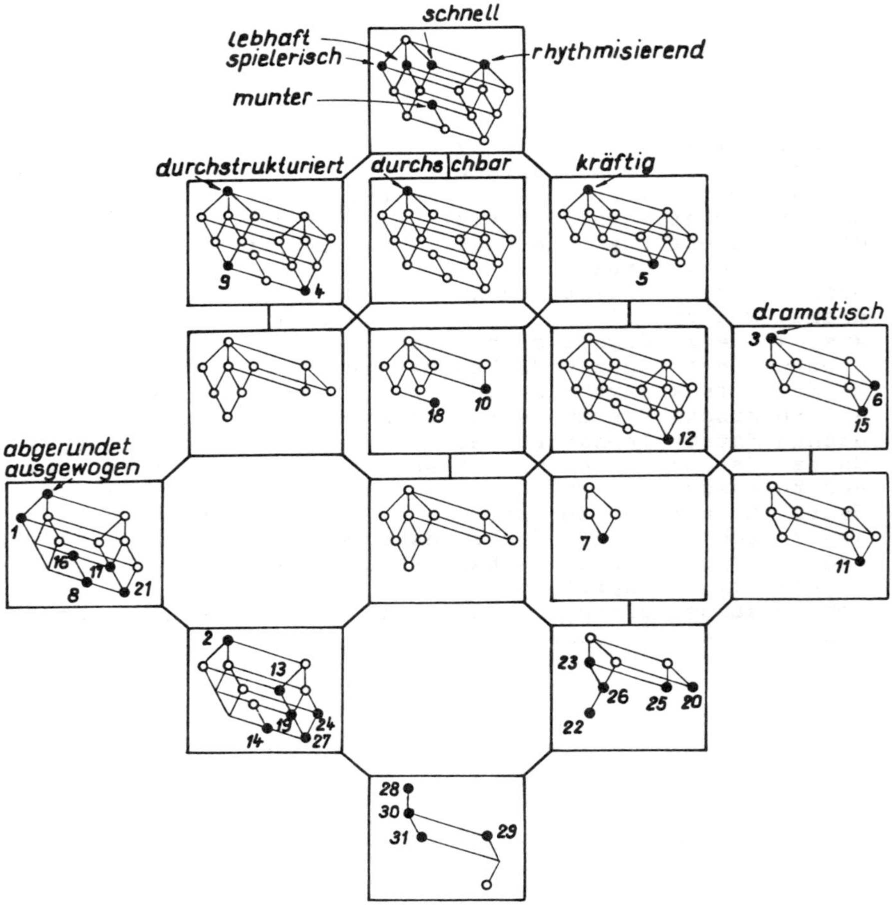

Figure 8: The concept lattice of the context of fig.6

3. Concept Exploration

The idea of *concept exploration* is already mentioned in [Wi82]; the scheme of its procedure is demonstrated in [Wi87] by an example. U.Klotz and A.Mann have worked out the concept exploration in [KM88] and have also implemented this method of knowledge acquisition as an interactive procedure. Let us also explain the concept exploration by an example. First we have to choose the universe $\mathsf{U} := (G_{\mathsf{U}}, M_{\mathsf{U}}, I_{\mathsf{U}})$: let G_{U} be the set of all countable relational structures with a binary relation R (up to isomorphism) and let M_{U} be the set of all universal sentences of the first order logic with the relational symbol R and equality (cf. [Gr68]); then $gI_{\mathsf{U}}m$ shall mean that the structure g satisfies the sentence m. We want to explore this universe by its concepts "similarity" "equivalence", "order", "orthogonality", "dominance", and "covering"; the extent of these concepts consist of the countable relational structures (S, R) for which the binary relation R is (cf. [Ro79])

reflexive and symmetric in case of "similarity",

reflexive, symmetric and transitive in case of "equivalence",

reflexive, antisymmetric and transitive in case of "order",

irreflexive and symmetric in case of "orthogonality",

asymmetric and negatively transitive in case of "dominance",

asymmetric and $(R \cup R^{-1}) \cap (R \circ \cdots \circ R) = \emptyset$ in the case of "covering".

The idea is to determine (up to isomorphism) the sublattice of $\mathfrak{B}(\mathsf{U})$ generated by the given concepts of the universe. So the main question will be whether two terms formed by the given concepts with meets and joins yield comparable concepts or not. The answer might be "yes" and "no"; if the answer is "no", the program asks for an object belonging to one of the two concepts and for an attribute of the other concept which is not valid for that object. To make this clear we present a short protocol of the concept exploration for the three irreflexive concepts "orthogonality" (o), "dominance" (d), and "covering" (c):

Question 1: Is "dominance" a subconcept of "orthogonality"?
Answer: No! $D_3 := (\{0,1,2\}, \{(1,0),(2,0),(2,1)\})$ is not symmetric.
Question 2: Is "orthogonality" a subconcept of "dominance"?
Answer: No! $K_2 := (\{0,1\}, \{(0,1),(1,0)\})$ is not asymmetric.
Question 3: Is "covering" a subconcept of "orthogonality"?
Answer: No! $P_3 := (\{0,1,2\}, \{(1,0),(2,1)\})$ is not symmetric.
Question 4: Is "covering" a subconcept of "dominance"?
Answer: No! P_3 is not as-transitive (i.e., $\neg yRx \wedge xRy \wedge yRz \rightarrow xRz$).
Question 5: Is "orthogonality" a subconcept of "covering"?
Answer: No! K_2 is not asymmetric.
Question 6: Is "dominance" a subconcept of "covering"?
Answer: No! D_3 is not as-atransitive (i.e., $\neg yRx \wedge xRy \wedge yRz \rightarrow \neg xRz$).
Question 7: Does"orthogonality" have the attribute "as-transitive"?

Answer: Yes!

Question 8: Does "orthogonality" have the attribute "as-atransitive"?

Answer: Yes!

Question 9: Is $c \wedge d$ a subconcept of "orthogonality"?

Answer: No! $P_2 := (\{0,1\}, \{(1,0)\})$ is not symmetric.

Question 10: Is "$c \wedge o$" a subconcept of "dominance"?

Answer: Yes!

Question 11: Is "$d \wedge o$" a subconcept of "covering"?

Answer: Yes!

Question 12: Is "$(c \vee o) \wedge (c \vee d)$" a subconcept of "covering"?

Answer: Yes!

Question 13: Is "$(c \vee d) \wedge (d \vee o)$" a subconcept of "dominance"?

Answer: Yes!

Question 14: Is "$(c \vee o) \wedge (d \vee o)$" a subconcept of "$(c \wedge d) \vee o$"?

Answer: Yes!

Stop! The sublattice generated by the concepts "orthogonality", "dominance" and "covering" consists of 10 concepts.

The acquired knowledge, as it is accumulating, is represented in a scheme of four contexts. For our example, the final state of this scheme is shown in fig.9: the upper

×										10				
×	×	×	×							9			×	×
×		×								8			×	
×			×							7				×
×				×						6		×		
×	×	×	×	×	×	×	×	×	×	5	×	×	×	×
×	×	×	×	×		×	×	×		4		×	×	×
×			×	×			×			3		×		×
×		×		×				×		2		×	×	
×	×	×	×						×	1	×		×	×
10	9	8	7	6	5	4	3	2	1		s	as	t	at
×		×		×				×		D_3		×	×	
×	×	×	×						×	K_2	×		×	×
×			×	×			×			P_3		×		×
×	×	×	×	×		×	×	×		P_2		×	×	×

1: o, 2: d, 3: c, 4: $c \wedge d$, 5: $c \wedge o = d \wedge o$, 6: $c \vee d$, 7: $c \vee o$, 8: $d \vee o$, 9: $(c \vee o) \wedge (d \vee o)$, 10: $c \vee d \vee o$,

s: symmetric, as: asymmetric, t: as-transitive, at: as-atransitive.

Figure 9: The resulting context of a concept exploration of binary relational structures

lefthand context indicates the order relation between all lattice terms determining the generated sublattice, the lower lefthand context and the upper righthand context inform which of the chosen objects and attributes belong to which of the concepts described by the listed lattice terms, respectively, and the lower righthand context is the subcontext of the universe formed by the selected objects and attributes; its concept lattice can be seen in fig.10. In our example the concept lattices of the four

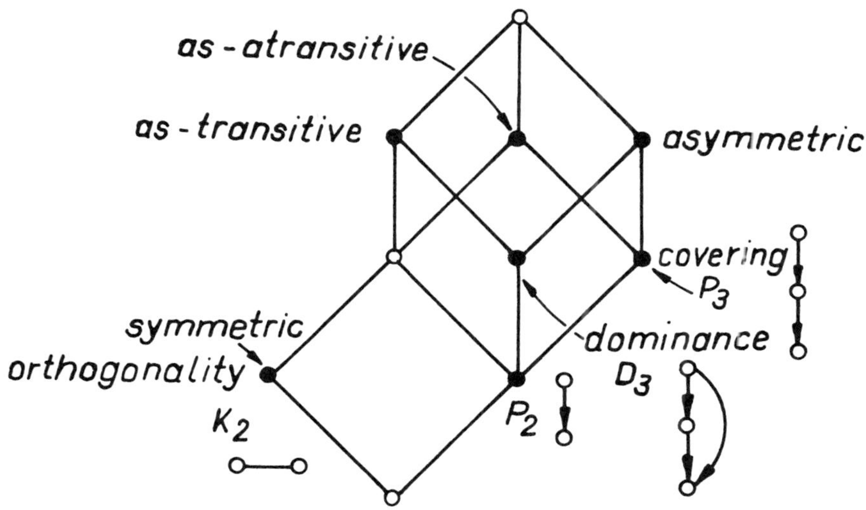

Figure 10: The concept lattice of the lower righthand context in fig.9

contexts are isomorphic which in general is not the case (cf. [Wi87]). We always have that the extents of the upper and lower lefthand contexts are extents of the upper and lower righthand contexts, respectively, and the intents of the upper lefthand and righthand contexts are intents of the lower lefthand and righthand contexts, respectively; in particular, the concept lattice of the upper lefthand context have an order-embedding into the concept lattice of the lower righthand context.

The resulting concept lattice for the exploration with all six given concepts is drawn in fig.11. For answering the questions of the program it is helpful to know characterization theorems joining syntactical and semantical descriptions of structure classes; such a theorem would be for our example a characterization of structure classes determined by universal sentences (cf. [Gr68], p.274). But in our case there is even a more effective method which helps to answer the questions: Notice that the intents of all six given concepts contain the universal sentence "$xRx \rightarrow yRy$" which yields that a relational structure in the extent of the supremum of the six concepts is either reflexive or irreflexive; furthermore, we can reformulate a universal sentence (e.g. symmetric: "$xRy \rightarrow yRx$") so that it becomes valid in all reflexive structures (i-symmetric: "$\neg xRx \wedge xRy \rightarrow yRx$") or valid in all irreflexive structures (r-symmetric: "$xRx \wedge xRy \rightarrow yRx$"). It follows that the resulting concept lattice of

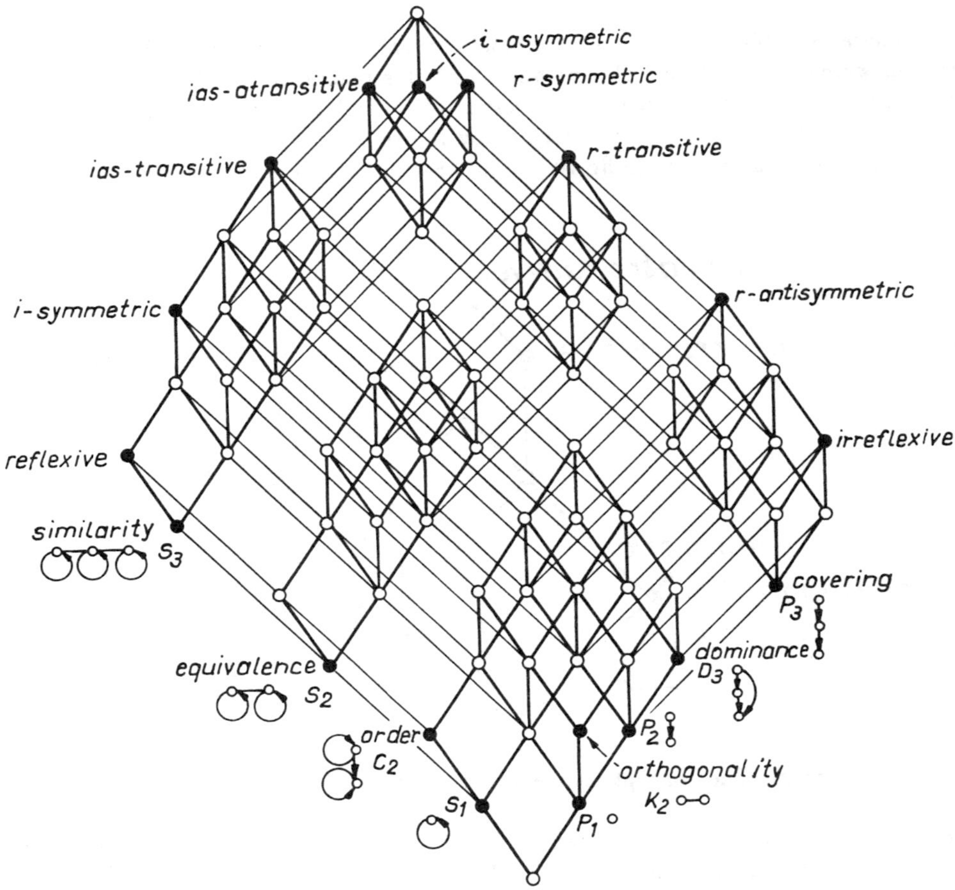

Figure 11: The resulting concept lattice of a concept exploration of binary relational structures

the exploration is a direct product of the sublattice generated by the three reflexive concepts "similarity", "equivalence", "order" and of the sublattice generated by the three irreflexive concepts "orthogonality", "dominance", "covering" (see fig.11). Such structure-theoretical reasoning cannot be performed by the program. It is an interesting research question how to include more structure theory into the program. But already now the mathematical background of concept exploration is quite extensive; in [KM88] the mathematical theory of concept exploration covers more than hundred pages which makes it impossible even to sketch this theory here.

References

[BGMT79] Z.M. Baloghné, J. Géczi, L.T. Molnár, V. Takács: INTEGRAF, Integrált Természettudomanyi Galois Relácioban Ábrázolt Filmek. OOK, Veszprem 1979.

[BM70] M. Barbut, B. Monjardet: L'ordre et classification, Algèbre et combinatoire, tome II. Hachette, Paris 1970.

[Ba76] G. Batel: Komponenten musikalischen Erlebens. Göttinger Musikwiss. Schriften 1976.

[Bi67] G. Birkhoff: Lattice theory, 3rd edition. Amer. Math. Soc., Providence/R.I. 1967.

[Bu87] P. Burmeister: Programm und Programmbeschreibung zur formalen Begriffsanalyse einwertiger Kontexte (unter Mithilfen von A. Rust, P. Scheich und N. Newrly). FB4, TH Darmstadt 1987, neueste Fassung 1989.

[Bu89] P. Burmeister: Merkmalsimplikationen bei unsicherem Wissen. FB4-Preprint Nr.1218, TH Darmstadt 1989.

[DG86] V. Duquenne, J.-L. Guigues: Familles minimales d'implications informatives resultant d'un tableau de données binaires. Math. Sci. Humaines 95 (1986), 5-18.

[Ga87] B. Ganter: Algorithmen zur Formalen Begriffsanalyse. In: B. Ganter, R. Wille, K.E. Wolff (Hrsg.): Beiträge zur Begriffsanalyse. B.I.-Wissenschaftsverlag, Mannheim/Wien/Zürich 1987, 241-254.

[GSW86] B. Ganter, J. Stahl, R. Wille: Conceptual measurement and many-valued contexts. In: W. Gaul, M. Schader (eds.): Classification as a tool of research. North-Holland, Amsterdam 1986, 169-176.

[GW89] B. Ganter, R. Wille: Conceptual scaling. In: F. Roberts (ed.): Applications of combinatorics and graph theory to the biological and social sciences. Springer, Berlin-Heidelberg-New York-Tokyo 1989.

[GW90] B. Ganter, R. Wille: Formale Begriffsanalyse. B.I.-Wissenschaftsverlag, Mannheim/Wien/Zürich (in preparation)

[Gr68] G. Grätzer: Universal algebra. D.van Nostrand, Princeton-Toronto-London-Melbourne 1968.

[KM88] U. Klotz, A. Mann: Begriffexploration. Diplomarbeit, FB4, TH Darmstadt 1988.

[Ma83] D. Maier: The theory of relational databases. Computer Science Press, Rockville 1983.

[Ro79] F. Roberts: Measurement theory. Addison-Wesley, Reading/Mass. 1979.

[Ta84] V. Takacs: Two applications of Galois graphs to pedagogical research. Manuscript of a lecture at the TH Darmstadt, Feb.8, 1984.

[Wa73] H. Wagner: Begriff. In: H. Krings, H.M. Baumgartner, C. Wild (Hrsg.): Handbuch philosophischer Grundbegriffe. Kösel, München 1973, 191-209.

[Wi82] R. Wille: Restructuring lattice theory: an approach based on hierarchies of concepts. In: I. Rival (ed.): Ordered sets. Reidel, Dordrecht-Boston 1982, 445-470.

[Wi84] R. Wille: Liniendiagramme hierarchischer Begriffssysteme. In: H.H. Bock (ed.): Anwendungen der Klassifikation: Datenanalyse und numerische Klassifikation. INDEKS Verlag, Frankfurt 1984, 34-51; engl. translation: Line diagrams of hierarchical concept systems. International Classification 11 (1984), 77-86.

[Wi87] R. Wille: Bedeutungen von Begriffsverbänden. In: B. Ganter, R. Wille, K.E. Wolff (Hrsg.): Beiträge zur Begriffsanalyse. B.I.- Wissenschaftsverlag, Mannheim/Wien/Zürich 1987, 161-211.

[Wi89] R. Wille: Geometric representation of concept lattices. In: O. Opitz (ed.): Conceptual and numerical analysis of data. Springer, Berlin- Heidelberg-New York-Tokyo 1989.

Forschungsgruppe Begriffsanalyse
Fachbereich Mathematik
Technische Hochschule Darmstadt

STATISTICS AND ARTIFICIAL INTELLIGENCE

Daniel Defays
Eurostat
Bâtiment Jean Monnet
L-2920 Luxembourg
Tel: 4301-2854 Telex: COMEUR LU 3423

The term "statistics" is inherently ambiguous. It implies both a scientific discipline essentially concerned with analysing data on a reference population (often on the basis of probabilistic hypotheses, in which case we speak of "mathematical statistics") and applied practice, which has not yet evolved into a discipline and covers the whole field of activities undertaken by statisticians in the public and private sectors to prime policymakers with information. It is the relationship between statistics in this second sense of the word and artificial intelligence which I have explored in this paper.

1. Public statistics

To define the work of the public sector statistician as "describing society and its economic context" would be erroneous in many respects. In fact, the statistician has an ambivalent relationship with power. He provides answers to questions which are often generated by the institution itself and its practices, and his figures are converted into instruments of planning, management and evaluation. His job is not only to propose methods of representation, but extends to supplying information which will provide a suitable basis for shaping policy. The resulting requirement for reliability and transparency and the enormous sums at stake can make his task highly demanding.

The statistician's real role lies in interpreting requirements and designing and setting up the information systems - for collection, validation, estimation, storage, analysis and dissemination - enabling these requirements to be met. Mathematical statistical techniques are only one of a whole series of methods involved. Measurement errors, for example, are far less likely to be random than a direct result of the source used, inconsistency of definition, errors in evaluation and the nature of proxies. Difficulties in coordinating surveys, or processing the results of censuses, and the sheer volume of data they entail often necessitate the development of specific techniques.

While complete automation of every area of the statistician's expertise may still seem to belong to some distant vision of utopia, a look at his job profile reveals one relatively systematic task after another which could benefit from being automated or at least computer-assisted.

2. Automation and statistics

It is no coincidence that statistics and informatics have a long common history of development. The systematic and repetitive nature of certain tasks and their scale (e.g. processing of census data) mean that even partial automation can often increase productivity significantly. Encoding certain replies to questionnaires can take up hundreds of man years; seen in this perspective, a saving of even 10% can be considerable.

As far back as a century ago, Hollerith was using punched cards to process census data in the United States. Automation of numerical calculation has since enabled statistical methods such as correspondence analysis or multidimensional scaling to be developed which would be unthinkable in a different technological environment. The advent of database management systems later helped facilitate processing of large volumes of data, with certain modifications. And progress shows no signs of halting. With the development of expert systems, the number of tasks benefiting from computer support seems set to continue to expand.

The actual and potential contribution to statistics of techniques based on artificial intelligence can be divided into different areas. Some projects seek to make existing statistical software more user-friendly, others to assist the data producer by automating certain tasks, while a third category seeks to expand the choice of statistical techniques available. In the following paragraphs, I have dealt with each aspect in turn, illustrating what has already been achieved and suggesting possible new avenues to be explored.

3. Making statistical software more user-friendly

Maintaining that current statistical software packages allow, if not encourage, the worst possible heresies has by now become a cliché. They are essentially non-intelligent since they do not use the data to check the validity of the hypotheses underlying the statistical procedures used. Various systems have been or are being developed in this area. The best known of these must surely be REX (Gale, 1986) which advises the inexperienced user on regression strategy; it examines the data supplied, suggests possible changes, explains its suggestions and, having completed its analysis, produces a short report (in heavy stereotype, no less). Nelder's work on the subject is also well known (Nelder 1989); GLIMPSE assists use of the GLIM software by, for example, posing questions and specifying possible answers and explaining the definitions of certain terms. Modification of forecasting models has also been the subject of several studies: Mandrake d'Azencott, FORECAST PRO, FOCA, to name but a few.

This field of application of artificial intelligence-based techniques is clearly the biggest growth area. Quite apart from its value to the end user, who does not always have a statistician at hand, it promotes investigation into the best data analysis strategy: identifying the problem, how best to solve it, how to get round sources of difficulty etc. Schemes involving the introduction of automation or computer support have the advantage of looking for ways of pin-pointing and encoding the heuristics used (although not necessarily universally) by statisticians and may in time contribute to improving their quality and organization. It has been suggested, for instance, that explicit statement of strategies is useful in identifying important gaps in statistical methodology.

4. Assisting the data producer

This field is perhaps of most direct interest to the public sector statistician. I have divided the topic into four sections for convenience:

- data collection and planning
- data processing and validation
- identifying and finding relevant statistics
- data analysis.

4.1 Collection and planning

Changes in the technological environment are progressively bringing about a change in methods of collecting information. The wide range of possibilities offered by telephone and minitel, direct capture of information on magnetic tape and real-time correction and verification have accelerated the growth of new, highly advanced systems. Representation techniques inspired by artificial intelligence are beginning to appear: W. Saris, for example, has developed an intelligent interview program for question selection, certain automatic consistency checks and automatic coding of certain answers to open-ended questions (Saris, 1989). M. Appel and T. Scopp have written an automatic coding program for the United States Census Bureau enabling the correct code for a given nomenclature to be selected on the basis of a description of an industrial activity or the occupation of a survey subject. At present the program has an average 90%+ success rate, and will be used to process the 1990 ten-yearly census (Appel and Scopp 1989). ISTAT (the Italian National Statistical Institute) is studying an automatic survey planning system using frames for representing information on samples, stages and strata (Barcaroli and Fortunato, 1989).

Other developments such as EDI (Electronic Data Interchange) are certain to widen the scope of opportunities open to the statistician still further. Methods of collecting information continuously rather than at a given point in time are now a possibility. Such techniques will, however, confront us with new problems (when to begin publishing the results, for example), which will have to be resolved.

4.2 Processing and validation

One of the major problems confronting the public sector statistician is how to deal with missing data. This is another area which has been tackled by introducing expert-type systems. At Eurostat (Statistical Office of the European Communities), for example, a system of data estimation for external trade statistics has been developed (Koszerek 1989). The software examines different databases and suggests a method of estimation on the basis of the characteristics of the information available (use of imports by Y from X to estimate exports from X to Y, data extrapolation, conversion of existing data by deflation etc.). B. Greenberg of the Census Bureau has developed an expert system for edit and imputation which may be applied to different surveys dealing with similar types of data (Greenberg 1989). Information specific to a particular survey is represented in the form of rules.

The sheer volume of data collected in statistical surveys also makes it difficult to check the information supplied. Visual examination is often impossible and automated procedures are therefore indispensable. Validity checks are, however, often based on very different areas of expertise. It involves knowledge of

a) the expected structure of individual records (in terms of relationships required or expected between various components);

b) the relationships required or expected between individual records;

c) the expected structures of various aggregates;

d) the relationships between individual records and information relating to the same entities collected on other occasions;

e) the relationships between aggregates and corresponding aggregates derived from other sources or collected on other occasions;

f) and, of course, ideally, the domain (industry, external trade or employment for instance).

Checking the consistency of information is not generally too difficult (items a to f), but estimating plausibility (item f) is a much trickier problem. What exactly would the machine have to know, for example, to question the quality of a statistic telling the user that the average age of widows in a particular region was 25 years? Further development is planned in this area (M. Talbot, for example, proposes using expert systems) (Talbot 1989), but there is a long way to go before checks can be completely automated.

Other problems arise in the context of preserving confidentiality of micro data where extensive cross-tabulations are generated from a data set. With increased computing power the variety of tabulations produced has multiplied and the risk of accidental disclosure from comparison of separate tabulations increased. This could be an area of application of expert systems with a potentially early application.

4.3 Identifying and finding relevant statistics

Here again, systems exist, but substantial progress remains to be made. Getting from the request stage, which is often formulated in very general terms, to retrieval of the right kind of quantitative information involves various operations: translating the problem into statistical terms - matching statistical requirements with the information available - accessing a relevant data base.

The level of support provided must, moreover, be adapted to the user profile of the requester. Existing bases in fact offer few possibilities for accessing information: these need to be extended to allow browsing, comparing alternatives, experimenting and scanning for information rather than simply entering an address.

Existing software generally uses data dictionaries and is inadequately adapted to dealing with the problems described. There is undoubtedly a need for a systematic study of the nature and methods of representing information (meta data) which different categories of user will need in order to access data bases more easily.

4.4 Data analysis

Most existing data analysis systems (many of which are already being brought onto the market) offer a user-friendly interface which helps the user to choose the most appropriate statistical method for his problem and his kind of data. Mention has already been made of this type of software in a previous paragraph. More integrated products of the Adèle type also exist (Gonzales and Hatabian, 1989) to guide the user through the various stages of survey analysis: choice of task, choice of type of processing, assisted analysis dependent on the characteristics of the information available and even assistance with interpretation. Products capable of generating reports based on a table of figures are also beginning to appear: SAM, developed at the INSEE (French Statistical Institute), can simulate the procedure followed by an economist analysing a table of economic indicators; each problem has a corresponding semantic information network, several sets of rules and, of course, numerical data (Ross, 1989).

5. Widening the choice of statistical techniques

It would be a pity to restrict advanced data processing techniques to automating tasks previously carried out manually. Results already obtained indicate that certain representation and handling techniques for symbolic information could considerably broaden the field of application of statistical techniques. All too often, current statistical methods can only use a very small part of the information available on a given subject. For a typological analysis on the regions of Europe, for example, traditional techniques involve defining the level of dissimilarity between regions on the basis of a certain number of variables. All our information on the regions, their wealth, their population structure and their industries must somehow be squeezed into one distance matrix. While this is not generally too much of a problem, it does not allow aspects of regional contiguity, such as belonging to the same country, to be taken into account. It may be that the regions neighbouring onto large metropolitan centres constitute a relevant group! Available techniques for integrating such information into a typological analysis are disappointing. New "conceptual clustering" methods are currently being developed which will enable relationships between units, as well as the units themselves, to be aggregated according to a certain number of variables. These methods combine clustering and machine learning techniques (e.g. B. Nordhausen, 1986). The classification strategy generally used in typology (aggregating similar units and separating different units) could be improved in this way, and other classification methods may be considered (such as definition of rules) (E. Diday, 1988, for example).

Neural networks could find a useful application in time series analyses. The learning system has been used to design different types of networks to predict time series (cf. A. Varfis, 1989 for example). The research is still in its infancy but promises to produce interesting results.

6. Conclusions

If expert statistical tools are going to be used in practice, they have to address problems that do exist and which by their nature can be solved through the application of principles of expert system techniques. There are, in fact, many points of contact between official statistics and artificial intelligence. This article has been limited to an examination of those situations in which the statistician could profitably apply artificial intelligence techniques. By the same token, artificial intelligence specialists could benefit from current developments in statistics. The two fields are intimately linked and cannot afford to develop in isolation.

By making available new information processing techniques, A.I. is capable of making an enormous contribution to the whole subject area of statistics and its methods. As this international conference organized by INRIA shows, it is a challenge which is being taken up by more and more statisticians.

7. References

APPEL, M.V., SCOPP, T. (1989), Automated industry and occupation coding. Proceedings of the seminar "Development of Statistical Expert Systems" held in Luxembourg in December 1987.

BARCAROLI, G., FORTUNATO, E. (1989), Expert system for defining statistical samples. Proceedings of the seminar "Development of Statistical Expert Systems" held in Luxembourg in December 1987.

DIDAY, E. (1988), Introduction a l'approche symbolique en analyse des données. Actes des "Journées numériques-symboliques". Université de Paris 9 Dauphine.

GONZALES, P., HATABIAN, G. (1989), Adelle or an aid in the analysis of surveys by expert software on a microcomputer. Proceedings of the seminar "Development of Statistical Expert Systems" held in Luxembourg in December 1987.

GREENBERG, B.V. (1989), Developing an expert system for edit and imputation. Proceedings of the seminar "Development of Statistical Expert Systems" held in Luxembourg in December 1987.

KOSZEREK, C. (1989), An expert system for gap filling in external trade statistics. Proceedings of the seminar "Development of Statistical Expert Systems" held in Luxembourg in December 1987.

NELDER, J.A. (1989), The user interface for the analysis phase. Proceedings of the seminar "Development of Statistical Expert Systems" held in Luxembourg in December 1987.

NORDHAUSEN, B. (1986), Conceptual clustering using relational information. Proceedings of the National Conference of Artificial Intelligence.

ROOS, J.L. (1989), SAM, a knowledge based system for modelling economist. Data Analysis, Decision Support and Expert Research in Economics, Hamburg.

SARIS, W.E. (1989), Intelligent computer programs for computer assisted data collection. Proceedings of the seminar "Development of Statistical Expert Systems" held in Luxembourg in December 1987.

TALBOT, M. (1989), Development of expert systems for routine data monitoring. Proceedings of the seminar "Development of Statistical Expert Systems" held in Luxembourg in December 1987.

VARFIS, A. (1989), Application des réseaux neuronaux aux séries temporelles. Perspectives d'une nouvelle technique. Ispra Research Center - Not published report.

RULES GENERATION

SYMBOLIC NUMERIC APPROACH FOR BIOLOGICAL KNOWLEDGE REPRESENTATION: A MEDICAL EXAMPLE WITH CREATION OF IDENTIFICATION GRAPHS

J. Lebbe (1, 2), R. Vignes (1, 2), S. Darmoni (3)

(1) Laboratoire de Phytologie quantitative, Université Pierre et Marie Curie, Bat C
7 quai Saint Bernard, 75230 Paris cedex 05, FRANCE
(2) INRIA, projet Clorec, Domaine de Voluceau, BP 105, 78153 Le Chesnay cedex, FRANCE
(3) Service de gastro-entérologie Hôpital Antoine Beclère, 141 rue de la porte de Trivaux 92140 Clamart, FRANCE

Summary

We study representation of concept descriptions. Particularities of these descriptions are disjunctions of values due to the internal variation in concept extension, and hierarchies of variables. This work refers to symbolic object formalism. We add typicality attribute to express the notion of relative frequency of a value. So we introduce the qualified symbolic objects.

We propose an algorithm (MAKEY) to create identification graphs from a set of qualified symbolic objects. The resulting graphs carry out identification with a value of typicality attached to each conclusion. Different parameters can be modified to adapt the algorithm to a particular application and its domain theory: function to compare descriptions, function to agregate typicality values, criteria to stop or to prune branches in the graph. This algorithm is applied to a real medical application on viral hepatitis.

INTRODUCTION

In biology and medicine, knowledge almost always appears as concept descriptions. The experts know sets of properties describing the concepts, but generally they do not know definitions or discriminating rules to identify one concept from the others. The concepts are classes (diseases, taxa, biocenosis, etc.) isolated as entities. The three aspects of a concept are: its extension (set of instances belonging to the concept), its name (to refer to it as an entity), and its definition.

Concepts may be represented by: production rules, samples, prototype and so on. This point was discussed in (Lebbe *et al*, 1989). In our application domain, a concept is known by its description: set of properties that an instance may have. In practice, the concept description results from an induction made by expert from observations and historical knowledge about the concept. These descriptions contain more information than samples or prototypes, and are easier to collect. The creation of identification graphs aims at producing discriminating strategies and identification rules.

To represent faithfully these concept descriptions we need to grasp internal variation of the extension. Indeed we don't want to reduce the information to average values, but we want to code all the range of values. To manage such concept descriptions with efficacy, a specific mathematical formalism is needed. Recently E. Diday has proposed symbolic object formalism well adapted for this purpose (Diday, 1987). Each concept description is represented by a symbolic object which is the conjunction of the concept properties. Each property (symbolic event) is the list, for one variable, of all the possible values in the extension. It appeared necessary for some applications to associate to each value in description a qualitative attribute which expresses whether this value is frequent or not in concept extension in comparison with the frequency of other possible values. To formalise this supplementary knowledge we propose the qualified symbolic objects (see also Diday 1988).

KNOWLEDGE REPRESENTATION

The first step in this work was to study the semantics of concept descriptions so as to develop a suitable formalism. Previously, Allkin (Allkin, 1984) has studied taxonomic descriptions. We insist below on two points: disjunction in description and notion of typicality. These points explain why we use qualified symbolic objects which are detailed in Lebbe et Vignes, 1988.

Disjunction

Disjunctions may have three completly different meanings, relating to context: uncertainty, imprecision, variation. If we describe a single case (specimen, example) disjunctions express uncertainty or imprecision (Zadeh, 1975 ; Shafer, 1976 ; Dubois & Prade, 1984). In our domain, when we describe classes (the concept extension), disjunctions more often indicate a range of values (variation). For example: "*Trifolium*: ... Flowers purple, white or yellow, rarely disposed in dense ears or solitary ...". This sentence is a translated extract of FLORE de la FRANCE (Coste, 1906). It means that some *Trifolium* have "purple flowers", some have "white flowers" and some others have "yellow flowers" ; but there is no *Trifolium* with another flower colour. So there is at least one trifolium with "purple flowers", one with "white flowers" and one with "yellow flowers". The disjunction is neither due to the doubt which the expert may be having nor due to the imprecision of the descriptors. It reflects a variation inside the taxon *Trifolium*.

It is not our purpose to describe here uncertainty and imprecision disjunctions. The formalism used here takes into account disjunctions of variation, but not imprecision and uncertainty. This variation is managed by accepting more than one value for each variable.

Typicality

The notion of typicality may have different meanings: specific value, most frequent, value, theoric value according to a model.

In biological descriptions, many terms express frequency and occurence of a value (state of character) in comparison with the other possible values. In the above example, "rarely" is an adverb refering to the frequency of the property "dense ears" but it does not code a precise frequency. It is an expert's subjective estimate of relative likelihood of the each alternative state: "dense ears of flowers" and "solitary flowers". The second state is more frequent than the first one. We incorporate this type of information giving a qualitative value of typicality to each possible value in concept description. It is generally impossible to have more precise information than a pre-order among character states. With such subjective estimates and without information about conditional likelihood, we cannot refer to probabilistic formalism (Gremy, 1987).

Note that it is also possible to qualify an instance of a concept extension with a typicality value. We say that an instance has a high typicality for its concept if this instance has many values associated with high typicality value in the concept description. For example: "this child has a typical viral hepatitis A" means that the disease of this child has typical properties of the concept viral hepatitis A.

Knowledge-base

In our applications, a knowledge-base includes a set of concept descriptions and background knowledge (domain theory). The list of concept descriptions must be exhaustive, and the concepts must be mutually exclusive: an example belongs to only one concept, even if the example observation is not sufficient to identify this concept without uncertainty. The concept descriptions are given by experts. These informations are considered certain, precise and true. All the variables are qualitative variables, and numerical data are split into ranges of values.

Sometimes, it is not possible to describe all the concepts with the same set of variables. For example: if an insect species has no wings, the variable "colour of wings" has no sense for it. If a variable has no sense in a concept, we code no value and we add in the backgroud knowledge a rule explaining why this variable has no sense: if "wing = absent" then "colour of wings has no sense". In a such case, the first variable (premise) is said "mother variable" and the second (conclusion) "daughter variable". A daugter variable may also have daugter variables. These logical relations and intervariable dependencies form hierarchies structuring descriptions. A no sense data is completly different from an unknown data which will be coded as the range of all the values.

MATHEMATICAL FORMALISM

We present some symbolic objects defined by E. Diday, adding possibility to formalize typicality. We take as example a knowledge-base about viral hepatitis diagnoses.

Ω = set of cases. For example: all the cases of viral hepatitis.
C = set of concepts. For example: all the viral hepatitis diagnoses.

To each $c \in C$ we associate:
- a symbolic assertion object which is its description (conjunction of properties).
- a subset of Ω which is the extension of c in Ω and hence we note: $|c|_\Omega$

Variable (= attribute or character)

A variable y_i is a mapping $\Omega \rightarrow O_i$.
O_i is the observation domain of y_i.

For a qualitative variable, O_i is a set of modalities (nominal values). For example: y_1 = "IgM anti HBc" and O_1 = {present, absent}.

Event (= property)

A qualified event eq_i is a mapping $\Omega \rightarrow T \cup$ {incompatible} and is noted:

$eq_i = [y_i = X_i]$ with $X_i \subset (O_i \times T)$
T = set of typicality values. In our application T = {typical, atypical, very atypical}.

We note V_i the subset of O_i present in X_i. It is the set of values of y_i possible in the concept extension.

An event is said incompatible for $\omega \in \Omega$ iff $y_i(\omega) \notin V_i$. In the contrary, the event takes the typicality value associated to $y_i(\omega)$ in X_i:

$\omega \in \Omega,\ eq_i(\omega) = t,\ t \in T,$ iff $(y_i(\omega), t) \in X_i$

Assertion (= description)

A qualified assertion object aq is a mapping $\Omega \rightarrow T \cup$ {incompatible} and is a conjunction of qualified events: $aq = eq_1 \wedge \ldots \wedge eq_k$

For example: c1 = viral hepatitis A
aq_1 = [IgM anti HAV = {(present, typical)}] ^ [IgM anti HBc = {(absent, typical), (present, atypical)}] ^ [Ag HBs = {(absent, typical)}]

It means that an instance of viral hepatitis A has always IgM anti HAV, never Ag HBs and sometimes IgM anti HBc; and IgM anti HBc absent is much more frequent than IgM anti HBc present. Note that events are linked by conjunctions, and values in an event by disjunctions.

$a^q(\omega) = fg(e^{q}_1(\omega), \dots, e^{q}_k(\omega))$
fg is a function which agregates the typicality values of the events of an assertion.
$fg : (T \cup \{incompatible\})^k \to T \cup \{incompatible\}$.

We propose some properties for fg:
1) $fg(t) = t$
2) $fg(t_1, t_2) = fg(t_2, t_1)$
3) if $t_2 > t_3$, $fg(t_1, t_2) \geq , fg(t_1, t_3)$ if $t_1 > t_2$, $fg(t_1, t_2) \leq t_1$

The function fg must be defined by the experts of the application domain (see application). It is an element of the background knowledge. For example if the agregation of several atypical values has a result less than atypical, there is negative enhancement. One may or not wish for this property depending on the application. If the expert thinks that some variables may be correlated, this negative enhancement can give too much weight to some information.

A qualified assertion $a^q = e^{q}_1 \wedge \dots \wedge e^{q}_k$ is said incompatible for $\omega \in \Omega$ iff $\exists i \in \{1, \dots k\} / e^{q}_i(\omega) =$ incompatible. We define the extension of a qualified assertion as the set of $\omega \in \Omega$ for which the assertion is not incompatible: $|a^q|_\Omega = \{\omega \in \Omega / a^q(\omega) \in T\}$

To each $\omega \in |a^q|_\Omega$, the qualified assertion associates an element of T which is the typicality of ω for the concept described by a^q.

$|a^q|_\Omega = \cup \{(a^q)^{-1}(t_i), \forall t_i \in T\}$

MAKEY ALGORITHM, IDENTIFICATION GRAPHS

An identification graph is a rooted, directed acyclic inheritance graph.

```
                    Ag HBs          <= (NODE = variable)
                      /\
           present /    \ absent  <= (BRANCH = modality or
                   /      \            disjunction of modalities)
                  /        \
       IgM anti HAV      NODE
                /\         /\
    present /    \       /    \
             /      \     ...    ...
            /        \
 IgM anti HBc       ...
          /\
        /    \ present
       /      \
      /        \
 ...      Cohepat. A B        <= (LEAF = concept or property)
            (typical)
```

A graph is defined by the pair (Π, Γ) where Π is a finite set of **points** and $\Gamma \in \Pi \times \Pi$. An element (a,b) of Γ is an **arc** whose initial point is a and terminal point is b. We name these arcs, or arrows, BRANCHS. We distinguish the LEAVES which are only terminal points of the graph, and the NODES which are all the other points.

Each terminal **leaf** represents one or more values of the variable to be explained. In our medical application, the explained variable is the name of viral hepatitis diagnosis (so each leaf is a concept name). Each **node** represents an explanatory variable. Each **branch** is one or more modalities (values) of the above variable. A path from the root to a terminal leaf describes a premise of an implication rule. The disjunction of all the premises reaching the same conclusion describes a premise of an equivalence rule. For example: Ag HBs present ^ IgM anti HAV present ^ IgM anti HBc present $\Rightarrow$ typical cohepatitis A B

MAKEY is a top down method (Vignes et Lebbe., 1987). It is stable, and is not dependent of data order: with the same data and the same parameters, MAKEY always creates the same graph. A heuristic selects step by step the variable with the maximum discriminant power in the context. This selected variable makes a node with a branch for each possible modality. In each branch the process comes back to the selection of the best variable, except if a criteria to stop is matched. The path ends with a leaf which is the list of the compatible conclusions. Same conclusion can appear in more than one leaf in the graph, due to the disjunctions in concept descriptions. The quality of the graph depends on the method of computing the discriminant power of the variables.

Discriminant Power, function of comparison

The discriminant power of a variable is the result of the comparison of its values for all the pairs of concept descriptions.

Assume two qualified events for the same variable: $eq_1 = [y_1 = X_1]$ with $X_1 \subset (O_1 \times T)$ and $eq_2 = [y_1 = X_2]$ with $X_2 \subset (O_1 \times T)$. The function cp for comparing two events is cp: $P(O_i \times T)^2 \rightarrow [0-1]$. This function is a dissimilarity index corresponding to the divergence between the two qualified events. If the events have no common modalities ($V_1 \cap V_2 = \emptyset$) the dissimilarity is maximal. Two events with many common and typical modalities have a low dissimilarity index. The total discriminant power of a variable is the sum of the function of comparison for all the pairs of concept descriptions. A great number of pairs correctly discriminated gives a high discriminant power to the variable. In MAKEY algorithm, to compute correctly the discriminant power at each step, it is necessary to take into account the precedent steps in the graph. In the general case the discriminant power is the weighted sum of the function of comparison of all the pairs of events belonging to assertion objects describing concepts always compatible at this step. A pair has the more weight as the corresponding concepts have high typicality.

Criteria to stop, to prune or to merge branches

A concept stays in a branch while the properties of the path are compatible with the concept description. A branch without any corresponding concept is pruned. The choice of the most discriminant variable at each step allows to obtain an identification graph with a short average length for the paths.

In all the cases, a path stops when there is only one concept compatible or when no variable is left. Other criteria may be added. For example:

- if in a branch, only one concept has maximal typicality, this branch is stopped.
- typicality threshold: if the typicality associated to a concept is less than a threshold, the concept is deleted from the branch.
- discriminant power threshold: when the largest discriminant power is less than the threshold, the branch is stopped.

Sometimes branches can be merged together with respect to some criteria. In the case of symbolic objects without typicality value, the following algorithm was proposed (Vignes et Lebbe, 1989) to merge branchs descending from the same node. For the variable of the node a graph is constructed. In this graph, the points are the modalities and the arrows link two modilities if merging these two is accepted. Merging two modalities is possible, if replacing them by a unique modality equal to their disjunction does not reduce the discriminant power of the variable. A good merging of branchs is given by a graph partition in cliques. Each clique forms a unique branch which is the disjunction of the modalities belonging to this clique.

Adding deduced variables

Each node is one variable. An option of MAKEY allows to add at each node the variables whose modalities are deduced. So a node is a list of variables, and the branchs conjunctions of modalities.

Suplementary knowledge

Hierarchical relations between variables dictate terms to construct the identification graph. It is impossible to use a daughter variable as long as mother variable is not known (selected or deduced). At each step the creation of the graph respects these hierarchical relations. A mother variable is sometimes misevaluated by its discriminant power because it does not take into account the discriminant power of the daughter variables which will appear if the mother is selected. So a parameter authorizes to select a mother variable which has many daughter variables with great discriminant powers even if its own discriminant power is low.

At each step the choice of the variable can be suggested by variable weights. These weights are given by the expert to code strategic knowledge. They induce a pre-oder between variables. The most discriminant variable is chosen among the list of variables of highest strategic interest.

Comparison and agregation functions are also parameters of the algorithm.

MEDICAL APPLICATION

A knowledge-base describing serologic profils (variables) of viral hepatitis (concepts) is realized. A previous knowledge-base was created in 1987 (Darmoni *et al*, 1987) and adapted later on MINITEL network (Darmoni *et al*, 1988). This first knowledge-base has no typicality values. Recently medical experts added typicality. Qualified assertion objects are given fig. 3. The knowledge-base concerns 14 concepts (fig. 1) described with 8 variables (fig. 2). The set T of typicality values is: T = {typical, atypical, very atypical} and the order: typical > atypical > very atypical.

Agregation function

Agregation function $fg(t_1, t_2)$ of two typicality values is done by the matrix:

	typical	atypical	very atyp.
typical	typical	atypical	very atypi.
atypical	atypical	atypical	very atypi.
very atyp.	very atypi.	very atypi.	very atypi.

and $fg(t_1, t_2, t_3) = fg(fg(t_1, t_2), t_3)$.

This function was chosen with a medical expert. It agrees with function minimum. It gives great importance to the less typical values whatever the number of typical values. Example: a^q = [Ag Hbs = {(present, typical), (absent, very atypical)}] ^ [Ag Hbc = {(present, typical), (absent, atypical)}]. If ω has Ag Hbs present and Ag Hbc absent, then $a^q(\omega)$ = fg(typical, atypical) = atypical.

Comparison function

Function cp to compare two qualified events is: rang(max{min(t1, t2)}) / card(T), $\forall$ (t_1, t_2), $\forall\, v_i \in (V_1 \cap V_2)$, $(t_1, v_i) \in X_1$ and $(t_2, v_i) \in X_2$, where rang() is a function returning for each value of T its rank.

For example:
e^{q}_1 = [Ag Hbs = {(present, typical), (absent, very atypical)}]
e^{q}_2 = [Ag Hbs = {(present, atypical), (absent, typical)}],
$cp(e^{q}_1, e^{q}_2)$ = rang(max(min(typical, very atypical), min(atypical, typical)) / 3 = 0,333.

Weighting function

Weighting function fp used in discriminant power computing is:

$$1 - (\mathrm{rang}(\min(t_1, t_2)) / \mathrm{card}(T))$$

If two pairs have the same value with function cp, the function fp gives more weight to the most typical pair: if $cp(eq_1, eq_2) > cp(eq_3, eq_4)$ then $fp(cp(eq_1, eq_2)) > fp(cp(eq_3, eq_4))$

Functions fg, cp and fp depend on the application and are parts of background knowledge. According to knowledge-base size we don't use pruning criteria in this application. Different identification graphs were tested with and without variable weight.

Result: Identification graph

One identification graph is presented fig. 4. Following a path, a case is identified to one or more diagnosis with typicality values allowing to evaluate the result. In a leaf, we keep only diagnoses which are compatible. If the case is very atypical for all diagnose in the conclusion, the user has an argument to doubt of the quality of its case description. The identification graph obtained with qualified symbolic objects is more precise and useful than this without typicality values.

This identification graph is considered good by medical experts. A real practical evaluation will be interesting: it will need a clinical trial comparing conclusions and diagnostic strategies of doctors versus those proposed by identification graphs.

CONCLUSION

Qualified symbolic object formalism seems well adapted to represent descriptions of classes (as concepts) with disjunctions of variation, and when data are sure and precise. The choice to estimate relative likelihood with qualitative attribute is a practical means when numerical values are difficult to obtain.

As there is no perfect objective way to choose comparison, agregation and weigting functions, we prefer to use parameters to help the expert to select the best ones for his application.

Representing one concept by one assertion object may introduce a bias. Only the coding of the description of each instance would be exact. But often, the complete extensions of concepts are inaccessible, and the expert knows concept descriptions but not concept definitions. With the symbolic object formalism we represent a concept description by only one assertion object, putting more than one modality to grasp variations. Unknown value are interpreted as all modalities are possible. Impossible values are explained by hierachies of variables. Due to their internal variation, concept descriptions may have overlapping extensions, even if we know that the concepts are exclusive. It is for this reason than some leaves contain more than one conclusion in the identification graph. The typicality values improve descriptions, and in graph conclusions they allow to evaluate the adherence of a case to each concept.

We need now adding correlations between properties. This is a difficult but necessary problem to avoid lost of information when representing variation. We are studying also new types of variables, and quantitative typicality values.

REFERENCES

ALLKIN R., 1984. "Handling Taxonomic Descriptions by computer". In Databases in Systematics, Allkin R. & BISBY F.A. (Eds): 279-290. London: Academic Press.

COSTE H., 1900-1906. "Flore descriptive et illustrée de la France". Librairie scientifique et technique Albert Blanchard.

DARMONI SJ., POYNARD T., MEMBRADO M., PARIGOT D., NAVEAU S., CHAPUT J.C., 1987. "Elaboration d'un système expert pour l'aide à l'enseignement du diagnostic sérologique de l'hépatite virale aigüe". Nouv. Press. Méd., 16, 1903-1906.

DARMONI S., VIGNES R., POYNARD T., LEBBE J. & CHAPUT J.C., 1988. "Diagnostic sérologique de l'hépatite virale aiguë sur MINITEL." in Systèmes experts en médecine, EC2 Ed., Nanterre, 255-267.

DIDAY E., 1987. "Introduction à l'approche symbolique en analyse des données", in premières Journées Symbolique - Numérique, Université Paris-Dauphine, Paris 8-9 décembre 1987, 21-56.

DIDAY E., 1988. "Introduction à l'analyse des données symboliques : objets symboliques mopdaux et implicites", Deuxièmes journées Symbolique - Numérique, LRI, Orsay 12-13 décembre 1988, 1-30.

DUBOIS D. & PRADE H., 1984. "Théorie des possibilités". Ed. Masson Paris.

GREMY F., 1987. "Aide à la décision : les méthodes classiques", in "Informatique médicale", Ed. Médecine-Sciences Flammarion.

LEBBE J. & VIGNES R., 1988. "Les objets symboliques qualifiés : application au diagnostic médical". Deuxièmes journées Symbolique - Numérique, LRI, Orsay 12-13 décembre 1988, 127-139.

LEBBE J. & VIGNES R., 1989. "Représentation des connaissances dans les sciences de la vie : l'approche des objets symboliques". XXI emes journées de statistiques, Rennes 22-26 mai 1989, 155-159.

SHAFER G., 1976. "A mathematical theory of evidence". Princeton University Press.

VIGNES R. & LEBBE J., 1987. "Un système générateur de graphes d'identification d'objets symboliques", in premières journées Symbolique - Numérique, Université Paris Dauphine, Paris 8-9 décembre 1987, 143-163.

VIGNES R. & LEBBE J., 1989. "Transformations de variables dans le cadre de la création de graphes d'identification d'objets symboliques". XXI emes journées de statistiques, Rennes 22-26 mai 1989, 279-283.

ZADEH L.A., 1965. "Fuzzy sets". Information and control 8: 338-353.

Fig. 1 : List of concepts

1 Viral hepatitis A
2 Viral hepatitis A and chronic hepatitis B
3 Viral hepatitis B
4 Viral hepatitis B detected with sensitive markers
5 Viral hepatitis A and B
6 Co viral hepatitis Delta and B
7 Co viral hepatitis Delta and B detected with sensitive markers
8 Delta and chronic hepatitis B
9 Viral hepatitis CMV
10 Viral hepatitis CMV and chronic hepatitis B
11 Viral hepatitis EBV
12 Viral hepatitis EBV and chronic hepatitis B
13 Viral hepatitis no A no B and chronic hepatitis B
14 Viral hepatitis no A no B or no viral hepatitis

Fig. 2 : List of variables

1 IgM anti HAV
2 IgM anti HBc
3 Ag HBs
4 Ac anti Delta
5 Sensitive marker of virus B
6 IgG anti EBNA
7 IgM anti VCA
8 IgM anti CMV

Fig. 3 : Descriptions = Qualified assertion objects

Each diagnosis is described by a qualified assertion. A variable is either positive (+) or négative (-). In an assertion, if a variable has two possible and typical states (+ and -), then, to simplify, we don't write the event. For typicality values: T = typical, AT = atypical, and VAT = very atypical. When a variable has only one state (+ or -) in an assertion, the typicality value is always T and hence we don't note it.

a1 = [HAV = +] ^ [HBc = (+, VAT), (-, T)] ^ [HBs = -] ^ [Delta = -] ^ [SensB = -] ^ [VCA = (+, VAT), (-, T)] ^ [CMV = -]
a2 = [HAV = +] ^ [HBc = -] ^ [HBs = +] ^ [Delta = -] ^ [SensB = -] ^ [VCA = (+, VAT), (-, T)] ^ [CMV = -]
a3 = [HAV = -] ^ [HBc = +] ^ [HBs = (+, T), (-, AT)] ^ [Delta = -] ^ [SensB = (+, T), (-, VAT)] ^ [VCA = -] ^ [CMV = -]
a4 = [HAV = -] ^ [HBc = -] ^ [HBs = -] ^ [Delta = -] ^ [SensB = +] ^ [VCA = -] ^ [CMV = -]
a5 = [HAV = +] ^ [HBc = +] ^ [HBs = (+, T), (- AT)] ^ [Delta = -] ^ [SensB = (+, T), (-, VAT)] ^ [VCA = -] ^ [CMV = -]
a6 = [HAV = -] ^ [HBc = +] ^ [HBs = (+, T), (-, AT)] ^ [Delta = +] ^ [SensB = (+, T), (-, VAT)] ^ [VCA = -] ^ [CMV = -]
a7 = [HAV = -] ^ [HBc = -] ^ [HBs = -] ^ [Delta = +] ^ [SensB = +] ^ [VCA = -] ^ [CMV = -]
a8 = [HAV = -] ^ [HBc = -] ^ [HBs = +] ^ [Delta = +] ^ [SensB = -] ^ [VCA = -] ^ [CMV = -]
a9 = [HAV = -] ^ [HBc = -] ^ [HBs = -] ^ [Delta = -] ^ [SensB = -] ^ [VCA = -] ^ [CMV = +]
a10 = [HAV = -] ^ [HBc = -] ^ [HBs = +] ^ [Delta = -] ^ [SensB = -] ^ [VCA = -] ^ [CMV = +]
a11 = [HAV = -] ^ [HBc = -] ^ [HBs = -] ^ [Delta = -] ^ [SensB = -] ^ [EBNA = -] ^ [VCA = +] ^ [CMV = -]
a12 = [HAV = -] ^ [HBc = -] ^ [HBs = +] ^ [Delta = -] ^ [SensB = -] ^ [EBNA = -] ^ [VCA = -] ^ [CMV = +]
a13 = [HAV = -] ^ [HBc = -] ^ [HBs = +] ^ [Delta = -] ^ [SensB = -] ^ [VCA = -] ^ [CMV = -]
a14 = [HAV = -] ^ [HBc = -] ^ [HBs = -] ^ [Delta = -] ^ [SensB = -] ^ [VCA = -] ^ [CMV = -]

Fig. 4 : Identification graph

This identification graph is read from top to bottom. The first node is Ag HBs. Two branchs start from this node (+) and (-); they have the same increment and follow number "1)". If marker HBs is present, the identification continues to second line (HBc), else the identification continues with the line following HBs (-). The process stops when a conclusion is reached (noted ⇒). Each diagnosis of the conclusion is associated with a typicality value.

```
1) HBs (+)
   2) HBc (+)
      3) HAV (+) ------------- ⇒ CoHep. A and B Typical
      3) HAV (-)
         4) Delta (+) -------- ⇒ CoHep. B and Delta Typical
         4) Delta (-) -------- ⇒ Hep. B Typical
   2) HBc (-)
      3) HAV (+) ------------- ⇒ Hep. A (Chronic B) Typical
      3) HAV (-)
         4) Delta (+) -------- ⇒ Hep. Delta (Chronic B) Typical
         4) Delta (-)
            5) VCA (+) ------- ⇒ Hep. EBV (Chronic B) Typical
            5) VCA (-)
              6) CMV (+) ----- ⇒ Hep. CMV (Chronic B) Typical
              6) CMV (-) ----- ⇒ Hep. nA nB (Chronic B) Typical
1) HBs (-)
   2) SensB (+)
      3) Delta (+)
         4) HBc (+) ---------- ⇒ CoHep. B and Delta Atypical
         4) HBc (-) ---------- ⇒ CoHep. B and Delta (Sens. mark.) Typical
      3) Delta (-)
         4) HAV (+) ---------- ⇒ CoHep. A and B Atypical
         4) HAV (-)
            5) HBc (+) ------- ⇒ Hep. B Atypical
            5) HBc (-) ------- ⇒ Hep. B (Sens. mark.) Typical
   2) SensB (-)
      3) HAV (+)
         4) HBc (+) ---------- ⇒ Hep. A Very atyp. or CoHep. A and B Very atyp.
         4) HBc (-) ---------- ⇒ Hep. A Typical
      3) HAV (-)
         4) VCA (+) ---------- ⇒ Hep. EBV Typical
         4) VCA (-)
            5) CMV (+) ------- ⇒ Hep. CMV Typical
            5) CMV (-)
               6) HBc (+)
                  7) Delta (+) ⇒ CoHep. B and Delta Very atypical
                  7) Delta (-) ⇒ Hep. B Very atypical
               6) HBc (-) ---- ⇒ Hep. nA nB or Hep. no viral Typical
```

AUTOMATIC KNOWLEDGE BASES CONSTRUCTION

Jean Gabriel GANASCIA
LAFORIA
UA 1095 - C.N.R.S.
Université Pierre et Marie Curie
Tour 45-46, 4 Place Jussieu
75252 Paris, CEDEX, FRANCE

Abstract : Designed for an operational prospect, the CHARADE system automatically learns consistent rule systems from a description language, a set of axioms reflecting the language semantics and a set of examples expressed in that language. The recommended technique is based on the use of Bolean lattices.Thanks their mathematical properties, constraints, reflecting the properties of the rule systems to be obtained, may be implemented to control the combinatorics inherent in such a research. The following article is a general presentation of the CHARADE system, the aim for which it was designed and the mechanisms used.

1. INTRODUCTION

In its conventional meaning, the notion of rule refers to the rational foundation of thought in two ways: events are self repeating and reasonings indefinitely reproducible..., equally, in the set of mirrors displayed by Artificial Intelligence on the spiritual scene, production rule systems must meet two requirements: to denote correlations between events and simulate semantic continuity reasonings with the support of these correlations. However, for complex and intelligible reasonings to be constructed, rules must elaborate into a system, i.e. they must be able of chaining together to form deductions and each of them, independently, must correspond to a readable and explicit atom of knowledge. Besides, in a pragmatic prospect, that of knowledge base construction and maintenance, the global properties of the rule system play a central part, they will allow to assist the specialists at the time of transfer of expertise, to prevent errors, and possibly to correct them. Yet there is a major gap between individual rule juxtaposition and a complete and efficient usable rule system. CHARADE, a program designed to automatically acquire rule systems from examples, proposes to bridge that gap. CHARADE has been designed to detect logical or statistical regularities existing in a set of examples and generate production rule systems which, reflecting such regularities, can operate on commercial inference engines.

2. CONSTRUCTION OF RULES BY GENERALIZATION.

To fully understand the incentives which have led to the implementation of the CHARADE system, it is useful to recall the characteristics of conventional systems of rule learning by generalization.

2.1. Generalization from the specific to the general.

Michalski INDUCE system [Michalski 83] is the most typical example of this type of approach and can be schematically expressed as follows:

Given

- A concept **C**
- A set of examples $E_1...E_n$ and of counter-examples $CE_1...CE_p$ of **C**

Stage 1: Construction of a generalization of $E_1...E_n$ discriminating $CE_1...CE_p$. Call it E_g

Stage 2: Construction of the rule $E_g \rightarrow \mathbf{C}$

Many papers (Cf. [Michalski 83] [Mitchell 82], [Plotkin 70], [Kodratoff & Ganascia 86], etc.) have shown the difficulties to elaborate a consistent theory of generalization, thus, we shall not dwell on this aspect, except to recall a few aspects of the generality relation, and, more specifically, their formal translation:

– **Constant variabilization:** if x is a variable, A a constant, P a predicate and $P(t_1, t_2, ..., t_n)$ an atom then $P(x, t_2, ..., t_n)$ is more general than $P(A, t_2, ..., t_n)$.
We shall note it as $P(x, t_2, ..., t_n) \leq P(A, t_2, ..., t_n)$.
– **Use of the dropping rule:** if P1 and P2 are two propositions, P1≤P1 & P2.
– **Climbing in a conceptual hierarchy:** given an elementary conceptual hierarchy to explicit relations of generality between plane figures (Cf. fig. 2). In such a hierarchy, POLYGON<SQUARE and thus the proposition (SHAPE = POLYGON) & (COLOR = RED) is more general than the proposition (SHAPE = SQUARE) & (COLOR = RED).

– **Extension of boundaries** in the case of valued attributes on ordered sets: if SIZE is an attribute which has values over the set of integers, (SIZE ≥ 7) is more general than (SIZE ≥ 37) and thus, (SIZE ∈ [-7, 7]) ≤ (SIZE ∈ [-5, 5]).
Also, whatever the formal theory of generalization chosen, an order of priority as to the potential generalizations must be indicated. In fact, exactly as man does not explore all the possibilities, and is able to limit his search to the point of only exploring a minute part of the space of generalizations, even with the risk of omitting some, a machine cannot explore the whole domain of possible, it is too vast. To do so, constraints are defined which, as they propagate in the space of generalizations, limit the research combinatorics. This set of constraints defines what T. Mitchell [Mitchell 82] has named the "learning bias". Without a "learning bias" there is, strictcly speaking, no real induction as the whole space of the potential generalizations has to be explored, on the other hand, if the "learning bias" imposes to many constraints to the exploration procedure, it is then possible that the good generalization be never studied.

Although this technique has been confirmed by major successes, it can be criticized on many respects. In fact, generally, the rules constructed cannot be used directly by an inference engine, and whenever they are, it is not without a major impoverishment of the resulting expert system. Among the causes of such an impoverishment, we may quote:

- Lack of rule chaining: A → B, B → C.
- Difficulties, or even impossibility, to introduce uncertainty.
- Lack of consistency and economy for the knowledge base: the rules are considered as isolated from one another.
- Last, the frequent use of heuristics during the generalization process which is not justified as these heuristics have no semantics.

2.2. Generalization from the general to the specific.

The building up of decision trees with the Quinlan method [Quinlan 83] and its numerous additions is a good representation of this approach. To summarize, we start from:

- A set of classes exclusive from one another: $\mathbf{C}_1...\mathbf{C}_k$
- A set of examples $E_1...E_n$ for each class $\mathbf{C}_i$

Stage 1 : Construct a decision tree to classify all the examples E_i
Stage 2 : Generate rules from the decision tree.

Just as we have criticized, on some respects, the strategy to construct rules by generalizing from the specific to the general, we could criticize the consequences of a generalization from the general to the specific:

- No rule chaining: A → B, B → C.
- No accounting for the description language semantics.
- Learning heuristics based on a numerical function.
- No accounting for the properties of the system of rules.

3. BUILDING RULES SYSTEMS

Modularity of production rule systems must not be deceptive: it is not enough for the rules, individually to make sense, they must also, as a whole, meet operational criteria as lack of redundancy and cycle, consistency, completeness, etc... To insure that such criteria are verified, knowledge acquisition assistance tools are built, and it is to be able to rid from such verifications that we hope to create systems to learn rules from examples. However, as stated above, in most cases, rule learning systems are limited to the acquisition of concept description from examples and counter-examples. To be convinced of it, let us study the similarity based learning system procedure in case of a bottom-up generalization: being a set of examples, $E(c_1) = \{e_1,e_2,\ldots,e_n\}$ and a set of counter-examples, $CE\ (c_1) = \{ce_1,ce_2,\ldots,ce_p\}$ of a concept c_1 a generalization is looked for: g_1 of $e_1,e_2,\ldots,e_n$ discriminating $ce_1,ce_2,\ldots,ce_p$. When g_1 has been found, the rule $g_1 \rightarrow c_1$ is generated.

In spite of the difficulty to introduce disjunction in the g_1 generalisation, (Cf. [Michalski 83], [Vere 80] etc.) the same operations must be done for all the concepts to be learnt. Moreover, once defined concepts c_1 and c_2, to define a concept c_3 so that $c_1\&c_2 \rightarrow c_3$, descriptors c_1 and c_2 must be introduced in the example descriptions, which means that rules chainings can be learnt only if they are predetermined and fixed. Thus, rules are isolated from one another during acquisition.

3.1. CHARADE : detection of regularities

Reversly, CHARADE does not refer to the notion of example or counter-example for separated concepts; it considers globally the set of examples, each of them being described by a logical formula, and explores the space of the elementary descriptors present in the examples conjunctions. Be $d_1\&d_2\&\ldots\&d_n$ one of those, CHARADE looks for descriptors $f_1, f_2, \ldots, f_p$ correlated to the set of examples covered by $d_1\&d_2\&\ldots\&d_n$, then it generates rule $d_1\&d_2\&\ldots\&d_n \rightarrow f_q, f_2, \ldots, f_m$, eliminating amongst $f_1, f_2, \ldots, f_p$ descriptors f_i already derived from $d_1\&d_2\&\ldots\&d_n$ through the rule system.

More precisely, to construct knowledge bases CHARADE [Ganascia 87] empirically looks for regularities which are present in the learning set. To do so, the set of parts of the learning set and the set of descriptor conjunctions are represented by two Boolean lattices. Besides the representation economy that they offer, the properties of the lattices are used in the learning process to facilitate the detection of regularities. In fact, a regularity corresponds to a correlation empirically observed in the learning set. If all examples that have a descriptor d_1 in their description also have the descriptor d_2, it is possible to induce that d_1 implies d_2 in the learning set. The principle of induction used in symbolic learning consists of a generalization of this relation to the whole description space. Thus, these regularities must then be detected.

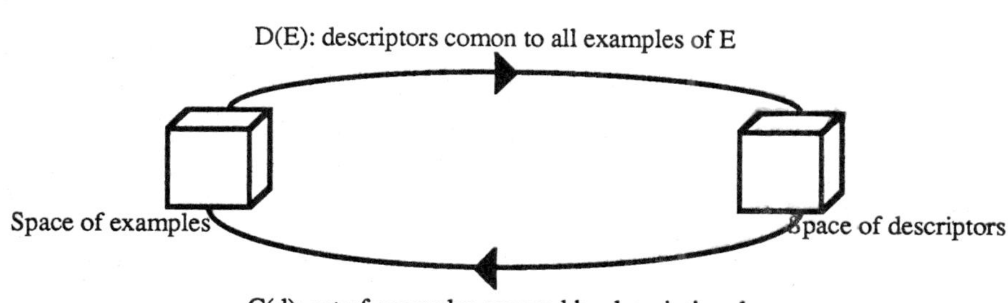

Two functions D and C are used (See figure). The first one, D, goes from the descriptors lattice to the examples lattice. To each description, it associates the set of examples of the learning set covered by this description. The other, C, relates to each subset of the learning set, i.e., to each element of the examples lattice, the set of descriptors present in the description of all the examples.

Example: Let us imagine a learning set made of three examples E_1, E_2, and E_3:
$E_1 = d_1\&d_2\&d_3\&d_4$
$E_2 = d_1\&d_2\&d_4\&d_5$
$E_3 = d_1\&d_2\&d_3\&d_4\&d_6$
An empirical regularity can be detected between $d_1\&d_2$ and d_4, in fact, all the examples described by $d_1\&d_2$ have also d_4 in their description. This is obtained with D and C as follows: $DoC(d_1\&d_2) = D(\{E1, E2, E3\}) = d_1\&d_2\&d_4$

This technique detects the certain rules; however, when building up a knowledge base, the uncertainty must also be detected and translated, such an uncertainty might be due to either the poor quality of the learning set or the rules which are uncertain by nature.
With a view to translating such uncertainties plausibility coefficients are frequently used. The latter are numbers comprised between +1 and -1 which alter the conclusions of the production rules, -1 representing the false and +1 the true.
Now, it is easy to translate statistical regularities as production rules with plausibility coefficients. It is sufficient to calculate conditional probabilities with frequencies, and then to translate the result from scale [0,1] to the scale [-1,+1]. So, if D is a descriptors conjunction, and if d' is a descriptor, the statistical correlation between D and d' is expressed by the following production rule: $D \rightarrow d'(\omega)$ where $\omega = 2*Pr(d'|D) - 1$

Example : to use once more the elementary example presented above, one obtains, among all the regularities detected, the following rules:
– Certain rule : $d_1\&d_2 \rightarrow d_4$ $[DoC(d_1\&d_2) = D(\{E1, E2, E3\}) = d_1\&d_2\&d_4]$
– Uncertain rule : $d_1\&d_2 \rightarrow d_3$ $(\pi = 0.33)$ $[\pi = 2*Pr(d_3 | d_1\&d_2) - 1 = 2*2/3 - 1]$

3.2. Exploration of the description space

Althoug we are able to detect and qualify regularities, it cannot be envisaged to detect all the possible regularities, as, on the one hand, the exploration procedure would be exponential (2^n conjunctions of descriptors, n representing the number of descriptors), on the other hand, the number of rules generated would be so large that they could not be managed by a conventional inference engine and, last, such rules would be redundant, which would lead both to useless duplications and to errors in the case of rules with plausibility coefficients. There are thus two problems:
– How to limit the number of rules ?
– How to limit the exploration ?

Without describing the detail of the mathematical formalism allowing the operations that we introduce, let us mention, in their general lines, the foundations on which the exploration procedure is based:
1) Use of the Boolean lattice structure of
– The set of parts of the learning set
– The description space (Set of descriptors conjunctions)
2) Exploration from the **GENERAL** to the **SPECIFIC**
3) Use of the properties of **rules**

As far as the properties of rules represent the equivalence between rules, their implementation permits to suppress useless rules. This avoids redundancies at the same time that it limits the exploration of the description space.
Among such properties some are related to exact rules, they reflect the properties of the logical implication, this is true for the two following properties:

– If $a \rightarrow b$ does not explore descriptions a&b&..., which implies that in our example, d_1&d_3, d_2&d_3, d_1&d_2&d_3... are **useless** as $d_3 \rightarrow d_1$&d_2

– If **C(a&b)** $\subseteq$ **C(a&c)** then descriptions of the type **a&b&c**... are **useless**.

Other properties are related to approximate rules, they reflect the properties of plausibility coefficients:

– If $a \rightarrow b(\omega)$ rules of the type a&... $\rightarrow$ b(ω'), are **useless** when $\omega \equiv \omega'$

4) Use of the properties of the **system of rules** to be constructed.

Just as we introduced the properties of rules to avoid exploration of descriptions which, a-priori, are not fruitful and to limit the creation of rules to those which are really useful, we use the properties of the system of rules that we want to build up. This is the most original aspect of CHARADE as, thanks to such properties, it establishes clear semantic for the generalization heuristics that are usually applied, at the same time that it introduces new heuristics. Thus, these properties become the operational characteristics of the expert system that we want to obtain. In fact, to build up a knowledge base, one must refer to the finality of the desired expert system. Otherwise, it would be a collection of rules, many of which useless, some corresponding to random coincidences, others having no practical interest whatsoever. These characteristics are used, sometimes, to create knowledge acquisition tools to verify the relevancy of the rules given by the user. The originality of the CHARADE system, is to translate these characteristics into constraints for the exploration procedure of the description space. As an illustration, we present below, some of these properties:

– **Goal** of the rules system (diagnosis, classification...). In the case of classification systems, if $\mathbf{C_1, ..., C_n}$ are classes, whenever a rule of the type $\mathbf{a \rightarrow C_i}$ is generated, descriptions more general than **a** (i.e. **a&...**) are **useless**.

– **Minimum number of examples** covered by the rule premise: if card(C(a))$\leq \nu$ then descriptions more general than **a** (i.e. **a&...**) are useless as **card**(C(**a&...**))$\leq \nu$

– **A-priori structure of the rule system:** when an expert system is constructed, the structure of the system of rules that we want to generate is known a-priori. Thus, for a therapeutical aid, whether in medicine or vegetal pathology, one knows beforehand that the rules go from the symptoms to the disease and then from the disease to the relevant therapy. This is what is reflected in the diagram below.

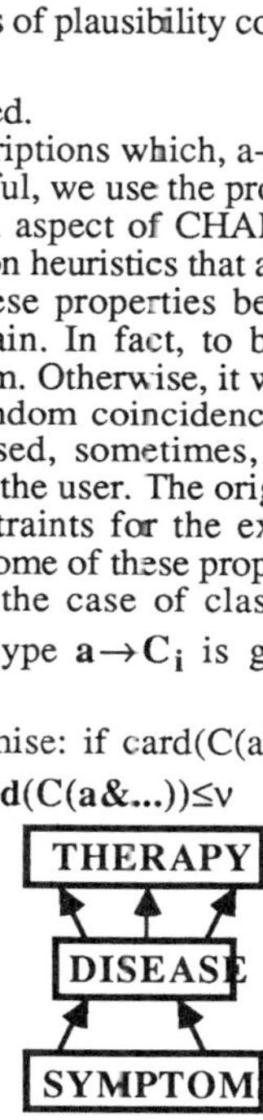

With this hierarchy, it is then possible to leave aside the exploration of the descriptions which correspond to a therapy and the study of regularities going from the diseases towards the symptoms of from symptoms to symptoms.

More generally, it is possible to introduce all the properties **P** such as when description **D** becomes useless because of the property **P,** then **D&...** is useless because of the same property.

4. REPRESENTATION FORMALISM

In CHARADE, the examples are represented as descriptors conjunctions. Each descriptor is a triplet (<Attribute><Selector><Value>).

The generalization can then be limited to the intersection of descriptors common to the various formulae, that is at the intersection of triplets (<Attribute><Selector><Value>) composing each formula. Working in propositional logics, there is no room for introducing the notion of constant variabilization in the generalization process, but one would still like to introduce the properties of descriptors or their mutual relations, which seems impossible though a mere use of elementary operations on sets.

For a greater clarity, consider the two expressions E1 and E2:

E1 = (Size = 3) & (Shape = Square) & (Couleur $\in$ {red, green})

E2 = (Size = 5) & (Shape = Triangle) & (Couleur ∈ {red, blue})

Although no descriptor belongs simultaneously to both formulae, the examples they refer to have many aspects in common: they are polygons, they both contain red, and their size is comprised between 3 and 5. One would like to emphazise these common characteristics in the generalization process. We shall demonstrate in the following paragraph that, if we take into account the description language axioms, the intersection of descriptors is sufficient to obtain such characteristics.

4.1. The description language

It can be considered as two consistent description sub-languages, the source language in which the examples are expressed, and the target language into which the generalizations are to be translated. The former example shows that, when reduced to the sole triplets (<Attribute><Selector><Value>) present in the examples, the target language is too limited to represent the generalization of E1 and E2. This means that the source language must be expanded and all the implicit relations between descriptors must be made explicit.

Now, all these relations derive directly from the attributes properties. It is then sufficient to express the properties of each attribute when defining the source language, to introduce them in the description of each example. But this, requires, as a prerequisite, an accurate definition of such properties and of the modalities according to which they can be introduced into the example descriptions. To do so, we chose to classify the attributes according to their type, the type being related to the *domain of values* which can be associated to an attibute, to a *set of selectors* defining the modalities of such associations and to a set of *axioms* creating logical relations between the various descriptors built up with the help of an attribute.

In the example described above, three types of attributes can be identified: the ordered type to which the attribute *Size* belongs, the hierarchical type to which the attribute *Shape* belongs and the set content type to which the attribute *Color* belongs. Obviously, it is possible to define as many types as desired, as the types enumerated-monovalued, Boolean, etc., on the condition that they are properly described, i.e., define for each of them the *domain of values*, the *set of selectors* and a set of *axioms*.

As an example, we give below the description of the type ordered mentioned above:

Ordered type :

domain : set provided with a complete order relation
Selectors : <,≤,>,≥.
Accepted selector : ≐
Syntactical transformation : (A = B) → (A ≥ B) & (A ≤ B)
Axioms :
[v1 < v2] (A ≤ v1) → (A ≤ v2)
[] (A < v1) → (A ≤ v1)
[] (A1< A2)&(A2 ≤v2) → (A1 ≤ v2) etc...
Lower boundary : [optional]
Upper boundary : [optional]

Once they are well defined, the types allow to characterize each attribute as a domain associated type, thus, *Size* is an *ordered type* for the integers or floating numbers accepting a lower and an upper boundary. Note that we establish in this case a distinction between the selectors <,≤,≥ and > and the accepted selectors, =. The first ones are related to the properties of attributes via the axioms, whereas the second ones are only considered as a writing convenience. Each accepted selector is associated to a syntactical tranformation which rewrites this selector containing descriptor into one or several descriptors containing the selectors of the corresponding attribute. For instance, the descriptor (*Size* = 3) will be rewritten as (*Size* ≤ 3) & (*Size* ≥ 3). It is only after that such tranformations have been made that it becomes possible to translate the properties of attributes with the axioms, in this case, the fact that the size of the two examples E1 and E2 is larger than 3 and smaller than 5.

In the hierarchical types, the axiom is labeled "addition", which means that one adds to the example description all the conclusions that can be drawn from the axiom. In other words, one goes up into the hierarchy to introduce in the example description all its properties inherited via the hierarchy. In

the example E2, the descriptor (*Shape* = triangle) will be translated into (*Shape* ≤ triangle) which will generate the descriptor (*Shape* ≤ polygon). As to the attribute *Color*, it is of the set content type. According to what we have just stated, the descriptor (*Color* ∈ {red, green}) will be rewritten as (*Color* = red)&(*Color* = green).
Taking advantage of these comments, one could be tempted to generate all the descriptors that are implicitly linked to one of the descriptors via an axiom, but this would be very cumbersome, or even impossible in some instances as that of *Size* where they are infinite.
Instead, we propose to generate only the descriptors likely to play a part in the generalization, i.e., those that play a part in other examples. Thus, if the two former examples, E1 and E2, are present at the same time, all the descriptors will be first tranformed by the syntactical transformations, so that the only selectors to intervene in the descriptors will be those that are specific to the attribute to which they are related. Simultaneously with this first transformation, the axioms labeled {addition} will be triggered.
After this transformation E1 and E2 are described by:

E1 = (*Size* ≤ 3) & (*Size* ≥ 3) & (*Size* ≤ 5) & (*Shape* ≤ Square) & (*Shape* ≤ Rectangle) & (*Shape* ≤ Quadrilateral) & (*Shape* ≤ Polygon) & (*Color* = green) & (*Color* = red)
E2 = (*Size* ≤ 5) & (*Size* ≥ 5) & (*Size* ≥ 3) & (*Shape* ≤ Triangle) & (*Shape* ≤ Polygon) & (*Color* = blue) & (*Color* = red)

For the latter, the descriptors common to E1 and E2 give a generalization gen(E1, E2) in agreement with the intuition one could have:

gen(E1, E2) = (*Size* ≥ 3) & (*Size* ≤ 5) & (*Shape* ≤ Polygon) & (*Color* = red)

Without going into the detail of the completion operations, it is possible to note that if limiting oneself to the descriptors present on the learning set, there is neither a risk of explosion nor of making a loop. The axioms are only used to define the set of relations existing between the target language descriptors, without increasing it indefinitely.

4.2. Learning heuristics

We have seen that CHARADE empirically detects regularities in an intelligent exploration of the description space. We have also demonstrated how the constraints on the exploration procedure derive from the properties of the rules and rules system that we wish to build up. On one hand, these properties represent a bias, in the meaning explained above, as they determine the nature of the results of learning at the same time that they constrive the induction procedure. On the other hand, they have a clear meaning as they describe the operational characteristics of the expert systems for which we intend to construct the knowledge base.
Thanks to the translation of the results properties as constraints on the exploration procedure, the number of descriptions to be studied is considerably reduced. However, to understand better the true incidence of these properties on the complexity of the induction procedure, and the exact role to be played by the language, let us take an example:
Be D the set of descriptors of the source language. Be $\{d_1, d_2, ... d_n\}$ n descriptors linked together by logical implications: $d_1 \Rightarrow d_2$, $d_2 \Rightarrow d_3$,... $d_{n-1} \Rightarrow d_n$. Such links are frequent whenever there is an order relation between the values given to an attribute, for instance if "height" is a numerical attribute, (height ≤ 2) ⇒ (height ≤ 3) ⇒ ...⇒ (height ≤ 85).
Be K the maximum number of descriptions to be explored to scan, in the source language D, all the descriptions susceptible of giving rise to a regularity. One would like to calculate K', the maximum number of descriptors to explore to scan the target language $D \cup \{d_1, d_2, ... d_n\}$.
In the case where no semantical relation would link $d_1, d_2, ... d_n$, one would have: $K'*2^n$, as on the basis of each description of the original language, it would be necessary to derive 2^n descriptions. But, as, according to the constraints derived from the rules properties (Cf. above), if a⇒b, it is unnecessary to study descriptions of the type a&b, none of the descriptions of the type $d_i \& d_j \& ...$

need be studied as $\forall(i, j) \in [1,n]$ $d_i \Rightarrow d_j \vee d_j \Rightarrow d_i$. Only the descriptions of the type A or A&d_i must be studied, A being a conjunction of descriptors of the source language D. As a consequence, the maximum size of the space to be explored is $K' = K*(n-1)$.
Thus, the structuration of the description space conjugated with the provision of the properties of rules to be acquired minimizes the size of the description space which must absolutely be explored.

5. CONCLUSION

CHARADE has been implemented on Macintosh Plus. Programmed in Le_Lisp, it was tested in several fields, from tomato pathology to first call at bridge. Moreover, as regards classification, the system returns the examples that it has not been able to classify which allows a feed back on the description language. Last, the introduction of a noise factor and the acquisition of approximate rules limits the ill effects of mistaken descriptions. One can note that the parameter adjustment not only depends on properties of the rule system to be obtained but also on the quality of the examples. It differs according to wether we are using a small quantity of prototypical examples or a large quantity of examples, some of them being noisy, to wether the example descriptions are similar or different, to wether the number of descriptors per example is large or small, etc...
In conclusion, this work opens onto two prospects: on one hand, continuing the formalization initiated, it should be possible to discover a new field at the border of data analysis and symbolic learning; on the other, after a systematic study of the sets of rules acquired and their adequation to the aim proposed, we should be in a position to assess the description language, and possibly to transform it. In the short run, this very ambitious goal could make way to another: to supply assistance for the classificaion and choice of relevant descriptors. This assistance capability is promoted by the set of axioms facilitating the enrichment of descriptions.

6. REFERENCES

[Davis & Lenat 81] Davis R. and Lenat D. B., *Knowledge Based Systems in Artificial Intelligence.* HPP 76-7 Memo and AI Memo 283, Stanford University, July 1976.

[Ganascia 87] Ganascia J.-G., "AGAPE et CHARADE: deux mécanismes d'apprentissage symbolique appliqués à la construction de bases de connaissance", thèse d'état, Université Paris-Sud, Mai 1987.

[Hayes-Roth & McDermott 78] Hayes-Roth F., McDermott J., *An Interference Matching Technique for Inducing Abstractions, CACM*, Vol. 21, PP. 401-411, 1978.

[Kodratoff & Ganascia 86] Kodratoff Y., Ganascia J.-G., *Improving the generalization step in learning,* in *Machine Learning : an artificial intelligence approach*, Vol. 2, Eds. R. S. Michalski, J. G. Carbonell and T. M. Mitchell, Pub. Morgan Kaufmann, 1986, pp. 215-244.

[Michalski 83] Michalski R. S., *A theory and methodology of inductive learning,* in *Machine Learning : an artificial intelligence approach* , Eds. R. S. Michalski, J. G. Carbonell and T. M. Mitchell, Pub. TIOGA publishing company, Palo Alto, California, 1983, pp.83-134.

[Mitchell 1978] Mitchell T. "Version Space: An approach to concept learning", PhD Thesis, Stanford University, 1978.

[Mitchell 1982] Mitchell T. "Generalization as Search", Artificial Intelligence **18**, 1982, 203-226.

[Plotkin 1970] Plotkin G., "A note on inductive generalization", in *Machine intelligence* 5, pp. 153-163, eds. Meltzer B. & Michie D., Edinburgh University Press, 1970.

[Quinlan 83] Quinlan R. J., *Learning Efficient Classification Procedure and their Application to Chess End Game,* in *Machine Learning : an artificial intelligence approach* , Eds. R. S. Michalski, J. G. Carbonell and T. M. Mitchell, Pub. TIOGA publishing company, Palo Alto, California, 1983, pp.463-482.

[Vere 80] Vere S. A., *Multilevel Counterfactuals for Generalizations of relational concepts and productions,* Artificial Intelligence J., Vol. 14, 1980, pp. 139-164.

SYMBOLIC APPROACH OF THE ANALYSIS OF CHRONOLOGICAL SERIES IN VIEW TO FORECASTING

M.O. Mennessier, CNRS and USTL, F-34060 Montpellier (France)
E. Diday, Univ. Paris XI and INRIA, F-78153 Le Chesnay (France)

The here presented analysis of chronological series wasconceived to allow us the learning of series corresponding to observational data unregularly spaced in time and even missing during a long time range. In this case the classical analyses of time series are difficult to apply. The available data on the light curves of variable stars have this specific character : the method is applied so as to forecast them.

I. DATA.

Let $L=\{l_1 \ldots l_n \ldots l_N\}$ be a set of time series. We only treat here the case in which each serie is represented by a pseudo-periodic curve. So each curve may be considered as an ordered sequence of pseudo-cycles c_i.

We define two data tables :

T_1 on L

T_2 on the set C of the cycles.

	$y_1 \ldots y_p \ldots y_P$
l_1	
.	
l_n	
.	
.	
l_N	

Table T_1

	$z_1 \ldots z_q \ldots z_Q \ldots z_n\, z_c$
c_1	
.	
c_m	
.	
.	
c_M	

Table T_2

Each curve L_n is described by P variables y_p that can characterize its global features and/or the object to which the curve is related (mean value of extrema, mean amplitude, class of a partition to which the object belongs in a classification independant of the variables of T2.mass or age of the star in the case of the example of stellar light variations....).

Each pseudo-cycle c_m is described by :

. Q variables z_q characterizing its peculiar shape (maximum, minimum, time-length, possible bump,...),

. the variable $z_n : C \rightarrow L$ indicating the curve l_n to which c_m belongs

Z the variable $z_c : C \rightarrow \mathbb{N}$ giving its row in the ordered sequence l_n

Let us remark that it is not necessary for all variables z_q to be defined for all pseudo-cycles. It is even possible to have curves with missing data on entire cycles.

II. PROBLEM.

1.Exposition

The full problem consists of classifying the curves of L according to rules able to forecast their later behaviour.

We only present here the first step of the work:*the search for rules for which the premise is a horde on cycles and the conclusion is an assertion on a later cycle, among subsets of L a priori defined.* Definitions of an assertion and a horde are just recalled after, much details can be seen in Diday (1988).

2.Searching for rules.

We consider the assertion $\mathcal{J} = \wedge_p [y_p = V_p]$ on the curves, defined from the table T1. Let us denote $|\mathcal{J}|_L$ the extension of $\mathcal{J}$ on L.

Let us recall that the assertion $\mathcal{J}$ is the application $L \rightarrow \{true, false\}$ such as $\mathcal{J}(l) = true$ iff $\forall p \ \ y_p(l) \in V_p \ = \ O_p$

The extension $|\mathcal{J}|_L$ is the subset of L in which the assertion $\mathcal{J}$ is true.

Let be $J_K = (j_1, ..., j_{K-1}) \in \mathbb{N}^{K-1}$ and $C_{J_K} = C^K$, the set of K-uples $c'_{j_K} = (c'_1, ..., c'_k)$ of cycles of the same curve $l \in |\mathcal{J}|_L$ such as the cycle c'_k is the j_{k-1} cycle after c'_{k-1} i.e. if $r(c'_k)$ denotes the rank of the cycle c'_k in the ordered sequence of cycles composing one curve:

$$r(c'_k) = r(c'_{k-1}) + j_{k-1}$$

Then the problem consists of searching sets of J_K, V_{ik}, V_j to satisfy rules of the form :

$$R_{J_K} : h_{J_K}(c_1, ..., c_K) \text{ true} \rightarrow a_c(c_{K+1}) \text{ true } \forall (c_1, ..., c_{K+1}) \in C_{J_{K+1}}$$

- h_{J_K} is a horde on C_{J_K} :

$$h_{J_K} = \bigwedge_{k=1,K} [\bigwedge_i [z_i(c_k) = V_{ik}] \wedge [z_n(c_k) = l_n \in |\mathcal{J}|_L]]$$

i.e. the application $C_{J_K} \rightarrow \{true, false\}$ such as :

$$h_{J_K}(c'_{j_K}) = \text{true iff } z_i(c'_k) \in V_{ik} \quad \forall i \in i_k = \{1, ..., P\}$$

and $\forall k = 1,\ldots,K$

- $a_c = \wedge [z_j = V_j]$ is an assertion on a cycle such as :

$$a_c(c') = \text{true} \quad \text{iff} \quad z_j(c') \in V_j \quad \forall_j \in j = \{1,\ldots,P\}$$

3. Validation of a rule.

Let denote:

- n_{R_K} the number of the $c'_{J_{K+1}} \in C_{J_{K+1}}$ which satisfy the implication

$$I_{J_K} : h_{J_K}(c_1 \ldots c_K) \rightarrow a_c(c_{K+1})$$

- n_{J_K} the number of the $c'_{J_{K+1}} \in C_{J_{K+1}}$ for which the premisse h_{J_K} is true.
- n_{T_K} the number of $c'_{J_{K+1}}$ in which $z_i(c_k)$ and $z_j(c_{K+1})$ are not missing

$h_{J_K} \rightarrow a_c$ will be kept as a rule if it has required qualities of:

- efficiency, i.e. if $\tau(I_{J_K}) = n_{R_K} / n_{J_K} > s$
- frequency, i.e. if $f(I_{J_K}) = n_{J_K} / n_{T_K} < \alpha$

The thresholds s and α depend on the application and are to be chosen by the user.

III. PRODUCTION OF THE RULES.

1. Algorithm

Let K_T be the maximum value for which rules are searched. It depends on the characteristics of the curves l_n and is to be chosen by the user.

We scan the space of all possible solutions calculating the efficiency and the frequency of the implications I_{J_K} successively with $K=1$, $K=2,\ldots K=K_T$. *At each K-level the results on lower levels are used to restrict the number of implications for which its quality is computed* and so to escape from the combinatory explosion and to remain as close as possible to the exhaustivity. For this we use pruning rules described after.

2.Producing law of rules.

At each K-level, we only keep the implications satisfying the selection criteria as rules.

3.Optimization by pruning rules.

We argue with a fixed conclusion a_c.

a) At K_{t+1}-level, i.e. when investigating the implications with a premise $h_{t+1} = h_t \wedge a_{k'}$

with $a_{k'} = \bigwedge_i [z_i(c_{k'}) = V_{ik'}] \wedge [z_n(c_{k'}) = l_n \equiv |3|_L]$

and denoting t the index J_{K_t}.it will be needless to compute the qualities of the implications h_{t+1} such as:

$\tau(I_t) = o$ then $\tau(I_{t+1}) = o$ (criterion Co)

and $f(I_t) = o$ then $f(I_{t+1}) = o$ (criterion C1).

It will also be needless to compute the quality of the implications such as (criterion C2) :

- the extensions $|h_t|_L$ and $|a_{k'}|_L$ are the same, then $f(I_{t+1})=f(I_t)$ and $\tau(I_{t+1})=\tau(I_t)$.
- $|a_{k'}|_L \equiv |h_t|_L$ then $f(I_{t+1})=f(I_{k'})$ and $\tau(I_{t+1})=\tau(I_{k'})$.
- $|h_t|_L \equiv |a_{k'}|_L$ then $f(I_{t+1})=f(I_t)$ and $\tau(I_{t+1})=\tau(I_t)$.

b) At K_T-level, we will not compute the qualities of the implications for which we can foresee that:

$f(I_{t+1}) < \alpha$ (criterion C3) or $\tau(I_{t+1}) < s$ (criterion C4)

$f(I_{t+1})$ and $\tau(I_{t+1})$ can be written with the help of the frequencies and the efficiencies of the implications $I_{t''}$ with $t' \leq t$, and the correlation functions of the assertions of I_{t+1} between themselves.(Diday and Mennessier, 1988). Let us remark that f and τ do not necessarily vary monotonously with t.

IV.APPLICATION TO LIGHT-CURVES OF VARIABLE STARS.

1.Data

Figure 1 shows such a typical curve. It represents the luminosity of a star as a function of time. It exhibites the irregularities of the observation dates and the long time range during which the data can be missing. We can see the variations of the pseudo-cycles between which the rules are searched: amplitude,time-length,shape variations.a. Table T1.

In this application, we consider assertions

$$\mathcal{3}_j = [y_1 = V_{1j}]$$

y_1 is characteristic of the cluster to which the star corresponding to the curve l_n belongs according to a previous classification of variable stars into three classes in relation with any physical properties (Mennessier,1985).Then rules are searched separately on three disconnected subsets of L (the extensions of the $\mathcal{3}_j$'s on L). The indices j=1 and j=3 correspond to physically homogeneous classes with

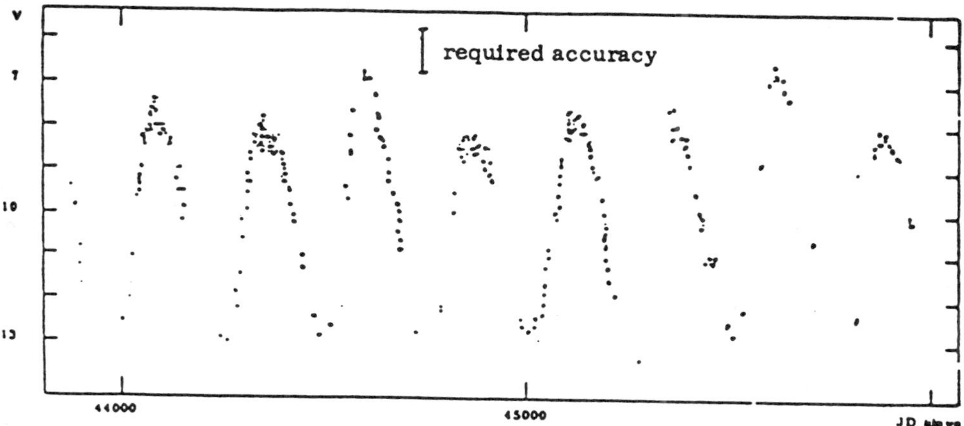

Figure 1. Light curve of a long period variable star T UMa- Data from AFOEV.

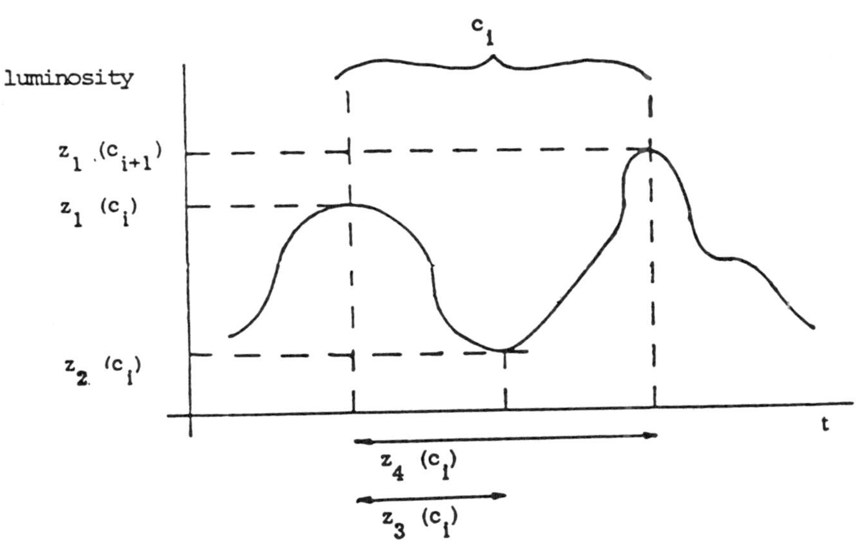

Figure 2.

quite distinct properties while j=2 corresponds to mixed physical populations.

b. Table T2.

In this first approach, we only use a limited observation number:dates and luminosities of successive observable extrema of 30 years light-curves (Campbell,1955).

The used z_q variables are shown on figure 2 : they are the luminosities of the extrema and the pseudo-period and pseudo-half-period of each pseudo-cycle.These variables are standardized on each curve :

$$z_q = (z_{q_o} - \bar{z}_{q_n}) / \sigma_{q_n} \qquad \text{with} \qquad \bar{z}_{q_n} = \sum_{c_i \in l_n} z_{q_o}(c_i)$$

σ_{q_n} denotes the standard deviation of z_{q_n} on l_n.

Concerning this application a good choise of V_q intervals is:

$$]-\infty, -0.5\sigma_{q_n}[\;,\; [-0.5\sigma_{q_n}, +0.5\sigma_{q_n}] \;,\;]+0.5\sigma_{q_n}, +\infty[$$

that leads to an equisharing between the intervals.

2. Algorithm of rules generation.

a.Choice of the thresholds.

The knowledge of the field of this application indicates that good values of the thresholds is s=0.8 and α=0.05. Moreover K_T=3 is a reasonable choice i.e. a maximum number of three pseudo-cycles will be considered in the premisse of a rule. So we obtain 4 variables with 3 modalities for each among the 3 pseudo-cycles considered in the premisse. There are 12 possible conclusions. Then there could be 432 (12x4x3x3) implications I_1, 7560 (12 C_{36}^2) implications I_2 and 85680 (12 C_{36}^3) implications I_3 .The quality of each of these implications must be computed if there is no optimization using pruning law.

b.Optimization.

The shortest computing time is obtained when the pruning criteria are applied beginning with the ones on the quantities that depend on the smaller number of variables (they will be tested in the most external loops). So the order of testing is: criterion C1, criterion C0, criterion C2, criterion C3, criterion C4. In this application with K_T=3:

- 1 % of the possible implications are eliminated by C1
- 1 %. of the remaining implications are eliminated by C0
- no implication is eliminated by C2
- 15 % of the remaining implications are eliminated by C3

- 5∅ % are eliminated by C4.

c.test

For a subset of curves, we have produced the rules by examining all the implications. The comparison with the case when we have used pruning rules shows that:

- all rules have been checked.
- the optimization reduced the computing time by a factor 2.

3. Obtained rules.

We find 126 rules for j=1, 11 for j=2 and 77 for j=3. Only one rule is common to j=)1 and j=3 ; all other rules are distinct. The number of rules for j=2 is low: this confirm the fact that the sample of stars defined by the extension $|3_2|_L$ is hetegeneous.

4. Forecasting.

As we have already mentionned it, the rules which have been obtained concern time series before 1955. We have tested these rules on time series between 1982 and 1987, by comparing the induced prevision to the observations (AFOEV,1983-1988). In case of a conflict we have chosen the majority rule; in case of equality there is an unsolvable conflict and no prevision can be made: this concerns only 1% of the tested cases.

All the obtained rules may not be applied because of missing data. In case of a value which has not been forecast, we take the most frequent one in the learning set.

Finally we have obtained the following results : 97% of forecast values are acceptable (i.e. 3% are completely false), 42% are exact.

The figure 3 shows the curve obtained by using the rules and simultaneously the observed values. The predicted smoothing curve is a sinusoide obtained by using the predicted extrema. It results that the curve predicted by the symbolic approach for these two stars is better than the one given by a classical numerical approach based on Fourier analysis.

The next work will consist in making a new classification by taking account of the rules obtained, the new classification will give more homogeneous classes and so more rules will be obtained and will improve the forecasting.

References.

AFOEV,1983-1988 Bulletin de l'Association Française des Observateurs d'Etoiles Variables n°23 à 44.

Campbell L.,1955 "Studies of long period variables" ed.AAVSO,Cambridge

Diday E.,1988 "The symbolic approach in clustering and related methods ofdata analysis : the basic choice" in "Classification and related methods of Data Analysis", ed. Bock, North Holland.

Diday E., Mennessier M.O.,1988 "Analyse symbolique pour la prévision de séries chronologiques pseudo-périodiques" 2° journées "Symbolique - Numérique" ,Orsay.

Mennessier M.O.,1985 "A classification of Miras from their visual and near-infrared light-curves : an attempt to correlate them with their evolution", Astron. Astrophys. 144, 463.

Mennessier M.O., Diday E.,1988 "Analysis of pseudo-periodic chronological series with irregularly time-spaced data in view to their prediction : comparison of several methods.", 4° International Symposium on applied Stochastic Models and Data Analysis ,Nancy.

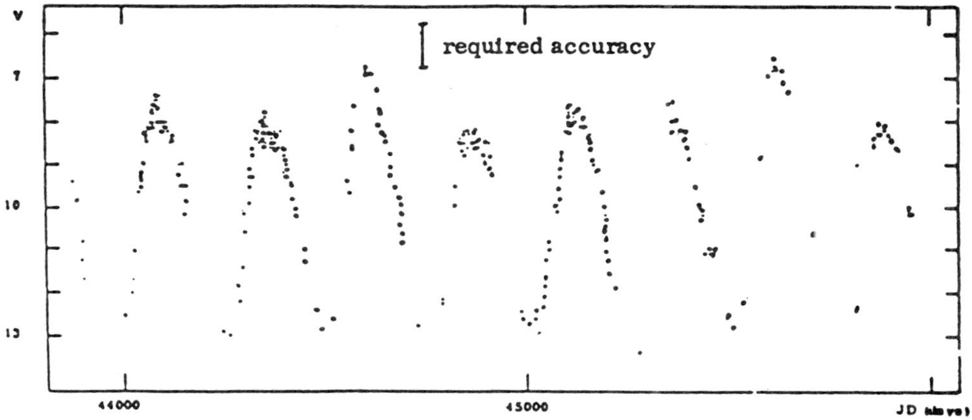

Figure 1. Light curve of a long period variable star T UMa- Data from AFOEV.

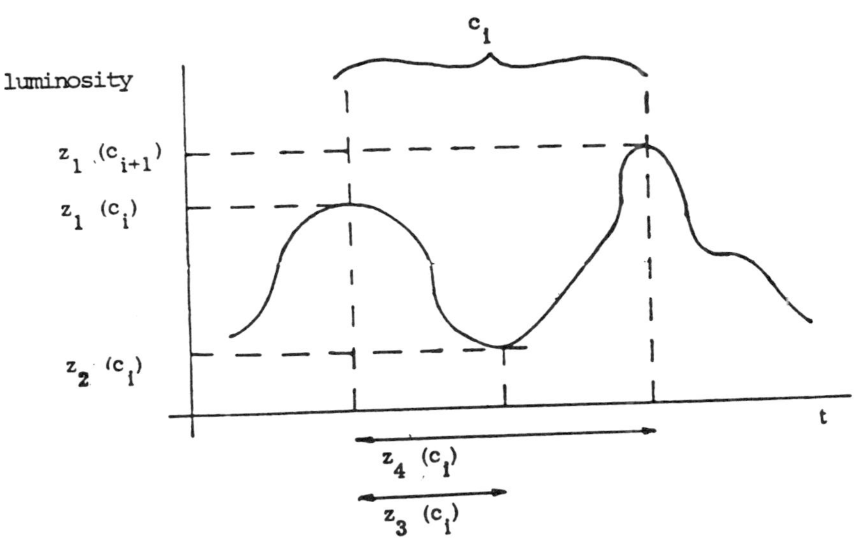

Figure 2.

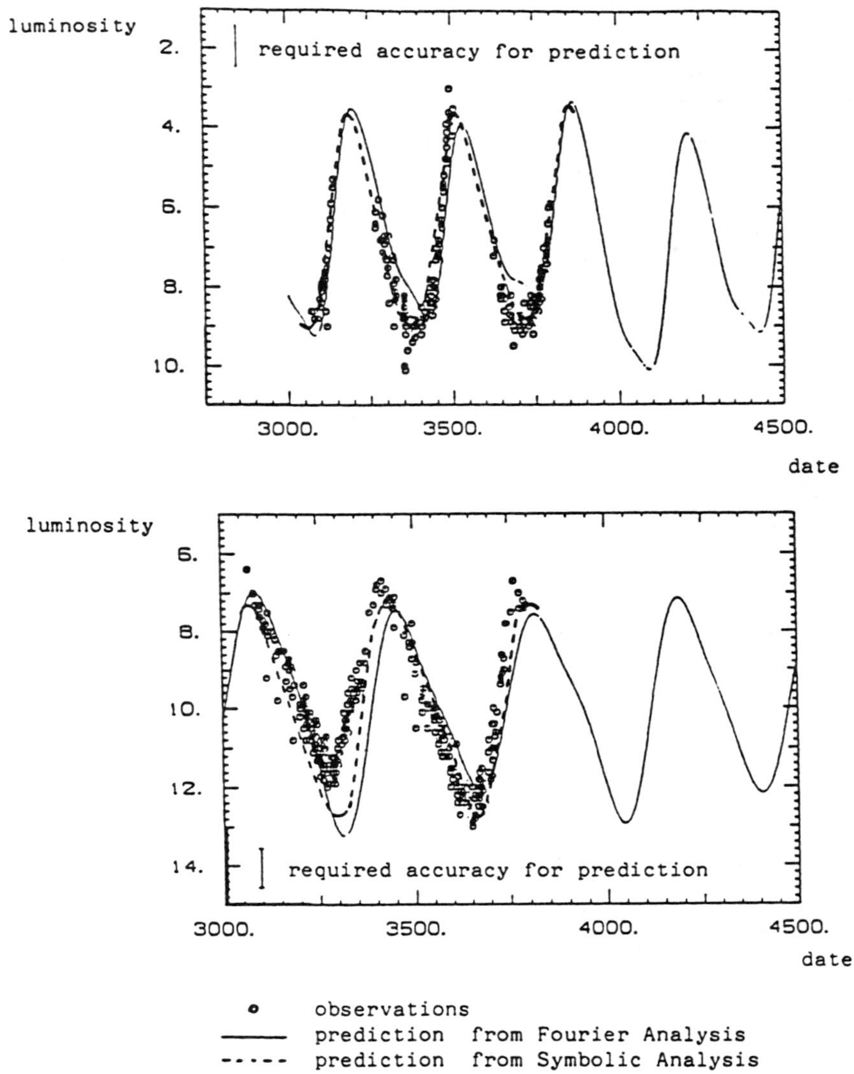
luminosity
2.
4.
6.
8.
10.
required accuracy for prediction
3000.
3500.
4000.
4500.
date
luminosity
6.
8.
10.
12.
14.
required accuracy for prediction
3000.
3500.
4000.
4500.
date
observations
prediction from Fourier Analysis
prediction from Symbolic Analysis

Figure 3

INDUCTIVE LEARNING, NUMERICAL CRITERIA AND COMBINATORIAL OPTIMIZATION, SOME RESULTS

Oliver Gascuel
CRIM, 860 rue de Saint Priest, 34100 - Montpellier (France)

1. Introduction

Several numerical learning methods are based on optimization. Given a set of examples, a parametric model, and a criterion, they determine the optimal values of the parameters according to the examples and the criterion at hand. The choice of the criterion is naturally based on its semantics. However, the mathematical properties of the criterion are also very determining. For example, convex criteria are often chosen because they are easy to minimize.

Several inductive (symbolic) methods [6, 10, 11, 12, 15 and 19; and in a certain way : 9 and 16] comply with a very similar schema. Given a set of examples, a descriptor space[1], and a(some) criterion(criteria) they search for the descriptor(s) which is(are) optimal, according to the examples and the criterion(criteria) at hand. The descriptor space is a discrete space which is partially ordered by the specialization-generalization relationship. The criteria are usually a combination of syntactical and numerical aspects. From an algorithmic point of view, these methods may be attached to combinatorial optimization.

Authors of the methods mentioned above looked into the meaning of the criteria they used rather than into their mathematical properties. They justified their choices by cognitive considerations, by the ease with which results are interpreted, by good results which have been obtained in particular domains etc ...

We adopt a different and complementary approach. We look into the links between the mathematical properties of numerical criteria and the algorithms which search the descriptor space. Our results show that, as in the numerical domain, "good" criteria allow "good" algorithms to be employed. More precisely, we show that criteria which possess certain properties, such as convexity, allow algorithms similar to the very classical branch-and-bound algorithms to be employed in a natural, efficient way.

This paper addresses the case where a unique numerical criterion is used (Part 3), and the case where two numerical criteria are used (Part 4). For brevity's sake, only two-class discrimination problems are considered here. However, most of the results presented here have a natural extension in n-class (n > 2) discrimination problems. This paper presents practical results rather than proofs. Complementary details on proofs, algorithms, and extensions to n-class problems are given in [14].

2. Background, definitions and notations :

2.1 Overview of the methods considered :

Inductive learning methods considered in this paper [6, 10, 11, 12, 15 and 19; and in a certain way : 9 and 16] have several common features :

- they deal with discrimination problems.
- they deal with noisy domains.
- they are all-at-once (rather than incremental).
- they use top-down search algorithms.
- they use numerical criteria for evaluating descriptors' performances.
- these numerical criteria are often combined with syntactical criteria which share a common aim : the descriptors, proposed as results, must be sufficiently general, or syntactically simple.

[1] By descriptor, we mean a rule left-hand side. The descriptor space is similar to the rule space [1].

Basically, numerical criteria are used for evaluating how good the trade-off is between goodness-of-fit and simplicity. Various solutions exist. Two solutions are given below (D is the descriptor, the responses of which on the training set are known, which has to be evaluated) :

- in PLAGE [12, 13], two numerical criteria are used : (i) a measure of simplicity, which is the observed frequency that D is true. The idea, which is of a heuristic nature, is that complex descriptors have a low probability of being true (see [5]). Therefor, if the observed frequency that a descriptor is true is sufficiently high, this descriptor has little chance to be complex. (ii) The Chi2 criterion, which is both a significance and a goodness-of-fit measure [2].

- in CN2 [11] two measures are used : (i) an entropy measure which is a pure goodness-of-fit measure; (ii) a significance measure : the likelihood ratio test [2].

Authors working towards various ends share in common optimization. For example :

- PLAGE searches for all minimal (the most general) descriptors which have a sufficient Chi2 and a sufficient simplicity.

- CN2 is based on the iterative use of an algorithm which searches for the descriptor which is best according to the entropy measure, which has a sufficient significance, and which contains less than K terms (descriptors are conjunctions).

In all cases (except [6]) the search is systematic in the sense that all descriptors of the space are either evaluated on the examples, or pruned because they may not obtain sufficient performances. Thus, these methods clearly depart from the methods based on the building of decision trees [8, 17, 18] which search a very small part of the descriptor space. In other words, in the case of binary data, decision tree methods examine few combinations of tests. Therefor, these methods are computationally more expensive than the decision tree methods, while they should provide best results. This has been shown by [10], on a real world application. Moreover, these methods (at least [6, 9, 12, 16]) deal with structural descriptions of examples, while pure decision tree methods seem to be more or less limited to attribute descriptions (see [7]).

2.2 Basic properties

Properties described in this paragraph are very general. They require only that the problems which are dealt with are discrimination problems and that they are approached in an all-at-once (non-incremental) manner.

Descriptors are boolean functions which are applied to examples. Descriptor spaces are partially ordered by the specialization-generalization relationship. This relationship verifies the following property :

(1) For every descriptor D, for every specialization SD of D and for every example e, we have : SD(e) => D(e).

This property is equivalent to the property :

(1)' For every descriptor D, for every specialization SD of D and for every set of examples SET, we have :
$|\{e \in SET / D(e) = TRUE\}| \geq |\{e \in SET / SD(e) = TRUE\}|$.

Let us consider the following notations :

L is the learning set, and S is the L's size;

L is partitioned in two classes : CLASS1 and CLASS2;

S1 and S2 are the CLASS1's and CLASS2's sizes (S1 + S2 = S).

Note that S, S1, and S2 are constant during the search (the method is all-at-once). The table of the responses of a descriptor D is called TBL(D) and is noted as follows :

	CLASS1	CLASS2	
TRUE	T1(D)	T2(D)	T(D)
FALSE	F1(D)	F2(D)	F(D)
	S1	S2	S

T(D) is the number of times D is TRUE; F(D) is the number of times D is FALSE; Ti(D) is the number of times D is TRUE on CLASSi; Fi(D) is the number of times D is FALSE on CLASSi.

(a) the Ti(D) and Fi(D) are bounded by the equality : Ti(D) + Fi(D) = Si;

(b) we have : Ø ≤ Ti(D) ≤ Si.

It follows from the constraint (a) that a table, TBL(D), may be unambiguously represented by the 2-dimensional vector with positive and integer coordinates : (T1(D), T2(D)). From now on, TBL(D) will be represented by this vector. The constraint (b) implies that TBL(D) belongs to the rectangle : [Ø, S1] x [Ø, S2].

When applying the property (1)' to the table TBL(SD) of a specialization of descriptor D, it directly follows :

(2) TBL(SD) belongs to the rectangle [Ø, T1(D)] x [Ø, T2(D)].

From now on, the rectangle [Ø, T1(D)] x [Ø, T2(D)] will be called Π(D). A graphic representation of the property (2) is given in Fig. 1.

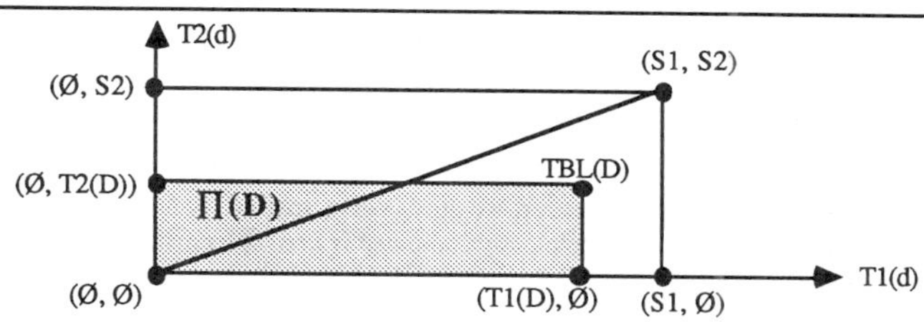

Every table TBL(D) may be represented by a point of the rectangle (Ø, Ø)(S1, Ø)(S1, S2) (Ø, S2). The point (S1, S2) corresponds to a descriptor which is always true (on the learning set); the point (Ø, Ø) corresponds to a descriptor which is always false; the two points (S1, Ø) and (Ø, S2) correspond to complete and consistent descriptors (see [16]). The points (with integer coordinates) of the two segments [(S1, Ø), (S1, S2)] and [(Ø, S2), (S1, S2)] correspond to complete descriptors. The points of the two segments [(Ø, Ø), (S1, Ø)] and [(Ø, Ø), (Ø, S2)] correspond to consistent descriptors. The diagonal of the rectangle, [(Ø, Ø)(S1, S2)], has for equation : T1/S1 = T2/S2. It represents descriptors the responses of which are rigorously independent of the class partition. In other words, it represents descriptors without any interest for the problem at hand, e.g. the Chi2 is equal to Ø along this diagonal. The property (2) indicates that for every specialization SD of D, TBL(SD) belongs to the grey rectangle Π(D) = (Ø, Ø)(T1(D), Ø)(TBL(D))(Ø, T2(D)).

Figure 1.

A numerical criterion C is a numerical function of the four variables : T1(D), F1(D), T2(D) and F2(D). Because constraint (a), C may be considered as a function of the two free variables T1(D) and T2(D). Below, for simplicity's sake, C(T1(D), F1(D), T2(D), F2(D)) will be noted C(D) or C(T1(D), T2(D)).

2.3 The pruning principle of branch-and-bound algorithms :

Inductive learning search algorithms may not be strictly identified with branch-and-bound algorithms. However, the pruning principle of the latter may be applied to methods such as those described in paragraph 2.1. Below, we will describe this principle using an inductive learning terminology. For simplicity's sake, we will assume, here and afterwards, that the search of the descriptor space is top-down. Principles and results presented herewith may be easily transposed to bottom-up search.

Let C be the numerical criterion, which is to be maximized[1]. The principle consists in having an "evaluation function of C", noted EC, which depends on C, and which verifies the following property :

(3) For every descriptor D and for every D's specialization SD, we have : EC(D) ≥ C(SD).

If, for example, the aim is to find the best descriptor of the space, the pruning is as follows :

(4) At each instant, we have for a current candidate : CANDIDATE. The descriptor space is searched in order to exhibit a descriptor D, such as C(D) > C(CANDIDATE). Each time such a descriptor is found, D becomes the new CANDIDATE. For every considered descriptor D, the search of the D's specializations is continued iff : EC(D) > C(CANDIDATE).

This pruning is safe in the sense that discarded descriptors may not be solutions. It must be distinguished from "probabilistic" pruning which prunes the parts of the space which have "little chance" to contain solutions.

An evaluation function E'C is better than an evaluation function EC when, for every descriptor D, we have : EC(D) ≥ E'C(D). When E'C is better than EC, the set of descriptors which may be pruned following E'C includes the set of descriptors which may be pruned following EC. When E'C is better than EC, the same search algorithm will run

[1] When the criterion has to be minimized, one considers the criterion - C.

faster[1] but it produces the same results. In the same way, the faster an evaluation function tends towards[2] Ø when specializing, the better it is. It has to be noted that an improvement of an evaluation function which is slight from a numerical point of view may lead to important run time gains.

3. A solution in the single criterion case

3.1 Principle

C is the criterion which has to be maximized. The problem is to exhibit an evaluation function (EC) which verifies (3). Let us consider the property (2) and Fig. 1. One may propose the following :

(5) EC(D) is equal to the maximum reached by C on ∏(D).

Clearly, this is a convenient solution. Moreover, we prove below that this is usually the best possible solution. However, a method allowing this maximum to be rapidly computed is needed so that this solution may be envisaged. Such a method is related to the mathematical properties of C. In the following paragraph, we examine the case where C is convex.

3.2 The case of convex criteria

In this paragraph, we consider that C is a convex function of the two variables T1(D) and T2(D)[3]. We will show that considering C as being convex is natural, that it leads to a solution which makes sense, and is quickly computed.

Considering that C is or may be convex is natural. C is a measure, and measure and convexity are often associated. This is because several measures are semi-norms and all semi-norms are convex. Among convex criteria, one finds the mutual information, the Chi2 criterion and the likelihood ratio test [2][proof in 14].

Convex functions have the following property [4] :

(6) If C is a convex function defined on a not-empty, bounded and convex polyhedron[4] ∏, then there exists at least a vertex V of ∏ such as C(V) is the C's maximum on ∏.

This property is on the basis of the simplex algorithm [3]. ∏(D) is a not-empty[5], bounded, and convex polyhedron. Thus, (6) may be applied to (5). We obtain :

(7) EC(D) = MAX(C(V); where V is one of the four vertices of ∏(D))

(7) may be interpreted simply. Let us consider Fig. 1. (7) indicates that the maximum of C on the grey rectangle ∏(D) is reached at one of the four points: (Ø, Ø), TBL(D), (T1(D), Ø) or (Ø, T2(D)). It seems that whatever C may be, the C's maximum will never be reached at (Ø, Ø) (a descriptor which is always false). If C is a measure of simplicity, such as the observed frequency that D is true (see 2.1), it is clear that its maximum is reached at TBL(D) and we have EC(D) = C(D). If C is a goodness-of-fit measure or a significance measure, one would think that its maximum is reached either at (T1(D), Ø) or at (Ø, T2(D)), which are the two "most consistent" and "most complete" points of ∏(D).

This last intuition (goodness-of-fit or significance measure => maximum reached either at (T1(D), Ø)) or at (Ø, T2(D))) may be easily shown in the case where C is convex and verifies the following property (proof in [14]) :

(8) C is constant[6] and minimal along the straight line T1/S1 = T2/S2.

---

[1] At least in the case where E'C and EC are computed in similar times.

[2] Or towards the C's minimum on [Ø, S1] x [Ø, S2], in the case where this minimum is different to Ø.

[3] One proves that, if C is a convex function of the 4 free variables T1, F1, T2 and F2, then it is also a convex function of the 4 variables T1, F1, T2 and F2 bounded by constraints (a) (see part 2.2) or equivalently a convex function of the 2 free variables T1 and T2. But the converse does not always hold. For instance, one proves that the mutual information and the Chi2 are convex when considered as functions of the 2 free variables T1 and T2, but are not convex when considered as 4 free variables functions.

[4] A convex polyhedron is the intersection of a finite number of half-closed spaces.

[5] Except in the trivial case where D is FALSE for all examples of the training set.

[6] Usually equal to Ø.

This line represents descriptors the responses of which are rigorously independent of the class partition (see Fig. 1). The most commonly used criteria (except measures of simplicity) verify (8), e.g. the Chi2, the mutual information and the likelihood ratio test.

3.3 Applying the solution (7) to non-convex criteria :

The solution (7) has such a simple interpretation that it leads one to think it may be applied to non-convex criteria. The following two results (proofs is [14]) go in that direction :

1- A sufficient condition for (7) to be applicable, is that C is convex with regard to each variables T1(D) and T2(D). When, as usual, C is twice differentiable, it is sufficient to prove that the two partial second derivatives : $\partial^2C/\partial T1^2$ and $\partial^2C/\partial T2^2$, are positive or null. This condition is less restricting than the convexity condition, and is often considerably easier to establish.

2- The solution (7) may be applied to criteria which are neither convex, nor convex with regard to each variable. For instance, the Mahanalobis distance which is used by Quinqueton et al. [19] is such a case[1].

Finally, we do not know any references concerning a criterion which makes sense in our framework, and to which (7) may not be applied. The convexity of the criterion is absolutely not a necessary condition. We point this out because several criteria are convex, and because convexity allows easy proofs. Moreover, the convexity is of a great interest in the two-criteria case.

3.4 Properties of this solution

Let us consider a criterion C to which (7) is applicable and a descriptor D, the responses of which on the training set are known.

Usually, when wondering about the necessity of searching the C's specializations, one does not have any information about the responses of these specializations (except the information reflected by the property (1)). Thus, the best possible evaluation function EC(D) is equal to the maximum reached by C on the points with integer coordinates which belong to Π(D). When considering that the four Π(D)'s vertices have integer coordinates, we deduce that (7) provides the best possible evaluation function of C.

Considering that for every D's specialization SD, we have : $\Pi(D) \supseteq \Pi(SD)$, we deduce : $EC(D) \geq EC(SD)$. Thus, the evaluation function EC decreases when specializing. For most of the criteria, EC is "almost always strictly decreasing" : if, as it almost always happens, we have : $\Pi(D) \supset \Pi(SD)$, then we also have : $EC(D) > EC(SD)$.

However, there are exceptions. For instance, the entropy measure used by CN2 [11], the Kullback's divergence and the Mahanalobis distance which are both used by Quinqueton et al. [19], present the following undesirable property : for every descriptor D, EC(D) is constant and equal to the maximum reached by C on [Ø, S1] x [Ø, S2]. In the case where C verifies this property, considering that (7) provides the best possible evaluation function of C, it follows that it is impossible to prune according to EC. In this case, C may not be the unique criterion, and the search has to be directed by one or more other criterion (as done in the papers mentioned above).

Finally, (7) considered as an algorithm, calls on the C's calculation, but is independent of it. Thus, it is very simple to implement a program in which several criteria are available and are chosen by users according to the problem at hand.

4. A solution in the two-criteria case :

4.1 Generalizing the pruning principle to the two-criteria case :

Let us consider now the case where the aim is to find descriptor(s) which is(are) "good", according to two numerical criteria C1 and C2, each having to be maximized. By looking at (4) it is clear that for each descriptor D the responses of which on the training set are known the basic question is as follows :

(9) Given two thresholds THR1 and THR2, is it certain that there does not exist any D's specialization SD such as : C1(SD) > THR1 and C2(SD) > THR2 ?

In order to answer this question, a first approach is to exhibit two evaluation functions EC1 and EC2, respectively attached to C1 and C2, which both verify (3). When $EC1(D) \leq THR1$ or $EC2(D) \leq THR2$, the answer to (9) is positive. However, this solution is not optimal. Let us consider the case where there exists two D's specializations SD1 and SD2, verifying C1(SD1) > THR1 and C2(SD2) > THR2, but there does not exist any D's specialization SD,

[1] However, the application of (7) to this distance produces a "null" evaluation function (see part 3.4).

verifying both C1(SD) > THR1 and C2(SD) > THR2. In this case, the answer to (9) will be inevitably negative, although it should be positive. The search on D's specialization will be continued despite its certain uselessness.

A better solution is the following : firstly, an evaluation function EC1 is available, attached to C1 and verifying (3); secondly, a "conditional evaluation function" is available, attached to C1 and C2, noted CondC1-EC2 (conditional on C1 evaluation function of C2) which verifies :

(10) For every D's specialization SD such as C1(SD) > THR1, we have : CondC1-EC2 (D, THR1) ≥ C2(SD).

If, for instance, the aim (similar to the one of CN2 [11]) is to find the descriptor verifying C1(D) > THR1, which is the best according to C2, the pruning is as follows:

At each instant, we have for a current candidate : CANDIDATE. The descriptor space is searched in order to exhibit a descriptor D, such as C1(D) > THR1 and C2(D) > C2(CANDIDATE). Each time such a descriptor is found, D becomes the new CANDIDATE. For every considered descriptor D, the search of the D's specializations is continued iff : EC1(D) >THR1 and CondC1-EC2(D,THR1) > C2(CANDIDATE).

When C1 and C2 have inverse roles : the aim is to find the descriptor, verifying C2(D) > THR2, which is the best according to C1, the condition for continuing the search is :

EC1(D) > C1(CANDIDATE) and CondC1-EC2(D, C1(CANDIDATE)) > THR2.

Finally, it appears that, whatever the C1's and C2's roles may be, what is important is the existence of an evaluation function for one of them (here, C1) and the existence of a conditional evaluation function for the other (here, C2). The choice of CondC1-EC2, rather than CondC2-EC1, will follow from their respective properties and not from their role.

The remarks of 2.3 concerning the evaluation functions' qualities may be extended to conditional evaluation functions.

4.2 Principle of the solution :

Let us consider the requirements which have been given in 4.1. EC1 has to verify (3). When possible, it will be determined by (7). CondC1-EC2 has to verify (10). As in the single-criterion case, the best possible evaluation function of C2, conditional on C1, is usually the maximum reached by C2, on the points P of Π(D) which verify C1(P) > THR1 and have integer coordinates. Unfortunately, except in particular cases (one of them is given below), this maximum may not be rapidly computed. It follows that constraints have to be somewhat relaxed[1], and we will consider that :

(11) CondC1-EC2 (D, THR1) is equal to the maximum reached by C2 on the points P of Π(D) which verify C1(P) ≥ THR1.

Still more than in the single-criterion case, the existence of a method for rapidly computing this maximum, is highly related to the mathematical properties of C1 and C2. In the following paragraph we consider a particular case where such a method does exist.

4.3 The case where a criterion is linear and the other convex

We will suppose that C1 is the linear criterion and C2 the convex one[2]. C1 is linear, thus it is convex, and (7) may be applied for determining EC1. The set of points P verifying C1(P) ≥ THR1 is a half-closed space. It follows that the set of points P which belong to Π(D) and verify C1(P) ≥ THR1 is a bounded convex polyhedron. It will be noted Π'(D, C1, THR1). In the case where Π'(D, C1, THR1) is not empty, (6) may be applied to (11) and we obtain :

(12) CondC1-EC2 (D, THR1) = Max (C2(V); where V is one of the vertices[3] of Π'(D, C1, THR1)).

In the case where Π'(D, C1, THR1) is empty it is consistent to stay CondC1-EC2(D, TRH1) = + ∞. A graphic interpretation of (12) is given in Fig. 2.

---

[1] The relaxation consists : (i) in replacing > by ≥; (ii) in suppressing the integer coordinate constraint.

[2] It is always possible to come at this case (see part 4.1).

[3] Π'(D, C1, THR1) has at most 5 vertices.

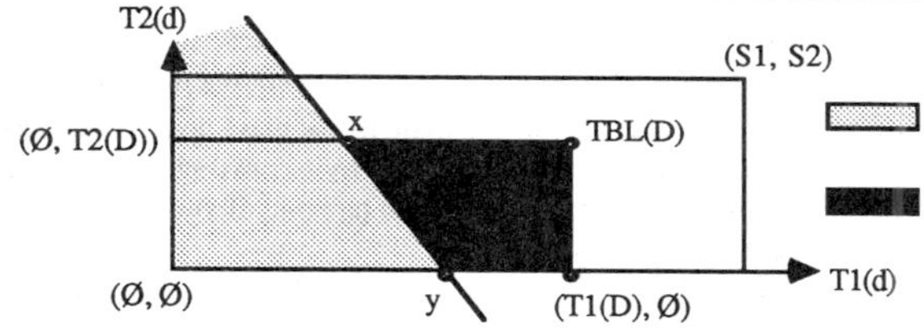

(12) indicates that CondC1-EC2 is equal to the maximum of C2(x), C2(y), C2(T1(D), Ø) and C2(TBL(D)).

Figure 2.

Two additional remarks may be done (proofs in [14]) :

1- In the case where C2 verify (8) (and is convex), the maximum of C2 on Π'(D, C1, THR1) is reached at one Π'(D, C1, THR1)'s vertex which belongs either to the half-open segment [(T1(D), Ø), TBL(D)[or to [(Ø, T2(D)), TBL(D)[. Such vertices are always exactly two[1] : x and (T1(D), Ø) in Fig. 2.

2- Let us consider the case where C1 is the measure of simplicity defined in 2.1 : C1(D) = observed frequency that D is true = (T1(D) + T2(D))/S. In this case, C1 is linear and we have :

when X.S is integer, all Π'(D, C1, X)'s vertices have integer coordinates;

when considering that T1(D) and T2(D) are integer, we also have :

(C1(D) > TRH1) <=> (T1(D) + T2(D) > S.TRH1) <=>

(T1(D) + T2(D) ≥ FLOOR (S.TRH1 + 1)) <=> (C1(D) ≥ (FLOOR (S.TRH1 + 1))/S)

where FLOOR (X) is, as usual, the largest integer not larger than X;

finally, by analogy with the result of part 3.4, it follows that the application of (12) to the polyhedron Π'(D, C1, (FLOOR(S.THR1 + 1))/S) produces the best possible evaluation function of C2, conditional on simplicity.

4.4 Properties of this solution

When considering that Π(D) ⊇ Π'(D, C1, THR1), it follows that the conditional C2's evaluation function is always better than (or equivalent to) the not-conditional C2's evaluation function.

In the case where C1 is the observed frequency that D is true, and C2 is convex, the pruning obtained by considering EC1 and CondC1-EC2 is the best possible. This means that the answer to (9) is negative (the search on D's specializations is continued) iff there exists at least a point P, which belongs to Π(D), has integer coordinates and verifies C1(P) > TRH1 and C2(P) > TRH2 (proof in [14]).

5. Conclusion :

This paper considers the links between the pruning of the descriptor space and the mathematical properties of numerical criteria. The basic idea is to use a pruning technique similar to the one used in branch-and-bound algorithms. This technique is based on the existence of an evaluation function attached to the considered criterion. It provides a safe pruning (discarded descriptors may not be solutions). A "good" criterion is a criterion for which a "good" evaluation function does exist, the computation of which is rapid. Our results may be summarized as follows :

- in the case where a unique numerical criterion is used, a method for calculating an evaluation function of the criterion does exist, which : (i) is computationally cheap; (ii) is independent of the criterion; (iii) is applicable to a wide class of criteria (making sense in our framework); (iiii) provides the best possible evaluation function of the criterion.

- however, there are criteria (making sense in our framework) for which the evaluation function, produced by this method, is "null", in the sense that it prunes no part of the descriptor space. Because this evaluation function is the best possible, these criteria may not be employed as unique criterion.

- in the case where two numerical criteria are used, of a particular interest are the couplings where one criterion is linear and the other convex. In this case, there is a way of pruning which : (i) requires few computations; (ii) is better than the pruning obtained by considering the two evaluation functions of the two criteria (and is at least equivalent). In the case where the linear criterion is the simplicity (observed frequency that the descriptor is true), the pruning which is obtained is the best possible.

Some of the results concerning the single criterion case, were probably more or less present in previous works. For instance, Clark et al. [11] speak of a look-ahead function (instead of an evaluation function) which is attached to the

[1] Except in the particular case where : C1(TBL(D)) = THR1.

likelihood ratio test they use. But they do not explain at all how this function is computed, and what its properties are. We have suggested [12], without any proofs, the solution which is proposed here for solving the single criterion case.

Practical impacts of these results are hard to evaluate. What is important is that these results often lead to the best possible, safe pruning. Therefor, they do not guaranty that a particular search may be envisaged but they guaranty that this search, when started, will be done as efficiently as possible, from the pruning point of view.

Two directions for further work seem to be important to us :

- extending the solution we propose for the two-criteria case, to a wider class of criteria (or proposing other solutions).

- searching for methods aimed at determining the "promise function" attached to a given criterion. This function would determine the "probability" that a solution may be reached by specializing a given descriptor. Such function would be useful in the case of a best-first strategy for choosing the descriptor to be first specialized, and also for pruning in a "probabilistic" way in the case of too large problems.

Acknowledgements : With sincere thanks to Gilles Caraux and to Joël Quinqueton for their helpful comments and plenty of meaningful discussions.

References :

[1] *The artificial intelligence handbook,* (1982), Cohen P. R. and Feigenbaum E. A. (Eds.), Vol. 3, Chapt. 14, Pitman, London.

[2] Kendall M. and Stuart A. (1976), *The advanced theory of statistics*, Charles Griffin, London.

[3] Papadimitriou C. H., Steiglitz K. (1982), *Combinatorial Optimization : Algorithms and Complexity*, Chapt. 2 and 18, Prentice Hall, New-Jersey.

[4] Roberts A. W., Varberg S. E. (1973), *Convex functions*, Chapt. 5, Pure and applied mathematics n° 57, Smith P.A. and Eilenberg S. (Eds.), Academic Press, New-York and London.

[5] Angluin D., Smith C. (1983), *Inductive inference: Theory and Method,* ACM Computing Surveys, 15(3), pp. 237-270.

[6] Bergadano F., Giordana A., Saitta L. (1988), *Automated concept acquisition in noisy environments*, IEEE Transactions on PAMI, Vol. 10, n° 4, pp. 555-578.

[7] Blythe J. (1988), *Constraining Search in a Hierarchical Discriminative Learning System*, ECAI 8, pp. 378-383.

[8] Breiman L., Friedman J. H., Olshen R. A., Stone C. J. (1984), *Classification and regression trees*, Wadsworth Inc., Belmont California.

[9] Buchanan B. G., Feigenbaum E. A. (1978), *DENDRAL and META-DENDRAL : their applications dimension*, Artificial Intelligence,Vol. 11, North-Holland, pp. 5-24.

[10] Cestnik C., Bratko I. (1988), *Learning redundant rules in noisy domains*, in ECAI 8, pp. 348-350.

[11] Clark P., Niblett T. (1987), *Induction in noisy domains*, in the proceedings of the second European Working Session on Learning, Sigma-Press, Wilmslow, pp. 11-30.

[12] Gascuel O. (1986), *PLAGE : A way to give and use knowledge in learning*, in the proceedings of the first European Working Session on Learning, LRI, Orsay.

[13] Gascuel O., Danchin A. (1988), *Data analysis using a learning program, a case study : an application of PLAGE to a biological sequence analysis*, ECAI 8, pp. 390-395.

[14] Gascuel O. (1988), *Apprentissage symbolique en présence de bruit, quelques résultats concernant l'exploration de l'espace des redescriptions*, Int. Rep. CRIM-61, Centre de Recherche en Informatique de Montpellier (CRIM).

[15] Goodman R. M. F., Smyth P. (1988), *Information-Theoretic Rule Induction*, in ECAI 8, pp. 357-362.

[16] Michalski R.S. (1983), *A theory and methodology of inductive learning*, Machine Learning, Vol. 1, Chapt. 4, Michalski R. S., Carbonell J.G., Mitchell T.M. (Eds.), Tioga publishing company Palo Alto California.

[17] Morgan J. N., Sonquist J. A. (1963), *Problems in the analysis of survey data, a proposal,* JASA 58, pp. 415-434.

[18] Quinlan J. R. (1986), *Induction of decision trees*, Machine Learning 1, pp. 81-86.

[19] Quinqueton J., Sallantin J. (1983), *Expansion and compression of binary data to build features by learning*, in IJCPR 6, pp. 641-644.

LEARNING DIFFERENTIAL DIAGNOSIS RULES FROM NUMERICAL KNOWLEDGE

Marie-Claude Perron
Electricite de France, D.E.R., I.M.A., T.I.E.M.,
1 Av du Général de Gaulle, 92141 Clamart Cedex (France)

Abstract

The problem studied lies in building decision rules from examples described by numerical attributes. Statistical knowledge based on the estimation of the significant descriptor domain is required in order to extract simple noise resistant filtering rules. Such generated decision rules are used to eliminate progressively unlikely diagnosis cases from a known list of possible candidate classes. This diagnosis help method has been successfully applied to ultrasonic data for defect characterization purposes in non-destructive testing.

1 Introduction [1,2,5,10,11]

In a discrimination problem between classes, the examples of which are described by a list of given attributes, the expected decision rules are intended to classify a new experiment from its attribute measurements. In most studied cases, the rule extraction process has to take account of the data noise. Although the methods that have been developed on the grounds of the Statistics are particularly well-adapted to noisy data, they are often poor at dealing with lack of data information. In contrast, some decision-tree based learning systems have been satisfactorily performed under noisy data and especially missing data conditions.

The simple way in which a multidimensional partitioning of the description space can be performed and, on the other hand, a Statistics background theory are used here in order to build easily negative diagnosis rules.

2 Learning decision rules from numerical knowledge.

2.1 Multidimensional ranges [6,7]

When the available descriptive information is entirely provided by numerical attributes related to physical measurements, discrimination rules based on the estimation of multidimensional ranges give a meaningful interpretation of the learning concepts.

The underlying principle consists in finding a formal definition of each example class against the set of counter-examples contained within the union of the other learning classes. A characteristic function of the class C can be obtained if the following condition is satisfied. It does not exist a learning example x, $x \notin C$, such that :

$$\forall\ j \in \{\ 1,\ na\ \} \quad xmin_j(C) \leq x_j \leq xmax_j(C)$$

where $xmin_j(C)$, $xmax_j(C)$ is the minimum, (resp. maximum) value of the attribute A_j for the clas C, x_j is the value of A_j for the event x and na is the total number of attributes used by the description language.

A characteristic function of C is obtained from the disjunction of sub-classes, included in C, and described by conjunctions of elementary assertions : <attribute> <operator of comparison> <numerical value>.

When dealing with noisy data, the conditions required to obtain characteristic functions are not always satisfied. Hence it is difficult to build decision rules according to the above-mentioned principle. One way to overcome this problem is to build negative rules the purpose of which consists in excluding applicant classes instead of classifying an experiment in a given class.

2.2 Noisy data and redundant rules [3,12]

A few authors put emphasis on the rule base redundancy to fight noise. While investigating real learning situations, M. Sebag and M. Schoenauer have shown that the problems of non-significant noise and lack of data lead to throw the purpose of concise and exact rule learning, based upon a consistency hypothesis, back into question. The proposed solution consists in a redundant rule network in which the evidence combination problems are solved by reiterating the learning step.

For the purpose of avoiding numerical problems encountered with the combination of redundant rule credibility, B. Cestnik and I. Bratko have suggested elimination rules obtained by exploring all conjunctions of up two attributes. This method is well adapted to situations where the descriptive variables have a finite set of values.

The elimination rules described here are intended to deal with examples described by continuous attributes. No completeness or consistency hypothesis is required to generate such rules. They are based on the estimation of description space constraints derived from statistical theory.

3 Construction of exclusion rules.

The proposed rules are based on statistical knowledge. They can take into account different kinds of noise among them we can mention missing attribute values, uncertain values due to lack of measurement accuracy, false information.

The first step towards rule generation consists in estimating various acceptable descriptor domains with respect to different probability coefficients.

Since the variables used as descriptors are captured from physical measurements, the value set of each attribute is potentially limited by experimental constraints. When local constraints on the feature space are not supplied by the expert, another way to constrain the domain extent consists in proceeding by extreme estimation. Evaluation of the extent in which the attribute values are meaningful for a given class, prevents the system from going further in investigations beyond these limits.

3.1 Estimation of extreme values [8]

The aim of theory of extreme values as defined by Gumbel (1954) is to "explain observed extremes arising in samples of given size or valid for a given period, length, area or volume and to forecast extremes that may be expected to occur within a certain sample size, time, area, etc". Since the procedures required to study extreme values are supposed to deal with statistical variables, it is assumed that the initial distribution remains constant from one sample to the next. Another condition is that the samples, from which the extremes are drawn, should be independent. When we are dealing with data issued from large samples, the independence hypothesis is currently applied. Moreover, in most cases the underlying data extraction process is imperfectly known and then the data analysis is based upon simple assumptions. If the developed theory seems to fit, we assume that the choosen method will not lead to substantial errors.

Exact distributions of extreme values may be approximated by numerical methods when the initial distribution is assumed. Otherwise, in many situations where the attribute values do not obey a known distribution, the extreme distributions can be estimated on the basis of the observed samples.

3.1.1 Limiting distribution

Let X(n) be the largest value observed from a sample of n independent observations of a statistical variable and F(x) the probability that the value of the variable is smaller than x. Then the distribution of the n-value maximum is given by :

$$F_{X(n)}(x) = \text{Prob}\{ X(n) < x \} = F^n(x)$$

The extreme value distributions are obtained as limiting distributions of maxima (resp. minima) in random samples of increasing size.

To obtain a nondegenerate limiting distribution, it is necessary to find the coefficients a_n and b_n such that :

$$G(x) = \lim_{n \to \infty} F^n(a_n x + b_n) \qquad b_n > 0$$

where G is a nondegenerate distribution function and F belongs to the attraction domain of G.

According to the stability postulate established by Frechet, Fisher and Tippett, the maximum value of N samples of size n is the same as the maximum value in a sample of size Nn. It follows that G(x) must satisfy the equation :

$$[G(x)]^N = G(a_N x + b_N)$$

This functional equation lead to the three following solutions which may be related to three types of initial distributions.

Type 1 : $a_N = 1$
Type 2 : $a_N \neq 1$ et $G(b_N / (1-a_N)) = 1$
Type 3 : $a_N \neq 1$ et $G(b_N / (1-a_N)) = 0$

3.1.2 Extreme estimation

In practice, according to our choice, the extremes are assumed to obey a Gumbel distribution (Type 1), which can be written as :

$G(x) = \exp(-\exp(-(x-\alpha)/\beta))$ where α and β ($\beta > 0$) are real

Let $x_{1,N}$, ..., $x_{N,N}$, be the maxima taken from N samples of size n, the largest value distribution is approximated by fitting a least square staight line to the discrete sequence of points defined by $(x_{k,N}, \log(-\log(f_k)))$, k = 1, N , where $f_k = k / N+1$.

Estimations of the distribution parameters α et β, are obtained from the computation of the real coefficients a et b minimizing the quadratic criterion :

$$\sum_{k=1,N} (a x_{k,N} + b + \log(-\log(f_k)))^2$$

Therefore $\beta = 1 / a$ et $\alpha = -b \beta$

The maximum value estimation corresponding to the quantile X_p associated with a probability p is given by :

$$X_p = \alpha - \beta \log(-\log(p))$$

Maximum values are estimated for high quantiles, that is to say for probabilities p close to p = 1. Estimation of the minima are obtained by similarly proceeding.

Exclusion rules generated by this process are expressed in the very simple following form :

If $(x_j > A_j max(C^i)(p))$ ou $(x_j < A_j min(C^i)(p))$

Then $x \notin C^i$ with an error probability (1-p)

where x is an event to classify, x_j is the value of the attribute A_j measured for x, $A_j max(C^i)(p)$ and $A_j min(C^i)(p)$ are the estimations of maximum (resp. minimum) value of A_j for the class C^i, corresponding to the p-order quantile.

3.2 Generation of rules

The set of generated rules may be important but its size is limited by the number of attributes used in the description language. A small set of rules can be built by choosing the most relevant attributes to the classes to be eliminated, in order to limit the space search.

Such rules can be expressed as many trees as pattern classes. Given a class C^{io}, a binary elimination tree is built from the root composed of the total training set X consisting of the union of nc disjoint classes, C^i, i=1,nc.

If a node N contains more than [Card(C^i) x THR] observations for at least one $i \neq i0$, it is divided into 2 subsets :

- events $\notin C^{io}$ with a probability p
- events about which no decision may be taken, that means events of either C^i or C^{io}.

Note that the threshold THR is choosen as a parameter dependending on the probability p and an arbitrary constant.

The subdivision criterion consists in selecting the attribute A_j such that either $A_j max(C^{io})(p)$, or $A_j min(C^{io})(p)$ implies the best adequation between the partition N = { N_1,N_2 } defined as :

$$N_2 = \{ x \;/\; x_j > A_j max(C^{io})(p) \}$$
$$N_1 = N - N_2$$

or
$$N_1 = \{ x \;/\; x_j < A_j min(C^{io})(p) \}$$
$$N_2 = N - N_1$$

and the partition $\{ C^{io} , N \cap \bigcup_{i \neq i0} C^i \}$

The most natural heurism consists in counting the examples on each side of the estimated bounds. The attribute selection is based on the maximization of the number of eliminated counter-examples.

The exclusion rules are built according to the following principle. The attribute selected at each node is such that the space partitioning is performed with respect to the highest probability assuring a maximum coverage of the counter-examples. So, the extreme values being estimated according to decreasing probabilities $p_1, \ldots, p_n$, when all the possible current node subdivisions lead to exclude a number of examples smaller than an arbitrary threshold for a given probability p_i, the procedure is repeated with p_{i+1}, $p_{i+1} < p_n$.

3.3 Decision process

A conflict arises when all the diagnosis cases are excluded. Each decision rule is supplied with a credibility coefficient equal to the quantile probability from which the attribute bound of the rule condition part has been computed. The generated rulès are ordered with respect to decreasing credibility coefficients. Thus, the last activated rule providing the conflict is not taken into account if its credibility coefficient is strictly smaller than the others. In a conflict between 2 rules of same probabilities, a second coefficient related to the rule efficiency estimated from the learning set may be required.

In many situations, a small set of simple filtering rules, which are the most efficient on the training set, may lead quickly to a diagnosis by minimizing the number of performed tests. When dealing with more difficult cases, where the description knowledge is very imperfectly assumed or for which the list of applicant diagnosis issued from the above-mentioned step is not satisfying, it can be useful to generate a greater number of redundant filtering rules, in a systematic way, in order to take into account all the conclusions of activated rules.

4 Application and results

4.1 Data

Data used for testing were collecting by ultrasonic inspection of ferritic specimens containing several types of flaws. Between them two main groups can be distinguished : crack-like defects (C^4,C^5) and the other ones (C^1,C^2,C^3). Material undergoing test is submitted to an excitation energy transferred by an appropriate transducer. The response echo resulting from the interaction with the reflector is then digitized and recorded. The feature extraction is based on our expert knowledge of the measurement process and the observed signal. The set of selected features consists of both time- and frequency-domain representations of the defect signatures.

The studied population is composed of about 230 signals belonging to 5 disjoint defect classes C^i, i=1,5. A set of 11 continuous attributes is used to describe the energy/defect interaction process.

The training set is composed of 50% of the total signals selected randomly from the respective pattern groups, 50% being used for testing.

4.2 Results

About 30 rules are generated according to the principle mentioned in §3.2, for probabilities choosen in the range [0.97, 0.999]. At most 4 descriptors are used to exclude one given class.

4.2.1 Learning set

The following table gives the percentages of learning examples, belonging to C^i (row), for which the diagnosis C_j (column) is excluded.

	C^1	C^2	C^3	C^4	C^5
C^1		100	100	16	100
C^2	100		100	80	100
C^3	100	100		100	100
C^4	100	100	100		72
C^5	100	100	100	72	

4.2.2 Test set

The results are satisfying with respect to the discrimination between cracks (modified or not) and non-cracks. Actually, the remaining defect list obtained after elimination lead to recognize correctly all the crack specimens. Moreover, an accurate diagnosis about the nature of the crack is achieved in 70% of cases. Concerning the non-crack group, 100% of the examples of C^1 and C^2 are correctly classified. Although only 20% of the objects belonging to C^3 are such that the final diagnosis is composed of a single classifying possibility, the diagnosis obtained for the other test samples is reduced to a defect list containing up to 2 candidates.

5 Conclusion

The construction of exact discrimination rules according to the example and counter-example coverage assume more or less strong hypotheses about the completeness and consistency of the learning set. In contrast, the building of filtering rules, the purpose of which does not consist in finding a single diagnosis, does not require such conditions.

It has been shown that a statistical theory based method can be used to generate quickly efficient diagnosis rules adapted to numerical data. In our experiments, a small set of rules has been sufficient to provide the user with a diagnosis help of interest with respect to the discrimination between crack and non-crack-defects. In more difficult cases, the activating of a larger set of redundant elimination rules could be useful to achieve a more reliable and accurate diagnosis.

6 References

[1] Breiman L., Friedman J.H., Olshen R.A., Stone C.J., "Classification and regression trees", Wadsworth Inc., Monterey, 1984.

[2] Celeux G., Lechevallier Y., "Méthodes de segmentation non paramétriques", Revue de Statistique Appliquée, vol.30, 4, 1982.

[3] Cestnik B., Bratko I., "Learning redundant rules in noisy domains", Proceedings of ECAI 88, 1988, Munich.

[4] Gondran M., "Diagnostic et maintenance assistés par systèmes experts", Note EDF, 1986.

[5] Gueguen A., Nakache J.P., "Méthode de discrimination basée sur la construction d'un arbre de décision binaire", Revue de Statistique Appliquée, vol. 36, no 1, 1988.

[6] Guénoche A., "Modélisation, algorithmes et informatique en sciences humaines", Thèse de doctorat d'état, Université d'Aix-Marseille, 1983.

[7] Guénoche A., "Propriétés caractéristiques d'une classe relativement à un contexte", Journées Symbolique-Numérique, Université Paris-Dauphine, 1987.

[8] Gumbel E.J., "Statistical theory of extreme values and some practical applications", Applied Mathematics Series, 33, 1954.

[9] Manago M., "Learning rules resilient to noise from noisy training data", Journées Symbolique-Numérique, Université Paris-Dauphine, 1987.

[10] Quinlan J.R., "Induction of decision trees", Machine Learning, 1, 1986, 81-106.

[11] Quinlan J.R., "Learning from noisy data", Machine Learning vol.2, Ed. Michalski R., Carbonell J., Mitchell T., Tioga publishing company, Palo Alto, 1986

[12] Sebag M., Schoenauer M., "Un réseau de règles d'apprentissage", Journées Symbolique Numérique, Université Paris-Dauphine, 1988.

SYMBOLIC CLUSTERING

INCREMENTAL CONCEPT FORMATION via a SUITABILITY CRITERION

Christine DECAESTECKER*[1]

ABSTRACT :

This article introduces a family of algorithms called ADECLU, which incrementally construct hierarchies of concepts ("Concept Formation"). The simplest version (ADECLU1) offers a performance level identical to that of COBWEB (Fisher, 1987), but uses a much simpler internal representation of concepts. The ADECLU family is based on an suitability criterion which permits the integration of new objects into the existing hierarchy which is itself restructured in the process.

KEY-WORDS : Concept formation, Non Supervised Learning, Incremental Processing, Suitability Criterion, Attribute Selection, Data Analysis.

1. INTRODUCTION :

The context of this paper is that of "Concept Formation" such as it has been defined by J.H. Gennari, P. Langley, D. Fisher [GENN88], and more particularly, incremental non-supervised learning.The aim of this paper is to construct a hierarchical classification using the descriptions of objects presented sequentially, where each class is given a definition which **summarises** its elements. The general aim is to **condense** (summarise) the information contained in the descriptions, with the possibility of **predicting** the unknown values of the attributes. Thus, the performance of the acquired hierarchies will be measured in terms of the ability to make predictions concerning the unknown attributes.

This paper presents ADECLU, a family of algorithms based on a general suitability criterion between an object and a concept in a hierarchy in the process of being formed. The first (simple) version (ADECLU1) already gives good results (slightly superior to the results from COBWEB [FISH87a][FISH87b]). A second version (ADECLU2) is proposed which uses association criteria from Data Analysis, the aim being to select the most appropriate combinations of attributes to describe a concept. Not only are there certain advantages concerning the description of concepts, early results show already that the hierarchies constructed by this second version give the best predictive power.

2. ADECLU : PRINCIPLES

ADECLU is a hierarchical, incremental "Conceptual Clustering" process, which extends ideas from UNIMEM [LEBO87], in the context of "Concept Formation" defined in [GENN88], following the work of Fisher [4]. This extension allows one to avoid the use of arbitrary (user specified) parameters, thanks to the procedure of optimising an evaluation function (inspired by COBWEB).

This section introduces ADECLU1 (the simple version) in detail. ADECLU2 follows the same principles, but differs in its knowledge representation of the concepts, namely in its selection of single and multi valued attributes to summarise information (cf section 4).

2.1 Knowledge Representation

The objects are described by a conjunction of **characteristics**, i.e. attribute-values pairs. ADECLU organises a series of objects presented sequentially, into a hierarchy of concepts, in which each node represents a concept and contains a **concept description**. Only the

---

* This work is supported by the Belgian National incentive-program for fundamental research in Artificial Intelligence. The scientific responsability is assumed by the author.

[1] CADEPS Artificial Intelligence Research Unit, Université Libre de Bruxelles (U.L.B.), Av. F.D.Roosevelt 50, C.P. 194/7. B-1050, Bruxelles, Belgium. Tel (+ 32 2) 642 2783; Telex 23069; Email CADEPS@BBRNSF11

descriptions intervene in the classification operation (and not the earlier objects previously classed). Note, for each concept C, the set of subconcepts generates a partition over the set of objects classified in C.

In ADECLU1, each concept is also defined by characteristics which are **common** to its objects. In short, we maintain two rules :

For each concept C in a hierarchy IH,

- the set of these subconcepts form a **partition** of the set of these objects of C (mutually exclusive) (rule 1).
- these characteristics, belonging to those **common to all its objects**, are those which do not appear in the parent concept of C (rule 2).

2.2 Suitability criterion : an evaluation function which guides the search

When a new object O is presented, a search is undertaken to find the most specific concept into which to class the new object, based on the descriptions of the different concepts (partial matching). The top level of the hierarchy is a set P of disjoint concepts, all having the root node as parent node (which has no specification) and forms a partition over all the objects already presented.There are several possibilities :

- integrate O into one of the concepts of P
- modify P by applying one of the operators, and integrating O into one of the concepts resulting in the set P'.

After having chosen one of these possibilities via an evaluation function, one recommences the process, for the set of sub concepts of the concept chosen at the previous step (the other branches are not investigated) (hill climbing search strategy).

We will elaborate the different aspects of this strategy later in this paper. The aim of this section is to define the evaluation function which determines the choice of the various options. This definition is presented in two steps; firstly, the case of a fixed partition of concepts, and secondly, when several partitions are to be compared.

1st step : fixed partition of concepts

Let be a set of disjunctive concepts (called a "partition" in the rest of the paper) P = (C1, C2,..., Cn). We define, for each concept Ci of P, an suitability score (positive or negative) between the new object O to be classified, and Ci : sc(O, Ci), which reflects the correspondence (or the lack of it) between the description of Ci and that of O. To this end, one verifies, for each characteristic (abbreviated as: charact.) of Ci, if it *"agrees" with O*, i.e. if it is found in the description of O.

Concerning the correspondence measure between Ci and O,

- each charact. of Ci which agrees with O, contributes positively to the measure
- disagrees negatively

The contribution of the charact. j of Ci is noted p^i_j and is evaluated at +1 if it is positive and at -1 if not. The terms p^i_+ and p^i_- are defined to be the number of charact. of Ci which match (do not match) the object O.

The definition of the suitability score includes the number of elements matched by the concept . This term was included as a result of several experiments. It appears to be intuitively preferable, for the purposes of generalisation, to classify an object O into a concept with a large card and an average number of common characteristics, than into a concept with a small number of objects and a larger number of common characteristics (for discussion see [DECA89b]).

The **score** between an object O and a concept Ci is defined as the **sum of the contributions** (+ve or -ve) of each charact. of Ci, weighted by |Ci| the cardinal of Ci (i.e. the number of objects in Ci) :

$$sc(O,C_i) = |Ci| \Sigma_j p^i_j = |Ci| (p^i_+ - p^i_-)$$

2nd step : comparison of several partitions

As mentioned at the beginning of section 2.2, we must compare different partitions resulting from the application of operators (described in 2.3) to the initial partition. The aim now is to evaluate the decision "classify the object O into the concept Ci of the partition P" where one cannot dissociate the concept from the partition to which it belongs, without risk of falsifying this evaluation. In fact, a concept is an entity which is defined not only by its own characteristics, but also by contrasting it with other concepts of the partition. (See [HANS86]). A good suitability measure between a new object O and a concept Ci of a partition P will thus be a function of O, Ci and P. It would thus be context dependent (in the style of Michalski [MICH84]).

The decision *"classify O in the concept Ci of the partition P"* is taken in favour of Ci and to the detriment of all the other concepts Ck ($k \neq i$), from which we have :

Definition : The **suitability** between an object O and a concept Ci of a partition P, is defined as follows : $\mathbf{S(O,Ci,P) = sc(O,Ci) - \Sigma_{k \neq i}\ sc(O,Ck)}$

In ADECLU1, one has : $\mathbf{S(O,Ci,P) = |Ci|\ (p^i_+ - p^i_-) - \Sigma_{k \neq i}\ |Ck|\ (p^k_+ - p^k_-)}$

2.3 Bidirectional operators :

The role of each new object is to launch the search process described in 2.2 and to modify the state space. In order to avoid an oversensitivity to the order of presentation of the objects, we will use bidirectional operators which will allow **backtracking** (at least cost) :

creation	<---->	suppression	of concepts
merging	<---->	splitting	of concepts

Three operators will be used (see COBWEB [FISH87a], CLASSIT [GENN88]) :

1. Merging of two concepts : creation of an intermediate concept (FIG.1)
2. Splitting of a concept : suppression of an intermediate concept (FIG.2)
3. Creation of a concept : demanded by a new object O, this new concept will have O (at its creation) as its only element and will be entirely characterised by it.

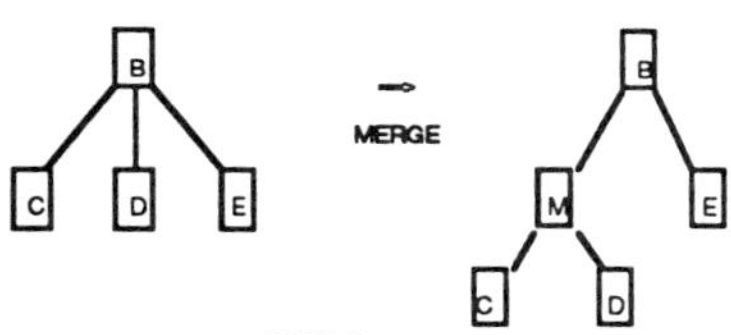

FIG.1

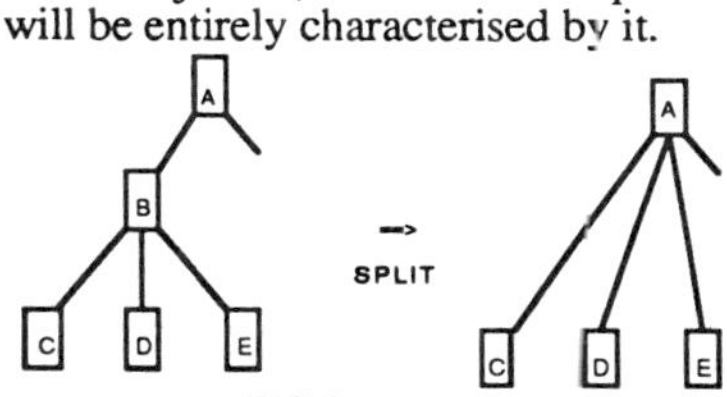

FIG.2

Each subsequent modification of the hierarchy is performed in accordance with the two rules mentioned in 2.1, which characterise the hierarchy produced by ADECLU1 (fore more details see [DECA89a&b]).

2.4 Control strategy

Each time we have to classify an object O into one of the concept's subconcepts (a partition P) (see 2.2), we will have to envisage the following possibilities :

- leave P as it is (operation L)
- modify P either by creating a new concept to receive O (operation C => partition Pc)
  or by merging two concepts into one (operation M => partition Pm)
  or by splitting a concept into its sub concepts (operation S => partition Ps)
  and integrating O into one of the concepts of the resulting partition (with the consequences that this implies for the description of the concept, see later).

The choice of the partition (i.e. the operation) and the concept into which O is to be integrated, will be that which optimises : MAX $S(O, C', P')$ (where $C' \in P'$)

To avoid combinatorial explosion problems (which would consist of testing all possible merges and splits), we restrict ourselves to: merging only two concepts (of P) (having the highest suitability scores), and splitting the concept (of P) with the highest score.

The suitability criterion S(O,C,P) has two general properties which are independant of the score function sc(O,C) and economical for the optimisation (see [DECA89a&b]).

3. A DETAILED EXAMPLE OF ADECLU1'S BEHAVIOUR

For example, assume that clustering occurs over the ordering animals (objects) given below (taken from [FISH87b]):

	BodyCover	HeartChamber	BodyTemp	Fertilization	Color
mammal-1	hair	four	regulated	internal	brown
reptile-1	corn-skin	imp4	unregulated	external	green
amphibian-1	moist-skin	three	unregulated	internal	red
bird-1	feathers	four	regulated	internal	red
reptile-2	corn-skin	imp4	unregulated	internal	gray
fish	scales	two	unregulated	external	red

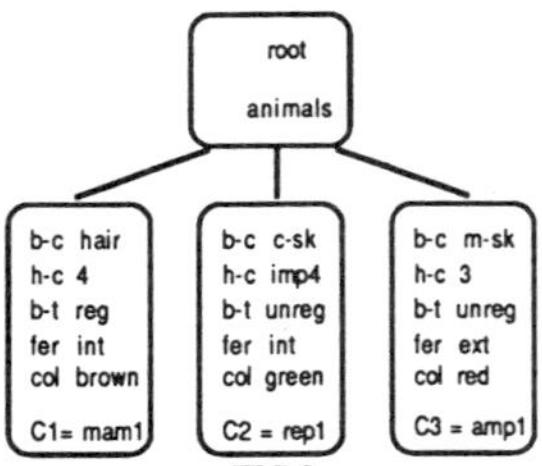

FIG.3

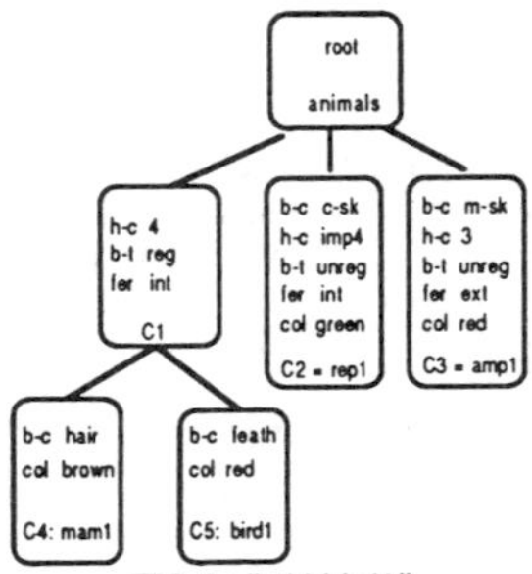

FIG.4 : "add bird1"

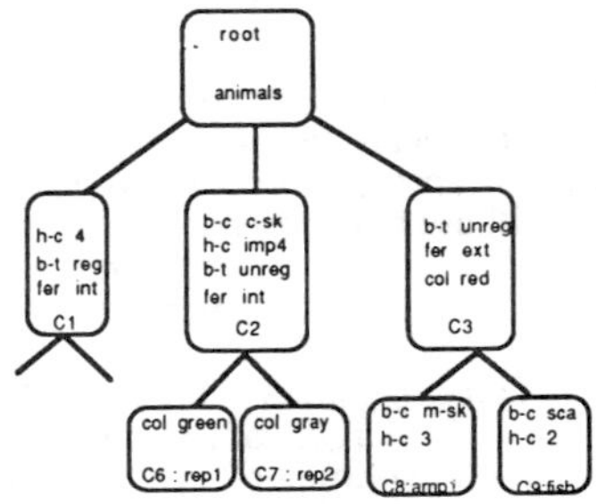

FIG.5 : "add rept2" & "add fish"

The successive integration of the first 3 objects causes (for each of them) the creation of a new concept (operation C) (FIG. 3) In fact, these first 3 objects have very few characteristics in common (if any). On the other hand, "bird-1" and the concept C1 ("mam-1") have 3 characteristics in common out of 5. C1 is thus judged to be more adequate for "bird-1". After the integration of "bird-1" into C1, the characteristics which differ between the two objects are used to create two sub-concepts of C1 (FIG.4). The process is the same between "reptile-2" and C2. For "fish", the most adequate concept at the first level is C3 ("amp1") (see FIG.5).

4. ATTRIBUTE SELECTION (ADECLU2)

We have met :

- processes such as ADECLU1 and UNIMEM, which remove an attribute from a concept description once it takes several values
- processes such as COBWEB and CLASSIT [GENN88],which keep all the attributes.

The aim of our research, at present, is to propose a selection of attributes (whether they take one or several values), which best describe a concept. We will present here ADECLU2, a variant of ADECLU1 which uses a selection strategy, based on a system of variable weighting of the attributes, and integrated into the suitability principle. ADECLU2 thus follows the principles described in section 2, so only the changes concerning the knowledge representation and the definition of the score function will be described below.

4.1. Concept description :

Each concept is described by a conjunction of characteristics, where :

characteristic = triplet : (attribute, set of values, weighting) noted (j , Vj , wj)

where V_j is the set of the values (of the attribute "j") taken by the objects integrated into the concept. V_j can be reduced to a singleton

w_j is a weighting ($0 \leq w_j \leq 1$) which is a function of the information supplied by the characteristics of the concept definition. This weighting <u>varies</u> during the incremental process.

We take the two rules characterising the hierarchies produced by ADECLU1: For each concept C of the hierarchy IH:

- the set of sub-concepts of C forms a **partition** of the set of objects of C (rule 1).
- a characteristic, for which Vj is a singleton (a value common to all the objects), appears in the description of C iff Vj is not a singleton in the parent concept of C (rule 2).

4.2. Choice of weightings :

The weighting wj in the concept C, must reflect the degree of association which exists between the attribute (or variable) "j" and the variable belonging to the concept C. Data Analysis offers us numerous **association criteria** between two nominal variable X and Y (See [ANDE73]). All of them refer to a **contingency table** between X and Y :

matrix $(n_{ij})_{pxq}$ where n_{ij} is the number of objects for which $X = x_i$ and $Y=y_j$

with marginal distributions $(n_{i.})$ and $(n_{.j})$ where $n_{i.} = \Sigma_j n_{ij}$ and $n_{.j} = \Sigma_i n_{.j}$

Goodman and Kruskal [GOOD54] suggested measuring the association between two variables X and Y by the predictive power of one variable as a predictor of the other (**"Lambda" index**) and is defined as the relative decrease in the probability of error (in the prediction of Y), due to knowledge of the value of X. If the prediction of the value of X from Y is as important as Y from X, then a symmetric relationship is considered (for more details see [ANDE73]). The **measure of the predictive power** between X and Y is :

$$L = \frac{\Sigma_i (\max_j n_{ij}) + \Sigma_j (\max_i n_{ij}) - (\max n_{.j}) - (\max n_{i.})}{2 n_{..} - (\max n_{.j}) - (\max n_{i.})}$$

Since we are in an incremental process, subject to numerous parameter updatings, we will calculate this Lambda index for a simplified 2x2 contingency table (see below), between the variable belonging to the concept (i.e. $O \in C$) and the variable belonging to the set of values Vj taken up by C for the attribute "j" (i.e. $j(O) \in Vj$). Furthermore, since we are in a hierarchical organisation, the set of objects which will be used as a reference to test the association criterion (and on which a contingency table will be constructed), will consist of objects integrated into the parent concept of the concept C being considered.

In conclusion, for each characteristic of the concept C, we will study the following contingency table :

$j(O) \in Vj$: $O \in C$:	yes	no	TOTAL
yes	$\lvert C\rvert$	0	$\lvert C\rvert$
no	$N1-\lvert C\rvert$	N-N1	$N-\lvert C\rvert$
TOTAL	N1	N-N1	N

This table is entirely defined by the quantities $|C|$, N, N1, where $|C|$ is the cardinal of C, N is the card of the parent concept of C and N1 is the number of objects of the parent concept which agree with $j = Vj$. One then defines **wj as the index L** applied to the 2x2 table above.

The choice of L among the different possible criteria has been guided by the following properties. As with the majority of association criteria, we have $0 \leq L \leq 1$ with

$L = 0$ in the case of independence and $L = 1$ in the case of perfect association[1].

Furthermore, L becomes zero in other cases as well. One can show for the 2x2 table above that: $L = 0 \iff$ independence of the two variables or $(|C| \leq N1-|C|$ and $N-N1 \leq N-|C|)$

This second condition expresses the following probabilistic fact :

$$P(O \in C \mid j(O) \in Vj) \leq P(O \notin C \mid j(O) \in Vj)$$

and $$P(j(O) \notin Vj \mid O \notin C) \leq P(j(O) \in Vj \mid O \notin C)$$

[1] That is, for each line i of the contingency table, a single nij is non zero; and similary for each column j.

Under such conditions, it is fully justifiable to "de-activate" the characteristic j=Vj in the definition of the concept C.

We can also remark that the use of rule 2 (cf 4.2) agrees with the use of L. In fact, one can show that L = 0 for each characteristic with a single value (singleton Vj) which also appears in this form in the parent concept.

The updating of the weights takes place after the integration of an object into concept C. The weightings of the characteristics of C must be adjusted, as well as those of the "brother" concepts (i.e. same direct parent in the hierarchy). In fact, the quantities N and N1 also vary for the brother concepts (due to integration into the preceding level).

During the process, the weightings allow an "activation" or "de-activation" of different characteristics, inside the definition of a concept. A zero weighting will mean that this characteristic must not be involved in the description of the concept. It will however, remain in memory, able to be rendered active at a later stage (incremental process). Thus the combination of selected attributes to describe the concept, may vary during the process.

4.3. Score function :

The contribution of each characteristic (j, Vj, wj) (of concept Ci) to the calculation of the score sc(O,Ci), is decomposed into a product of two factors : the weighting wj, and the factor p^i_j, which takes also the value +1 or -1, depending on the attribute value of O (i.e. j(O)) and the attribute value of Ci (i.e. Vj),

agrees : $j(O) \in Vj$ or disagrees : $j(O) \notin Vj$

The **score** between an object O and a concept Ci is defined as the sum of the contributions of each characteristic of Ci, weighted by |Ci|. This is written as :

$$sc(O,Ci) = |Ci| \Sigma_j p^i_j wj$$

4.4. An example of ADECLU2's behaviour

Let us take the example used by ADECLU1, supplemented by this several new objects:

	BodyCover	HeartChamber	BodyTemp	Fertilization	Color
mammal-2	hair	four	regulated	internal	gray
amphibian-2	moist-skin	three	unregulated	external	green
bird-2	feathers	four	regulated	internal	black
fish-2	scales	two	unregulated	external	red

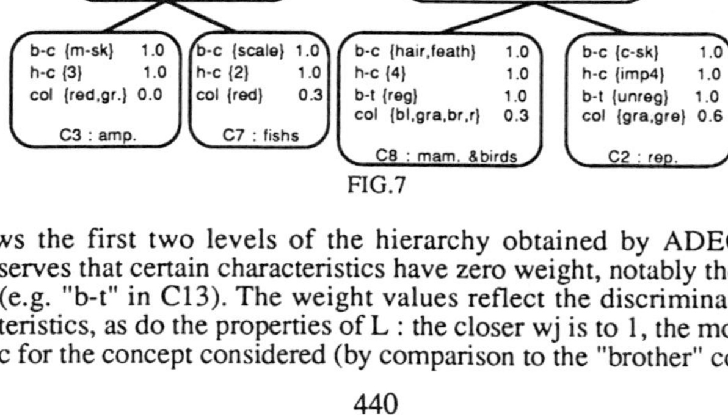

FIG.7

FIG.7 shows the first two levels of the hierarchy obtained by ADECLU2 for the 10 objects. One observes that certain characteristics have zero weight, notably those which take all the values met (e.g. "b-t" in C13). The weight values reflect the discriminating aspect of the different characteristics, as do the properties of L : the closer wj is to 1, the more discriminant is the characteristic for the concept considered (by comparison to the "brother" concepts).

5. RESULTS (ADECLU1 & ADECLU2)

Recall (see section 1.) that the aims of concept formation as in [GENN88] are to **condense** (summarise) information and to **predict** unknown values. For comparative purposes, ADECLU(1&2) was tested, amongst others, on a data base used by Fisher [FISH87b]; a collection of 47 Soybean disease cases taken from 4 categories. Each case is described by 35 attributes, which does not include the name of the disease. This data base is, in fact, extracted from a larger data base, (introduced by Michalski in [MICH80]) which we will refer to later.

The evaluation procedure, inspired by the methods of supervised learning, now follows. (see [FISH87a][FISH87b][GENN88]). ADECLU incorporates a number of cases (the socalled training set), which include the name of the disease (as the value of the 36th attribute). The hierarchy of concepts thus created, will be used later to classify (but not to incorporate[1]) the remaining cases (the socalled test set), which do not include the name of the disease. The value of this attribute is inferred from the result of the classification, and takes as value the name of the disease, taken from the concept into which the test case was classified. One can then calculate the percentage of correct predictions as a function of the number of cases present in the training set (TABLE 1). The same experiment was performed for the other 35 attributes. TABLE1 presents also the average (over all the attributes) of the percentages of correct predictions, as a function of the card of the training set.

card of training set :		5	10	15	20	25		
correct diagn.:	COBWEB:	88	100	100	100	100	%	
	ADECLU1:	97	100	100	100	100	%	
	ADECLU2:	**100**	**100**		**idem**		%	(100% predict.)
average of correct predict.:	COBWEB:	82	87	87	87	87	%	
	ADECLU1:	85	87	87	88	88	%	
	ADECLU2:	**86**	**90**	**90**	**90**	**90**	%	
	(% predict.)	(77)	(87)	(92)	(89)	(90)		
	ADECLU2 forced :	86	87	88	88	88	%	(100% predict.)

TABLE 1

ADECLU2 allows us to detect those cases for which **prediction cannot be inferred from the classification**. In fact, this value to be predicted can be in a characteristic of zero weight. In this case, no value can be proposed, because the predictive power between this characteristic and the fact of belonging to the concept, is judged to be zero. We will therefore have to distinguish the cases of "**non-prediction**" from those of "prediction", where the percentage of correct predictions can be calculated. In TABLE 1, (% predict.) gives the percentage of predictions (correct or not) actually achieved. As an example, and to permit a comparison with ADECLU1, the latest result has been added, which was obtained by forcing ADECLU2 to predict the missing values, whatever the weight of the characteristic.

By comparing the results obtained from COBWEB and ADECLU(1&2), one notices that ADECLU's results are better than COBWEB's. COBWEB retains all of attributes in each concept, and for each attribute, the distribution of the different values met inside the concept. It seems, at least with this data and in a qualitative context, that this extreme and costly specification does not permit any improvement of the result.

We will perform experiments on 4 classes, extracted from the full and more complex data base of Michalski (4 x 17 cases of 50 variables). One notices (see [DECA89a&b]) that the good results are maintained, thus justifying the suitability criterion and reinforcing the idea that it is not necessary to keep everything (i.e. attributes, distributions of values) at each node. An appropriate selection can be more useful to reach the fixed goals with reasonable success.

We have also measured (see [DECA89a&b]) the performance of ADECLU as a hierarchical classification algorithm of objects. The method proposed also uses techniques from Data Analysis and in particular from Multiple Correspondence (Factorial) Analysis (MCA). This technique allows the display of natural clusters in the initial data. One would expect that an

[1] Classification without modification of the hierarchy constructed from the training set (i.e. without merging, splitting nor creation of concepts).

algorithm which performs well would find these clusters as subsets of objects defined by the concepts. ADECLU2 performs very well in this capacity.

As Fisher has already shown [FISH87a&b], the bidirectional operators reduce the dependency of the order of presentation of the data. Hence, the experiments using the same data but presented in a different order (ensuring that each class is represented) give similar results.

6. CONCLUSION AND PERSPECTIVES :

The use of the suitability criterion in "concept formation" is justified by the results obtained in the simple version (i.e. ADECLU1). Using this simple form, ADECLU has a predictive power equivalent to that of COBWEB, while using a much simpler concept description.

The second version, (i.e. ADECLU2) attacks the problem of attribute selection, and suggests the use of criteria of association of variables (taken from Data Analysis). The index of Goodman and Kruskal possesses properties appropriate to the problem. Other criteria can be envisaged, notably the parameters whose calculation and updating would be made easier in the incremental context of the problem. This second version, whose complexity is comparable to that of COBWEB (one must also take into account the distributions of the values of the variables), does offer however several important advantages over COBWEB:

- better predictive performance is achieved with detection of those cases where the prediction cannot be inferred from the classification (thanks to the "Lambda" index).
- selection of a combination of (single or multi-valued) characteristics, to define a concept
- calculation of the evaluation function reduces to the characteristics with positive weight.

The suitability criterion, because of its generality, offers a wide flexibility of application, and we suggest adopting it to numerical attributes, so as to be able to treat **mixed data**. We note that there are similarities between the formalism of ADECLU2 and neural networks. A possible third version of ADECLU, expressed in connectionist terms is envisaged. This version would permit using a learning law for the weights, which would solve the problem of the association criteria updating (mentioned above).

REFERENCES

[ANDE73] Anderberg, M.R. (1973). Cluster analysis for applications. Academic Press. New York.

[DECA89a] Decaestecker, C. (1989). Formation incrémentale de concepts par un critère d'adéquation. Proceedings of "4èmes Journées Française d'Apprentissage" (May 1989) published by INRIA Rocquencourt (France). (english version available)

[DECA89b] Decaestecker, C. (1989). ADECLU : A Incremental conceptual clustering system. Technical Report IA-06-89, CADEPS, Université Libre de Bruxelles.

[FISH87a] Fisher, D.H. (1987). Knowledge acquisition via incremental conceptual clustering. *Machine Learning,* 2, 139-172.

[FISH87b] Fisher, D.H. (1987). Knowledge acquisition via incremental conceptual clustering. Doctoral dissertation, Department of Information and Computer Science, University of California, Irvine.

[GENN88] Gennari, J.H. , Langley, P. and Fisher, D. (1988). Models of Incremental Concept Formation. Technical Report 88-16, Department of Information and Computer Science, University of California, Irvine.

[GOOD54] Goodman, L.A. and Kruskal, W.H. (1954). Measures of association for cross classifications. *J.of the American Stat. Assoc.,* vol. 49, 732-764.

[HANS86] Hanson, S.J and Bauer, M. (1986). Machine Learning, Clustering and Polymorphy. In L.N. Kanal & J.F. Lemmer (Eds), *Uncertainty in Artificial Intelligence*. North-Holland.

[LEBO87] Lebowitz, M. (1987). Experiments with incremental concept formation : UNIMEM. *Machine Learning,* 2, 103-138.

[MICH80] Michalski, R.S. and Chilausky, R.L.(1980). Learning by being told and learning from examples: an experimental comparison of the two methods of knowledge acquisition in the context of developping an expert system for soybean disease diagnosis. *Policy analysis anf Information Systems,* vol.4, No 2, June 1980.

[MICH84] Michalski, R.S. and Stepp, R.E. (1984). Learning from observation : Conceptual Clustering, in "Machine Learning, An Artificial Intelligence Approach", Michalski, Carbonell, Mitchell (eds.), Springer Verlag.

CONCEPTUAL CLUSTERING IN BIOLOGY: APPLICATION AND PERSPECTIVES

J. Lebbe*, I.C. Lerman+, J. Nicolas+, P. Peter# and R. Vignes*[1]

ABSTRACT

In this work we strive towards two complementary goals. First, we look for a cooperative clustering process of symbolic and numerical methods. Second, we want to stay as near as possible the expert knowledge. Our choices are to apply **existing** symbolic and numerical approaches as much as possible and to manage fidelity using a rich and complex knowledge representation: the **taxonomic preordonnance with multiple choice variable**. A symbolic method seems more fitted to represent the data, the knowledge and the result of a clustering algorithm, and a numerical method seems far more efficient to build clusters.

In order to test our proposal we construct a classification of medical interest insects, the species of phlebotomine sandflies of French Guiana, with the Likelihood Linkage Analysis.

We show also how we plan to carry on this work in order to improve the data acquisition and the clusters explanation.

1 INTRODUCTION

Sandflies are small blood-sucking insects widely distributed all over the world. More than 700 different species have been observed. They are important from the medical and veterinary point of view due to their role in the transmission of various infectious illness by protozoa, bacteria or virus. One of them, the leishmaniasis, is prevalent throughout the world (about 400,000 new cases each year). Only 20 of the species of sandflies are proved vectors of human leishmaniasis. The accurate identification of these sandflies is of vital importance for epidemiological studies and for programs of disease control. But data on sandflies are widely dispersed and not readily available. Identification ability is shared only by few specialists and is acquired by a single biologist only through huge effort. In order to alleviate this unsatisfactory situation, during the last four years, a Computer Aided

[1]* Laboratoire de Phytologie quantitative Université Pierre&Marie Curie Paris VI.

+ IRISA Rennes Tel 99 36 20 00 Telex UNIRISA 95 0473 F # IRESTE Nantes FRANCE

Identification (C.A.I.) system was created with XPER (an expert system generator, [LEBB 84]) for the identification of the phlebotomine sandflies of French Guiana [LEBB 87][2]. This identification system is used by many entomologists with success and will be extended to South and Central America.

Starting from the knowledge base of this expert system, we are now interested in creating and explaining clusters of sandflies species. Such classifications are very important for the biologist to structure the different informations collected about species. They try to discover natural affinity from the morphological attributes of the species. The most informative natural affinity is the phylogenetic[3] relationship between species [HENN 66].

Many methods of numerical clustering algorithm are used to construct a phylogenetic tree. But they show some commonly admitted drawbacks [FARR 72], among them being their unability to distinguish clusters on absolute values and to assign different values of distance between a mother node and its daugthers.

New methods of clustering, called cladistic methods, are now widely used by biologists, trying to solve these problems [FARR 70], [ESTA 72]. The global philosophy of these methods is to seek a parsimonious tree which minimizes the necessary hypotheses about attributes to "explain" the phylogeny. Unfortunately these methods are time consuming and are in fact unapplicable to our data. In particular they need the phylogenetic orientation of each modality in each variable and this information is not available. They are also unable to manage multivalued attributes (character polymorphy) and the relations between variables (dependent characters) necessary in the description of the sandflies.

Consequently, we have developped a method able to represent all the knowledge available about sandflies, and a clustering process avoiding the problem of classic numerical methods.

2 METHOD AND RESULTS

2.1 KNOWLEDGE REPRESENTATION

A knowledge base about phlebotomine sandflies of French Guiana and an expert system was created in order to manage identifications [LEBB 87]. The knowledge base covers a matrix of objects / variables describing the 77 species of phlebotomine, and a background knowledge including dependency relations between variables.

---

[2]This system is produced by the Pasteur Institut with the financial backing of the World and Pan-American Health Organization.

[3]Two species have close *phylogenetic relationship* if they share a recent common ancestor in the evolutionnary history.

The biological descriptions are very complex. Each species is a class of specimens and its description must represent not only a prototype but all the possible variations in the concerned species [LEBB 88]. Therefore, for each variable, the description of a species often requires a disjunction of values.
Example: Variable 4, which describe a part of male genitalia (*course of the terminal portion of genital filaments*), presents to possible modalities
1) straight or curved
2) twisted or spiraled
The disjunction does not reflect an uncertainty on the ral value of the variable but instead describes an **observed** set of values occuring for the given variable.
In our application, the variables are qualitative and the numeric values dealt with range classes. Some variables depend on values of other variables. These relations between variables explain why some species have no value for some variables.
Example: Variable 19, *Distribution of insertion of four style spines*, is relevant only if Variable 18, *Number of style spines*, takes value 4.

In this work, we try to translate this knowledge base using a new type of descriptive variable, called *taxonomic preordonnance with multiple choice* [LERM 88].
The notion of such qualitative variable provides the most accurate representation of the expert knowledge.
A **preordonnance variable** is a qualitative variable with a total preorder on the set of pairs of modalities. The higher the rank of a modality pair, the bigger the perceptual similarity between the two modalities. This preorder constitutes an important part of the knowledge supplied by the expert. It improves the accuracy of the similarity measure between species, bringing it its semantical foundations.
Example: Consider variable 33 of the knowledge base describing a part of female genitalia *Aspect of individual ducts* and its modalities
1) smooth non-sclerotized
2) smooth sclerotized
3) transversally striated or annulated
4) with small prominent tubercles
Starting from most similar pairs to least similar ones, the expert has provided the following preordonnance: $11 \sim 22 \sim 33 \sim 44 > 12 \sim 13 \sim 23 > 14 \sim 24 \sim 34$, where jh ($1 \leq j \leq h \leq 4$) represents the pair (j,h).
This notion is general enough to represent hierarchical relations between variables.This relation leads to a synthetic **taxonomic variable** which is a particular case of a preordonnance variable A taxonomic variable is a set of variables with a hierachical relation.A **taxonomic preordonnance with multiple choice variable** is the result of numerous variables having a hierachical relation together and a preorder. Furthermore, the "value" of a qualitative variable on a given object is defined by a disjunction on the set of its modalities. The preorder takes into account the preorders defined on each variable of the taxonomy.

Example: In order to represent the original descriptive variable 14, *Number of setae on coxite tuft*, we use mother variables 12, *Ornementation of coxite* (with modality 3, *One tuft without longitudinal row of setae*, and 4, *One tuft without longitudinal row(s) of setae*), and 1, *sex* (with modality 1, *male*), to build a new taxonomic variable:

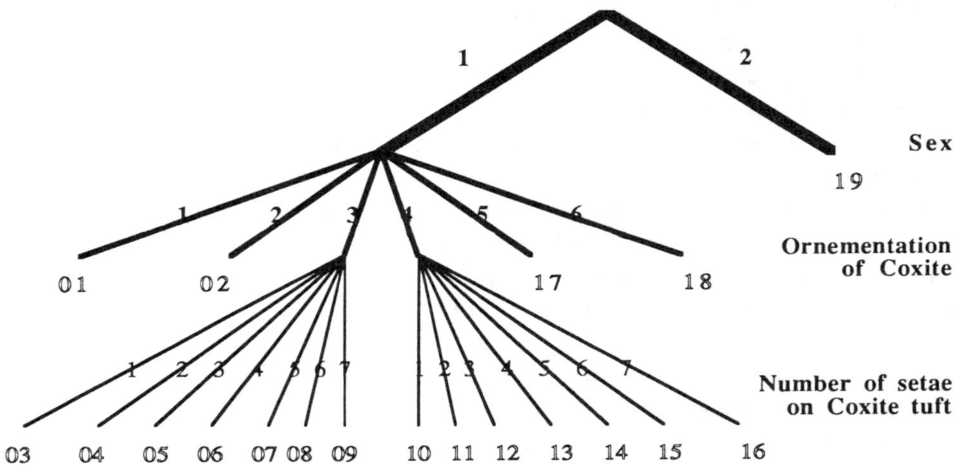

Each modality may be a disjunct of values. For example, modality 08 of the taxonomic variable (i.e. modality 6 of variable 14) is *eight to ten* (i.e 8 *or* 9 *or* 10).

2.2 THE NUMERICAL METHOD

Our method is a hierarchical classification method (**Likelihood Linkage Analysis** [LERM 81]) founded on a combinatorial and statistical approach. The method leads to significantly relevant classes on both sets, the object set and the variable set, for all types of descriptive variables considered in the method as relations.

In this method two classes are quickly merged if their proximity is likely big with respect to a random no relation hypothesis. This algorithm enables the structuration of the family of species one by one or by couple (male, female) of each species.

The similarity index between species is consistent with the Likelihood Linkage Algorithm. In short, for each preordonnance variable, the rough index is the average of the rank (defined in a precise way) on all the modality pairs respectively possible for the two species to be compared. This index is statistically standardized and summed for all the variables.

Finally, a method based on statistical criteria allows to detect significant steps and nodes.

2.3 RESULTS AND INTERPRETATION.

A classification tree of the phlebotomine sandflies has been produced. This tree matches quite well the empirical entomological classification. This demonstrates that the similarity index evaluates the likeness between species in a way close to the biologist appreciation. However, some groups are sparsely distributed among the clusters. Are the standard entomological groups not real (natural) groups or is the classification not perfect ?

Two sources of these problems may be pointed out.

1) The behaviour of the similarity index between disjunctions of modalities is not fully satisfying for discrimination of very distant species or aggregation of very close species. This is due to the specific interpretation of the biological (not logical, see §2.1) **or** and we plan to test new similarity indexes, better suited to this interpretation, including notions of typicality.

2) In the classification algorithm, some characters may have an arbitrary important weight due to the choice of variables to describe these characters.For instance, all the numerical variables about length of a part of the phlebotomine are correlated with the total length of the specimen. So this last character has an important weight. We also deal with numerical variables for lengthes and ratio of these lengthes The ratio (A/B) is given for its practical descriptive power (a ratio is much more easily measured than absolute values) but we do not want to attribute a more important weigth to the length of A and B due to the variable A/B.Moreover, the way to describe each character has importance. The choice of taxonomic variables or of variables with conjunction of values for the same character introduces a bias on the similarity between species and gives different results.

For instance Variable 11, *Shape of paramere*, presents three possible modalities with a conjunction of values:

1) simple unilobed without stout differenciated seta

2) simple unilobed with stout differenciated seta

3) complex and multi lobed

This descriptive choice has to be equivalent to the selection of two "mother-daugther variables":

Variable 11', *Shape of paramere* (modalities *1 simple 2 complex*), and Variable 11", *Setae on a simple paramere* (modalities *1 stout differenciated 2 other cases*).

Generally speaking, we have to make a clear distinction between the **description level**, where variables are designed for their ability to measure relevant data, and the **taxonomic level**, where variables are designed to partition the description space without interacting. A possible restricted solution would be to exclusively create trees where groups are stable for sets of randomized weights assigned to each variable. Our view is rather to organize original variables and their modalities on the basis of their formulas and the background knowledge of the application.

To sum up, our method does not give directly a phylogenetic tree (cladogram) but is probably a good way to summarize morphological (phenetic) relationships of our species. The phenetic tree can be used to enforce our opinion about empirical classification and moreover, allows to make hypotheses on characters evolution. However, this interpretation requires a better knowledge representation and an automated production of cluster explanations. The next section outlines some solutions that we plan to develop.

3 PERSPECTIVES

It is well known that a conceptual clustering problem cannot be solved with a sequential application of a single numerical classification task and a single symbolic characterization task [FISH 85]. The issues raised by our application require obviously more diversified numerical and symbolic tasks and complex connections between these two types of methods. The purpose of this section is to make clear these tasks and links in the framework of a general method we are currently developping.

3.1 RESEARCH TOPICS

Perspectives concern three basic themes.

Knowledge Acquisition:

In order to provide to the expert a language in which he can easily express all his knowledge and to produce rich and shaded characterizations of classes, we extend the internal language in two directions. First, descriptions are expressed in a restricted first order predicate calculus - functions are not allowed for decidability -. Second, modal operators of typicality are introduced to modulate some specific parts of the description of species and of the description of classes.

However, representation is just one component of the knowledge acquisition problem. We have to ensure the consistency and the adequation of the system knowledge base with the knowledge of the expert. That is, we must provide some tools, able to control the knowledge acquisition process. For this purpose, variables and their modalities are organised on both combinatorial and logical criteria. These organisations are then submitted to and/or compared with the own expert knowledge. The expert must provide new knowledge until a consensus is reached.

Definition of a Quality criterion for a partition:

The originality of conceptual clustering is to found the quality of a partition on both the structure and the characterization of each node of this partition. We don't think that the usual simplicity criterion on formulas provides a satisfying specification of suitable generalizations. Our approach is rather to favour a **relative** simplicity of the formulas, that is, to found the evolution of a characterization in a hierarchy on the evolution of the structure itself of this hierarchy.

Extension of the base of sandflies:

We have a specific group of species available to validate our approach (sandflies of French Guiana). Consequently, we have at our disposal too few species for some natural groups. A significative and really interesting study of sandflies requires more data to be collected. A planned extension of the project, taking into account all the species of South and Central America, seems promising towards the achievement of this goal .

3.2 OVERVIEW OF THE PROJECT

We outline now the implementing of the previous general project. The main steps of our method follow:

1) Knowledge acquisition : A little parser translate a "pseudo-natural" language external form into a logical internal one. We have chosen first order predicate calculus, excluding functions but including negation and disjunction Three modal operators of typicality are added : Typical, Atypical, Very_Atypical. The knowledge of possible combinations of these operators is application dependent and has to be provided by the expert.

2) Classification of variables : During this numerical step, we build a hierarchical clustering of the set of variables, using the L.L.A. method.This leads to new potential structered descriptors.

3) Organisation of variables and their modalities: The next symbolic step search for a logical explanation of the previously observed clusters of variables. More accurately, the formula describing a "mother" variable has to be implied by the formula describing a "daugther variable.If it is not the case, the expert may either reject the discovered link between data (it is a fortuitous link) or extend the knowledge base in order to explain the link. On the contrary, if the implication holds, the expert is always allowed to reject the link (it is a non-interesting link).

In a similar way, modalities of each variable are organised. That is, based on their formulas and on the logical similarity between these formulas, a preordonnance between modalities is established and compared with the preordonnance supplied by the expert. Thus, the system is able to detect inconsistency (the expert preordonnance is false) or incompleteness (the system preordonnance is false) of its knowledge base.

4) Production of species classifications: This numerical step may be splitted itself into two subprocesses. The first one is the classification task of objects we have already

described. A promising extension of the method concerns the treatment of typicality informations within data. The second one is a statistical characterisation of the significative nodes of the previous classification, expressing the influence of a same variable on an organized set of variables.

5) Production of local characterizations : A symbolic algorithm of generalization [NICO 87, 88b] considers significative clusters of objects produced during the previous step and generalize formulas associated with objects of each cluster to produce its characterization. The originality of the method lies in its capacity to fully treat negation (and hence disjunction) [NICO 88a]. The combinatorial cost of the method is reduced with a focalisation of the characterization process on small "significative" subgroups. An interesting research direction is to use effectively statistical cues as a means to further focus the generalization process and to make profit of typicality informations to produce more interesting disjunctive formulas.

6) Global detection of the quality of the partition, with a branching to step 4 if a sufficient level of quality is not reached. The set of local characterizations is globally logically tested. A good partition must ensure a certain evolution of the formulas associated with nodes on a same branch of the tree.

4 CONCLUSION

We have established first encouraging results on a natural classification task. We propose and develop a general plan, closely combining symbolic and numerical methods, in order to deal with this problem thoroughly. Much work remains to be done, but we hope that this new approach within a practical domain will favour further research in the challenging conceptual clustering field.

REFERENCES

[ESTA 72] "Cladistic methodology: a discussion of the theoretical bases for the induction of evolutionnary history" G.F. ESTABROOK
Ann. Rev. of Ecol. System. Vol 3 pp 427-456

[FARR 70] "Methods for computing Wagner trees"
J.S. FARRIS
System. Zool. Vol 19 pp 83-92

[FARR 72] "Estimating phylogenetic trees from distance matrices"
J.S. FARRIS
American Nature Vol 106 pp 645-668

[FISH 85] "Approaches To Conceptual Clustering"
D. FISHER P. LANGLEY
Proc. of IJCAI 85

[HENN 66] "Phylogenetic: Systematics"
W. HENNIG
University of Illinois Press. Urbana Champaign.

[LEBB 84] "Manuel d'utilisation du logiciel XPER"
J. LEBBE
Micro Application Rueil Malmaison 1984

[LEBB 87] "Identification assistée par ordinateur des phlébotomes de la Guyane Française"
J. LEBBE, R. VIGNES, J.P. DEDET
Institut Pasteur de la Guyane Française Rapport Juillet 1987

[LEBB 88] "Base de connaissances sur les pollens et spores atmosphériques de l'Europe du Nord et Tempérée : Identification et application taxonomique" J. LEBBE, R. VIGNES, M. HIDEUX
Masson Paris A paraitre

[LERM 81] "Classification et analyse ordinale des données"
I.C. LERMAN
Dunod, Paris 1981 Chap.3.

[LERM & PETE 88] "Classification en présence de variables taxonomiques à choix multiple. Application à la structuration des phlébotomes de la Guyane Française" I.C. LERMAN, P. PETER
IRISA, Publication interne n° 426 Rennes sept.1988

[NICO 87] "ALLY, un système logique pour la généralisation en apprentissage automatique" J. NICOLAS
Thèse d'Université de Rennes I Mai 1987

[NICO 88a] "Consistency and Preference Criteria for Generalization Languages handling Negation and Disjunction" J. NICOLAS
Proc of ECAI 88, München August 88

[NICO 88b] "Constructive proofs for Concept Learning"
J. NICOLAS
Proc of AIMSA 88, Varna Septembre 88

LEARNING SYSTEMS

THE MEFISTO* MODEL: AN OBJECT ORIENTED REPRESENTATION FOR STATISTICAL DATA MANAGEMENT+

G. FALCITELLI[1], L. MEO EVOLI[2], E. NARDELLI[3], F.L. RICCI[2]

1: Istituto di Medicina Sperimentale, CNR, V.le Carlo Marx 15, 00156, Roma, Italy.
2: Istituto di Studi sulla Ricerca e la Documentazione Scientifica, CNR, via C. De Lollis 12, 00185 - Roma, Italy
3: Istituto di Analisi dei Sistemi e Informatica, CNR, V.le Manzoni 30, 00185 Roma, Italy

ABSTRACT

Statistical Databases can support user activity by means of queries on aggregated data (macrodata). At the logical level, specific models to represent the different aspects of macrodata are necessary; to achieve this, the authors propose Mefisto*, an object oriented model. At the database definition stage it is not possible to predict which macrodata-generating functions will be required during the database's lifecycle. An object based model allows the knowledge required to describe the function to be acquired as such functions become necessary.
After analyzing the characteristics of statistical tables, the authors first indicate the role of data type as a bridge between the closed world of databases and the open world of statistics. Then they illustrate the Mefisto* model and go on and discuss the object oriented representation of statistical information.
KEY-WORDS : data model, statistical databases, object oriented representation

1.INTRODUCTION

One of the problems currently investigated in database theory is the logical representation of query languages on Statistical Data Bases (SDBs). The characteristics and problems related to statistical and scientific databases, in particular macro-SDBs, are discussed in [SH85]. Macro-SDBs refer to SDBs containing macrodata; they are obtained by applying aggregation functions to records of individual entities or events (such as tables and/or charts of mortality by causes). This kind of macrodata can be represented by a complex data structure, which we shall call **Statistical Table** (ST). The kind of data structure consists of [FO86]:

a) a single summary attribute (quantitative data) representing a property of the statistical phenomenon described in the ST; its instances (summary values) are the numeric values contained in the table;
b) the unit of measure of the summary attribute;
c) a set of category attributes (qualitative data), that is, the variables that characterize the summary attribute univocally;
d) a set of values that we shall call the table variable domain, corresponding to each category attribute; these values (modalities) are generally strings of alphanumeric characters;
e) one specific "data type", depending upon the particular aggregation function from which the summary attribute originates [FO86]; for example, the aggregation function "percentage" produces, as a data type, "rate value"; the aggregation function "count" produces, as a data type, "absolute value", and so on.

Therefore, user searches carried out on macro-SDBs usually include two activities mutually orthogonal: 1) Table management, or manipulation of the descriptive elements of the ST (variables and modalities), 2) Data analysis, or processing of the instances of summary attributes.
Here only the first phase will be discussed, i.e. the one in which appropriate operators are required in order to build up new STs starting from existing STs in the database.

The table manipulation generally means changing their characteristics both at the level of category attributes (e.g. by eliminating one or more variables) and at the table variable domain level of category attributes (e.g. by selecting or compacting modalities). As a result, this process generally requires the corresponding summary values to be recalculated. This recalculation is strictly dependent on the ST data type.
Account must be taken of the fact that it is impossible to know a priori which data types must be taken into consideration in a macro-SDB insofar as there is always the possibility of defining new ones. In fact, statistical activity involves identifying new aggregation functions to generate indicators which are valid for specific areas of investigation. It follows that the universe of the data type is an open one and that a macro-SDB must therefore be able to acquire the knowledge needed to manage these new data types.

+ This work has been supported by CNR under project MULTIDATA.

Our approach to data type management is related to management transparent to the user, without having to define a special procedure for each table considered. Furthermore, the database administrator, for the new data types, provides the system with the information (expressed in general terms and not bound to the individual table) needed to build up the procedures for recalculating the values of the summary attribute.

2. THE MODEL

In a database composed of macrodata, the activity of table management involves the generation of new tables starting from stored tables (i.e. without having access to the microdata) by operating on the descriptive elements of the starting tables [BR84].
The relational algebra operators are not sufficient because they cannot be used to recalculate the summary values [FO87]. It is precisely in this recalculation that statistical experience plays a fundamental role. If the operators is in terms of manipulation of the descriptive elements and the user is not required to have any knowledge of statistics; this kind of experience must be supplied to the system. This has the advantage that the user can ask the system for a table without having to specify how to obtain it from among those present in the database. Problems about representation of statistical knowledge and implemmentation of statistical expertise are discussed in [GA86].

2.1 The data structure

An example of an ST is shown in Fig. 1 (the information source for this table is the W.H.O. Cancer Register). The summary attribute is "cancer cases" and its type is "rate" (the aggregative activity in this application is the function "count" on the above microdata set and the subsequent function "ratio"). The unit of measure is the number of cases per 100000. The set of category attributes comprises "year", "sex", "country", and "site" of the cancer; the table variable domain of year is {1981}; the table variable domain of sex is {M(=male), F(=female)}, the table variable domain of site is {oral, lung}, and the table variable domain of country is {Australia, Austria, Brazil, Canada, Denmark}.
The Mefisto* model is based on a functional approach and represents an ST by means of a complex data structure described by an ordered pair (R,g), where [FO86]:
R is a relation, known as table space, whose attributes are category attributes describing an ST to which the pair refers, and whose extension is the cartesian product of the table variable domains;
g is a function which maps from the elements describing the macrodata to the macrodata themselves.
The aggregation function we apply to the microdata to obtain the ST generates the data type.
In the preceding example the scheme of the relation is {year, sex, country, site}, the relation is the cartesian product of {1981}, {M, F}, {oral, lung}, {Australia, Austria, Brazil, Canada, Denmark}, i.e. {(1981, M, oral, Australia), ... , (1981, F, lung, Denmark)}, while the function g is g(1981, M, oral, Australia) = 5.1, ... , g(1981, F, lung, Denmark)=18.7.

site	oral		lung	
sex	M	F	M	F
Country				
Australia	5.1	1.7	71.6	21.4
Austria	5.3	0.6	65.9	9.7
Brazil	N.A.	N.A.	22	4.3
Canada	5.2	1.5	68.4	16.6
Denmark	4	2	70.6	18.7

Cancer cases by selected sites in 1981, in 5 countries, by sex(rates per 100000 population)
Source: World Health Organ., Annual Statistics 1984
Note: N.A.="Not Available"

Fig. 1

The table scheme could thus be written as:
cancer_cases_by_selected_sites (year, sex, country, site) (rate per 100.000).

2.2The algebra operators

The operators of interest here are those which involve the recalculation of the summary values (summarization, classification, restriction, enlargment) [FO86]. Even if disaggregation operator involves recalculating the summary values, it is not included among the table management type operators because the fundamental role in its application is played by the user's statistical experience. This experience is necessary to express the law of disaggregation.

SUMMARIZATION
For instance, let us consider the table "fruit production by year_region" (T_1 of Fig. 2), characterized by "region" and "year" as category attributes. If we are interested in knowing the total fruit production for the years 1981-1983, we perform the summarization of T_1 with respect to year and obtain the table of Fig. 3.

FRUIT PRODUCTION BY YEAR_REGION
(10^3 tonns) (absolute value)

T_1	Years		
Regions	1981	1982	1983
Lazio	87	92	127
Piemonte	58	63	61

Fig. 2

FRUIT PRODUCTION BY REGION IN 1981-1983 (10^3 tonns)
(absolute value)

Regions	
Lazio	306
Piemonte	182

Fig. 3

CLASSIFICATION

For instance, let us consider the table "car production by year_model" (T_1 of Fig. 4) characterized by "model" and "year". If we are interested in having this production for "HP" and "year" (the link between the "model" and "HP" is the relation r of Fig. 5), we perform the reclassification of T_1 by r and obtain the table of Fig. 6.

CAR PRODUCTION BY YEAR_MODEL (10^5 cars) (absolute value)

T_1	Years				
Model	1981	1982	1983	1984	1985
Panda	127	132	139	100	102
Fiesta	68	73	71	150	200
Alfa	70	80	247	252	269

Fig. 4

	models	HP
	Panda	12
r	Fiesta	12
	Alfa	15

Fig. 5

CAR PRODUCTION BY YEAR_HP (10^5 cars) (absolute value)

	Years				
HP	1981	1982	1983	1984	1985
12	195	205	210	250	302
15	70	80	247	252	269

Fig.6

RESTRICTION

For instance, let "occupied in building by sex_region" be the table T_1 of Fig. 7, the data type is "percentage with respect to the total of sex" (i.e., %sex), and let r_1 be the set of Fig. 8. The restriction of T_1 by r_1 gives the table of Fig. 9, this table represents the occupied in building in the Veneto and Piemonte regions, broken down by "sex".

OCCUPIED IN BUILDING BY SEX_REGION (%sex)

T_1	Sex	
Regions	M	F
Lazio	14,6	19,4
Piemonte	26,8	29,3
Veneto	46,6	41,4
Sicilia	12	9,9

Fig. 7

r_1	region
	Veneto
	Piemonte

Fig. 8

OCCUPIED IN BUILDING IN VENETO_PIEMONTE BY SEX (%sex)

T_1	Sex	
Region	M	F
Veneto	63,5	58,6
Piemonte	36,5	41,4

Fig. 9

The computation formula of summary values is described in following section.

ENLARGEMENT

Let us consider two tables of "occupied in building in Lazio_Sicilia by sex" (Fig. 10) and "occupied in building in Veneto_Piemonte by sex" (Fig. 9), where the data type is "%sex". If we need the table "occupied in building" by region_sex", if we are interested to the "union" of these tables, we perform the enlargement of these tables and obtain the table of Fig. 7.

OCCUPIED IN BUILDING IN LAZIO_SICILIA BY SEX (%sex)

T_1	Sex	
Region	M	F
Lazio	54,9	66,2
Sicilia	45,1	33,8

Fig. 10

The computation formula of summary values is described in following section.

3. THE OBJECTS

In the definition of these four operators the function appears which allows the values associated with the summary attribute to be recalculated. The definition of this function varies from operator to operator. The application of the same operator on the STs of a macro-SDB can involve different algorithms for recalculating the summary values involved in the manipulation which are dependent on the ST data type. From the way the operators are defined, the user can know how the operator acts at the table description level and need not know how the summary values are actually computed. This means for example that if the user wants a table to be described by a smaller set of category attributes he has to use the summarization operator.
Let us consider the case of a usee intrested in houses distribution. The starting table is described in terms of "municipality" (with table variable domain "Acquanegra, Asola, Bagnolo"), "occupation" (with table variable domain "occupied, unoccupied") and "year" (with table variable domain "1981"). If the user is interested in the distribution of the houses by "municipality" and "year", he performs a summarization on "occupation". Let us take the starting statistical table as T_1 in Fig. 11 ("number of houses"), the data type of which has an absolute value. Its summarization gives table T_2 (fig. 12) by eliminating the attribute "occupation".

Number of houses by municipality and occupation in 1981
(abs. values)

occupation / municipalities	unoccupied	occupied
Acquanegra	922	130
Asola	2768	296
Bagnolo	1454	195

Tab. **T_1** / fig. 11

Number of houses by municipality in 1981
(abs. values)

municipalities	
Acquanegra	1052
Asola	3064
Bagnolo	1649

Tab. **T_2** / fig. 12

In this case the values of T_2 must be computed by summing the corresponding values of T_1 for every element of the table variable domain associated with "municipality", applying to T_1 the computation function:

$t_i^2 = \sum_j t_{i,j}^1$ where:

- t_i^2 is the generic instance of the summary attribute of T_2
- i is the generic element of the variable "municipality"
- $t_{i,j}^1$ is the generic instance of the summary attribute of T_1
- j is the generic element of the variable "occupation".

If the original statistical table is T_3 in Fig. 13 (where the data type is arithmetic mean), the same summarization gives T_4 in Fig. 14 as its result.

Average number of rooms per house by municipality and occupation in 1981
(arithmetic mean)

occupation / municipalities	unoccupied	occupied
Acquanegra	5,5	4,7
Asola	4,8	4,2
Bagnolo	5,4	5,3

Tab. **T_3** / fig. 13

Average number of rooms per house by municipality in 1981
(arithmetic mean)

municipalities	
Acquanegra	5,4
Asola	4,7
Bagnolo	5,4

Tab. **T_4** / fig. 14

It can be seen that, in this case, the values of T_4 cannot be computed as in the preceding case. In order to obtain the table T_4 from table T_3 the data contained in table T_1 must be involved, according to the following computation function:

$$t_i^4 = \frac{\sum_j t_{i,j}^3 \, t_{i,j}^1}{\sum_j t_{i,j}^1}$$ where:

- t_i^4 is the generic instance of the summary attribute of T_4
- i is the generic element of the variable "municipality"
- $t_{i,j}^3$ is the generic instance of the summary attribute of T_3
- $t_{i,j}^1$ is the generic instance of the summary attribute of T_1
- j is the generic element of the variable "occupation".

3.1 The data types

In order to allow transparent data type management, for each data type, the database administrator must provide the system with the information (expressed in general terms and not linked to individual tables) that the system needs in order to construct the procedures for recalculating the summary attribute values for the reclassification, restriction and enlargement operators (summarization is a special case of reclassification).
The various algorithms for calculating the summary values as a function of data type are set out below. For the sake of simplicity, an essential data typology has been identified which includes the most common data types, but without any claim to being exhaustive. This typology is as follows: absolute value, arithmetic mean, percentage, fixed base index number, moving base index number, ratio between tables or rate.

For each of these data types we shall indicate:
- the aggregation function that generates the statistical table when applied to micro-data;
- preferential variable, if there are any variable types for which a different behaviour is required;
- condition of applicability, i.e. which tables must be present in the database in order to allow the summary values to be computed;
- formulae for computing the summary values with respect to the Mefisto* operators.

For the sake of semplicity we assume that the modalities have null intersection ($t_i \cap t_j = \emptyset$).

absolute value: originating function: count, sum
preferential variable: none
condition of applicability : none

formulae:

summarization / classification $t_i = \sum_j t_{i_j}^1$

restriction $t_i = t_i^1$

enlargement $t_i = t_i^1$

arithmetic mean originating function: average
preferential variable: none
condition of applicability: the availability of a table with the same descriptive elements obtained by means of an originating function of the type count (T_c) or sum (T_s)

formulae: summarization / classification:

- with count table (Tc): $t_i = \frac{\sum_j t_{i,j}^c \, t_{i,j}^1}{\sum_j t_{i,j}^c}$ with sum table (Ts): $t_i = \sum_j t_{i,j}^s \Big/ \sum_j \frac{t_{i,j}^s}{t_{i_j}^1}$

restriction $t_i = t_i^1$

enlargement $t_i = t_i^1$

percentage (respect to the total)
originating function:
1. count / sum in order to obtain the table of absolute values Tp
2. division by the value of the corresponding element of the table TN obtained by summarizing on Cj
3. multiplication of the single value by 100

preferential variable: a variable Cj
condition of applicability: it is necessary to know table TN

formulae: summarization respect to Cj: has no meaning

classification respect Cj $$t_i = \frac{\sum_k t_k^N t_{i,j}^1}{\sum_k t_k^N}$$

classification respect any category attribute other than Cj $$t_i = \sum_k t_{i,k}^1$$

restriction $$t_{i,j} = 100 * t_{i,j}^1 \Big/ \sum_k t_{k}^1$$

enlargement: if is performed along any variable other than Cj $$t_i = t_i^1$$

otherwise, in presence of the tables TN1 and TN2 $$t_{i,j} = t_{i,j}^1 * t_j^N \Big/ t_j^{N_1} + t_j^{N_2}$$

fixed base index originating function: 1 count / sum in order to obtain the table of absolute values Tp
2 ratio to reference value (base value)
3 multiplication of the single value by 100

preferential variable: a time variabile Ct
condition of applicability: it is necessary to know table Tp

formulae: summarization respect to Ct : has no meaning
classification respect to Ct in presence of table Tp: ti,t=1 = 100

$$t_{i,t\neq 1} = 100 * \frac{\sum_{t_j \to t} t_{i,t_j}^P}{\sum_{t_j \to t=1} t_{i,t_j}^P}$$

summarization/classification respect to any category attribute other than Ct, in presence of table TR obtained by selecting the prime value of Ct : ti,t=1 = 100

$$t_{i,t\neq 1} = \frac{\sum_j t_{i,j,t}^1 * \sum_j t_{i,j,t}^R}{\sum_j t_{i,j,t=1}^R}$$

restriction having the same base: $t_i = t_i^1$

enlargement having the same base: $t_i = t_i^1$

moving base index: originating function: 1 count / sum in order to obtain the table of absolute values Tp
2 ratio with the previous value
3 multiplication of the single value by 100

preferential variable: a time variabile Ct
condition of applicability: it is necessary to know table Tp

formulae: summarization respect to Ct : has no meaning

classification respect Ct in presence of table Tp: ti,t=1 = 100, $t_{i,t\neq 1} = 100 * \frac{\sum_{t_j \to t} t_{i,t_j}^P}{\sum_{t_j \to t-1} t_{i,t_j}^P}$

summarization/classification respect to any category attribute other than Ct, in presence of table Tp; ti,t=1 = 100;

$$t_{i,t\neq 1} = 100 * \frac{\sum_j t^p_{i,t,j}}{\sum_j t^p_{i,t-1,j}}$$

restriction having the same base $t_i = t^1_i$

enlargement having the same base if performed respect Ct $t_i = t^1_i$

otherwise it is necessary to know Tp1 and Tp2 $t_i = t^1_i$

except for the first value referring to T2 (with the index t2 of Ct equal to 1)

$$t_{i,t_2} = 100 * \frac{t^2_{i,t_2}}{t^1_{i,t_T 1}}$$

ratio or rate: originating function: 1 construction of first table TI

2 construction of second table TII

3 division betwenn the corresponding element of TI and TII

4 multiplication of the single value by 100

preferential variable: none

condition of applicability: it is necessary to have the two tables TI and TII

formulae:

summarization / classification: $$t_i = 100 * \frac{\sum_j t^I_{i,j}}{\sum_j t^{II}_{i,j}}$$

restriction $t_i = t^1_i$

enlargement $t_i = t^1_i$

3.2 Object oriented scheme

On the basis of these examples it can quite naturally be observed that:

1) the computation function of the summary values may be <u>simple</u> (e.g. absolute value) or <u>complex</u> (e.g. arithmetic mean) according to whether values from other tables are needed. Furthermore, there may be one (e.g. percentage of total value of a variable) or more (e.g. rate) additional tables, each having its own role;

2) in the case of complex functions, the additional tables over and above the one we are operating on can have the same table space (e.g. arithmetic mean, rate) or a smaller one (e.g. percentage of total value of a variable);

3) in the case of complex functions, the operators for which the presence of additional tables are required can be computed only if the tables are present and, moreover, the computation function must depend on the function generating the tables (e.g. in the case of arithmetic mean the additional table may be generated by count or sum);

4) linked to the type of function is the possibility that an operator may not be allowed (e.g. summarization along the variable on which the percentage is computed);

5) not all category attributes have the same weight for <u>non homogeneous</u> computation functions, in which the computation function varies according to the category attribute with respect to which the operation is performed (e.g. enlargement for tables with percentage data type with respect to the total of a variable).

Clearly, then, the same operator can have several different types of computation function and further information, in addition to the algorithm, is required, such as reference tables and the preferential variable. All this is a function of data type.

These problems coincide with the advantages offered by the object oriented models [BA88]. In fact, this type of model has the following characteristics: 1) encapsulation, in order to model data and operations at the same time; 2) inheritance, in order to give a concise and accurate description of the world; 3) overriding, in order to allow the same name to be used for different operations.

The data type permits the organization of hierarchy of classes. The semplified object oriented model of macrodata of Figs. 11 and 13 is shown in Fig. 15.

The "table scheme" class represents the scheme of the relation describing table space, while the "table space" class represents the table space as the cartesian product of the modalities. The table corresponds to two classes: the first "table structure" is used to define the data and methods common to all tables; the its subclass (typed

table) is used to describe the information which varies according to data type. For example, in the case of "data type=arithmetic mean", the additional data entails reference to the table of absolute values having the same table space. At the level of the "table structure" class, the data refer to the summary values and the generating function. At this level also the operators used to manipulate the statistical tables are defined, at least as the name of the method. In fact the data type independent operators are fully specified (for instance, disaggregation), while the data type dependent operators are specified in the subclass of "table structure". There are two inheritance hierarchies (with transmission of data and methods): "subclass of" among classes and "subinstance of" among instances. In this way a tree structure can be used to represent the set of tables having the same table space and therefore to give a more abstract definition of the algorithms for computing the summary values. In the case in which the user studies the distribution of houses by means of the tables in Fig. 11 and 13, it is necessary to construct first the instance of the "table scheme" in which the names of the variables are contained as well as the domains for each of them, then the instance of the "table space "in which the various modalities are specified and, finally, the two tables (absolute values and arithmetic mean) which are interrelated because it is necessary to know the table of absolute values in order to compute the summarization of the table of data type aritmetic mean. It should be noted that both these instances refer to two instances of the "matrix" class because these instances incorporate in the relative methods the access method to the mass storage.
From this figure it can be seen that for each data type introduced into the scheme it is necessary to define the lower-level subclass in terms of the computation algorithms of the summary values for enlargement, restriction and summarization/reclassification and, if necessary, of additional information (reference tables and preferential variable). In this way a framework is built up which can be used as a guide in describing the new data types with which the system must be provided.

Current research topics include:

- defining methods and thus a module with a user-friendly interface for the new data type acquisition phase;
- studying the model extensions in order to include more sophisticated operators and statistical functions;
- extending the Smalltalk 80 language [GO83] to manage the inheritance among instances;
- implementing the model and integrating it with the macrodata management system already constructed [MA88].

4. CONCLUSIONS

The object oriented model for statistical table representation allows characteristics from the outside world of statistics to be integrated with the closed world of databases.
Although such a model cannot be directly utilized by the user, it allows the latter to express his table management type queries without worrying about the algorithms required to compute the summary values.
This model represents the starting point of statistical knowledge capture in order to simplify user interaction with the system as the user's attention is focused on the semantics of the statistical operator and not on the procedures for implementing it.

Acknowledgement

The authors wish to thank E. Fortunato and M. Rafanelli for the precious discussions.

REFERENCES

[BA88] F. Bancilhon "Object-oriented database systems", Proc. VII ACM SIGACT/SIGMOD/SIGART symposium on PODS, 1988.
[BR84] F. Bry, "Un modèle de données statistique", Rapport IRT, september 1984.
[FO86] E.Fortunato, M.Rafanelli, F.L.Ricci, A.Sebastio, "An algebra for statistical data", Proceed. of the III° Internat. Workshop on Statistical and Scientific Database Management, Grand Duchy of Luxembourg, 1986 .
[FO87] E.Fortunato, M.Rafanelli, F.L.Ricci, "The statistical functional model for the logical representation of a statistical table", Technical Report n. RT11/87, CNR-ISRDS, 1987.
[GA86] W.A. Gale (ed.), "Artificial intelligence and statistics", Addison-Wesley, Princeton, 1986.
[GO83] A.Goldberg, D.Robson, "Smalltalk 80: The language and its implementation", Addison-Wesley, 1983.
[MA88] L. Malavenda, L. Meo-Evoli: "La gestione di basi di dati statistici", Technical Report n. RT17/88, CNR-ISRDS, 1988.
[SH85] A.Shoshani, H.K.T.Wong, "Statistical and scientific database issues", IEEE Transactions on Software Engineering, Vol.SE-11, N.10, October 1985.

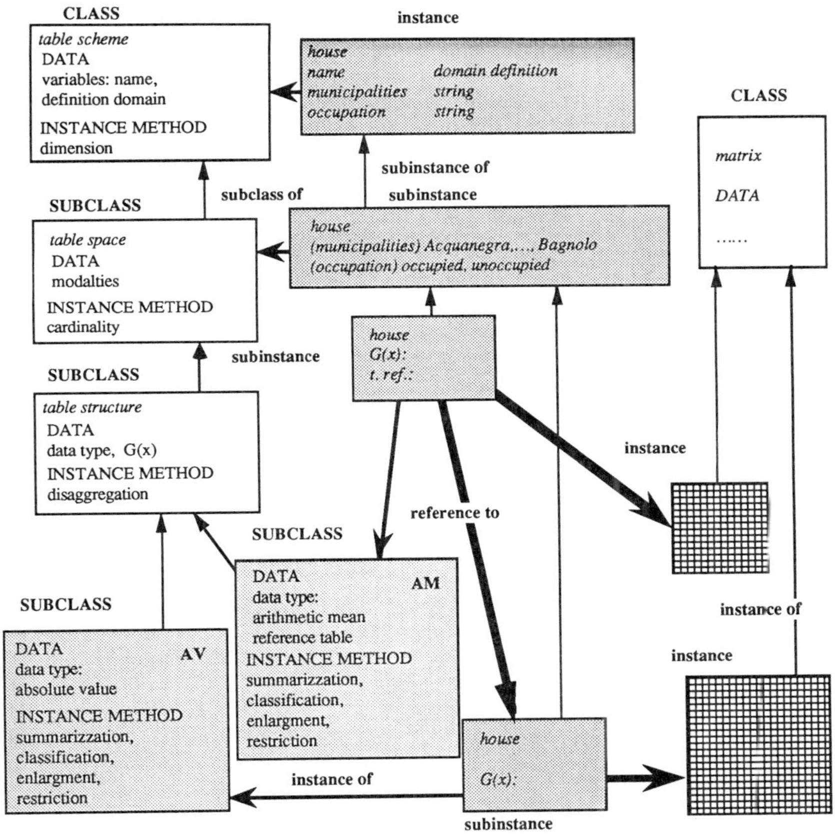

G(x) is a function from table space to summary value

Fig. 15

PINOCCHIO: AN AUTOMATIC CONCEPT LEARNING SYSTEM

Anna Giannetti* and Mimma Lucchelli**
*SIGMA-TAU Foundation for Scientific Research & IP/CNR
**Artificial Intelligence Lab., SELENIA S.p.a & IP/CNR
National Research Council, Institute of Psychology
Rome, (Italy)

Abstract: We present Pinocchio, a system which taxonomizes objects and associates complex descriptions to them. This system learns incrementally by simulating direct input from the external environment. Its major task is to link each concept with the others and to create "fictitious" classes that cover significant sets of given instances. It stores all of the information about concepts in a graph . It is able to perform deductive and inductive reasoning tasks and to answer a variety of questions about the concepts and their relationships..

1.Introduction

There is a growing interest in the research on automatic knowledge acquisition. The major goal of this research is to build intelligent autonomous systems, which should be able to categorize input data, to deduce new information, to organize the knowledge base so as to make the retrieval task more efficient and simpler. Following the taxonomy of Machine Learning research presented in [Carbonell,Michalski,Mitchell 83], we can characterize our system PINOCCHIO as follows:

- It is an automatic concept learning system, i.e. a system which is able to generate taxonomies of concepts and complex descriptions associated to each concept;
- The underlying learning strategy goes under the name of learning from observation and discovery. PINOCCHIO discovers regularities in the external world by means of passive observation and , sometimes, interacts with an expert (or a tutor) which should be considered as part of the complex environment;
- The general task of concept formation is accomplished by adopting an original conceptual clustering algorithm [Fisher and Langley 85], which is able to handle many concurrent hypotheses at time.

Among other systems which make use of a conceptual clustering method to build classifications we have: UNIMEM [Lebowitz 87] and COBWEB [Fisher 87]. Both systems are able to create complete hierarchies of concepts, learning incrementally and without the intervention of a supervisor. Contrary to PINOCCHIO, they allow to handle only one hypothesis at time and use operators in order to move inside the search spaces; moreover they build their concept hierarchies preferably in a top-down fashion whereas PINOCCHIO is a genuine data-driven model, adopting preferably but not exclusively a bottom-up, agglomerative approach [Hanson and Bauer 86].

2.PINOCCHIO

2.1 Goals of the research

We wanted to build a system which, when exposed to an unknown word, was able to optimally insert it into the knowledge base (we take for granted the existence of a module that correctly inserts the new word into a lexicon , see [Parisi and Castelfranchi 87]).

bIn the system, concept refers to a concrete thing, i.e. something which can be denominated by a concrete noun. Our choice is motivated by the fact that the concrete thing domain appears to be the most tractable one. At the beginning of learning, PINOCCHIO considers any object to be classified as a "potential" class (or concept).

At the end of a learning cycle, our system supplies a hierarchical classification of the given objects.

We consider the intensional meaning of a concept such as a collection of properties that can be ascribed to it, both positive (properties that a concept has) and negative (properties that a concept doesn't have).

Each property ascribed to a concept performs the same role in the description associated with a concept; the only important thing is the properties that concepts have in common. The concepts are of a general type : "a cat meows" means <each entity of "cat" kind meows (sometimes)>.

We can inspect the knowledge base both during the learning period and at the end of the period; we can also use the question/answering module in order to check out the results.

2.2 Task description

The main task of the system is to taxonomize concepts, by grouping those with properties in common into classes which can either be supplied by the external environment or can be internally built by the system.

Our learning task has some interesting qualities which should be investigated in more detail:

-it occurs by observation, i.e. the concepts' properties are presented one after another, in mixed order.

-it is strongly incremental. The system "learns" every new concept without wasting time waiting for the associated description to be complete and all data to be supplied. This means that the system continually changes the built-in representation, giving , sometimes, some strange intermediate results, in absence of discriminating information.

-it can be either supervised or unsupervised. Information about concepts can be fed to the system in two ways: one can supply a property (positive or negative) for a concept. In this sense we refer to "unsupervised learning", because the external environment supplies information which is not based on previous classifications. One can also supply a membership of a concept to its superordinate .In this sense we can speak of "supervised learning", i.e. there is a tutor,or an expert, who already knows a part of the classification, or is a result of a previous inductive reasoning).

Let's begin to examine each allowed input:

Positive Evidence:

"the cat miaows" : The system registers from the external world the information regarding an action usually performed by a certain set of elements, named "cats".

"the animals sleep": This input can be seen as a shortcut for a complex induction task. The concept of "animal" derives from the individual concepts "cat","dog","fish"..., and the property "sleep" should be first ascribed to all the instances. This input could also be seen as a result, obtained by the system, which in turn can be fed into the system.

"the mouse is an animal": Like the previous example, this input presupposes the existence of the two concepts and it is a way of abbreviating the description of a new concept, deducing properties from the superordinate to the subordinate concept.

"to run is a type of moving": This input refers to the classification of properties, along the SPEC link.

Negative Evidence:
"the cat does not bark": This input can be used to contradict a previous presupposition about the fact that "the cat barks" (like a "dog"). The negative evidence is weaker than the positive because, if an instance of cat doesn't happen to bark, this doesn't mean that each entity of "cat" type is never allowed to bark (unless one can check out over a long period that that is the case).

"animals cannot break down": The system should already know the concept of "animal".

"not every animal has scales":

"some animals have scales":

These inputs represent other shortcuts for a complex inductive task. With this input we exploit linguistic constraints in order to express two observations regarding the fact that there is at least one instance of the "animal" concept which doesn't have scales.

"chair is not an animal" : The system should already know both the concepts and their reciprocal differences.

"chair is not a dog": This is a very unusual input; it is evident that two different instances of the same level, having different nouns, are different things.

"small is the opposite of large": This input refers to the classification of properties where the negative label means "the opposite of" inside the PROPERTY table.

We want to remark that input of the "supervised" type, could be substituted, as it has indeed happened in some experiments, by a list of input, of "unsupervised" type, coming directly from the external environment.

The question/answering module is part of the system and it works by extracting information from the graph in order to give the requested information.
We briefly present a list of questions to which the system is able to reply:

" Has a concept X the property Y (or does it perform an action Y) ?"
The system is able to reply in different ways:

YES,SURE (if the information about the positive evidence was given by the external world);

NO,SURE (if the information about the negative evidence was given by the external world);

SOMETIMES (if the system learns from external or internal evidence, that only some subconcepts of X have the property Y);

MAYBE (if the system has inferred, from previous knowledge, the positive evidence);

THINKNOT (if the system has inferred the negative evidence);

DON'TKNOW (if the system has no information about that)..

"Is a concept X an instance or a superordinate of the concept Y ?"

The system replies:

YES,SURE (if knowledge about membership comes from the external world);

NO,SURE (if the negation of a membership comes from outside);

MAYBE (if the system has inferred the existence of that relationship);

THINKNOT (if the system has inferred the non existence of that relationship);

DON'TKNOW (if the system cannot infer anything).

"Which (how many) properties does (doesn't) the concept X have?"

The system gives as output the list (number) of the properties that are ascribed (positively or negatively) to the concept X, labelling, in the same way as we have described before, the answer coming from an inference.

"Which (how many) concepts are characterized by the property Y ?"

The system gives as output the list (number) of the concepts that have (or don't have) that property, labelling the answers as usual.

2.3 Data Representation

The system's knowledge base is represented by a graph whose nodes are concept names (concept-node), property names (property-node), or nodes indicating 'fictitious classes' i.e. classes created by the system on the basis of the regularities discovered in the context (XX-node). The edges represent different relationships between nodes:
a) we have IS_A edges which indicate membership;
b) either concept-node or XX-node can be connected with the property-nodes by means of HAS_PROP edges;
c) each property could be a specialization or the contrary of another property, and this fact is represented by a SPEC edge between the two property-nodes. The edge may represent different kinds of relationship (namely, specialization and opposition) depending on a positive or negative label.
All edges may have different labels: they can be the results of an inference of the system or can correspond directly to input from outside.
Learning happens by labelling or modifying already existing labels on edges. We classify labels according to their "meanings", i.e. external or internal origin of the corresponding information. If a) there is explicit information from outside that a concept belongs (or doesn't belong) to another concept; b) a concept has (or doesn't have) a property;or only some

instances of a concept have a certain property; c) a property is the contrary or a specialization of another, then Pinocchio attaches to the corresponding edges a label of "confident" type: EXT_EV (external evidence) if the information is positive, STR_NEG (strong negation) if the information is negative, WEA_NEG (weak negation) if the information pertains only to a subset of instances.
These labels cannot be subject to modifications.
On the contrary, if the system **infers** (according to rules) some information, it labels the corresponding edge with a label of "non confident" type:

INT_EV (internal positive evidence),
INT_NEG (internal negative evidence),
IS_A_XX (membership in a fictitious class),
INT_DED (internal deduction, if that property has been inferred from a superordinate concept)
XX_PR (presence of a property in the description of a fictitious class).

The "non-confident" labels can be modified during learning. As we shall see, some rules are very unwary because their major goal is the maximization of the information about each concept, regardless of the correctness of the classification.
Picture n. 1 shows a simple graph created by the system.

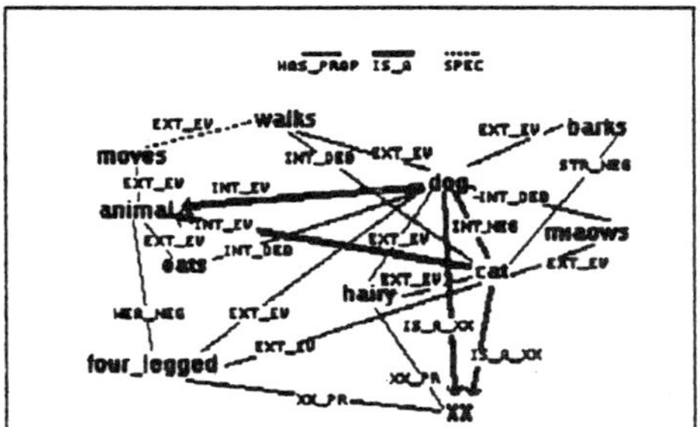

Picture 1. A graph created by the system

Let us now illustrate how the input is encoded: it consists of a pair: a) <concept-name,property-name>; b) <concept-name,concept-name>; c) <property-name, property-name>, preceeded by a code (indicating the type of information).
Picture n.2 is a brief example of the meaning and use of these codes.

		CODE	PAIR
cats miaow	⟶	1	cat miaows
cats do not bark	⟶	2	cat barks
only some animals have four legs	⟶	3	animal four_legged
snakes are animals	⟶	4	snake animal
vegetables are not animals	⟶	5	vegetable animal
crawling is a way of moving	⟶	6	crawls moves
small is the opposite of big	⟶	7	small big

Picture 2. The various kinds of code-pair input

During a learning session as well as at the end of the learning process we can examine the Hierarchy Table, the Concept-Property Table and the Property Table.

In Picture n.3 we present Pinocchio's data structures and output.

HIERARCHY TABLE (IS_A EDGES)

	snake	animal	hen	dog	cat	XX
snake		EXT_EV				
animal	INT_NEG		INT_NEG	INT_NEG	INT_NEG	
hen		EXT_EV				
dog		INT_EV		INT_NEG	INT_NEG	IS_A_XX
cat		INT_EV				IS_A_XX

CONCEPT-PROPERTY TABLE (HAS_PROP EDGES)

	eats	walks	moves	four_legged	hairy	miaows	barks
snake	INT_DED		INT_DED				
animal	EXT_EV		EXT_EV	WEA_NEG			
hen	INT_DED		INT_DED	-INT_DED			
dog	INT_DED	EXT_EV	INT_DED	EXT_EV	EXT_EV	-INT_DED	EXT_EV
cat	INT_DED	INT_DED		EXT_EV	EXT_EV	EXT_EV	STR_NEG
XX				XX_PR	XX_PR		

PROPERTY TABLE (SPEC EDGES)

	walks	moves	crawls	four_legged	two_legged
walks		EXT_EV			
moves					
crawls		EXT_EV			
four_legged					STR_NEG
two_legged				STR_NEG	

Picture 3. The tables represent different links between entities

2.4 The learning algorithm

As soon as the input is given to the system, the immediate consequences are: a) the insertion of the new nodes (if they don't already exist) into the knowledge base; b) the recognition of the kind of input pair and the generation of the correct edge between nodes; c) the labelling of an edge with a "confident" label (this occurs only if the label doesn't contradict a previous label on the same edge).
After these operations the system performs either a deductive or inductive task, depending on the kind of input. If the input is a <concept,concept> positive pair (code 4), the system deduces that all properties ascribed to the parent concept will also describe the child.
If the input is a <concept,property> pair (codes 1-3), both of positive and negative type, the knowledge base is modified according to some rules, which are applied cyclily to each current pair.
These induction rules, together with the deduction rules mentioned above, can be categorized as following: **preliminary rules** (they try to connect a

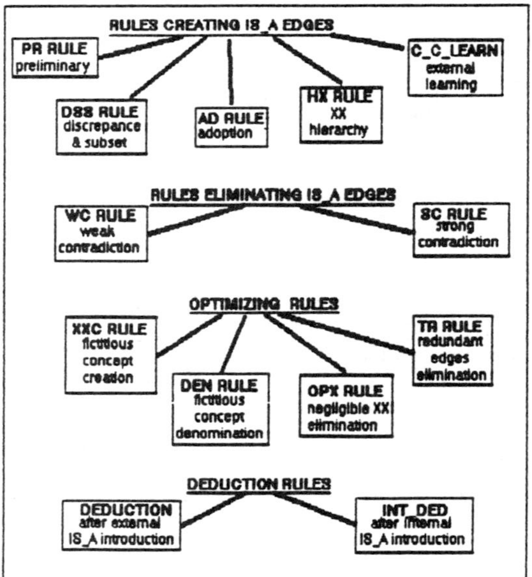

Picture 4. Rules' classification

new concept with the most general class, if it exists), **global rules** (they examine the whole graph in order to retrieve or create other relationships), and **reorganization rules** (they optimize the graph, eliminating redundancies).
A different classification for these rules is shown in Picture n.4.

PR-RULE: It is the first rule to be applied as soon as a new concept with only one property ,is introduced: this rule tries to link the single-property concept with another, choosing the most general concept with the same property (i.e. the highest in the hierarchy,if it exists).
If the concept in the current input pair is already known (i.e. the input adds only a new property to its description), then the system examines all other concepts in order to create every new legal relationship with the current concept.
Let us have a quick look at the rules, given in the same order in which they are applied.

SC RULE: if a concept has a property which cannot be shared by another concept, then the IS_A edges between the two concepts are both labelled INT_NEG.

WC RULE: if a concept has a property which the system knows to be valid only for a single subset of istances of a superordinate concept, then this could be considered (by default) as another instance of that superordinate concept (the IS_A edge from the current concept to the superordinate is labelled INT_EV and the opposite edge is directly labelled INT_NEG).

DSS RULE: a) if two concepts share common properties, but they also have some characteristic properties, then they are considered siblings (they mutually belong to each other; both IS_A edges are labelled INT_EV); b) if the set of properties of a concept is a subset of the properties of another concept, then the former is parent of the latter (the IS_A edge coming from the latter is labelled INT_EV and the opposite edge is deleted).

AD RULE: if two concepts are siblings, and one of them has a parent., which shares at least one property with the other concept, then the parent adopts the other sibling. (the IS_A edge is labelled INT_EV).

Each time the system puts a new INT_EV label on an edge, it performs an INTERNAL DEDUCTION: the child concept inherits all the parent's properties (the HAS_PROP edges from the parent's properties to the child concept are labelled INT_DED or -INT_DED, depending on the existing labels on the edges from which they have been inferred).
At the end of each learning cycle, the system optimizes its knowledge, creating new classes, grouping concepts or eliminating redundant information.
The optimization rules are:

CXX RULE: if at least two concepts have some properties in common, then a new class will be created, which is made up of the same properties of the two concepts and whose members are exactly all those concepts that share (or maybe will share) those properties. This new class is called a **fictitious concept or class**, because it has no name initially.
So a new XX-node is introduced in the graph, the edges going to its properties , labelled XX-PR, and the edges coming to it from the concepts IS_A_XX.

DEN RULE: if there exists a **named** concept whose properties are the very same as those of a **fictitious concept**, then the fictitious concept is **denominated.** The fictitious concept is deleted and the **named** concept becomes the new parent of the fictitious concept's members. The XX-node is eliminated as well as all edges from and to it (the IS_A edges from the members to the XX-node are erased and new INT_EV edges to the new concept are created).

At the end of a learning session, Pinocchio tries to reorganize the knowledge base through these rules:

HX RULE: (XX hierarchy rule) If a **fictitious class** has a set of properties which contains as a subset of the properties all of the properties of another **fictitious class**, then the former becomes a <u>parent</u> of the latter . The relationship between the two fictitious classes is labelled IS_A_XX.

OPX RULE: (XX optimization rule) If a **fictitious class** has either one single member or one single property, it can be deleted. All edges are eliminated too.

TR RULE: (triangle rule) If a concept is directly connected (i.e. by a single edge) with an <u>ancestor</u> (i.e. a concept with which it is already connected by a sequence of edges), then the most direct edge is deleted.

3. Experiments with Pinocchio

3.1 Vertebrate classification

We have chosen 11 animals among the vertebrate to test the classification power of the system.
The chosen animals are:

man, cat, cow, zebra,
hen, canary,
shark, trout,
adder, boa,
frog.

The properties are:
drinks, sleeps, eats, moves, breaths, has_blood, reproduces (general properties),
has_lungs, has_branchiae (breathing properties),
swims, flies, jumps, crawls, walks (moving properties), two_legged, four_legged, has_wings, has_fins (limbs properties),
hairy, feathered, scaly (covering properties)
cold_blooded, hot_blooded (temperature properties)
oviparous, ovoviviparous (reproduction properties),
speaks, miaows, bellows, chirps, croaks, hisses, clucks, dumb (sound properties),
big, small, long, short, tall (dimension properties),
domestic, aquatic, terrestrial (environmental properties),
useful, dangerous, edible, friend, enemy (man relationship properties).

The pairs in input were only 1-3 code pairs i.e. <concept,property> pairs.
Pinocchio has created some interesting fictitious classes which partially correspond to the five well-known classes among which we had initially chosen the concepts. It also show some other implicit regularities discovered by the system. There are no direct relationships between concepts; they are grouped in a hierarchy of fictitious classes whose properties (and membèrs) are the following:
xx1: ovoviviparous, four_legged, hairy, has_lungs, moves, sleeps, eats, drinks, gas_blood, hot_blooded. big. (cow, zebra).
xx2: ovoviviparous, four_legged, hairy, has_lungs, moves, sleeps, eats, drinks, has_blood, hot_blooded. (cat, xx1).
xx3: edible, useful. (hen, cow).
xx4: has_lungs, moves, sleeps, eats, drinks, has_blood, hot_blooded. (hen, xx2).
xx5: has_lungs, eats, drinks, has_blood, hot_blooded. (canary, xx4).
xx6: ovoviviparous, hairy. (man, xx2).
xx7: oviparous, two_legged, has_wings, feathered. (hen, canary).
xx8: moves, sleeps. (trout, shark, xx4).
xx9: eats, has_blood. (man, trout, shark, frog, xx5).
xx10: aquatic, has_branchiae, has_fins. (trout, shark).
xx11: cold_blooded, scaly, crawls. (adder, boa).

3.2 An experiment on learning natural kinds and artifacts

We fed the system 60 pairs of input of different types. The domains of classification are natural kinds (dog,cat...) and artifact categories (chair, table...) [Gelman 88]. This experiment was designed to test the classification ability of the system in these two poorly correlated subdomains.
The input pairs were the following:

1 chair has_back	1 chair made_man	1 liv-thing live
2 artfat live	1 phy_obj exist	3 phy_obj live
3 liv_thing breath	1 liv_thing exist	1 artfat exist
2 cat bark	1 cat meow	1 animal live
1 dom_an s_stroke*	3 animal s_stroke	1 artfat made_man
3 artfat furnish	1 table furnish	1 chair furnish
4 chair artfat	1 building concrete	5 animal artfat
1 lion roar	1 lion sleep	2 lion s_stroke
1 cat s_stroke	1 dog s_stroke	1 building man_made
1 building ine_mat	2 building is_small	1 chair is_small
1 table is_small	1 table ine_mat	3 furnit has_back
1 furnit furnish	1 furnit is_small	1 furnit is_portab
2 building is_portab	2 building furnish	1 dom_an live_w_hum
1 dom_an breath	1 animal breath	1 animal sleep
1 cat breath	1 dog feed_itself	2 table has_back
1 artfat ine_mat	1 furnit ine_mat	1 furnit hard
1 chair hard	1 dom_an live	1 table man_made
5 building animal	4 building phy_obj	2 dog meow
3 dom_an meow	1 dog bark	1 dog sleep
1 cat sleep	2 lion s_stroke	1 furnit man_made

*someone can stroke it.

After the learning period the system has converged on the hierarchy, represented in Picture n.5.

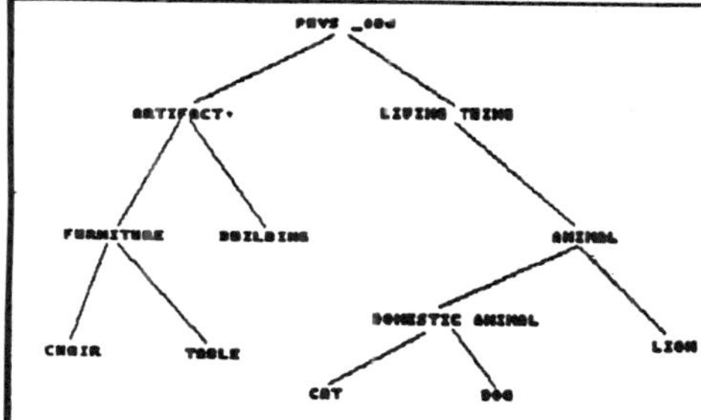

Picture 5. Natural Kind-Artifact Categories Experiment: the output

at the system appears not to be sensitive to the order of presentation of the input pairs in the sense that whereas the inductive/deductive path could be different, the resulting hierarchy is always the same anyhow you feed the system. However, during the experiments, we have noticed that by grouping input pairs in such a way that for each object were first given all its properties, the convergence of the system was much faster;
b) the resulting hierarchy doesn't contain any "fictitious" class of XX type. This depends on the nature of the experiment and on the fact that all those XX classes, created by the system during learning, which have not been eliminated , have become coincident with the already known concepts and have been subject to the denomination process.

4. Validation of the Pinocchio system

Contrary to what is claimed by [Fisher and Langley 85], [Michalski and Stepp 83] and many others, we don't view the aggregation problem (i.e. the formation of a set of classes) and the characterization problem (i.e. the assignment of a conceptual description to each new formed class) as two distinct issues. Rather we believe that the characterization occurs while aggregation takes place. The obtained classes are subject to dynamical changes in their descriptive content as new entities or new properties are presented to the system. So we have a gradual, incremental process of augmentation of the intensional description associated with the classes, both "named" and "fictitious", such that each class will cover all its potential instances or subclasses.
In our system we can have more than one class for each single instance (or set of instances): as the knowledge of the system increases, some of these classes become more and more marginal, until they completely disappear (as in the case of a single-attribute or single-member class), while other classes reinforce themselves through a **process of denomination**. This process testifies that we have a particular linguistic expression only for the most frequent or useful classes. To attach a linguistic label (a name) to a set of properties serves to the purpose of obtaining a fast and optimal classification and can be considered as a distinct and later cognitive process [Nelson 78].
At the beginning of the learning process, all the linkable concepts are connected ; as the system's knowledge increases, some of the connections are modified or eventually deleted. Each membership hypothesis among two (or three) concepts, together with the "fictitious" class, is handled in a parallel fashion.The system is prone to oscillate at the beginning of the learning period, when the knowledge base is poor, but becomes more and more stable as new pieces of knowledge are added.
PINOCCHIO does not use particular strategies for moving inside the space of hierarchies in order to find the optimal one, because each concept-node is immediately inserted into the hierarchy by means of concurrent hypotheses; only the next input will decide which of those hypotheses will survive, i.e. the most consistent with the next input. Anyway, we tend to maintain all those hypotheses which are not discarded by the next coming input, because they could be useful in **relevant contexts** [Stepp and Michalski 86].
One interesting result of our experiments is that, if the system knows a large number of properties of a concept, it will try to compare all the remaining concepts to that concept, with the result that this concept will be connected to many other concepts. Conversely, if the system knows only a few properties of a concept, this concept will not have many shared properties and it will be partially disconnected from all the others. These experimental results show that the performance of PINOCCHIO reflects two observed phenomena in early categorization: overextension and underextension.
The learning rules are transparent from a cognitive point of view [see Merriman 86] ; the PR (peliminary) rule has its cognitive counterpart in the fact that every new concept, of which only ine property is known should belong, at the beginning, to another concept which has the very same property and is at a higher hierarchical level (if it exists); the XX-C rule gives an account of the use of properties in order to classify instances, before a noun is associated with the class.
Most non-uniformity in the hierarchy is hidden in the input structure. As we have seen, some input operates on the current hierarchy, creating or deleting membership hypotheses; some other presupposes the intervention of a tutor or a domain expert, or an inductive pre-processing, and operates augmenting the number of levels of the hierarchy. This non-uniformity is due to the difference between information sources (observed world/tutor).
The system allows the execution of a single rule for each couple (or triple) of concepts, the rule which better satisfies the constraints imposed by a) the new input, b) the internal criterion of the sharing of attributes and c) the state of the knowledge base. For instance, the SD rule (of discrepancy and subset) is more frequently applied to basic level (it gives an account of the well-known phenomenon of the overlapping among concepts' intensions) and generates a mutual membership hypothesis between the two concepts, instead of the WD rule (of weak contradiction) that is only applied to higher levels.

5.Limitations and future directions

The most limiting aspect of our system resides in the choice of the "classical" representation model of the concepts [Smith and Medin 81]. We claim that this kind of representation of concepts can better model the process of acquisition of general conceptual knowledge. Anyway, in order to give more flexibility and reliability to the system, we want to assign to each property a typicity value, representing the role of that property in the intensional description of the concept. We want also to order concepts according to a typicality scale, representing the role of a particular instance in the extensional representation of the concept.
A technical limitation regards the computational cost of the embodiment of each single observation. The unique "heuristic strategy" adopted by the system is that two concepts can be compared if and only if they have at least one shared property and one specific property.
The membership hypotheses, which link concepts, are handled simoultaneously and our system will take advantage of a parallel computer architecture, in which every concept can act as a "demon", sending messages to every other concept sharing common properties.
Some other future directions of development are a) the refinement the object description language, by introducing more structured knowledge in input, like complex relations among concepts with different roles, b) the implementation of constructive induction techniques so that the system is able to generate new properties.

6. Conclusions

Pinocchio can be considered, even with its obvious limitations, a cognitive model of human concept learning. It works with a linguistically structured input and its core is the process of assignment of nouns to intensional descriptions based on the comparison and grouping of concepts according to internal criteria.

The system could also be integrated in more complex systems, such as natural language text understanding systems or educational systems. In the latter case PINOCCHIO would support people in the development of concepts' descriptions and hierarchies in a variety of useful domains.

Acknowledgements

The authors thank Prof. Domenico Parisi for his precious suggestions and helpful comments. The first author was supported by a fellowship from the Sigma-Tau Foundation for scientific research, and the second author by a fellowship from the AI Lab. Selenia , S.p.A.

They also thank the Institute of Psychology of National Research Council, Rome , Italy for the support given to the research.

References

Carbonell, J.G. Michalski, R.S. Mitchell, T.M. (1983) "An Overview of Machine Learning" in Michalski, Carbonell, Mitchell (eds) Machine Learning: An Artificial Intelligence Approach, Tioga, Palo Alto, CA, 1983.

Fisher,D.H. (1987) "Knowledge Acquisition via Incremental Conceptual Clustering", Machine Learning, Boston MA 1987, vol.2 n.2 , p.138-175.

Fisher,D.H. (1987) "Improving Inference through Conceptual Clustering", Proceedings of AAAI. Fisher, D. Langley, P. (1985) "Approaches to Conceptual Clustering", Proceedings of 9th IJCAI, Los Angeles CA 1985, vol. 1 , p.691-697.

Gelman,S.A. (1988) "The Development of Induction within Natural Kind and Artifact Categories", Cognitive Psychology, 20 1988, p.65-95.

Hanson, S.J. Bauer, M. (1986) "Machine Learning, Clustering and Polymorphy", Uncertainty in Artificial Intelligence, L.N. Kanal and J.F. Lemmer (Eds), Elsevier Science Publ.,1986.

Keil,F.C. (1987) "Conceptual Development and Category Structure", Concepts and Conceptual Development: Ecological and Intellectual Factors in Categorization, U.Neisser(ed), CUP 1987.

Lebowitz,M. (1987) "Experiments with Incremental Concept Formation: UNIMEM", Machine Learning, Boston MA 1987, vol.2 n.2, p.103-137.

Medin, D.L. Wattenmaker, W.D. Michalski, R.S. (1987) "Constraints and Preferences in Inductive Learning: An Experimental Study of Human and Machine Performance", Cognitive Science, 11 1987, p.299-339.

Merriman, W.E. (1986) "How Children Learn the Reference of Concrete Nouns: a Critique of Current Hyphothesis", The Development of Word Meaning, S.A. Kuczaj II and M.D. Barrett (Eds), Springer&Verlag, 1986.

Michalski, R.S. Carbonell, J.G. Mitchell, T.M. (1983) Machine Learning: An Artificial Intelligence Approach, Tioga, Palo Alto, CA,1986, vol. 1.

Michalski, R.S. Carbonell, J.G. Mitchell, T.M. (1986) Machine Learning: An Artificial Intelligence Approach, Morgan Kaufmann Publ, 1986, vol. 2.

Michalski, R.S. Stepp, R.E. (1983) "Learning from Observation: Conceptual Clustering" in Michalski, R.S. Carbonell, J.G. Mitchell, T.M. Machine Learning: An Artificial Intelligence Approach, Tioga, Palo Alto, CA, 1983.

Nelson, K. (1978) "Semantic development and the development of semantic memory" in K.E. Nelson (ed) Children's Language (vol.1), Gardner Press, NY.

Parisi, D. Castelfranchi, C. (1987) La macchina e il linguaggio Boringhieri, Torino, 1987.

Smith, E.E. Medin, D.L. (1981) Categories and Concepts, Harvard University Press, 1981.

Stepp, R.E. Michalski, R.S. (1986) "Conceptual Clustering: Inventing Goal-Oriented Classification of Structured Objects" in Michalski, R.S. Carbonell, J.G. Mitchell, T.M. Machine Learning: An Artificial Intelligence Approach, Morgan Kaufmann Publ, 1986, vol. 2.

THE GENERALIZATION STEP IN A STATISTICS INTERPRETER

W. Klösgen
Institute for Applied Information Technology,
German National Research Center for Computer Science (GMD)
D-5205 Sankt Augustin (FRG)

Abstract

The statistics-interpreter EXPLORA is a system for the discovery of interesting facts in statistical data. EXPLORA integrates statistical and knowledge-based methods and applies not only statistical but also domain-knowledge for its exploration and interpretation. The system uses three knowledge-processes. They exploit the semantical context-knowledge about the interesting objects and relations of the domain which has produced the data. First EXPLORA *systematically searches* for statistical results *(facts)* to detect relations within the data. A second knowledge-process consists in *discovering messages* about the data by condensing the numerous facts. Approaches of inductive generalization which have been developed for machine learning are adapted. These identify common values of attributes of the objects to which the facts relate. In a third process EXPLORA uses methods of discourse-organization to produce a coherent analysis report. The system formulates the messages in terms of the domain, groups and orders them and even provides flexible *navigations* in the *fact-spaces*.
This paper gives an overview about these knowledge processes implemented in EXPLORA. Some examples of its beta-test applications in the area of survey analysis are presented. Special emphasis is laid on describing the methods of inductive generalization used for the condensation of facts.

Keywords: data exploration, AI for statistics, semantic interpretation, machine learning, data navigation.

1. The generalization step in learning and in EXPLORA

Learning systems studied in Artificial Intelligence (AI) incorporate two distinct steps: a learning step and a performance step. In the performance step the system applies the information it has gained in its learning step. By applying the "learned" knowledge the system solves a task in the problem area it was designed for. One possible performance step would be to propose and to execute an acquired strategy for solving a data analysis or a graphical design task. In this case the learning step of the system consists in machine acquisition and compilation of strategies (of data analysis or of graphical design) by watching and questioning experienced users. Another performance step would be to identify a disease of a patient by classifying his symptoms and medical data. In the associated learning step some classification rules are derived, analysing a lot of cases for diseases given as examples.

There are various frameworks for learning approaches in AI (Michalski83 and Michalski86). One major area is to generalize from examples. Generalization involves analysing a set of positive and negative training examples of some general concept, identifying the essential attributes and relations common to these examples, and then formulating a concept description based on these common features.

Generalization techniques are used in EXPLORA to condense the numerous facts. These facts are generated by systematically searching for statistical results to detect relations within the data which could possibly be overlooked by a human analyst. On the other hand EXPLORA will contribute to overcome the bulk of information which is generally produced when presenting the data. Therefore the role of generalization in EXPLORA consists in concentrating the findings about the analysed data, in particular as a method of redundancy reduction.

In this way EXPLORA embodies only the generalization step of the learning paradigm. That means that there is no machine performance step following the generalization step. The informations derived by generalization techniques are presented to the user as messages and arranged in a report of analysis.

Statement-types which represent generic statements about the data in a special application area are fundamental to the process of searching for facts (compare 3). Facts are defined as true instances of these statement-types. They are verified in the data by statistical methods. The facts are the examples of the concept relating to the statement-type. Wrong instances (not verified in the data) are the counter-examples submitted to the generalization techniques.

A simple example of such a statement-type would be the concept of a *record-sale*. Instances of this statement-type relate to individual objects like firms, sales-districts and fiscal-years. The positive examples are the instances (triples of these objects) which represent a record-sale. Negative examples would be given by non-record instances (characteristic description) or by only these instances which represent a negative record (discriminant description).

The quality of a derived generalization is influenced mainly by the following topics: How representative of the concept are the examples? How adequate can the examples and the generalizations be represented in the associated description languages? How homogeneous is the concept?

Therefore it is first of all a matter of the quality of the knowledge-base that was established in EXPLORA for an application field that determines the usefulness of the output of the generalization techniques. This requires that the statistical methods used to verify the instances of a statement-type must be statistically sound, that the domain model must include adequate attributes and relations to describe the objects and that the degree of randomness inherent in the concept may not be too high (it surely makes no sense to generalize instances which were drawn randomly).

The EXPLORA-project is part of a GMD key Project *Assisting Computer (AC)*. The vision of this key project is the paradigm of the computer as a *personal assisting machine*. The goal of this long-term activity is to create intelligent, active support systems which have the abilities to complete user commands, to render them preciser and to correct them if necessary. Furthermore they will learn and adapt themselves to their users, consult and critizise them and explain themselves. The final result of the AC will be an ensemble of machine assistants adaptable to specific requirements and consisting of universal office assistants, of domain assistants (like EXPLORA) and of communication assistants (Hoschka89a).

2. Overview about knowledge-based techniques of EXPLORA

The interpretation process in EXPLORA consists of the phases for the generation of facts, messages and reports. First a systematic search for *facts* is performed on the data. This is done by exploiting a statement-pool. This pool has to be established, so that it includes all types of statements which in principle are interesting for the interpretation of the data within the domain (see figure 1). Those statements which are *true* in the given data are then located. When searching for true statements the system takes into account that the statements may partially overlap and may be redundant.

Since not every fact is interesting and worth mentioning, the identified facts are in a second step condensed to *messages*. To formalize the vague notion of interesting statements, EXPLORA offers several approaches. In particular, methods of inductive generalization are applied to discover common values of the attributes of the objects to which the facts relate.

EXPLORA uses semantic relations of the domain not merely to speed up the search process and to identify messages. Moreover, in a third step the discovered messages are structured into a *report* consisting of natural language and graphical representations of the statements. Furthermore, these relations are available to offer flexible navigations in the spaces of facts.

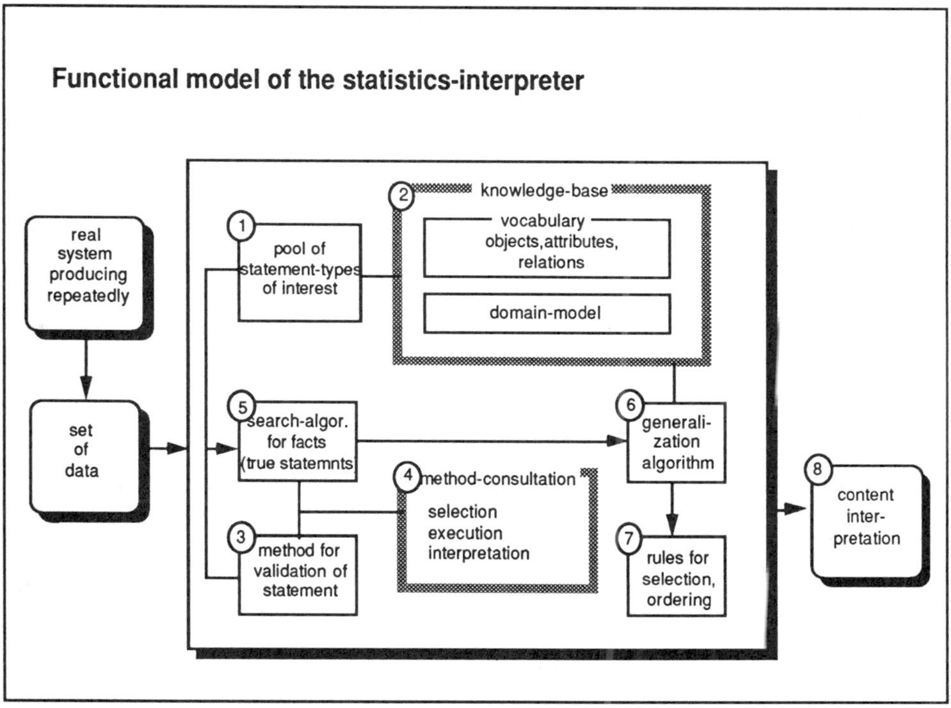

Figure 1: Functional model of content interpretation

These processes are described in more detail in (Hoschka/Klösgen89b). The proceeding of EXPLORA can be characterized as a semantic approach for the exploration of statement-spaces. Formally the problem can be stated as follows (Latocha89):

Given:
- A descriptional model of a domain using objects and relations
- Data D about the domain organized by these objects
- A formal language L; statements expressed in L can be verified in D by using (statistical) methods
- Constraints C (e.g. redundancy-, selection-, ordering-rules).

Wanted:
All true statements about the domain (and the data D) expressed in L satisfying C.

See (Hajek77) for a similar, more specific formulation.

3. Statement-types, object-structures and search-algorithm

An example of a statement-type for the application-class *aggregate data analysis* (see 5) is given by the following template:

Outstanding-behaviour of *result-variable* for *target-group.*

This statement-type has three arguments: *outstanding-behaviour, result-variable, target-group*. An *object-structure* is associated to each argument of a statement-type. Such an object-structure consists of a *set* of objects and of a *partial ordering* on this set. The object-structure *outstanding-behaviour* has as elements several patterns of behaviour like *above the average.* A stronger-relation supplies a partial ordering on this set of behaviour-patterns. The argument *target-group* refers to a set of basic units of analysis like persons or firms. In this example we are not interested in statements about the individual units of analysis but in statements about some classes of units (aggregates or target-groups). These target-groups are constructed by using the attributes which are available for the units of analysis. A partial ordering is defined on the set of all possible attributive generalizations of the units of analysis. A target-group is more general than another, if the attribute-values in its description are more general (referring to the structure defined on the value set of attributes) and/or if its description holds only a subset of the attributes which are used for the other class. Therefore a target-group which is more general than another includes the other more particular target group.

Further a *verification-method* is associated with each type of statement. This method includes statistical criteria used to exclude fortuitous results and to indicate the degree of interest of a concrete *instance* of a statement type (Gebhardt88). An instance of a statement-type is given by an application of the statement-type, where for each of its associated object-structures an element of its object-set is taken as an argument. Some instances of the above statement-type:

Penetration-rate for *Computers* in *Manufacturing and Process* is *below the average.*
Penetration-rate for *General Purpose Computers* in *Manufacturing and Process* in *Northern-Regions* is *much below the average.*

The verification-method checks whether an instance of a statement-type is true in the given data. A method which verifies the above instances could be based on statistical analyses of contingency-tables. Then a value of the result-variable for a target-group corresponds to a cell in an associated table. This value is compared with its expected value. Depending on the deviation between the value and its expected value in terms of multiples of the standard-deviation of the cell, the value is classified according to the specified qualitative patterns of behaviour like *much below the average.*

A statement-type therefore defines a search-space with dimensions corresponding to the associated object-structures. In the above example we have a three dimensional search-space spanned by the arguments *result-variable, target-group, behaviour.* A *fact* is now defined as a true instance of a statement-type in this search-space.

To *eliminate redundancy* in the space of instances of statements a partial ordering may be defined on this set of instances. This ordering is induced by the partial orderings existing in the associated object-structures. It is defined as the product of the adapted partial orderings of the associated object-structures. To adapt a given partial ordering of an object-structure one has to define whether it is also essential for the ordering of instances and if so, in what direction (stronger, weaker) it is essential.

As an example of such a redundancy relation let us consider the above statement-type. There are basic partial orderings on the object-structures of behaviour-patterns (a pattern may be stronger than another pattern), of result-variables (let us assume that the result-variables are arranged in a tree-like structure) and of target-groups. Then one instance of this statement is stronger than or equal to another, if the pattern of the first instance is stronger than or equal to the pattern of the second instance, if the result-variable appearing in the first instance is on a higher hierarchical level than the result-variable of the second (or is equal) and if the first target-group is more general than the second (or is equal). For the tree-structure we can apply one of several possible redundancy rules. In our example this is a heuristic redundancy rule of the form:

If a statement is true for an element in the tree structure, then for all subordinate elements the statement is not interesting.

These are local rules for the individual dimensions (object-structures) in the statement space. They are combined (product-ordering) to deliver a partial ordering in the statement space. A search-algorithm uses this partial ordering to reduce the search for true statements and to present only these non-redundant statements to the user. If this algorithm for instance finds a true statement, it looks for the strongest statements (compared to this true one) which are also true and then can eliminate all still stronger ones (which are false) and all weaker ones (which are redundant).

The two instances of the statement-type presented above are non-redundant. The target-group (Manufacturing and Process) and the result-variable (Penetration-rate of Computers) of the first instance are more general than these of the second instance (Manufacturing and Process in Northern Regions and Penetration-rate of General Purpose Computers) but the behaviour-pattern of the second instance (much below the average) is stronger than the pattern of the first (below the average). Therefore according to the product partial ordering there is no relation between these instances in the sense that one is stronger than the other, so that both are selected by the search-algorithm.

The direction which has to be specified for the adaptation of a component partial ordering for the product partial ordering depends on the semantics of the statement-type. Consider the following type:

The *n* units with the *largest* values of a *result-variable* are located in *target-group*.

In this case a statement about a more general target-group is weaker than a statement about a more particular target-group. Therefore the partial ordering on target-groups here is used in an inversed direction by the search algorithm.

4. Generalization of facts

As a simple example (see 1) for generalization let us assume that a record-sale was achieved in the sales-district Munich. Then it is checked if perhaps there was a record sale in every southern German sales-district in the analysed time-range. If so, then the fact about Munich surely is not interesting, but the message about southern Germany is. Another example: if in successive periods a certain fact has been discovered, then this can be summarized into a single message about this space of time. Such generalizations are also possible if only few exceptions exist: record sales in *nearly* all reorganized districts.

Let us consider again the argument *target-group* (see 3). Now we assume that we have basic statements about single units of analysis like firms (e.g.: *The best value of result-variable since n fiscal-years for firm A*). We want to find target-groups of firms with a high density of true statements of this type. Generalization-techniques use the partial ordering on the target-groups and apply evaluation criteria (e.g. LEF-functions, compare Michalski83) to identify interesting target-groups.Figure 2 summarizes the connections between the data, the objects, the facts and the generalizations.

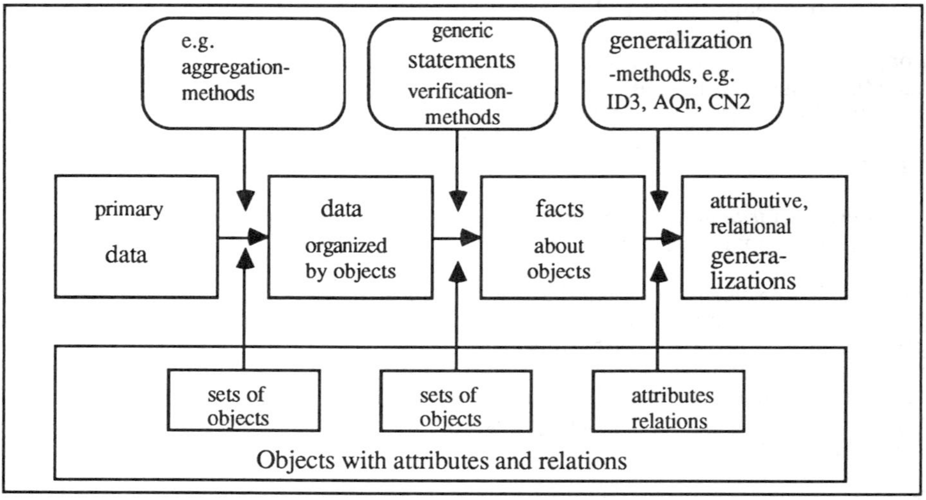

Figure 2: Application of the generalization approach in EXPLORA

Meanwhile the methodology of machine learning is quite advanced. Some methods which are relevant for the generalization in EXPLORA are described in (Michalski87, Quinlan87, Clark/Niblett89). The generalization problem can formally be stated as follows (compare Mitchell82):

Given:
- A language to describe instances. In our case the instances are the discovered facts; these facts are described in terms of the objects (elements of the sets of the object-structures).
- A language to describe generalizations. Attributive and relational descriptions are possible. Attributive descriptions use the attributes defined for the object-structures. Example: region(sales-district)=south; *region* is an attribute defined for the structure *sales-district*. Relational descriptions use relations, defined on several structures. Example of a binary relation: special-advertizing(product,sales-district) is a relation between products and sales-districts, which states that a special advertizement measure was performed for a product in a sales-district. Values of attributes can be structured (e.g. a tree-structured value-set). Generalizations can also contain internal disjunctions of objects or attribute-values and formations of intervals in case of linear object-structures.
- Evaluation-functions to judge the quality of generalizations. We use the "density of facts" within a generalization, the degree of generality of a generalization and the syntactical complexity of the descriptions by which generalized facts are represented to evaluate a single generalization and its description.
- Generalization-aims which relate to overall properties of the wanted descriptions. Example: disjunctive descriptions *overlap* the given set of facts. Global evaluation-criteria are associated to evaluate a disjunctive set of single descriptions. Another overall property refers to the distinction between characteristic and discriminant generalization (see 1).

Wanted:
All generalizing descriptions. Each description condenses a part of the set of facts and satisfies the evaluation-criteria. All descriptions together satisfy the generalization-aims.

Within this framework the user can supply several specifications of the description language, the evaluation criteria and the generalization aims in order to direct the generalization procedure. Default settings may be changed, e.g. by selecting subsets of facts for which a generalization is demanded or by choosing object-structures, attributes and relations which will be used for generalization.

5. Application examples of EXPLORA

The examples are chosen from the practice of one of the selected pilot-users in the area of survey analysis. They belong to the application-class *aggregate data analysis*. Homogeneous applications of the same structure are gathered in an application-class to facilitate the incorporation of an application into the EXPLORA frame.

The application class *aggregate data analysis*

Data consist of records about units of analysis such as persons, companies or equipment. Variables are collected for a total or a sample of these units. There are result-variables and other variables which characterize the units. Other types and structures of variables are important for selecting relevant statement-types, appropriate statistical tests and parameters. Statements about remarkable patterns of the aggregated result-variables for classes of units are of primary interest. These classes are constructed by the characterizing variables. The files, structured by records (units) and variables, are covered regularly, so that an additional time-dimension is available.

Examples of this application-class are present in marketing research where interviews of establishments concerning their equipment or of persons concerning their buying-behaviour constitute the samples of units. Another example of this type is the analysis of political elections, where the units are given as a total of regional districts; the result-variables are not interview-values, but the actual results for political parties. This example was used as an alpha-test for our first prototype (figure 3).

Results are typically presented in the form of tabulation. Two or three characterizing variables are selected to structure the rows and columns. The aggregated values of a result-variable appear within the tabulation. This is how the notorious bulks of information arise: the users are drowned by tabulations (corresponding to the large number of possible combinations of variables) in which they have to identify remarkable values, e.g. by comparing the values with the margins.

EXPLORA will overcome this situation, so that its user obtains the most interesting messages concerning the data. Interactively he may focus the themes of the exploration and browse through the presented messages. These messages are constructed on the basis of statement-types referring to longitudinal or cross-section analyses.

For the longitudinal analysis some elementary time-series characteristics are applied, such as records and series. The following questions are answered:

Is the present value of a result-variable for any target-group a record, compared to the past values? Is the present value a further link in a series or is any series terminated?

Several partially ordered patterns of records or series are available, so that the redundancy formalism produces statements about a less general target-group only if it shows a stronger pattern.

Cross-sectional analyses can answer the following questions where the redundancy- and generalization-formalisms again reduce the number of generated statements:

For what target-groups are any patterns of behaviours remarkable for any result-variables? What competing objects possess the best (worst) ranks for any result-variable in any target group? In what sections are any cumulative values remarkable for any result-variable? What functional relations exist between any result-variable and any characterizing variables in any target groups?

The following figure presents an example of the results of our first prototype.

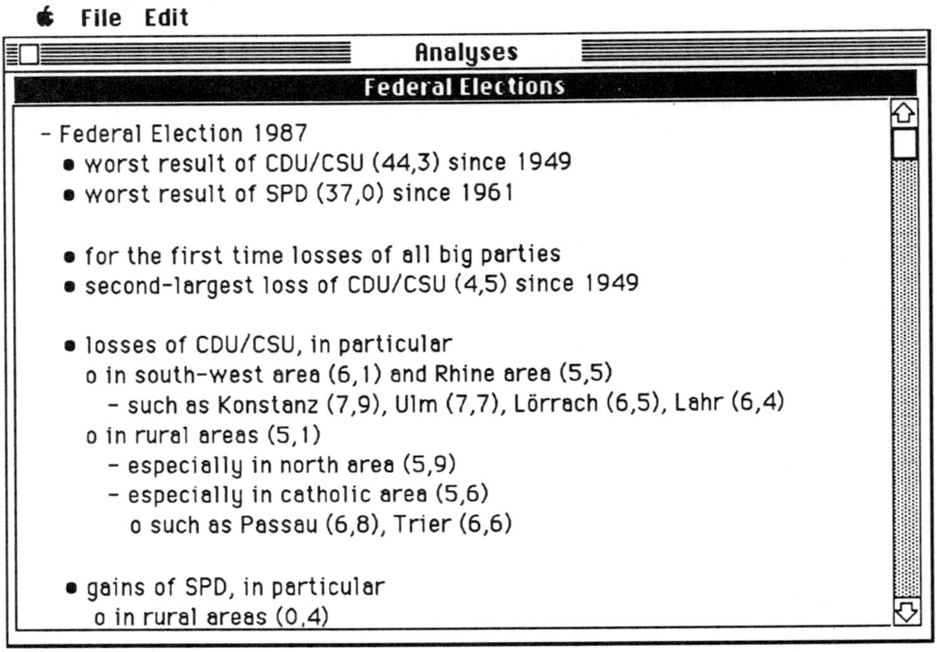

Figure 3: Analysis of the Federal-Election FRG in 1987

Conclusion

The interpretation of data is a task of human experts who are versed in the domain and who know how one has to analyse the data to discover their meaning. The aim of our EXPLORA-system is to support parts of this interpretation process by a knowledge-based system for data exploration. A first prototype of EXPLORA was presented in 1988 at the Hannover fair for information technology, a beta-test version is now delivered to an international marketing research institute and a German car-manufacturer.

References

Clark, P. and Niblett, T.: The CN2 induction algorithm, Machine Learning, in press.

Gebhardt, F. (1988): Statistische Fragestellungen bei einem Expertensystem zur explorativen Datenanalyse. Sankt Augustin, GMD, (GMD-Studien 137).

Hajek, P. and Havranek, T. (1977): On generation of inductive hypotheses, Int. Journal Man-Machine Studies, 9 .

Hoschka, P. and Klösgen, W. (1989a): Assistenzcomputer als Entdeckungshilfe. Sankt Augustin, GMD, (Aus der wissenschaftlichen Arbeit, Jahresbericht 1988).

Hoschka, P. and Klösgen, W. (1989b): EXPLORA: Content Interpretation of data. Prepared for: Journal of Artificial Intelligence and Mathematics, 1.

Latocha, P. (1989): Exploration von Aussagen-Räumen. Sankt Augustin, GMD, (GMD-Studien, in press).

Michalski, R.S. et al. (ed., 1983 and 1986): Machine Learning I and II. Morgan Kaufmann, Los Altos.

Michalski, R.S. (1987): How to learn imprecise concepts, in: *Proceedings of the fourth international workshop on Machine Learning*, ed. P. Langley, Morgan Kaufmann.

Mitchell, T.M. (1982): Generalization as search, Artificial Intelligence, 18.

Quinlan, J.R. (1987): Induction of decision trees, Machine Learning, 1.

NEURONAL APPROACH

A NEURAL MODEL FOR HIGHLY MULTIDIMENSIONAL DATA ANALYSIS

Alain Lelu
CNRS - Institut de l'Information Scientifique et Technique
26 rue Boyer, 75971 Paris Cedex 20 (France)

Abstract : In this paper, we present a neural model implementing an associative mapping of highly multidimensional data - as encountered in information retrieval systems - upon a structured set of half-axes. Our non supervised learning rule uses a local reconstitution of the data by the network, and an analytic expression of the gain coefficient. We will show first that our model can be configured for a stochastic approximation of principal component analysis and correspondence analysis. Then we will discuss two variations of our model : spherical k-means, and local component analysis.

Our starting point is an application problem : how to enhance information retrieval systems in order to allow for "convivial" browsing in documentary data bases ? How does one analyze those dormant information stocks in order to identify relevant trends and make recommendations about the development of science and technology ?

Existing studies (9) show that many association substructures between limited amounts of terms and documents do exist in databases, but identifying them is generally not within the scope of our classical data analysis methods, both for qualitative, as well as quantitative reasons. The minimum size of a data table in our field is several hundred terms multiplied by several thousand documents.

We will present here a method qualitatively suited to our problem, and whose storage requirement is roughly equivalent to the number of terms multiplied by the number of derived clusters, or axes. Furthermore, its "neural" formalism intrinsecally adapts it to highly parallel computers.

1 - Factors and neurons

Principal components analysis consists in reducing the representation space of a set of observations : the goal is to shift from a description of each object by numerous describers to a description by a few factorial values only, i. e. to reduce the number of dimensions in the representation space (the axes of such a reduced space are known as the "factors") .

Formally it is possible to represent simultaneously the rows and columns of a data table in this space ; the factorial value u_j of the j-th "pattern" of data (for example a row of numbers in a table containing the values of descriptive variables for a particular "individual") is given by the classical transition formula between the normalized factorial values w_i of the columns and those of the rows :

$$u_j = (1/\sqrt{\lambda}) \Sigma_i x_{ij} w_i$$

where λ is a singular value of the covariance matrix XX^T.

Let us denote m a "synaptic weight" vector of arbitrary norm $|m|$: $m/|m| = \{w_i\}$

Thus the transition formula is in fact a very simple transfer function, i.e. the inner product :

$$\eta = \langle m, x \rangle \qquad \text{where} \quad \eta = u_j |m| \sqrt{\lambda}$$

This leads us to consider the first p factors as being the responses of p simply defined neurons (with linear response and without threshold), all subjected to the successive input rows $\{x_{ij}\}$, of "synaptic weights" m_K, each creating η_K in output, this output being the sum of the inputs weighted by the synaptic weights (see figure 1).

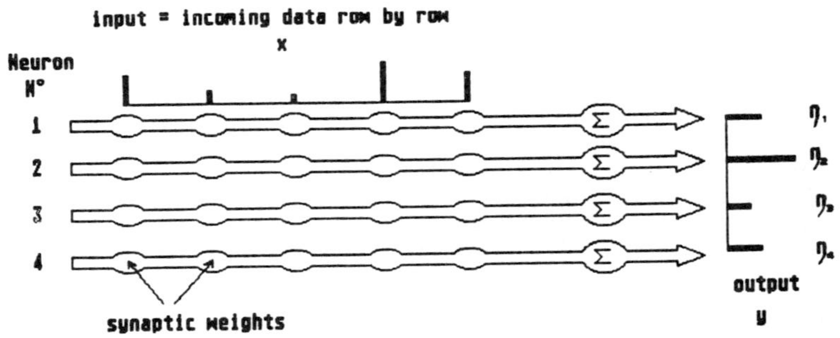

figure 1

2 - **Learning process and normalization**

If we want to enhance this model in order to produce real factorial analysis, fast dynamics must coexist with slow dynamics, i. e. output derived from inputs must coexist with the process of upgrading synaptic weights which converge towards the factors. Such a process is in fact achieved by the stochastic approximation method of factor analysis (3, 4), a method that inspired us our model for one factor :

Starting from the "iterated power" algorithm applied to the data matrix $X = \{x\}$ we obtain the following formula for updating the weight vector m_k :

$$m_t = m_{t-1} + (1/\tau_t)\, \eta\, x - (1 - \tau_{t-1}/\tau_t)\, m_{t-1}$$

where $(1/\tau)$ is a gain coefficient. Under the constraint of norm $|m|$ to be a constant with a desired value $|\underline{m}|$, and under second order approximation, it yields :

$$\left|\begin{array}{l} \delta m = (1/\tau)\, \eta\, (x - \eta\, m / |\underline{m}|^2) \\ \\ \tau := \tau + \eta^2/|\underline{m}|^2 \qquad\qquad \text{with} \quad \tau_{(\text{initial})} = 0 \end{array}\right.$$

This algorithm, like all those written in the form $\delta m = \alpha\eta x - \beta\eta^2 m$, allows m to converge stochastically (5) towards the first eigenvector of the matrix XX^T, which norm is $\sqrt{\alpha/\beta}$, and here $|\underline{m}|$. In our model, the gain $(1/\tau)$ is expressed analytically in order to optimize the convergence process. The scalar τ is an estimate of the sum of the squared projections of the data vectors upon the first principal component, i.e. τ is the "objective function" maximized in Principal Component Analysis (PCA). Note that we have proved (2) this formula to be a one-factor approximation of an "instantaneous", one-pass, PCA algorithm.

As presented here, the algorithm cannot supply uncorrelated factors : our aim will therefore be to introduce into the network structural couplings between several neurons. The result of this action may then be equivalent to orthogonalization, depending on the configuration of those couplings.

3 - **Orthogonalization and lateral inhibition**. We shall expose first an enhancement of our model by a feedback mechanism :

3 - 1 - A model with a predictor network :

Using principal component analysis, given K factors, the original data can be reconstituted according to the formula :

$$\underline{x} = \Sigma_k u_k w_k \sqrt{\lambda_k} \qquad ; k=1, \dots K$$

With our notations, and when the convergence of the $|m|$ is achieved, it yields :

$$\underline{x} = (1/|\underline{m}|^2)\Sigma_k m_k \eta_k$$

Thus $\underline{x}$ can be interpreted as the output vector of a "dual" network whose input vector is $y = \{\eta_k\}$; in fact the weights of this network are the same as those of the "direct" net.

We can improve our model with the following feedback "delta" learning rule :

$$\left| \begin{array}{l} \delta m_k = (1/\tau_k)\ \eta_k\ (x - \underline{x}) \\ \\ \tau_k := \tau_k + \eta_k^2/|\underline{m}|^2 \qquad \text{with } \tau_{k(\text{initial})} = 0 \end{array} \right.$$

which includes our original law, in the case of a network of a single neuron.

3 - 2 - A "neural" interpretation of the Gram-Schmidt orthogonalization procedure :

It is easy to verify that the Gram-Schmidt orthogonalization procedure can be obtained with our model under the following conditions :

- each weight vector is introduced successively as a data vector - so $\tau_k = \eta_k$ - and updates only itself.

- the reconstitution of the data used when updating the factor k is limited to the contribution of this factor and that of its predecessors.

This exact procedure can be approximated stochastically in the process of updating the weight vectors by using a filtered vector $\tilde{y}$ instead of the original one : if η_k is the "dominant" component of this vector, the components η_{k+1} to η_K are set to 0.

Another procedure is more accurate but has less generalization potential : it consists in updating the first factor during as many data "sweeps" as it needs to converge, then updating the second one, and so on.

Our simulations with the first procedure on real data are encouraging : orthogonalization is good, and even if the behaviour of the algorithm is sensitive to the initial conditions and to the presentation order of the data, a structure which can be interpreted emerges as the data are introduced. Our aim, with this first procedure, is not to design a new principal component algorithm but rather to derive a more local method adapted to the description of very large data tables.

4 - Metric and weights suited to correspondence tables

Good results for qualitative data are obtained from correspondence analysis of co-occurence matrices, given the underlying chi-square distance measure used ; this distance indeed has been demonstrated (6) to be strongly related to Shannon's entropy. In the scope of our model, we can simulate a correspondence analysis (7) :

- transforming our raw data ($_{i.}$, $_{.j}$ and $_{..}$ are respectively the sums of columns, rows, and general sum of the data matrix X) :

$$\xi_{ij} := (\xi_{ij} - \xi_{i.}\xi_{.j}/\xi_{..}) / \sqrt{\xi_{i.}\xi_{.j}/\xi_{..}}$$

- and then weighting our factors : $f(i) = m(i,j) \sqrt{\xi_{..}/\xi_{i.}}$

$$g(j) = \langle m, x \rangle \sqrt{\xi_{..}/\xi_{.j}}$$

This transformation of the data can be interpreted as some sort of "contrast enhancement", or orthogonalization, of the input patterns.

5 - Generalization : a neural model for "partial half-factors"

5 - 1 - Cooperative/competitive learning for non-linear mapping

It is possible to experiment with other interconnection structures between cells : if each cell "excites" its nearby neighbors and "inhibits" more distant ones ("mexican hat" interaction), a network of numerous cells geographically arranged in space creates a global "map" of learned data ; this map may be a highly non-linear projection of the multidimensional data structure on a portion of the surface. This approach is inspired by Kohonen (8), who implemented simple versions of such associative mapping : "positive" mexican hat, $\alpha(x-m)$ learning rule.

We have designed a model in which $\underline{x}$ is a kind of "local estimation" of the input vector x, which tends to generalise the previous pseudo-Gram-Schmidt orthogonalization method beyond hierarchically derived axes :

$$\underline{x} = \left(\Sigma_k \ddot{\eta}_k m_k / |m|^2 \ ; \ k \in D \right)$$

where D is the domain of influence of the "dominant" neuron, whose output is maximum when input vector x is read, and $\ddot{\eta}_k$ is a "filtered" version of the component η_k of the output vector $\ddot{y}$, transformed by the "mexican hat" coefficients (see figure 2).

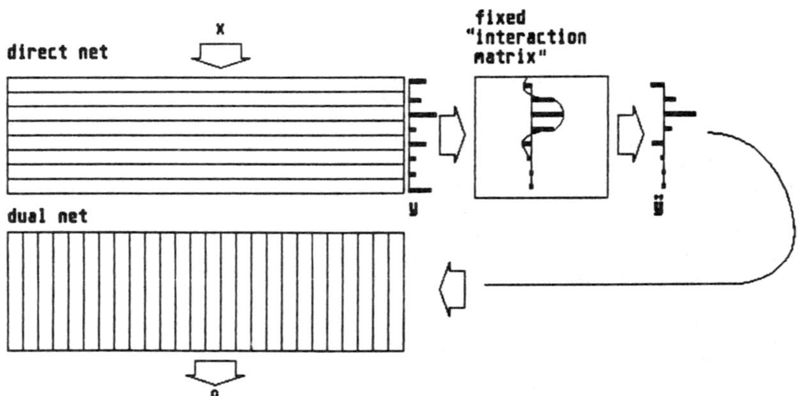

figure 2

5 - 2 - Strict orthogonalization or not ? Axes or half-axes ? Preliminary remarks :

Our practice of data analysis techniques suggests us a few remarks :

- Strict orthogonalization of the factors is an artificial constraint imposed on data representation ; principal component analysis is not the unique solution to the factor analysis problem : see the Oblimax, Varimax,... methods ; the relevant point is to "catch" local areas with a relative higher density of data vectors x ;

- Factor representation tends to privilege couples of oppositions. Observed structures of data are often different : for example, three poles may induce unstable and arbitrary rotations between two eigenvectors with neighbouring eigenvalues. In our opinion, the most adequate representation is a set of projections upon half-axes issued from the centroïd of the data cloud. This type of representation can also be considered as a set of <u>overlapping</u> clusters of descriptors and described objects (see figure 3). It is in accordance with the fuzziness of most of natural classes, proven by the well-known instability of clustering techniques with respect to small variations in their control parameters (number of classes, thresholds, metrics, ...).

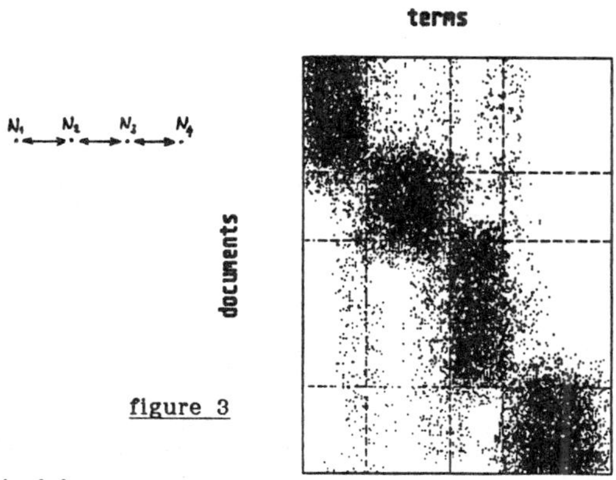

figure 3

5 - 3 - Spherical k-means :

We have applied the "mexican hat" interaction rule to various documentary data sets with acceptable results (1), but the mathematical understanding and control of the process is very hard to obtain. If we replace this rule by the "winner takes all" minimum interaction rule, where the interaction matrix is diagonal, of course we no longer have a global mapping, but we empirically found that the weight vectors were the same.

Then the process looks very much like the well-known "k-means" adaptive clustering method, where each new data is aggregated to the nearest centroïd, and then the position of this centroïd is modified slightly. But there is a difference : as stated above, our objective function is to maximize the sum of the squared projections of the data vectors on the weight vector m ; here, for the subset of data aggregated to the cluster k, this criterion is equivalent to minimizing the sum of the squared distances to the <u>axis</u> m_k, whereas in k-means the sum of the squared distances to the <u>point</u> m_k is minimized. This method converges, like k-means, toward local minima of the total objective function of the network ; these local minima depend on the initial conditions.

So, if we suppose that the data vectors are normalized, we have, as far as we know, a new type of k-means which we propose to call "spherical k-means". It provides some advantages :

- for lexical applications, <u>angles</u> and not euclidean distances are relevant : when two document descriptions have no word in common, all you can say is that they are orthogonal,

- you dispose not only of the projections of the elements of each cluster on its "axoïd", but also of the projections of all the other elements : in this way, each axis can be directly interpreted as a <u>fuzzy</u> cluster (see Table 1).

The convergence process is fast : in our example of Table 1, three passes upon the data were sufficient to stabilize the content (list of described objects) of each cluster.

5 - 4 - Local component analysis :

For the purposes of clustering data vectors, an interesting "structuring objective function" (10) that should be maximized is the partial inertia defined by the sum of the squared projections of the data vectors upon an axis m, when these vectors have an angle with m which is inferior to a threshold parameter Θ_t :

- when $\Theta_t = 0$ there are as many clusters as data vectors.

- when $\Theta_t = 180°$ there is one single cluster, and the objective function is that used in PCA.

- when $\Theta_t = 90°$ - this case is equivalent to fixing a threshold on the transfer functions of the neurons below 0 - the objective function is a sum of squared cosines, drawing a smooth, continuous landscape (fig. 4) along which our gradient-climbing learning law leads to local maxima. In this case, data patterns are centered, for example with our transformation rule of § 4 ; neurons have no interactions at all, and they are initialized at random.

Simulations on a small data set (see Table 1) show that if we freeze all the gains $1/\tau$ after the first sweep upon the data, and then carry out a dozen extra sweeps, the system will converge towards a set of stable "important" half axes ; these axes are :

- clearly interpretable, and in our opinion clearer than in any other method,
- independent of the order of the patterns, and of the initial random values of the synaptic weights,
- independent of the interaction structure.

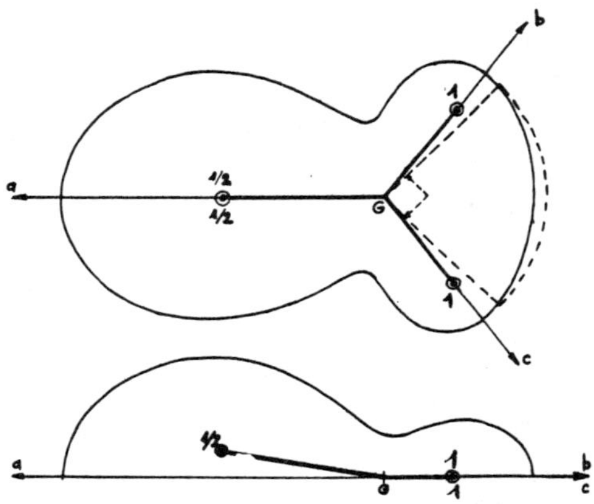

Inertia landscape along half-axes issued from the centroïd G : example of 4 points (weighted 1 and ½) in the 3D space ; Ga, Gb, Gc are the half-axes with maximal inertia.

figure 4

A difficult problem with our model is to know how many local peaks should be found in a given set of data, and how to be sure not to "miss" any of them ? For the moment we only have :

- a theoretical answer, unworkable on large data sets : if we build a network with as many neurons as data vectors, each neuron being initialized with a data vector, we will not miss any local peak, and we will derive as well the exact "basin of attraction" for each peak.

- an empirical answer : if we create a "sufficient" number of neurons, initialized by applying spherical k-means to the centered data, few local peaks will be missed. They can be identified by selecting data vectors closest to the orthogonal to all other peaks, then re-initializing "free" or redundant neurons if necessary, and letting the convergence process go on.

Ideally, an analytically exact one-pass learning rule would be desirable, but we have not achieved it as yet... However we feel that we are on the way : we have solved the problem in the linear, PCA case (2).

4 - CONCLUSION AND FURTHER DEVELOPMENTS

Highly multidimensional documentary spaces are an entangled mesh of associations between overlapping objects containing both global and local structures. Their representation can be achieved using a global map of spatially interconnected neurons "switched on" in a low number for each pattern of data. This map - which will as a rule not bė linear - will yield an optimal representation of data properties by defining the appropriate projection for a global interpretation of the documentary space. Each half-axis emphasizes the positions of a fuzzy subset of terms and documents, and the half-axes need not be orthogonal.

The process of documentary indexing will result in real time "learning" of each new documentary unit by the simulated network at each step. Assimilation of each unit will result in a slight modification of the various "points of view" embodied in the neurons. This operation of "synaptic weights" updating is, of course, intrinsically adapted to massively parallel data processing architectures.

Neural modelling is generally applied in two distinct situations : when a great many objects are described by a few dimensions (associative mapping), or on the contrary when a few patterns are described by a great many dimensions. Our particular case is different, in that we have highly multidimensionnal data and very large data sets. Our problem is consequently one of determining the right groups of dimensions, rather than to identify groups of objects within these dimensions. We hope that we have shown that this particular neural formalism provides a general conceptual framework for the interpretation, hybridation and generalization of classical data analysis methods.

References :

(1) **A. Lelu, D. Rosenblatt** - "A neural network model for information retrieval" - proceedings of N'Euro 88 - ESPCI/IDSET, Paris, June 6-9, 1988

(2) **D. Rosenblatt, A. Lelu, A. Georgel** - "Learning in a single pass : A neural model for instantaneous principal component analysis and linear regression" - forthcoming in proceedings of the IEE conference on neural computing, London, Oct. 18-20 1989

(3) **J. P. Benzécri** - "Approximation stochastique dans une algèbre non normée commutative" - Bull. Soc. Math. France N° 97 - Paris - 1969

(4) **L. Lebart** - "Sur les calculs impliqués par la description de certains grands tableaux" - Annales de l'INSEE N° 22-23 - Paris - 1976

(5) **E. Oja** - "A Simplified Neuron Model as a Principal Component Analyzer" - Journal of Mathematical Biology, 15 (1982)

(6) **J. P. Benzécri et coll.** - "L'analyse des données" - Dunod - Paris - 1973

(7) **M. Volle, D. Domengès** - "Analyse factorielle sphérique : une exploration" - Annales de l'INSEE, N°35 - Paris 1979

(8) **T. Kohonen** - "Self-organization and associative memory" - Springer-Verlag - New York - 1984

(9) **J.P. Courtial** - "Qualitative models, quantitative tools and network analysis" - forthcoming in Scientometrics, Special Issue : Amsterdam Workshop of Dec. 1987

(10) **H. Emptoz** et coll. - "Fonctions structurantes et classification automatique" - proceedings of the 3rd conference Reconnaissance des formes et intelligence artificielle - AFCET, Nancy, Sept. 16-18, 1981

(11) **F. Marcotorchino** - "Une approche unifiée des problèmes de sériation par blocs" - 5èmes journées internationales Analyse des Données et Informatique - Versailles, Octobre 1987 - INRIA

Table 1

DATA TABLE
(from (11))

		a	b	c	d	e	f	g	h	i	j	k	l	m	n	o	p	q	r	s
		Size			Weight			Speed			Intelligence			Affection		Aggressivity		Company	Hunting	Guard
		−	+	++	−	+	++	−	+	++	−	+	++	−	+	−	+			
1	Beauceron	0	0	1	0	1	0	0	0	1	0	1	0	0	1	0	1	0	0	1
2	Basset hound	1	0	0	1	0	0	1	0	0	1	0	0	1	0	0	1	0	1	0
3	German shepherd	0	0	1	0	1	0	0	0	1	0	0	1	0	1	0	1	0	0	1
4	Boxer	0	1	0	0	1	0	0	1	0	0	1	0	0	1	0	1	1	0	0
5	Bulldog	1	0	0	1	0	0	1	0	0	0	1	0	0	1	1	0	1	0	0
6	Bullmastiff	0	0	1	0	0	1	1	0	0	0	0	1	1	0	0	1	0	0	1
7	Poodle	1	0	0	1	0	0	0	1	0	0	0	1	0	1	1	0	1	0	0
8	Chihuahua	1	0	0	1	0	0	1	0	0	1	0	0	0	1	1	0	1	0	0
9	Cocker	0	1	0	1	0	0	1	0	0	0	1	0	0	1	0	1	1	0	0
10	Collie	0	0	1	0	1	0	0	0	1	0	1	0	0	1	1	0	1	0	0
11	Dalmatian	0	1	0	0	1	0	0	1	0	0	1	0	0	1	1	0	1	0	0
12	Doberman pinscher	0	0	1	0	1	0	0	0	1	0	0	1	1	0	0	1	0	0	1
13	German mastiff	0	0	1	0	0	1	0	0	1	1	0	0	1	0	0	1	0	0	1
14	Breton spaniel	0	1	0	0	1	0	0	1	0	0	0	1	0	1	1	0	0	1	0
15	French spaniel	0	0	1	0	1	0	0	1	0	0	1	0	1	0	1	0	0	1	0
16	Foxhound	0	0	1	0	1	0	0	0	1	1	0	0	1	0	0	1	0	1	0
17	Fox terrier	1	0	0	1	0	0	0	1	0	0	1	0	0	1	0	1	1	0	0
18	Gascony blue hound	0	0	1	0	1	0	0	1	0	1	0	0	1	0	0	1	0	1	0
19	Labrador retriever	0	1	0	0	1	0	0	1	0	0	1	0	0	1	1	0	0	1	0
20	Greyhound	0	0	1	0	1	0	0	0	1	1	0	0	1	0	1	0	0	1	0
21	Mastiff	0	0	1	0	0	1	1	0	0	1	0	0	1	0	0	1	0	0	1
22	Pekingese	1	0	0	1	0	0	1	0	0	1	0	0	0	1	1	0	1	0	0
23	Pointer	0	0	1	0	1	0	0	0	1	0	0	1	1	0	1	0	0	1	0
24	Saint Bernard	0	0	1	0	0	1	1	0	0	0	1	0	1	0	0	1	0	0	1
25	Setter	0	0	1	0	1	0	0	0	1	0	1	0	1	0	1	0	0	1	0
26	Teckel	1	0	0	1	0	0	1	0	0	0	1	0	0	1	1	0	1	0	0
27	Newfoundland dog	0	0	1	0	0	1	1	0	0	0	1	0	1	0	1	0	0	0	1

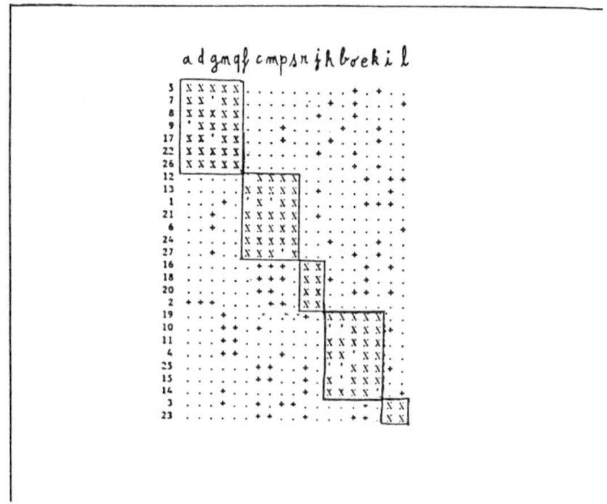

Data table re-ordered with the Block-Seriation method (11)

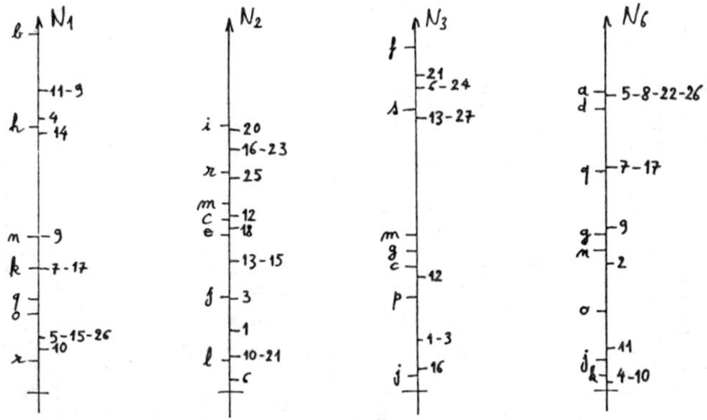

Local Component Analysis

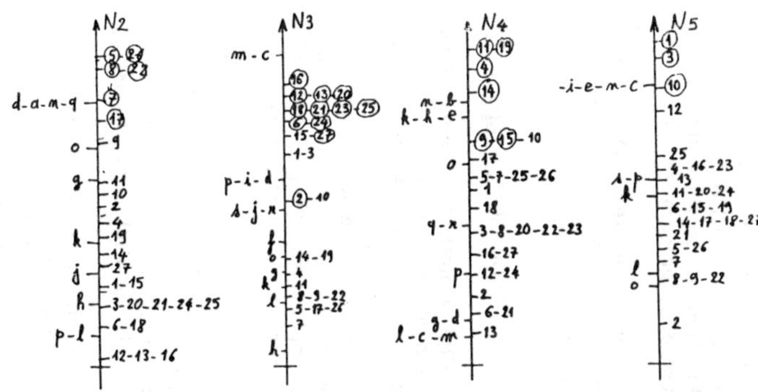

Spherical k-means

STRUCTURING AND GENERALIZATION KNOWLEDGE

METHOD GENERALIZATION BASED ON SPECIALIZATION

Sukhan lee
EE-Systems, Electrical Engineering Department,
University of Southern California, Los Angeles, CA 90089 (USA)
Bin Wang
Computer Science Department, University of Southern California
Los Angeles, CA 90089 (USA)

ABSTRACT: Most machine learning research in the last two decades has been focused on concept formation, that is the induction of non-procedural descriptions of concepts from examples [1,2]. Yet the learning of problem-solving methods has not been extensively investigated [3,4]. This paper investigates how to learn problem solving methods through *method generalization.* Method generalization is to develop the computational approaches and techniques for the generalization of problem-solving methods. In this paper, we investigate a restricted class of method generalization problems: given a concept hierarchy and a set of methods associated with individual concepts, the methods applicable to more general concepts are constructed from the methods associated with more specific concepts. A case study is presented which shows how methods for computing the properties of geometric entities in a hierarchy can be generalized to compute the properties of higher level entities in the domain of Euclidean geometry.

1. INTRODUCTION

Developing general methods for a given problem domain is considered difficult, and it depends largely on the domain and developers' experimental knowledge. On the other hand, instead of building generic algorithms to solve domain problems, it would be a good idea to develop approaches and techniques that generalize existing problem solving methods to solve new problems in the domain. As it is observed in real life, human beings always start with simple problems and find methods to solve them, then attempt to generalize those methods to solve more complicated problems based on their heuristic knowledge for generalization. This important aspect of human cognitive behavior motivates us the idea of building a machine by modelling this process. So method generalization is to study the computational approaches and techniques and the corresponding implementation issues for the generalization of problem-solving methods.

Method generalization involves a broad range of investigations. It relates closely to some AI areas such as machine learning and cognitive modelling. Basically, method generalization can be characterized as a mixed process of problem solving and learning. Problem solving is to construct a method for a specific problem based on domain knowledge, while learning is to generalize the method obtained for the specific problem by extracting invariant features and verify the generalized method based on generalization heuristics.

This paper is organized as follows: section 2 describes the method generalization problem in general; in section 3, we propose the problem in the restrict domain of polygon family in Euclidean geometry; section 4 presents a formal model for the restricted method generalization problem; section 5 illustrates the approach to cope with the restricted method generalization problem and provides a functional procedure to carry our the approach; section 6 presents a complete example to show the step-by-step application of the method generalization procedure; finally, section 7 concludes the paper.

2. THE METHOD GENERALIZATION PROBLEM

Whenever trying to find a method applicable to a complicated problem, people often investigate the methods applicable to the restricted versions of the problem and attempt to generalize those methods. For instance, the method for computing the area of a polygon may be obtained by generalizing the methods for computing the areas of the simpler geometric entities such as triangle and quadrilateral. A *method generalization problem* is to construct a method for a more general problem based on domain knowledge and methods for the less general problems previously solved. Therefore, *method generalization* is to develop computational approaches and techniques for the generalization of problem-solving methods.

However, method generalization in general is considered difficult to achieve and it is very much dependent on the application domain as well as heuristic knowledge [6,7]. It is impossible to accomplish it in a predictable way using a generic approach for all the problems from various application domains. Consequently, this paper only investigates method generalization under some restrictions. Given a concept hierarchy and a set of methods associated with individual concepts, a method generalization problem is to construct a method for a higher level concept based on the methods for the lower level concepts.

Problem solving and learning are considered as two phases of method generalization. The problem solving is to decompose a problem into subproblems and aggregate the solutions to the subproblems, while learning is to extract how to decompose the problems and how to aggregate the subsolutions. In general, problem solving and learning process problems in two opposite directions: 1) from general to specific and 2) from specific to general. Problem solving is based on the recognition that a specific problem is a specialization of a general problem, and learning is based on the characterization of problem solving. Specific-to-general may be considered more difficult to achieve than its counterpart, general-to-specific, and requires deeper understanding of domain knowledge and generalization techniques. For example, if we know how to compute the area of polygon, we can easily apply the method to computing the area of a triangle as long as we realize a triangle is a polygon of three edges; not obviously, however, if we only know the method of computing the area of a triangle and try to compute the area of an arbitrary polygon.

To be more specific and further investigate the problem, we constrain in the domain of Euclidean geometry and restrict the problems to the computation of geometric properties.

3. A RESTRICTED METHOD GENERALIZATION PROBLEM

In Euclidean geometry, some metric properties of geometric concepts (entities) are defined, such as length of a segment, area of a triangle and perimeter of a polygon. We take polygon family as our domain, the members of the family are polygon, triangle, quadrilateral, parallelogram, rhombus, ..., etc. These concepts form a structural hierarchy (See Figure 1).

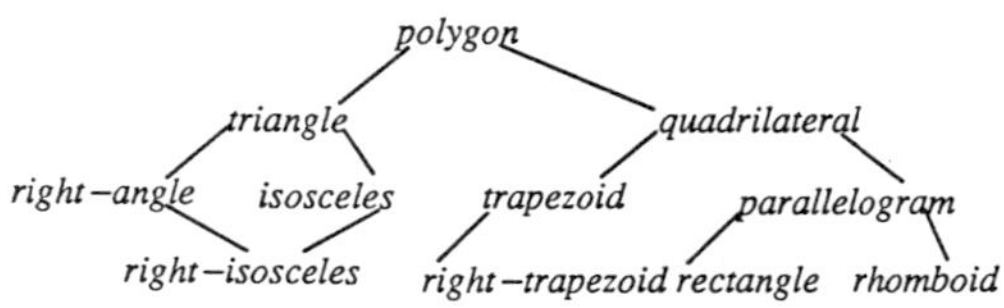

Polygon Class Hierarchy

Figure 1

We consider the generalization of geometric methods based on the polygon concept hierarchy. The problem instances are to construct methods of computing some geometric properties of the concepts in the hierarchy. Let us first investigate a simple example, the problem instance is how to compute the perimeter of a polygon.

Assume we have a method P_3 to compute a triangle's perimeter, that is to add three edge lengths together.

```
P3(e1,e2,e3);
int: e1, e2, e3, perimeter, i;
{
        perimeter = 0;
        perimeter = e1+e2+e3;
        return(perimeter);
}.
```

It is easy to see that this method can be generalized to a polygon. The generalization here is intuitive. Instead of using three edge parameters as the inputs to the procedure. We use a list to hold the lengthes of all the edges of a polygon and add every element of the list one by one. Then, P_3 is generalized to compute the perimeter of a polygon. This generalization process is easy to be validated. We have constructed the method P_n for polygon:

```
Pn (length,n);
int: length[0..n-1], n, perimeter, i;
{
        perimeter = 0;
        for i = 0 to n-1 do
        perimeter = perimeter+length[i];
        return(perimeter);
}.
```

Consider another problem instance, that is to how to compute the area of a polygon. Again, assume we know a method A_3 that compute the area of a triangle.

```
A3(height, base);
int: height, base, area;
{
        area = height*base/2;
        return(area);
}.
```

However, it seems not obvious how A_3 can be generalized to compute the area of a polygon. A natural question is that "is there any rule we can follow to make a method generalization from a simple concept to a more general one? ".

Some observations about the method generalization problem under this context are made below:

- The development of methods in application domains is always carried out ad hoc. The lack of deep understanding of the domain problems and their relationships causes the methods developed for individual problems hard to be understood and generalized to other problems in the same domain, let alone other domain problems. We need to investigate approaches and techniques that facilitate development of methods in a systematic way. As in human cognitive process, method generalization is an important approach to systematic method development.
- Concept hierarchy may be used to guide the systematic method development. We observe the bottom to top generalization is a promising and elegant way to solve some complicated problems. The reasons are 1) simple cases are easy to handle, 2) building from bottom up, complicated problems can be solved by divide and conquer 3) methods developed are more robust and 4) repeated work can be avoided.
- The concept hierarchy actually defines a *isa relation* between members of polygon family. When we say x *isa* y, it means x is a specialization of y, and x is a subconcept of y. For example, a triangle is a special polygon with three vertices. Intuitively, lower level concepts can inherit methods of its ancestors in the hierarchy. For instance, if we know the method for computing area of a polygon, we can apply it to compute area of a pentagon due to the fact that pentagon *isa* polygon. However, the top-down view of the hierarchy does not help with method generalization. But if we view it from bottom-up, we can use concept hierarchy to guide method generalization from lower level concepts to higher level ones.
- Go further with the *isa* relation, we investigate the condition for each specialization for x isa y, that is the sufficient condition which makes y to be a x. Intuitive, this information is useful for method generalization from x to y because it implies the inherent connection between two concepts.

In following section, we formalize the above restricted method generalization problem.

4. FORMAL REPRESENTATION OF RMG PROBLEM

Before the formalization of the restricted method generalization (RMG) problem, let us define the following terms to be used in the formal representation of RMG problem. A *method* is a predefined sequence of operations that operate on problem specifications. A method represents a problem solving procedure including mathematical functions, algorithms and procedures. A method can be a simple mathematical function, or can be a complicated procedure. For example, a method to compute the perimeter of a triangle is a simple function **area**(*triangle*)=$e_1+e_2+e_3$, where e_1, e_2, e_3 represent the three edges of the triangle; while a method to classify whether a point is inside or outside a polygon is a complicated procedure that takes a number of steps. A *concept* defined in a domain is an entity that represents a class of objects that share some common features. For example, the concept *triangle* represents all polygons of three vertices. A concept is characterized by its *properties* described as its attributes. A *value* is assigned to each property to measure the property of the concept.

The RMG problem is now formally defined as a 6-tuple <C,H,S,M,G,D>:

A *concept space C* is a set of concepts defined in a domain.

A *concept hierarchy H* is a relation defined by

$H=\{<x,y>|x \textit{ isa } y$, i.e. x is a *specialization* of y, $x,y\in C\}$

A set of sufficient conditions *s* is assigned to each <x,y> in H to define <x,y>. A *specialization space S* is the set of all the tuples composed by *s* and <x,y>:

$S=\{<s,<x,y>>|$ s is a set of sufficient conditions for $<x,y>$, $<x,y>\in H\}$.

A *method tuple* is a triple <x,y,z> where x is a concept, y is a property of x, and z is a method to compute the value of y. Let us denote the result computed by z, i.e. the property value of y, as z(x). A *method space M* is a set of method tuples in the domain:

$M=\{<x,y,z>|<x,y,z>$ is a method tuple$\}$.

A *method generalization* of a method tuple <x,y,z> is a tuple <u,y,f(z)>, where <x,u> is in H, f(z) is a method that computes the value of the property y of u; f(z) is obtained by generalization of the method z that computes the value of the property y of x. The *generalization space G* is the set of all possible method generalizations that can be generated from *M*:

$G=\{<u,y,f(z)>|<u,y,f(z)>$ is a method generalization of $<x,y,z>$, $<x,y,z>\in M\}$.

The *domain knowledge D* consists of a set of domain inference rules. Domain knowledge is organized and represented in form of inference rules. The problem-solving heuristics are the core of domain inference rules.

To make the above formalization clear, we present in the following more detailed interpretations of the tuple elements based on examples:

- The *concept hierarchy* is defined by the *isa relation* among concepts in the concept space. Examples include "an isosceles triangle is a triangle" and "a triangle is a polygon". This represents a special relation defining a subclass relations: *x isa y* implies that x is a specialization of y; therefore x can inherit methods and properties defined for y. The specializations of concepts are generated based on concept hierarchy.
- A set of sufficient conditions is required to define specialization relation <x,y> in H. For example, the specialization relation, <triangle,polygon>, is defined by the condition {number_of_vertices(polygon,3)}. The set may be empty as indicated by a special case, x isa x. The set of sufficient conditions for y to be x is closely related to generalization of a method from x to y.
- A *method tuple* associates a concept with a property and a method that computes the value of the property. For instance, the method tuple <triangle,perimeter,"$P_3(e_1,e_2,e_3):=e_1+e_2+e_3$"> means that a triangle has a perimeter property which is computed by a function "$P_3(e_1,e_2,e_3):=e_1+e_2+e_3$". A concept may have many properties, and each property may have multiple methods for computing its value.
- A *method generalization* constructs a function f(z) for computing the value of the property of concept u based on the specialization relation, x isa u, and on the function z that computes the value of the same property of x. For instance, given a method tuple <triangle, perimeter, $P_3(e_1,e_2,e_3)$> and "triangle isa polygon", the *method generalization* results in <polygon, perimeter, P_n(lenght,n)> (P_3 and P_n are method procedures mentioned in last section).

- The *domain knowledge D* in Euclidean geometry can have rules that represent geometric theorems, such as to represent "if line l_1 and line l_2 are both parallel to line l_3, then l_1 parallel to l_2" as an inference rule (in Prolog code):

```
parallel(l1,l2):-
        parallel(l1,l3), parallel(l2,l3).
```

An example of problem-solving heuristic rules is "some properties of polygon can be computed by triangulation". This rule suggests the triangulation as a possible method to compute the properties of a polygon.

The RMG problem can be defined as follows: given <C,H,S,M,D>, how to construct G; specifically, how to construct each individual method generalization <u,y,f(z)> from method tuple <x,y,z>?

5. AN APPROACH TO THE RMG PROBLEM

With the formal representation of the RMG problem, a constructive approach to the solution of the RMG problem is proposed. The proposed approach is based on the following heuristics fundamental to the studies of the RMG problem:

Restricted Method Generalization Heuristics:

Given the RMG problem to construct a method f(z) of computing the property y associated with u from the method z of computing the property y associated with x, with x **isa** u, there exists a transformation that decomposes u in terms of x's, such that the property of u can be computed by collecting the properties of x's. The construction of such a transformation is based on specialization constraints that makes u be x and on problem-solving heuristics for the given domain.

Description of the approach:

The proposed approach is based on a constructive method composed by four steps: 1) the formation of a specific collection rule from an example generated by specialization based on domain problem-solving heuristic rules; 2) the generalization of the specific collection rule formulated for the selected example and the operationalization of the heuristic rule; 3) the verification of the generalized collection rule and the operationalized domain knowledge using additional examples by specialization; 4) the formation of desired method by composition of verified collection rule and heuristic rule.

To derive the method f(z) of computing property y of u (given problem instance <u,y,f(z)>), first we generate x as a specialization or an example of u, where x isa u. Then we derive the collection rules that aggregates the property y of u from the problem-solving process of the sample problem. Note that domain problem-solving heuristic rules suggest how to transform u into x's as defined in RMG heuristics, and the collection rules generally indicate how to collect distributed properties imposed by RMG heuristics. The collection rule derived from an example will then be generalized to formulate the method f(z) of computing the property y of u, while the problem-solving heuristic is operationalized into a more specific rule. The f(z) obtained from the generalization should be further verified for its validity based on additional examples.

For example, to solve the problems of deriving some polygon property, we first select a domain heuristic rule related to the problem, say "some properties of a polygon can be computed by triangulation", which suggests the transformation of polygon in terms of triangles. Then we generate an example by making a specialization of the polygon, say, a rectangle. Assuming that we are to find a method of computing the area of a polygon, the selected heuristic rule suggests the decomposition of the area of a rectangle by the area of triangles. Then, the specific collection rule for a rectangle is derived from the known formula of computing the areas of a rectangle and a triangle. The collection rule thus obtained for a rectangle will be generalized based on the climb-tree heuristic such that the derived collection rule applicable to a specific concept can also be applicable to a high-level concept in concept hierarchy.

Of course, the selected heuristic rule or the derived collection rule from an example may either be unable to lead to a valid generalization or insufficient to solve a RMG problem. This is because either the transformation hueristic or the collection rule may result in an invalid solution to a RMG problem. Therefore, the verification of a generalized collection rule is necessary.

The verification can be carried out in two different ways: 1) the example-based verification and 2) the proof-based verification.

The example-based verification keeps generating examples by specialization and verifying the generalized collection rule until some credibility criterion is satisfied. One creditability criterion may be the number of positive examples tested in verification. This method does not give a complete proof of the generalized collection rule but provides creditable generalization.

The proof-based verification formally proves the validity of the generalized collection rule based on axioms or theorems available to the system. This method provides a perfect verification but poses difficulty in implementation. We suggest that proof-based verification be conducted by human experts as an external process.

We present the example-based verification procedure step by step:

Step 1. Select a new example by specialization.

Step 2. Apply the generalized collection rule to the selected example.

Step 3. Verify the validity of the generalized collection rule.

Case I: the rule is verified valid, check if creditability criterion is satisfied, if satisfied, verification is successful and stop; otherwise, put the example into positive example bin for the particular generalized collection rule and go to Step 1.

Case II: the rule is verified invalid, invoke the process of deriving the collection rule for the selected example and obtain a new generalized collection rule by generalizing the specific collection rule derived for the selected example.

Apply the new generalized collection rule to every example in the positive bins for other generalized collection rules and verify their validity. If the new collection rule is valid for each example in a bin, then substitute the collection rule for the bin by the new collection rule. If the new collection rule is valid for two bins, the two bins are merged into one.

Step 4. Repeat the process until the verification criterion is satisfied or no more example can be generated.

At the end of this process, if only one bin exists, the verification process is successful; if several bins exist, the collection rule is valid by cases.

Example-based Verification Procedure:

```
example_based_verification(c,p,decomp_proc,grule,status);
    BEGIN
      L1: generate_specialization(c,csp);
          /* generate a new example of "c" as "csp" */

          decomp_proc(csp,subcsps);
          /* invoke the same decomposition procedure "decomp_proc" that
             decomposes "csp" in terms "subcsps" */

          collection_rule_application(csp,subcsps,grule,process);
          /* apply the generalized collection rule "grule" on "subcsps"
             and record the process as "process" */

          external_verification(process,result);
          /* verify the application "process" using external expert and
             report the result of verification as "result" */

          CASE result
          VALID:    {
                check_criterion(c,grule,value);
                IF value=SATISFIED THEN return(YES)
                      ELSE { put_in_bin(csp,grule,bin); GOTO L1}
                /* in case of verification result is VALID, then check whether
                   creditability is satisfied, if SATISFIED, return YES to the
                   main procedure, otherwise, put "csp" into the "bin" of
                   "grule", then back to beginning of the procedure */
                };
```

```
            collection_rule_formation(csp,subcsps,p,crule);
            collection_rule_generalization(c,crule,grule);
            example_based_verification(c,p,decomp_proc,grule,status);
            /* in case of verification result is INVALID, solve "p" of
               "csp" and extract a new collection "crule", then
               generalize "crule" to "grule", involve example-based
               verification procedure recursively */
            };
    END.
```

Since the example-based verification procedure is easier to implement, we give a detailed procedure for the restricted method generalization problem based only on the example-based verification:

Example-based Method Generalization Procedure:

```
GIVEN: problem domain = <C, H, S, M, D>;

        C: concept space;
        H: concept hierarchy;
        S: specialization space;
        M: method space;
        D: domain knowledge;

INPUT: <c, p, m>
/* construct a method "m" that computes property "p" of concept "c" */
        BEGIN
          L1: generate_specialization(c,csp);
            /* generate an example of "c" as "csp" based on H */

            find_decomposition_procedure(csp,p,c0,decomp_proc);
            /* find a proper decomposition rule to divide "csp" in terms of "c0"
               and the corresponding procedure is "decomp_proc" based on domain
               problem-solving heuristic rules */

            decomp_proc(csp,subcsps);
            /* invoke procedure "decomp_proc" that decomposes "csp" into "subcsps" */

            collection_rule_formation(csp,subcsps,p,crule);
            /* solve "p" of "csp" based on domain knowledge and extract the
               collection rule "crule" that aggregates property "p" on "subcsps" */

            collection_rule_generalization(c,crule,grule);
            /* generalize the specific collection rule "crule" to "grule" based
               on generalization techniques */

            example_based_verification(c,p,decomp_proc,grule,status);
            /* verify the decomposition rule "decomp_proc" and the generalized
               collection rule "grule" based on examples and report the "status" */

            IF status=YES
                THEN method_composition(c,p,decomp_proc,grule,m)
                ELSE GOTO L1;
                /* if the status of the verification is YES, then compose
                   "decomp_proc" and "grule" into expected method "m";
                   otherwise, repeat the process from beginning */

        END.
```

6. A CASE STUDY

A complete example is illustrated to show various steps of our constructive generalization approach.

Problem Space Description:

- Concepts Space C = { polygon, triangle, parallelogram, quadrilateral,... };
- Concept Hierarchy H = { <triangle,polygon>, <parallelogram,polygon>, <quadrilateral,polygon>, ... };
- Specialization Space S = { <{#_of_vertices(polygon,3)},<triangle,polygon>>,... }
- Method A_3 is associated with *triangle* that computes the *area* of the *triangle*.

  M = { <triangle,area,A_3>, ... }
- Domain knowledge: (assumption: plane geometry, Euclidean space)

  [1] definitional domain knowledge, such as definition of polygon, area of a geometric entity, triangulation procedure, ..., etc;

  [2] domain problem-solving heuristics, such as "some problems of polygon can be solved by triangulation";

Generalization Problem:

Construct a method An that computes the *area* of *polygon*:

G = { <*polygon,area,*A_n>, ... }

Here we are concerning about the generalization problem of constructing the method A_n based on the approach we proposed in last section.

To obtain the method A_n for polygon, we first look at the domain problem-solving heuristic which suggests the transformation of polygon in terms of its specializations. We find the heuristic rule "polygon can be divided into triangle areas" from domain knowledge D and the fact "triangle isa polygon" from the concept hierarchy H. Both the heuristic rule and the fact provide the decomposition of polygon into triangle areas such that the property of polygon can be obtained by computing and collecting the triangle areas. So an instance of RMG heuristics is formed.

Next we choose a specialization of polygon, say, triangle again. Since we know the method for computing the area of a triangle and a single triangle should not be triangulated, this specialization can not help the generalization process because it is a trivial application of RMG heuristics.

Another specialization of polygon is generated, say, parallelogram. Apply the selected heuristic rule and formulate the collection rule.

Formation of collection rule from an example:

We know how to compute the area of parallelogram and triangle given their base and height.

Express the methods in expressions:

Area(parallelogram)=a*b;

Area(triangle)=c*d/2;

Goal:

1. Find the function f such that Area(parallelogram)=f(Area(triangle)).
2. Generalization of function f.

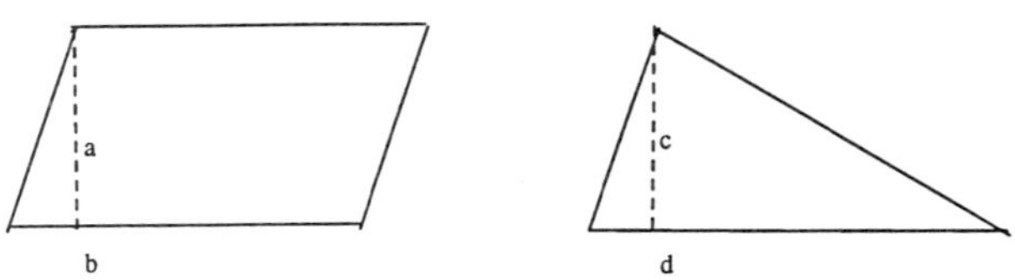

Figure 2

Since we have Area(parallelogram)=a*b, make the unification of a*b with the right side of Area(triangle)=c*d/2, we get:

a/c and b/d ------- (1)

From (1), we have

Area(parallelogram)=2*Area(triangle) provided c=a and d=b. ------(2)

Now let's see what is the geometric interpretation of (2): The area of a parallelogram is equal to two times the area of a triangle which has the same lengths of height and base as of the parallelogram. Note that unification is regardless of the geometric meanings of a,b,c,d; where a and b are height and base of a parallelogram respectively, and c and d are triangle's.

From above unification process, we obtain the collection rule "addition in terms of two triangles with same base and height".

Generalization of the collection rule and operationalization of the heuristic rule

1. *Domain knowledge is used to extract features from the example*;

For instance, the relative features drawn from above example could be

a) the area of parallelogram is the areas of two triangle with same height and base;

b) the triangles are not overlapping;

c) the number of triangles are two for a parallelogram;

d) the diagonal partitions the parallelogram into two triangle areas;

2. *Generalization techniques are used to generalize the features;*

Some commonly-used domain-independent generalization rules [10] may be used to generalize these features:

1) drop the conditions of the collection rule; we get the area of parallelogram is equal to the sum of two triangle areas; if we further generalize parallelogram to quadrilateral (since parallelogram is a quadrilateral, isa hierarchy is used to guide the concept generalization). The generalized collection rule is that quadrilateral area is equal to the sum of two triangle areas, that is both features of parallelogram and triangle in the example are generalized (features 1.a and 1.c).

2) generalize the feature 1.c by induction: number of triangles resulted from decomposition of N-vertex polygon is N-2; (from 1.c)

3. *Generalization of collection rule*

It is easily seen that generalized collection rule for polygon implied in above example is the *addition of N-2 triangle areas.*

Verification of the generalized collection rules

Generate another specialization of polygon, say, a pentagon. Apply the triangulation procedure and collection rule. Obviously this example is valid. Many other specializations may be generated and tested and the results are valid too.

Formation of method A_n:

The required method A_n is constructed by composition of the operationalization of the heuristic rule and the generalized collection rule:

A N-vertex polygon area can be computed by triangulating the polygon into N-2 triangle areas and adding these triangle areas.

7. CONCLUSION

In this paper, we formalize the method generalization problem and present the computational approaches and techniques as well as the corresponding implementational issues for generalizing problem-solving methods. We focus on the method generalization problem in a restricted domain of Euclidean geometry and propose a constructive approach to solve the problem. A case study is presented to illustrate the approach in a step-by-step manner. As it is observed, method generalization in general is a difficult problem and deserves extensive research efforts. We hope this paper stimulate readers to initiate more research in that direction. Instead of drawing any conclusions about the practical applications of the proposed approach, we would like to build a prototype of the system that supports our proposed approach, even though in a restricted manner. We believe that experimenting the results of theoretical research will greatly help understand better the problem and proceed the research in a right direction.

REFERENCES

[1] Mitchell, T.M., Keller, R.M., Keder-Cabelli, S.T., "Explanation-Based Generalization: A Unifying View", Machine Learning 1: 47-80, 1986.

[2] DeJong, G., Mooney, R., "Explanation-Based Learning: An Alternative View", Machine Learning 1: 145-176, 1986.

[3] Michalski, R.S., Carbonell, J.G., Mitchell, T.M., "Learning Problem Solving and Planning" in Machine Learning, Volume I, Tiogo Press, Palo Alto, CA. 1983.

[4] Mitchell, T. "Learning and Problem Solving", in proceeding IJCAI83, August, 1983, pp1139-1151.

[5] Mahadevan, S. "Verification-Based Learning: A Generalization Strategy for Inferring Problem-Reduction Methods", in proceedings of IJCAI85, LA, 1985, pp616-623.

[6] Kant, E. "Understanding And Automating Algorithm Design", in proceedings of IJCAI85, LA, 1985, pp1243-1253.

[7] Steier, D., Kant, E. "Symbolic Execution In Algorithm Design", in proceedings of IJCAI85, LA, 1985, pp225-231.

[8] Haussler, D. "Learning Conjunctive Concepts in Structural Domains", in proceeding AAAI87, Seattle, 1987, pp466-470.

[9] Carbonell, J.G. "Learning By Analogy: Formulating And Generalizing Plans From Past Experience", in Machine Learning, Volume I, Tiogo Press, Palo Alto, CA. 1983.

[10] Winston, P.H., Artificial Intelligence, Addison-Wesley, 1984.

GENERALIZATION AND CONCEPTUAL CLASSIFICATION: INDICES AND ALGORITHM

A. Guénoche
G.R.T.C.-C.N.R.S. 31 Ch. J. Aiguier 13402 Marseille Cedex 9 (France)

In this text we tackle the generalization problem of a set of objects described with boolean variables. In learning theory, objects are called *instances* and belong to *concepts*. We will build generalizations that are logical formulas of present or absent, characteristics that are complete formulas for set of instances; they give the value "true" to any instance. We only use positive instances and compute simililarity between them, so it is a contribution in *similarity based learning*.

We will illustrate this text with animals described by the following boolean variables; answers "yes" and "no" are respectively coded 1 and 0 :

A : Does it lay eggs ?

B : Has it feathers ?

C : Has it scales ?

D : Has it teeth ?

E : Does it fly ?

F : Does it swim ?

G : Does it breathe in the air (1) or in the water (0) ?

For the considered animals, we get the following array that is a useful representation but not necessarily a data structure for computer treatments.

	A	B	C	D	E	F	G
1 Ostrich	1	1	0	0	0	0	1
2 Canari	1	1	0	0	1	0	1
3 Duck	1	1	0	0	1	1	1
4 Shark	1	0	0	1	0	1	0
5 Salmon	1	0	1	0	0	1	0
6 Frog	1	0	0	0	0	1	1
7 Crocodile	1	0	0	1	0	1	1
8 Barracuda	1	0	1	1	0	1	0

If we give to *generalization* the intuitive meaning of a characterization of the objects in the terms of the descriptive system, it seems "natural" to resume these data with the property that all our animals share : To lay eggs. It constitutes a fist level of generalization with a complete formula. Now if we look at descriptions with a little more attention, we have on one hand animals with feathers, no scales, no theeth, that live in the air, and on the other hand, animals that do not have feathers, that do not fly, but live in the water. Thus we have made a partition of our animals in two classes, each one can be characterized by its shared features. When we say that animals belong to one of these classes, each having a cohesion greater than the whole set, we formulate a generalization more specific than the only one common characteristic. We can begin again this operation on one or the other class to obtain a larger part of description possessed by each subclass. This underlines relations between *taxonomical generalization* as we say in learning and *classification* as we say in data analysis.

The aim of this communication is to give a formal framework to this process, to present parameters to evaluate goodness of generalization, and to give an algorithm to build formulas that are disjunctions of conjonctions of properties.

1. Notations and terminology

Variables may note presence or absence of characteristics (coded 1/0) or may be dichotomic with modalities arbitrary coded 0 or 1 (as to breathe in the air or in the water). This distinction is necessary because if in the first case we can only be interested by present attributs, in the second case, it is convenient to give to both modalities the same role. That is why we use boolean algebra, as we did previously [Guénoche 88]. We first recall its terminology.

Let X be a set of n objects described by m variables $\mathbf{A} = \{A, B, C, .. \}$, each one, for example A, can takes values a (*direct form,* coded 1) and a' (*complemented form*, coded 0). We can consider X as a 0-1 array with n rows corresponding to X elements and m columns corresponding to **A**. We pose $\mathbf{a} = \{a, b, .. j, .. \}$, $\mathbf{a}' = \{a', b', ..j', ..\}$ and $\mathbf{a}^* = \mathbf{a} \cup \mathbf{a}'$ is the set of the 2.m values of the m variables. We call a *monomial* a conjunction of variables in any form, that is a subset of $\mathbf{a}^*$ where each variable appears at most once. A monomial is noted by a concatenation of these forms. Its length, noted $l(\mu)$, is the number of variables (or forms) that it contains. For instance μ = ab'd is a monomial of length 3. A monomial is *complete* if its lenght is $m = | \mathbf{A} |$. We remark that X is a subset of $H = \{0,1\}^m$ that constitutes our *events space.*

We note $X(\mu)$ the part of X of which elements have all the variables of μ with the same form as μ. We shall say that they *contain* μ. For instance X(ab) is the set of rows that have 1 in first and second columns; For our animals this set is {ostrich, canari and duck}. In the same way we define $H(\mu)$ as the subset of vectors of $\{0,1\}^m$ that contain μ. As there are 7 attributs, fixing the two first, we get 2^5=32 elements in H(ab).

Let us call *disjonctive formula* a disjunction of monomials noted $\Phi = \mu_1 \vee \mu_2 \vee ... \vee \mu_q$. For any element x of H, $\Phi(x)$ = "true" if and only if there exists i such that x belongs to $H(\mu_i)$; else we have $\Phi(x)$ = "false". We define

$$X(\Phi) = \cup_{i=1,..q} X(\mu_i).$$

In the same way $H(\Phi)$ is the subset of $\{0,1\}^m$ for which Φ takes the value "true". If we note μ_i complete monomial corresponding to the i-th element of X, the function

$$\underline{\Phi} = \vee_{i=1,..n} \mu_i$$

is equivalent to X. We have $X(\underline{\Phi}) = X$ and $\underline{\Phi}$ takes the value "true" only for the elements of X.

2. Characteristic monomials of a class

For boolean algebra, both forms have the same importance; 0 and 1 have the same weight. Frequently in Artificial Intelligence, especially for present attributs, only direct forms have to be considered. Putting in evidence that a set of objects does not possess an attribut may have no sense. To avoid this difficulty, we shall say that complemented forms are *relevant* if they must be taken into account. We admit that direct forms are always relevant. We note $\mathbf{r}^*$ the set of relevant forms of the m variables; We have $\mathbf{r}^* \subseteq \mathbf{a}^*$.

Let Y be a subset of X and $\mathbf{c}^*$ the set of values of the variables, relevant or irrelevant, that are constant on Y. We call *characteristic monomial* (resp. *relevant characteristic monomial*) of Y, noted $\mu(Y)$ (resp. $\nu(Y)$), the conjunction of the elements of $\mathbf{c}^*$ (resp. $\mathbf{c}^* \cap \mathbf{r}^*$); $\nu(Y)$ is the conjunction of the relevant shared characters of Y. We can considerate $l(\nu(Y))$ as a measure of the cohesiveness of the characterization of Y by $\nu(Y)$. Y is as well characterized as $\nu(Y)$ is longer. We can extend this notion by giving some weights to the variables, proportional to the importance that they have in the description. If we note p_A the weight of variable A we define the weight of a characteristic monomial :

$$p(\mu(Y)) = p(\nu(Y)) = \Sigma_{K \in \nu(Y)} p_K$$

If all the forms are relevant, $\mu(Y) = \nu(Y)$ and if all the weights are equal, also are $p(\mu(Y))$ and $l(\mu(Y))$.

3. Generalization

Consider now a partition of X in q classes $X_1, .. X_q$. By definition we have :

$$\cup_{i=1,q} X_i = X \quad \text{and} \quad \text{for all } i \neq j \;\; X_i \cap X_j = \emptyset.$$

Each class X_i can be generalized by its relevant characteristic monomial. We shall now express a generalization Φ of X starting from monomials $\nu(X_i)$.

A generalization of X is a formula $\Phi = \nu(X_1) \vee \nu(X_2) \vee ... \vee \nu(X_q)$, disjunction of relevant characteristic monomials of the classes. Φ is a complete formula for X. The previous function $\underline{\Phi}$ is a particular generalization of X since it is true only for the X elements. To each partition corresponds a generalization. The question is how largely X has been generalized with that function Φ. For instance any formula $\Phi_A = a \vee a'$ is also a generalization since any element of X presents variable A with direct or complemented form. In fact for any element of $\{0,1\}^m$, Φ_A is true and we can say that we have generalized X to H.

3.1 Measures of quality for generalizations

In this way, there is a large number of generalizations. Each time we subdivide a class, we create a new formula Φ more specific in that sense that it covers less elements of H, and it uses more variables to characterize both new classes. We are going to quantify this approach with two indices. We give their expressions first for a monomial and then for a generalization.

3.1.1 Measures for a monomial

Characterizing Y by $\nu(Y)$, we generalized it to $H(\mu(Y))$, that is, all the elements of $\{0,1\}^m$ that contain $\mu(Y)$. They are the $|H(\mu(Y))|$ *potential* instances generated from Y. This operation can be evaluated by a *generalization index* $G[\nu(Y)] = |H(\mu(Y))| / |Y| = 2^{m - l(\mu(Y))} / |Y|$. It is the same notion that Michalski & Stepp quantify by "the relative sparseness of a complex", a complex being equivalent to a monomial in a discrete quantitative variable space.

Characterizing Y by $\nu(Y)$, we also use a part of the description of Y, variables included in $\nu(Y)$. We can say that we have resumed Y to its common relevant characters, using only a part of the knowledge. We can quantify this part by an *explanation index* $E[\nu(Y)] = p(\nu(Y)).|Y|$.

Examples

If all the Y elements begin with 01, a' and b are relevant forms with weights 1 and there is no other variable constant on Y, we have $\mu(Y)$ = a'b and $p(\nu(Y)) = 2$; a'b is a generalization on Y and $|H(a'b)|=2^{m-2}$. We also have $G[a'b] = 2^{m-2} / |Y|$ and $E[a'b] = 2.|Y|$.

For the X animals, we get $\mu(X) = \nu(X) = a$, monomial of length 1. When we say that animals that belong to X lay eggs, we make no restriction on values of the other characteristics. Therefore there are 64 possibilities generated with 8 animals; we get $G[a] = 64/8 = 8$. The explanation index take the same value : $E[a] = 8$.

If **c*** is the empty set - Y has no constant variable - we pose $\mu(Y) = \nu(Y) = \emptyset$. We get $p(\mu(Y))=0$ and $H(\emptyset) = \{0,1\}^m$, so $G[\emptyset] = 2^m / |Y|$ et $E[\emptyset] = 0$. In fact we have generalized Y to the hypercube $\{0,1\}^m$.

3.1.2 Measures for a generalization

We can extend without problem the explanation index to generalizations since it is already an additive formula. We pose :

$$E[\Phi] = \Sigma_{i=1, .. q} E[\nu(X_i)].$$

For the generalization index, we will make first the following remarks :

- A formula $G[\Phi] = |H[\Phi]| / |X|$ is "hard" to compute, because we must enumerate $\{0,1\}^m$ to evaluate $G[\Phi]$. An additive formula $G[\Phi] = \Sigma_i G[\nu(X_i)]$ is wrong since the intersection of $H(\mu(X_i))$ and $H(\mu(X_j)$ is not necessarely empty.

- Some subdivisions are arbitrary, specially those that we can do without any reason, that do not bring new information. We do not want to improve the generalization index with this kind of segmentation. Particularly :

- For any variable, for instance A, we can build a bipartition on X with two classes X(a) and X(a'). This choice is arbitrary if there is no other variable constant on one of the classes.

- We always can separate one element x from X and build the bipartition $(\{x\}, X\backslash\{x\})$. For the class with a single element, its characteristic monomial is $\mu(x)$ and $G[\mu(\{x\})] = 1$. So we cannot define the generalization index on $\Phi = \nu(\{x\}) \vee \nu(X\backslash\{x\})$ with the average of its values for the two classes.

In consequence of these two remarks, we will pose for X partitioned in q classes X_i with n_i elements and $l_i = l(\mu(X_i))$ is

$$G[\Phi] = \max_{i=1,..q} G[\mu(X_i)] = \max_{i=1,..p} 2^{m-l_i} / n_i.$$

Proposition

We do not improve the generalization index with arbitrary subdivisions.

Proof

We will suppose, without lost of generality, that X has no constant variable and that all the forms are relevant. So $G[\mu(X)] = 2^m / n$. We calculate G in the two arbitrary subdivision cases above :

Subdivising X with variable A, we make appear two classes $X_1 = X(a)$ and $X_2 = X(a')$, with n_1 and n_2 elements. If both classes have no other constant variable, $\mu(X_1) = a$, $\mu(X_2) = a'$ and $l_1 = l_2 = 1$. So $G[\mu(X_1) \vee \mu(X_2)] = \max_{i=1,2} 2^{m-1} / n_i$. Necessarily one of these classes has a cardinality lower or equal to n/2, and

$$G[\mu(X_1) \vee \mu(X_2)] \geq G[\mu(X)].$$

So we do not improve G with a subdivision from one variable, if it does not make appear other constant variables, particularly in the class with less elements.

If we choose arbitrarily one element x from X, we get the bipartition $X_1 = X\backslash\{x\}$ and $X_2 = \{x\}$, and we have $G[\mu(X_2)] = 1$. If there is no constant variable on X_1, we have :

$$G[\mu(X_1) \vee \mu(X_2)] = 2^m / (n-1) > G[\mu(X)].$$

4 Algorithm

As we present it, the generalization problem of a set X, can be exposed as an optimization one. We seek for an optimal partition with a fixed number of classes. Let $(X_1, X_2, .. X_q)$ be a partition on X with q classes, and $\Phi = \nu(X_1) \vee \nu(X_2) \vee ... \vee \nu(X_q)$ the corresponding generalization. We can pose the two following problems :

P1 : To build a partition on X with q classes such that $G[\Phi]$ is minimum.

P2 : To build a partition on X with q classes such that $E[\Phi]$ is maximum.

These optimization problems are distinct (a partition can minimize $G[\Phi]$ without maximizing $E[\Phi]$) and unsolved. But we can observe that $G[\Phi]$ is as small, and $E[\Phi]$ is as great, as monomials $v(X_i)$ are longer and so the elements of X_i are similar. We find again the classification aim : To build classes with similar objects. We remark that similarity between objects is mesured with the part of description they shared. The method described under is a method of *conceptual clustering* [Michalski & Stepp 83] since we also get a characterization of the classes.

4.1 Ascending method

As in the ascending methods of classification, we consider initially that each element is alone in its class. At each step, we join two most similar classes, until it remains one single class X. We start with a partition with n classes and stop with a partition with one class, building a hierarchy. We just have to define similarity values between classes, and how to choice among maximum values the pair of classes to be joined.

In consequence of the preceding, we will say that two classes are as similar as they share a large part of relevant description. So we pose :

$$s(X_i, X_j) = p(v(X_i \cup X_j)).$$

This similarity function often give several maximum values. Rather than to choice a pair of classes arbitrarily, we propose to join at each step many pair of classes. The maximum similarity values, s, link several classes that are vertices of a threshold graph G_s. To maximize the explanation index, we must join a maximum number of classes, that correspond to a matching with maximum cardinality of G_s. The maximum cardinality matching algorithm, classical in graph theory [Berge, 1970], uses an alternated tree, and will not be detailed here. It is sufficient to know that it has a polynomial complexity.

Example

We suppose that maximum similarity values link classes (1,2), (2,3), (3,4). If we choose to join classes 2 and 3, we cannot make other joins with classes 1 and 4, even when if we join 1 and 2 on one hand, and 3 and 4 on the other hand, the explanation index will be maximized.

This method is very closed to the single linkage one [Sokal & Sneath 63]. But at each step, we must calculate similarity values between the new class and all the unchanged ones, even when for the single linkage one, they are evaluated from the initial values. The tie maximum values imply to join vertices that belong to the same connex component, even when our method imply to calculate a matching. It is, in the boolean algebra frame, similar to the *monomial clustering* method, exposed by Pitt & Reinke [1988] as an example of a conceptual classification method with polynomial complexity.

4.2 Example

We deal now with initial data considering that variables B and C are only relevant in direct form, the other ones being relevant in both forms. We give to all variables the same weight equal to 1. We start with the 8 relevant characteristic monomials :

1 : abd'e'f'g 2 : abd'ef'g 3 : abd'efg 4 : ade'fg' 5 : acd'e'fg' 6 : ad'e'fg 7 : ade'fg 8 : acde'fg'

The computed similarities are :

	1	2	3	4	5	6	7
2 :	5						
3 :	4	5					
4 :	2	1	2				
5 :	3	2	3	4			
6 :	4	3	4	3	4		
7 :	3	2	3	4	3	4	
8 :	2	1	2	5	5	3	4

We obtain 4 maximum values equal to 5, that are s(1,2), s(2,3), s(4,8), s(5,8). We chose the perfect matching {(1,2), (4,8)}, but we could have chose one of the three others, {(1,2), (5,8)}, {(2,3), (4,8)}, {(2,3), (5,8)}. We join into a class numbered 9, with characteristic monomial abd'f'g, classes 1 and 2 and into class 10, with monomial ade'fg', classes 4 and 8. The new similarity array is :

	3	5	6	7	9
5 :	3				
6 :	4	4			
7 :	3	3	4		
9 :	4	2	3	2	
10 :	2	4	3	4	1

The six maximal values alloow perfect matching {(3,9), (6,7), (5,10)} :

Classes	3 and 9	give classe	11	with monomial abd'g
	6 and 7		12	ae'fg
	5 and 10		13	ae'fg'

The resulting similarity is :

	11	12
12 :	2	
13 :	1	3

Classes 12 et 13 are joined in classe 14 with monomial ae'f. The last iteration joins class 11 and 14 into class 15 with monomial a. It remains one class and the algorithm stops. Instances are sorted by a depht first search path in the classification tree, and just like a dendrogram we can draw this hierarchy :

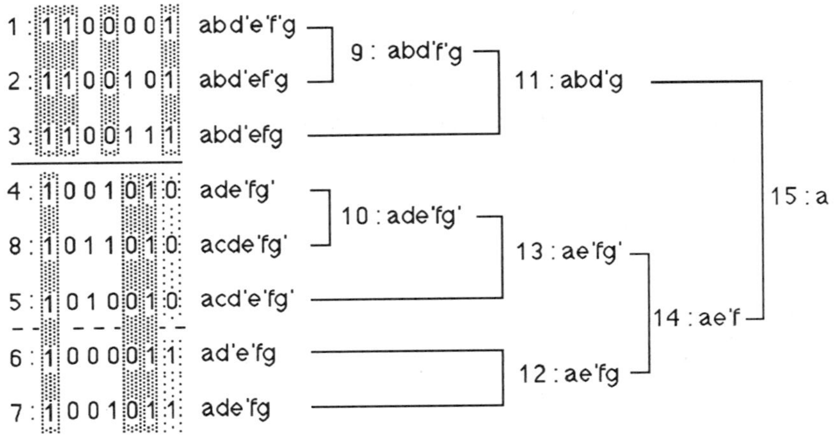

The generalization in two classes separated by a continuous line gives the logical function :

$$\Phi_2 = abd'g \vee ae'f.$$

$G[\Phi_2] = \max(2^2/3, 2^3/5) = 1.6$; $E[\Phi_2] = 4x3 + 3x5 = 27$; the explained part is dark-shadowed. The generalization in three classes, separated by a line (continuous or dotted), gives the function :

$$\Phi_3 = abd'g \vee ae'fg' \vee ae'fg.$$

$G[\Phi_3] = \max(2^2/3, 2^2/3, 2^1/2) = 1.33$; $E[\Phi_3] = 4x3 + 4x3 + 4x2 = 32$; The explained part is shadowed (dark and light).

As we see, the explanation index increases, since at each step the explained part is larger, and the generalisation index decreases. If fact $G[\Phi]$ does not depend on relevant or irrelevant variables, since G is computed from $\mu(X_i)$, but the joins of classes are computed from $\nu(X_i)$.

<u>Références</u>

C. Berge.- *Graphes et Hypergraphes*, Dunod, Paris, 1970.

A. Guénoche.- Méthodes booléennes pour la caractérisation des exemples par les contre-exemples, *Actes des Journées Françaises de l'Apprentissage*, Cassis, 1988, p.75-90.

R.S. Michalski, R.E. Stepp.- Learning from observation : Conceptual Clustering, *Machine Learning : an Artificial Intelligence Approach*, Tioga, Palo Alto, 1983, p. 331-363.

T. Mitchell.- Generalization as search, *Artificial intelligence*, 18, 1982, p. 203-226.

J. Nicolas.- Généralisation en logique des prédicats, in *Actes des Journées nationales PRC-GRECO I.A.*, Toulouse, Mars 1988, p. 255-274.

L. Pitt, R.E. Reinke.- Criteria for Polynomial-Time (Conceptual) Clustering, *Machine Learning*, 2, 4, 1988, p. 371-396.

R.R. Sokal, P.H.A. Sneath.- *Principles of numerical taxonomy*, Freeman, San Francisco, 1963.

ITERATIVE LEARNING

and

REDUNDANT GENERALIZATIONS

M. SEBAG[1], *M. SCHOENAUER*[1]

Abstract :

We present a strategy of learning from examples and counter-examples, adapted to some characteristics of real-world data (noisy, insufficient, mixed numerical and nominal data).

Our aim is the discrimination of classes, building a set of discriminant rules generalizing the examples.

The main steps of the presented approach are the following :

1- A rule is allowed to be inexact, i.e. cover some counter-examples because the data is assumed to be noisy.

2- The global reliability of a set of rules is then greatly increased by redundancy, that is, strong overlapping between rules.

3- Conflicts arise, as contradictory rules cover further examples. The learning algorithm itself is applied to solve those conflicts : the related data are obtained by changing the representation of the problem, according to the set of rules previously determined.

A multi-layer generalization is so defined, enabling incremental refinement of the learning.

Keywords :
Artificial Intelligence, Machine Learning, Learning from examples and counter-examples, Inductive Learning, Learning by discovery.

---

[1] *Centre de Mathématiques Appliquées, Ecole Polytechnique, 91128 PALAISEAU Cedex.*

ITERATIVE LEARNING

and

REDUNDANT GENERALIZATIONS

INTRODUCTION

We present a discriminant generalization strategy, adapted to learning rules from examples and counter-examples [Michalski 84] [Diday 87]. The accuracy criterion is the result obtained on a test set.

Our approach is concerned with industrial problems ; it is adapted to the very specific problems of learning in some real-world applications:

- the data are mixed, both numerical and nominal, and the background knowledge is poor. Examples and partial knowledge expressed as rules, are represented by attribute-value pairs, in 0+ logic.
- the aim is the discrimination of classes ; those classes are often identified by the experts, and cover some complex disjunctive concepts. This in turn will penalize numerical approach.
- the learning base is sparse, compared to the possible cases of the domain (if not, the application exceeds the memory size of the machine). The data is noisy too.

The problems of noise and sparseness are considered simultaneously (see also [Schlimmer-Granger 86], [Bratko Kononenko 87]). We try to improve the reliability of a rule set, given that each rule must admit an error rate, due to the initial noise. The redundancy, in a biological way, allows the whole to be more reliable than its parts.

An algorithm, leading to minimally discriminant rules is parameterized with error and redundancy rates. But the redundancy raises the problem of conflicts, when contradictory rules are simultaneously triggered. Two strategies are proposed to solve these conflicts :

First, the rules are weighted. The final conclusion of a term T is given according to the weighted sum of the rules fired par T.

Second, the learning algorithm itself is applied :

- we consider new descriptors, computable as the premises of the first set of rules. A term T takes the value True or False for the attribute R, wether T fires R or not.

- The same learning algorithm is then applied on the new base, and leads to "second tiered" maximally general rules, refining and specializing the "first tiered" ones.
  The main interest of this iterative technique is to automatically refine the rule set.

We first introduce some hypotheses : the learning base includes elementary examples as well as rules ; a weak coherence is defined, according to the general knowledge about diagnosis problems.

Second, we state some implicit choices of intelligent learning, such as conciseness, and their consequences in a poor-knowledge domain. Some heuristics are defined according to the quality of initial data.

We finally review the problems raised by redundancy, and detail the iteration of learning proposed to solve them.

I BACKGROUND KNOWLEDGE AND WEAK COHERENCE.

In some industrial problems, the available data is composed of elementary examples or counter-examples, together with partial rules.
The partial rules are provided by the experts, stating some general relations between (attribute-value) pairs and diagnosis, or sometimes, excluding parts of the domain as never encountered.

The chosen formalism is the following :

- Set Ω the domain of learning.

- Nominal descriptors y_i, $i = 1..P$ are defined on Ω :

  $y_i : \Omega \longrightarrow O_i$, where O_i is finite.

  Numerical attributes are previously discretized ; see [Lebowitz 85].

- Nominal conclusion y_0, characterizing the exclusive k classes to discriminate :

  $y_0 : \Omega \longrightarrow O_0 = \{ 1, .. k \}$

- a term T, elementary example or rule, is described by

  $\{ y_0(T), y_1(T), .. y_P(T) \}$

  $y_j(T) \in O_j$ or $y_j(T) = X \Leftrightarrow y_i$ *is not pertinent in T.*

  $y_0(T) = j \in O_0$

  T is a j-example, a k-counter-example for $k \neq j$.

Due to the sparseness of the data, a missing value will also be encoded by X. The error is then translated, not altered (which is the case when *Unknown* is considered as an attribute value, or replaced by a probabilistic value [Bhatnagar-Kanal 86]).

A partial order on the data is given by the relation of generality / particularity [Michalski 84] [Diday 87].
Rules and elementary terms can overlap. We cannot assume a strong coherence, that is, disjoined examples and counter-examples. But the

following assumptions, characterizing the weak coherence of a domain, are made :

-*1*- We are interested in conclusions which have a default value;
A failure is by default False.
-*2*- Counter-examples can cover examples;
A failure is specific compared to normal-state.
-*3*- But examples cannot cover counter-examples;
But normal-state is not a specific instance of failure.

II GOALS, ALGORITHM AND HEURISTICS.

Under those assumptions, likely in diagnosis problems, the goals are defined (for details, see [Sebag-Schoenauer 88]) :

II.1 Algorithm.

- From a term *T*, one (or several) maximal generalization(s), is (are) to be selected as rule(s) ; maximally general, because :

  1. The maximally specific rules would lead to conclusion *'I don't know'* on most further cases : the learning set contains very few information, compared to all possible cases;
  2. We don't know the actual level of generality relevant to the data [Mitchell 82];
  3. Maximally general rules are easy to refute, and counter-examples are easy to find.

- Optimizing some evaluation function,

  This evaluation generally refers to the number of elementary terms covered or discriminated. In the following, the score of a generalisation R denotes the number of examples covered by R, and the error denotes the number of covered counter-examples.

- With respect to counter-examples.

  If a term T and a counter-example C do not intersect, then a generalization of T must not intersect C.

For an example T, set

$$L_T = \{i = 1..P \;/\; y_i(T) \neq X\}$$

For a counter-example C, set

$$\Delta(T,C) = \{i \in L_T \;/\; y_i(T) \neq y_i(C)\}$$

One is looking for the minimal sets L_R, *such that :*

$$L_R \subset L_T \;\; ; \;\; \forall\; C \text{ counter-example} \quad L_R \cap \Delta(T,C) \neq \emptyset$$

A maximal discriminant generalisation R of T is then given by :

$$y_i(R) = y_i(T) \qquad \forall\, i \in L_R\,;$$

$$y_i(R) = X \qquad \forall\, i \in \{1,..\,P\} - L_R.$$

$$y_0(R) = y_0(T)$$

The pruning of the sets $\Delta(T,C)$ and $L(T)$ is based on the following remark, using the near-misses counter-examples information [Kodratoff 86] [Sebag 88] :

If $\Delta(T,C_1) \subset \Delta(T,C_2)$, then $\Delta(T,C_2)$ can be pruned.

We will consider first the problem of redundancy / conciseness.

II.2 Redundancy / conciseness.

The generalization is supposed to compress the initial information. An intelligent learning would build a concise set of generalizations, more comfortable for the mind and for the experts.

But, in our context, conciseness leads to unstable results ; the more frequent conclusion is *I don't know*, on further examples.

Remark : Some intelligent reasonning would reasonnably do the same, given a noisy and insufficient training set.

The redundancy then refers to the quality of available knowledge : the experts must know how many significant guesses can be extracted from an example.

We choose to control the ratio of allowed rules generalizing a given term, and not a global threshold such as the number of hypotheses per rule, or average level of generality, because the classes to characterize are not likely to be homogeneous.

II.3 Noise.

Second, the problem of noisy data is not easily handled by symbolic learning : a cleaning process derives from some distance on the domain [Diday & Al 85], and a meaningful distance should be a result of the learning phase.

Clearly, in our context, a rule must not be inhibited by a false counter-example, neither generalize only false examples.

These controls over the learning process are achieved through 3 parameters :

II.4 Some parameters of generalization.

The minimal size α is the amount of initial data a selected generalization must at least cover.

$Score(R) \geq \alpha$, $\alpha = 1, 2, \ldots$

One can observe that isolated terms generate many generalizations, only covering these terms.

When α is superior to 2, those generalizations are forgotten, that is, those isolated points are forgotten too except possibly as counter-examples : the cleaning of the initial data is done *a posteriori*.

The error rate ϵ $(0 \leq \epsilon \leq 1)$ is the ratio of covered counter-examples covered examples allowed for a generalization.

The exact discrimination algorithm is now considering the counter-examples with a probability of $1 - \epsilon$. The approximatively discriminant rules obtained are selected if

$$Error(R) \,/\, Score(R) \leq \epsilon \,,$$

The number of rules decreases when ϵ increases ; first results show a degradation on the training set, and a slight improvement on the test set. Numerical results are given in [Sebag 89].

The redundancy rate τ $(0 \leq \tau \leq 1)$ is a local parameter : a maximal generalization R of a term T is selected if

$$Score(R) \geq \tau \quad \underset{R \text{ generalization of } T}{Max} \; (Score\,(R))$$

The redundancy allows to correct the effects of the error rate : as it increases the number of generalizations, the results nearly reaches 100% correct on the training set, and are significantly improved on the test set.

III CONFLICTS.

A term T is called

- *certain*, if all the rules fired by T have the same conclusion (which in turn can be true or wrong);
- *unknown*, if T fires no rule;
- *ambiguous*, if T fires contradictory rules.

Without redundancy ($\tau = 1$), the terms of the test set are mainly *unknown*. As redundancy increases (that is, τ decreases), more and more terms become *ambiguous*.

III.1 Weighting.

Conflicts can be solved by weighting the rules : the class C associated to a term T is defined if it exists, by

$$\forall j \neq i \qquad \sum_k a_{ik} R_{ik}(T) > \sum_k a_{jk} R_{jk}(T) \,,$$

where a_{ik} is the score of R_{ik}, k^{th} i-example of the initial data.

We so obtain an "elastic" knowledge, more reliable than its components : each generalization may be slightly wrong, but the set of generalizations is redundant.
Then, if a term T is globally missed by the rules, T is a singularity being either an isolated point compared to the initial data, or a point whose neighbours are contradictory. Significant improvement of the results is observed on real-world data, compared to exact and concise approach ($\epsilon = 0$, $\tau = 1$).

Those coefficients a_{ik} could be optimized ; but weighting cannot solve all conflicts (for instance, circular conflicts cannot be solved in such a hierarchical way).

III.2 Change of representation.

On the other hand, conflicts can be solved by the learning process itself, if appropriate data is avalaible, that is examples of triggered rules, together with the associated conclusion.
This appropriate data is obtained by changing the domain representation, according to the given rule set : a rule is now considered as a new boolean descriptor, computable as the premises of the rule.

$$R(T) = \begin{cases} \textit{True if T fires R} \\ \textit{False if not.} \end{cases}$$

The initial learning base A, or other parts of the initial data, are then encoded, using the rule set B learnt from A. A term T is represented by :

$$T : \{y_0(T),\ R_j(T)\ \},\ R_j \in B.$$

The same discrimination algorithm is then applied (with same parameters values or not ; likely, the redundancy rate is to decrease), and leads to "second-tiered" generalizations, such as :

If R_i-hypotheses are verified,
And R_j-hypotheses are not,
Then Class k.

Notice that second-layer rules implicitly handle disjunctions, referring to the initial representation of the learning domain.
This change of representation allows to specialize and refine rule sets, iteratively building multi-layers generalizations. The number of layers is controlled by the error on the test set; this error first decreases, then become steady, as the information is compacted. See [Sebag 89] for results and complete discussion.

CONCLUSION.

The discrimination is iteratively performed through 2 steps :

- A rule set is learnt from a learning set, allowing redundant and approximatively discriminant generalization;

- A rule set defines a change of representation on the learning domain : new descriptors are computed as the rules premises. The new datas are then considered as input for the generalisation process.

This procedure performs in an automatic fashion,

- the specialization and refinement of an approximatively discriminant rule set, solving conflicts, and detecting exceptions;

- the handling of a concept's negation, and disjunctions;

- a partial treatment of data, allowing fragmentation of the learning base : an incrementality "by blocks" of the learning process is achieved.

Furtherly, these multi-layers rules will be compared to a connectionnist network [Le Cun 87] : as the generalization iteratively builds a neuronal-like network, some weights on the rule network can be optimized using connexionnist techniques.

BIBLIOGRAPHY

[Bhatnagar-Kanal 86] R.K. BHATNAGAR, L.N. KANAL, *Handling uncertain information: A review of numeric and non-numeric methods.* In Uncertainty in Artificial Intelligence (L.N.Kanal, J.F.Lemmer ed.) North Holland 1986.

[Bratko Kononenko 87] I. BRATKO, I. KONONENKO. *Learning diagnostic rules from incomplete and noisy data.* Interaction in A.I. and statistics. Eds B.Phelps, London, Gower technical press.

[Diday & Al. 85] E. DIDAY, J. LEMAIRE, J. POUGET, F. TESTU. *Elements d'analyse de données,* Dunod 1985.

[Diday 87] E. DIDAY. *Introduction a l'approche symbolique en analyse de données.* Journées symbolique/numérique, Paris 9-Dauphine, Décembre 87.

[Kodratoff 86] Y. KODRATOFF *Leçons d'apprentissage symbolique.* Capadues Ed, Toulouse 1986.

[Lebowitz 85] M. LEBOWITZ, *Categorizing Numeric Information for Generalization.* Cognitive Science 9, pp 285-308.

[Le Cun 87] Y. LE CUN, *Modèles Connexionnistes de l'Apprentissage.* Thèse Paris 1987.

[Michalski 84] R.S. MICHALSKI, *A theory and methodology for inductive learning.* In Machine Learning : An Artificial Intelligence Approach, (R.S. Michalsky, J.G. Carbonnel, T.M. Mitchell Eds), Palo Alto, 1984.

[Mitchell 82] T.M. MITCHELL, *Generalization as Search.* Artificial Intelligence N° 18, 1982.

[Schlimmer-Granger 86] J.C. SCHLIMMER, R.H. GRANGER, *Incremental Learning from Noisy Data.* Machine Learning, Vol 1, pp 317-354.

[Sebag-Schoenauer 88] M. SEBAG, M. SCHOENAUER, *Apprentissage par réseaux de règles, redondance et fiabilité.* Journées symbolique/numérique, Orsay, Décembre 88.

[Sebag 89] M. SEBAG. *Apprentissage multi-couches,* Thèse d'Université, Paris 9 Dauphine, to appear.

ANALYSIS, SYNTHESIS AND PARALLEL PROCESSING OF LARGE DATA AND KNOWLEDGE BASES

Boris E. Polyachenko, Filipp I. Andon and Oleg L. Gunko
Institute of Cybernetics, USSR Academy of Sciences, Kiev USSR and
International Basic Laboratory for Artificial Intelligence
Institute of Technical Cybernetics
Dubravska cesta 9
842 37 Bratislavia
Czechoslovakia

Abstract. Mathematical models of data and knowledge representation for very large data and knowledge bases and their processsing at parallel computing systems with various architectures are suggested. Mathematical models base on the introduced algorithms' algebra and the external informational environment communications are taken into account. Various programs and parallel computations processes properties are examined. The approach for data processing program synthesis and dynamic parallelizing are disscussed too. These techniques and means provides promoting the efficiency of this class of programs and based on the equivalent transformations of programs and their paralleling. This approach maximally corresponds to the class of large sets of (numeric, symbolic, etc.) data processing programs and facilitates the high computational efficiency when servicing the programs in the multiprocessor environment of various architectural structures. The approach for knoweledge base constructed as production set parallel processing are examined too. Parallel scheme for sequential optimization techniques(branch-and-bound, dynamic programming methods) for patterns recognition problems is suggested. The high level programming language ACORD and the technology for data processing programs designing, optimization and parallel processing are discussed. The structure of software packages is described which supports the proposed approach. The experimental studies show that these programs servicing time decreased several times as compared to using the standard operation facilities. The structure of parallel hardware tools is suggested too.

Please mail to Dr. B.POLYACHENKO: BRATISLAVA

We can demonstrate all described in this paper Software(IBM/370,IBM PC, PDP 11/70 compatible computers).

1. INTRODUCTION

Traditional computing systems and standard operational environment is not effective for large sets of symbolic and numeric data, Data Bases (DB) and Knowledge Bases processing. Besides the most techniques and software packages for parallel computing process organization is oriented to computational problems and not taken into account DB and Knowledge Bases specific features[EVANS82],OZKARAHAN86],[ULLMAN80].

The present paper proposes mathematical models of important class of programs - data processing programs, which works with very large DB, as well as mathematical models of computing system. These models are based on algorithms' algebra [MICHAL84] and extends this algebra with respect to communications with DB resided at external memory. By means of suggested models new approaches for promoting the efficiency of DB processing system is proposed based on the equivalent transformation of the program and its parallelizing. These approaches support two-level paralleling and are accordingly classified as vertical and horisontal paralleling techniques.

Data analysis problems need very large DB creating, supporting and processing with great amount of exchanges with external memory devices This is a 'bottleneck' of the computer systems, as well as system and applied software. In present paper we'll propose some approaches to solve this problem and examine efficient parallel computation organizations methods and parallel optimization scheme for patterns recognitions.

The vertical paralleling technique extends the data flow [EVANS82] and macropipeline [GLUSHKOV74],[MICHAL84] computation organization to the class of DB processing system. This approach was originally presented in [ANDPOL80].

The horisontal paralleling technique [ANDMARPOL88],[POLYACHENKO89] provides the program statements' transformation into the set of asynchronously executed statements; each of them processes its 'own' data portion(data substructure).

The program processing overhead function is considered as a program servicing time in the multiprocessing computing system(MCS) environment. The optimization problems are introduced providing the incremental increasing of overhead function while both suggested vertical and horisontal techniques are in usage.

Some of important properties of various class of DB processing programs are considered.

Mathematical model for matching problem is considered and parallel optimization algorithm for patterns recognition problem solving is suggested for exact and quaranteed approximate solutions generating.

The programming language syntax and semantics for DB processing programs designing, as well as technology for such programs construction, paralleling and optimization are suggested.

Knowledge-based approach is using now for classification and other pattern recognition problems[CHANDGOEL88]. And therefore we discussed some ideas of parallel computation organization for AI production systems.

The structure of software packages and hardware tools is described which supports the proposed approaches to paralleling and optimization. The experimental studies show that program processing time decreased several times as compared to using the standard operating facilities.

Existing applied software may be processed by means suggested techniques and tools without any modifications. These applied programs may be written at any sequential programming languages but will processed in parallel mode at multiprocessor environment under our system control.

2.MATHEMATICAL MODELS FOR LARGE DATA ARRAYS PARALLEL PROCESSING

2.1.Multiprocessor environment

The large experimental data arrays(DB) processing program servicing techniques on the multiprocessor computing system(MCS) are disscussed assuming the following MCS architecture. Let consider MCS with distributed memory, distributed control(distributed operation system) and universal switching system.

Let N processor (N>=1) be available, each of these possesses its own main memory; the shared external memory is available and the shared main memory can be available. Any hardware components of MCS can communicate with each other by means of switching system(processor). Such MCS obviously spreads over the wide range of MCS architectures, as static, as well as dynamically reconfigurable ones[EVANS82],[GLUSHKOV74], [MICHAL84].

2.2.Parallel data processing techniques

Here we present new techniques for data(DB) processing programs parallelizing and parallel processing at MCS with various architectures. These techniques are rather general approaches and can be used for wide range of applications.

The present paper proposes new paralleling approaches to data and knowledge bases processing programs. These techniques provide high degree of parallelizm at two levels of paralleling, referred here as vertical and horisontal techniques.

2.2.1.Vertical paralleling technique

The vertical program paralleling covers the principles of data flow [EVANS82] and macropipeline [GLUSHKOV74],[MICHAL84] implementations in relation to DB processing and consists in the following[ANDPOL80], [POLYACHENKO89]. To each array we'll correspond the buffer in the 'fast'(main) memory for one or more records(tuples) residing. The asynchronous pipeline execution of the MMP components is provided when the computability of a component is determined by the accessability of its data required for its operation and resided in the appropriate buffers. Thus the component (statement) is considered to be computable if required records(tuples) of the source arrays are stored in the buffers and the buffers' memory space is available for storing the resulting arrays' records(tuples). Otherwise the component(statement) is incomputable.

The execution of the components is reduced to the sequence of computable component chosen, assigning to the processor and computable cómponent execution itself continuing by processor until the component becomes incomputable and to cause the component to be suspended until the buffer data become accessible. This technique provides paralleling of the execution of the MMP(of infomational graph, of program graph) consisting of the set of components which are mutually dependent informationally by their array(relations). While implementing the vertical paralleling technique in MCS environment, each processor is assigned for servicing the part of MMP(the subgraph of MMP initial informational graph). The buffers for this subgraph component are allocated within the main memory space of that processor. The interprocessor (intersubgraph) communication is also realized asynchronously via the buffers allocated within the shared memory space (if not available - via the external storage).

The details of implementing the above technique on MCS of various architectures as well as the optimization algorithms for computations organization which can be reduced to solving the special discrete optimization problems, are examined in [ANDPOLGUN87],ANDKUKPOL80].

2.2.2.Horisontal paralleling technique

The horisontal paralleling technique[ANDMARPOL88],[POLYACHENKO89] provides the MMP components' transformation into the set of asynchronously executed components. The idea of this technique lies in that the component(statement) Z=s(X,Y) with the source arrays X,Y and the resulting array Z is transformed into the asynchronous cycle statement which specifies the asynchronous execution of the instances of s statement on 'own'(indexed) data structures relatively to the source arrays X and Y of the source component s. Each of this instances processes its 'own' data set portion. The construction of these 'own' data set portions provides by means of special indexing procedure and array partitioning. For the whole MMP by means of suggested equivalent MMP (informational graph) transformation, horisontal and vertical paralleling techniques, the optimal method for its realization at MCS is chosen.

This technique is rather general one, studies by us for various class of data bases processing MMP and may be used for different applications.

2.3.Data processing program models

2.3.1.Algorithms' algebra

For formal programs representation we'll use the algorithm's algebra A(U,Y,Z) under the memory with

U- the set of elementary conditions,

Y- the set of elementary functional statements,

Z- the set of elementary statements for communications with external informational environment. This algebra is based on Glushkov's algebra [MICHAL84] and extended by statements Z. This model of DB may describes various data structures for different applications(numeric and character processing etc.).

Let us consider following statements for communication with external informational environment which corresponds to MAJAK languaqe [MICHAL84]:

```
WRITE     array element,
READ      array element,
LOCK      array element,
UNLOCK    array element,
SEND      data,
RECEIVE   data,
CALL      program.
```

In informal way READ(WRITE) statements provides reading(writing) of array elements from(to) external informational environment(external memory). The LOCK statement usage for array element locking in K component following to MAJAK langage semantics will be mean that this element may be processed by this component program only. The UNLOCK statement provides array element be accessable for other component programs. The SEND(RECEIVE) statements will be give data communications between two components by means of (synchronous or asynchronous) 'rendez-vous'-like principle which similar to one of Hoar's CSP system[EVANS82].

The CALL statement provides the transferring of control to any program which will be executed (a)synchronously with calling program.

The multimodular program G with arrays (MMP),which consists of the set P of sequential components(programs), is described as the set of pairs

$G=\{(K_1,P_1),\ldots,(K_N,P_N)\}$, where K_i - components` names, $P_i$ - corresponding components` programs. Let us $I(K_i)$ and $O(K_i)$ are the set of source arrays(DB,relations) and the set of resulting arrays of the K_i components correspondingly.

Grafically MMP may be represented as a oriented multigraph (P,U) in which the nodes set P corresponds to the components` set, and the arc defines by infofmational relations between the components(by arrays).

The set of source arrays(relations) resides on external informational environment(external devices). The execution of the MMP(of the graph (P,U)) will be followed to reside resulting arrays at external informational environment. The equivalent transformation means that from equility of source array follows the equality of resulting array. Later in this paper we`ll consider MMP equivalent transformations only.

During vertical paralleling technique usage for MCS each processor executes any subgraph of initial graph.

2.3.2.Discrete dynamic systems

Discrete dynamic system(DDS)[MICHAL84],[ANDPOLGUN87] is defined as the states set S and the system of transitions rules {==>} which are given on the set S. The sequence $pr=s_0,\ldots,s_n,\ldots$, where $s_j \Longrightarrow s_{j+1}$ for each $j=0,1,\ldots,n-1,\ldots$, will be referred as process in DDS given, and s_0 -its initial state. There are the subset {s*} of final states in S, to which, by definiton, the transitions rules not applicable. The deadlock states will be refered as states which are not final states and to which the transitions states not applicable. The process which is terminated by final state will be terminatable and the process which to go to deadlock state will be deadlock one.

We described in [ANDPOLGUN87] traditional computation process organization with synchronization by means of LOCK, UNLOCK and asynchronous (vertical and horisontal paralleling) computations and proved their features for different important data processing programs` classes, examined deadlock conditions.

3.PATTERN MATCHING OPTIMIZATION PROBLEM

Let`s consider mathematical model of patterns recognition problem [CHANDGOEL88],[DUDAHART63],[SAKOE79],[SOKSNEATH63].

To examine dynamic programming and branch-and-bound methods for patterns recognition problems solving in relation to parallel computations organization.

Let's consider dynamic programming-based pattern matching algorithm based on [SAKOE79]. In this paper time-normalization procedure proposed for speech recognition problem.

Let

$$A=a_1,\dots,a_i,\dots,a_I$$

$$B=b_1,\dots,b_j,\dots,b_J$$

be two patterns. Each of them is represented as a sequence of feature vectors with quasi-continious time axis i or j. The time-normalized distance measure between these two patterns is defined as a function of vector distance d(i,j) between vectors A and B:

$$D(A,B)=\min_{j=j(i)}\left[\sum_{i=1}^{I} d(i,j)\right]$$

Warping function j(i) is subjected to monotonic and continious conditions, etc. This minimization problem is very efficiently solved by the use of dynamic programming.

Variants` graph (P,U) for dynamic programming(as well as branch-and-bound) method is defined as oriented tree with the set P of nodes which corresponds to the set of subtasks and the set of vertices U. Dependencies between nodes correspond to relation `parents-childrens` and sequention given at desicion rules set and present as data arrays. Usually these arrays stored at external memory devices of given computing system.

Dynamic programming algorithm can be used for only low-dimensional optimization problems. For large-scale optimization problem with number of constraints branch-and-bound method [HORSAHNI78] can be used as for exact as well as approximate solutions (with a posteriori error estimation) generating. Such approach suggested in [KUKSAPOL84] for some network scheduling problems solving. This algorithm possesses the same (exponential) computational complexity and above mentioned advantages.

Such graph (P,U) can be transformed, optimized, parallelized and processed by means of above described techniques and tools. Actually, horisontal paralleling technique can be used for each task pcP parallelizing, vertical paralleling technique - for asynchronous(pipeline) executing of the graph with buffering. However, arrays will not reside at external memory and will transfer from one task to outhers via buffers in main memory in asynchronous mode. This approach for computations organization provides high efficiency due to parallel processing, minimization of data exchanges between memory levels and uniform resources loading.

Computation optimization algorithms should be used for optimal utilization of resources of MCS with given architecture. It`s true for MCS with static(unchanged) architecture. If we want to process graph (P,U) at MCS with dynamically reconfigurable architecture suggested tools will control as tasks processing, as well as MCS configuration in dynamic mode.

4. PARALLEL OPTIMIZATION ALGORITHM

The scheme of branch-and-bound algorithm for any minimization problem solving will be briefly described by means of following key parameters[HORSAHNI78], [KUKSAPOL84]: subtasks, decomposition(branching), strategy, estimations(bounds), records, test.

The set of *subtasks* is defined as $Z=\{z^t, t=0,1,...\}$, where z^0 is initial optimization problem.

The *decomposition* is the following procedure of subtasks-childrens attaching from subtask z^t:

$$d(z^t):z^t \dashrightarrow z^{t+1},...,z^{t+1}.$$

The *strategy* S for subtask choice from the set of active subtasks will be LIFO(Last-In-First-Out).

Low $G(z^t)$ and high $R(z^t)$ *bounds* for any subtask z^t are defined in a such way that $G(z^t)<=F^*(z^t)<=R(z^t)$, where $F^*(z^t)$ is optimal cost function value for this subtask.

The $REC(z^t)$ and REC are *records*(record solutions) for subtask z^t and the best solution of the initial problem at this step of the algorithm correspondingly. The main *test* is the following relation:

$$G(z^t) >= REC.$$

If this condition is true then subtask z^t is deleted from the set of active subtasks, else decomposition of this subtask will be provided.

For parallel branch-and-bound algorithm supporting let's organize two lists. The first one is the fixed size list of active subtasks L(Z) and the second one is the N elements list of free processors L(PR).

The parallel branch-and-bound algorithm consists of the following steps.

At each time period(algorithm step) we'll assign any subtask $z \in L(Z)$ to any processor $p \in L(PR)$, $L(Z)=L(Z)\setminus\{z\}$. Such action (as well as loading into the buffer and into the list) will be defined as $z==>p$.

Let RECBUF and TASKBUF are one element and fixed element buffers correspondingly in main ('fast') memory of MCS for record REC and informations of active subtasks storing. Subtasks processing and communications via these buffers will be provided by means of above described vertical paralleling technique. The horisontal paralleling technique can be usec too for each subtask parallelizing.

To consider any subtask z^t executed by any processor p. Following procedure should be done.

```
         t
  IF REC(z  ) < REC THEN
BEGIN;
                   t
       REC=REC(z  ); REC ==> RECBUF;
               t
  EXECUTE d(z  );

    t+1      t+1
  {z   ,...,z    } ==> TASKBUF;
END;
```

This branch-and-bound algorithm is more general than one for above mentioned pattern recognition problem and can be used for various (discrete, scheduling theory, etc.) problems solving.

5.DATA PROCESSING PROGRAM SYNTHESIS FROM MODULES

Let's consider the set of parametrized modules which forms the library(base) of modules.

Each module S_j is any mapping(transformation) of its set $I(S_j)$ of input arrays(DB) into the set $O(S_j)$ of output arrays(DB). Each array corresponds to large DB resides at external memory.

We'll consider only logical DB and their logical structures. The correspondence between logical and physical structures must be provided by supporting systems(which will be referred as SYSTEM).

One(SYSTEM user) could formulate a nonprocedural request to the SYSTEM in the following way by means of very high level problem-oriented language[POLYACHENKO89]. This request, for example, may be followed to any array M* (with defined logical structure) creating:

CREATE ARRAY <name> WITH STRUCTURE <structure>

SYSTEM in interactive mode will choose such module S* from base of modules which could produce given array M*, M*cO(S*).

After that SYSTEM will define such modules formed input arrays for S* , etc. By recursion graph G will be constructed.

During graph G building user could (and must if it's necessary) introduce additional(unexisting) modules, arrays, procedures by means of SYSTEM help and control.

Constructed graph G can be stored in base of graphs or immediately used for transformations, optimization or processing by means of our suggested tools.

6. LANGUAGE FOR ASYNCHRONOUS DATA PROCESSING PROGRAMS DEVELOPMENT

Here we suggest simple and convenient high level parallel programming language ACORD(Asynchronous COnstruction of pRograms for Data processing), which oriented to compex data processing programs designing from data processing(DP) statements and reflected asynchroous style of programming. This style feature is such that all described in the program actions are supposed as potentially parallel, independent, but their scheduling are produced by language (system) supporting means. We'll defined by DP statement rather complex procedure, which consists of DP statement source arrays set into its resulting array set. This procedure may be written by means of any (sequential or parallel) programming(query) language, supported by the operational environment. These statements may be the statements of relational algebta,statements for hierarchical, network and others data structure manipulation [OZKARAHAN86],[ULLMAN80].

The DP program is the expression, consists of DP statement and similar to the expression in algorithms algebra under memory[MICHAL84], [ANDPOLGUN87].

The base of control semantic in ACORD language is the model for parallel processing [ANDPOLGUN87],[POLYACHENKO89] for the DP programs class. ACORD language corresponds to not von Neumann (functional) type.

For ACORD, exept defined above, following features are characterized:

1)effective realization of constructed DP program by means of high paralleling and automatization level, uniform computing resource loading and minimization of exchange volume between memory levels[4,10];

2)independency from computing system architecture;

3) indirect parallelizm expression by means of single assignment principle[EVANS82] extended to DP systems;

4)modularity and portability (mobilty) of constructing DP programs, simplity of extensions;

5)programs are free from side-effect[EVANS82];

6)lack of nessesity of DP statements(modules) interfacing and synchronization description.

The main logical data structure is array(relation), defined on the set of homogeneous records(tuples). One have a possibility of data structure 'array of arrays' like organization on the set of records attributes values. This possibility extends programming means and DP task features are taken into consideration. Besides it's can allow computation parallelizing increasing at arrays level, as well as array record level and maximally effective multiprogramming and/or multiprocessor system means usage. In some cases for maximally asynchronous realization the computation possibility principle is used (for example, parallel initiation of conditional DP statement alternative branches, etc.).

During language construction choosen modern programming languages traditions are taken into account. As above emphasized, the main ACORD peculiarity is simple and convenience realization of complex, effective and portable DP programs designing from DP statements, in which all means of languages by means of these statements are written, may be used. ACORD language isn't attached to any operational environment and can be easily realized within any system.

ACORD language can be used as interface with other software packages (for example, CIMOD ,CPM [ANDMARPOL88], etc.), DBMS, various access methods. For example, realization of control system for language supporting will potentially do by means of ACORD language means usage.

7. TECHNOLOGY OF SOFTWARE DEVELOPMENT AND COMPUTATIONS OPTIMIZATION

To consider the technology for data processing software packages development and exploitation. This technology consists of all technological steps of program life.

1. Each software component is described as expression in algebra A. These components may be any (sequential) programs or relational, hierarchical, network data manipulation statements, which are written at corresponding (query) lanquaqes[OZKARAHAN86],[ULLMAN80].

2. The ACORD language intends for development of complex(multimodular) data processing programs from components. This language reflects asynchronous way of programming. These means provide following features: program mobility, indirect parallelizm description by means of single assignment principle [EVANS82], which spreaded to data processing systems; extending simplicity; side effect-free nature; components interface and synchronization description nessecity absence.

The informational unit in ACORD is array notion - abstract data structure, which describes, for example, relations, hierarchical and network data structures, files, matrices, etc. The semantic of control and components communication in data processing program based on the suggested model of parallel computations. The components communications with each others provides only by means common external arrays (elements) usage, i.e. components `see` their own source and resulting data, which are arrays in external memory. However, this isn`t a constraint and give some advantages, because:

vertical paralleling technique provides components pipeline processing with their communications not via external memory,but via special supported buffers at main memory; scalar is one element array; any interfacing agreement for communication between components and complex debugging of the program aren`t need.

During components complexing into set of programs, which of these will be processed at one (logical) processor, the processor workload and other characteristics are estimated. Then the decision, how many and which components will be complexed into one program of software complex, is chosen.

Exept data processing program itself, program informational graph and its description tables, as well as components synchronization program are generated too.

3. Optimization and paralleling of obtained MMP is being by means of both vertical and horisontal paralleling techniques. There are two alternative ways to optimal MMP construction and realization at MCS:

1) MCS for MMP realization is given. The load program must maximally take into account MCS architecture(number of processors, memories, communications facility, etc.) by points of view of MMP asynchronous degree and computations organization possibilities at MCS;

2) there is dynamically reconfigurable MCS for MMP realization. Then one must to construct the program and choose optimal MCS architecture for this program realization. Besides one must consider limitations on maximal MCS configuration(maximal number of processors, volumes of memories, etc.)and constraints of reliability and/or cost characteris-
4. The load program [illegible] ics steps solutions and optimization and realization methods used taken into account. Load program is the software complex with generated means for computation process organization, which are already adapted for execution at chosen MCS configuration.

5. The final stage of the technological cycle is chosen of realization method from methods set, relization of constructed software complex and forming of resulting data for input data given.

8. KNOWELEDGE BASE PARALLEL PROCESSING

Knowledge-based approach for patterns recognition is widely used now. In [CHANDGOEL88] such approach is examined for classification task. Classification is one of the formulations of the more general recognition problem. In [CHANDGOEL88] knowledge-based approach for classification focused on the issues of how to represent knowledge in symbolic form, how to organize and access this knowledge, how to use this knowledge for solving problems, and how to control the problem solving process.

The problem of great amount of searching and looking out is very important for system of knowledge representation and other Artificial Intelligence(AI) system. Here we'll propose an approach for promoting the efficiency of production-based AI systems.

Let us consider the production system with set U of productions. Each production $u_i \in U$ is the following expression E_i = IF u_i THEN P_i, $i=1,\dots,N$, and corresponds to the elementary condition in algorithms' algebra A(U,Z,Y) under memory[MICHAL84],[ANDPOLGUN87],[POLYACHENKO89].

However, we can consider the set $S=\{P_1,\dots,P_n\}$, where $P_i \in S$ is any expression in condition algebra A(U), n<N. Component P can also inference to the external environment state.

To correspond to any problem Q which solved by means of production system a set P(Q) of productions related to problem Q. This set P(Q) may be built by means of experience during this production system exploitation. This information is expressed by set M of metaproductions, each of which describes productions set P(Q) for each problem Q.

Both any condition u_i and component P_i, $i=1,\dots,N$, have some source and resulting relations or arrays reside on external memory(may be $I(P_i)=0, O(P_i)=0$). Therefore, for production system paralleling techniques and means, proposed in present paper and in [ANDPOLGUN87], [ANDMARPOL88] can be used for its realization at MCS, namely, vertical paralleling technique for production set processing, horisontal paralleling technique - for single production $u_i \in U$ paralleling and computation optimization for production set efficient processing. During the production system exploitation one can extend the set of metaproductions M, greatly decreasing look out amount.

Above examined vertical as well as horisontal paralleling techniques can be used for parallel computations organization in AI system.

9. SOFTWARE AND HARDWARE TOOLS

Described paralleling techniques have been used in design of some projects conducted in the Institute of Cybernetics of the Academy of Sciences of the Ukrainian SSR in Kiev,USSR, and at the International Basic Laboratory for Artificial Intelligence with the Institute of Technical Cybernetics of the Slovac Academy of Sciences in Bratislava, Czechoslovakia, namely in software package for computing process organization at ES(IBM/360,370) computers and their multimachine systems [ANDMARPOL88], in the operational system for MCS ES-1766 with macropipeline organization [MICHAL84],[ANDPOLGUN87],[WOLCGOOD88], in the relational data manipulation system [ANDMARPOL88], in the software package for optimization programs generating[POLYACHENKO83] and project of DB MCS.

The MCS ES-1766 with macropipeline organisation is the MIMD-type computing system of the reconfigurable architecture, distributed control, diertibuted memory and versatile switching system. The multimodular data handling program consisting of a number of components serves as a subject for application of the proposed at this paper techniques. Each component is a sequential program running on the processor of one of thefollowing types: arithmetic, control and peripheral. The component of MMP is represented as a statement with a set of source external arrays and a set of resulting external arrays (data bases). The supporting software for paralleling techniques are implemented as a separate control procesor which controls the multiprocessor configuration during MMP execution. The main memory of a single or several arithmetic processors functioning in the `memory` mode only is used for buffers allocation. The components are written in the programming languages of the mentioned MCS and require no modifications to be made. On the presented components the multimodular program and the control program(for initiating, rescurce allocation, communication between components and external storage, etc.) are generated using the vertical paralleling technique.

The MCS performance characteristics are obtained as well as the servicing overhead function dependence on the number of processors, array and buffer capacity values, the topology of informational graph, the ratio of the component processor/communication operations, etc. The real and test tasks of MCS data bases handling are examined. The processing time of such tasks is reduced 2-7 times as compared to the usage of the traditional computations organization facilities.

Hardware implementation of suggested methods could improve the efficiency of parallel computations organization. For these reasons we designed the architecture of higly parallel multiprocessor DB machine [POLYACHENKO89] with associative parallel processing, active hardware buffering, etc. Described vertical and horisontal paralleling techniqùes, program transformations and optimization at this MCS are provided by means of hardware tools.

AKNOWLEGMENTS. The authors wish to thank Academician V.S.Michalevich, Academician I.Plander, Professor N.G.Zagorujko, Dr. J.Mikloshko and Dr. K.Richter for their support of this work and valuable discussions.

REFERENCES

[ANDPOL80] ANDON F.I.,POLYACENKO B.E. On a Technique of Data Processing in MIS.-Cybernetics,Vol.16,1980,No.2,pp.65-69.

[ANDGUNPOL87] ANDON F.I.,POLYACHENKO B.E., GUNKO O.L. Models for Organization of Asynchronous Computations in Multimodular Data Processing Programs.-Cybernetics,Vol.23 1987,No.5,pp.57-63.

[ANDKUKPOL80] ANDON F.I., KUKSA A.I.,POLYACHENKO B.E. Effective Implementation of One Class Data-Related Programs with a Computer.-Programming and Computer Software,Vol.6,1980, No.3,pp.57-63.

[ANDMARPOL88] ANDON F.I.,MAR`JANCHIK A.M.,POLYACHENKO B.E. DBMS Query Paralleling Techniques.-Proceedings of the 11th International Seminar on DBMS (Seregelyes,1988), Hungary, Statistical Publishing House,pp.131-140.

[CHANDGOEL88] CHANDRASEKARAN B.,GOEL A. From Numbers to System to Knowledge Structures: Artificial Intelligence Perspectives on the Classification Task.-IEEE Transactions on Systems, Man, and Cybernetics,Vol.18,1988,No.3,pp.415-424

[DUDAHART73] DUDA R.O.,HART P.E. Pattern Classification and Scene Analysis.-John Wiley, New York,1973.

[EVANS82] EVANS J. Parallel Processing Systems.-Cambridge Univ. Press,1982.

[GLUSHKOV74] GLUSHKOV V.M. et al. Recursive Machines and Computing Technology.-IFIP Congress-74(Stokholm,1974),1974, pp.80-85.

[HORSAHNI78] HOROWITZ E.,SACHNI S. Fundamentals of Computer Algorithm. Computer Science Press, Rockvill, Maryland,1978.

[KUKSAPOL84] KUKSA A.I.,POLYACHENKO B.E. Branch-and-Bound Algorithm for Two Network Scheduling Problems Solving.-Cybernetics, Vol.20,1984,No.1,pp.73-76.

[MICHAL84] MICHALEVICH V.S. et al. Organization of Computations in Multiprocessor Computer Systems.-Cybernetics,Vol.20, 1984,No.3,pp.1-10.

[OZKARAHAN86] OZKARAHAN E.A. Database Machines and Database Managemet.-Prentice-Hall, Englewood Cliffs, N.Y.,1986.

[POLYACHENKO83] POLYACHENKO B.E. Interactive Software System for Solving of the One Class of Dyscrete Optimization Problems.- Control Systems and Machines,1983,No.6, pp.95-98(in Russian).

[POLYACHENKO89] POLYACHENKO B.E. Models, Techniques and Tools of Databases Parallel Processing.-To appear in Computers and Artificial Intelligence,1989.

[SAKOE79] SAKOE H. Two-Level DP-Matching-A Dynamic Programming-Based Pattern Matching Algorithm for Connected Word Recognition.-IEEE Transactions on Acoustics, Speech, and Signal Processing,Vol.ASSP-27,1979,No.6,pp.588-595.

[SOKALSNEATH63] SOKAL R.R.,SNEATH P.H.A. Principles of Numerical Taxonomy.-Freeman, San Francisco,1963.

[ULLMAN80] ULLMAN J.D. Principles of Database Systems.-Stanford Univ., Computer Science Press Inc.,1980.

[WOLCGOOD88] WOLCOTT P.,GOODMAN S.E. High-Speed Computers of the Soviet Union.-Computer IEEE,Vol.21,1988,No.9,pp.32-41.

SUBJECT INDEX